D1223401

STUDIES IN THE HISTORY OF STATISTICS AND PROBABILITY

Some Griffin books on statistics

The advanced theory of statistics (3 volumes)
M. G. Kendall & A. Stuart

An introduction to the theory of statistics
G. U. Yule & M. G. Kendall

Rank correlation methods M. G. Kendall

Games, gods, and gambling: A history of probability and statistical ideas F. N. David

Combinatorial chance F. N. David & D. E. Barton

Style and vocabulary: Numerical studies C. B. Williams

Biomathematics: C. A. B. Smith

Volume II: Numerical methods, matrices, probability, statistics

Griffin Statistical Monographs

Geometrical probability M. G. Kendall & P. A. P. Moran

Statistical models and their experimental applications P. Ottestad

The linear hypothesis: A general theory G. A. F. Seber

Families of bivariate distributions K. V. Mardia

Statistical tolerance regions: Classical and Bayesian I. Guttman

Mathematical methods in the theory of queueing
A. Y. Khintchine

The fundamentals of statistical reasoning M. H. Quenouille

Analysis of variance A. Huitson

STUDIES IN THE HISTORY OF STATISTICS AND PROBABILITY

A series of papers selected and edited by

E. S. PEARSON, C.B.E., M.A., D.Sc., F.R.S.
M. G. KENDALL, M.A., Sc.D.

1970
HAFNER PUBLISHING COMPANY
DARIEN, CONN.

Charles Griffin & Company Limited
42 Drury Lane, London, WC2B 5RX

$10\frac{1}{4}'' \times 7\frac{1}{4}''$, ix +481 pages
33 illustrations
ISBN 0 85264 193 1

PRINTED BY LITHOGRAPHY IN GREAT BRITAIN
BY BUTLER & TANNER LTD, FROME AND LONDON

CONTENTS

PREFACE

It was decided in 1954 to publish a series of historical articles in *Biometrika* under the general title: "Studies in the history of probability and statistics". The series was launched with the two long articles, by Florence David and Maurice Kendall, No. 1 and 2 in the present volume: it was hoped that when sufficient contributions had accumulated, the whole series might be reissued together in a single volume. Twenty-one of these articles have now, by 1969, been received and published.

As the time for taking action approached, we realised that there were additional articles, mainly published earlier in *Biometrika*, which it would be of value to include in our volume. The majority of these are concerned with the development of statistics in this country in the years since 1890, that is covering a period through a part of which their authors had lived themselves. While as a result the characterization is bound to be to some extent subjective, we are confident that contemporary records of this kind will be of value in any future assessment of this important and formative period.

An interest in the history of our subject was fostered over a number of years in the Department of Applied Statistics at University College London, where Karl Pearson gave a series of lectures in the 1920s, mainly concerned with the history of statistics in the 17th and 18th centuries. Three of the authors of articles in the present volume attended at least some of these courses and are indebted to the stimulus which the lecturer conveyed. Pearson had written out these lectures carefully in long-hand and may have hoped to revise and publish them later. With pressure of other work he never found time, however, to do this and the lectures remain just as he gave them, week by week, to a student audience. We include as an Appendix a summary of their contents. In view of requests received from a number of present-day students of history, who have heard of the existence of the MSS., it is hoped before long to have available typed copies in mimeographed form for limited circulation.

Interest in the history and philosophy of science has been growing steadily for a number of years, but there have been very few *books* published on our subject. The classical volume on probability is, of course, Isaac Todhunter's *History of the Mathematical Theory of Probability* published in 1865. But Todhunter's account ended with the work of Laplace and says little or nothing about statistics. In more recent years we have had Harald Westergaard's *Contributions to the History of Statistics* (1932) which is largely concerned with descriptive statistics; Helen Walker's (1929) *Studies in the History of Statistical Method*, a series of sketches, and finally Florence David's *Games, Gods and Gambling* (1962).

The articles in the following pages are clearly no substitute for a unified historical appreciation, but all of them, in greater or less degree, represent studies which have involved a considerable amount of patient research, and together, we believe, they give a very fair idea of the whole domain. They have been printed in approximate historical order of subject matter. We hope that by publishing them together we shall give the mathematical statistician some hours of pleasant browsing and suggest to the more serious student of history and historical continuity some of the gaps which still remain to be filled in.

For permission to reproduce we are indebted to the various living authors of these papers; to the Royal Statistical Society in respect of Kendall's obituary of Yule; and to John Wiley & Sons in respect of E. S. Pearson's article on the Neyman–Pearson story.

REFERENCES

Todhunter, Isaac (1865) 1949 reprint. *A History of the Mathematical Theory of Probability from the Time of Pascal to that of Laplace*, New York: Chelsea Publishing Company.

Walker, Helen M. (1929). *Studies in the History of Statistical Method, with Special References to Certain Educational Problems*, Baltimore: Williams & Wilkins.

Westergaard, Harald (1932). *Contributions to the History of Statistics*, London: P. S. King & Son Ltd.

E. S. PEARSON
M. G. KENDALL

November 1969

DICING AND GAMING (A NOTE ON THE HISTORY OF PROBABILITY)

By F. N. DAVID

'See, this is new? It hath been already of old time.' (Ecclesiastes i. 10.)

1. A cynical archaeologist remarked recently that a symptom of decadence in a civilization is when men become interested in their own history, and he added that in the unlikely eventuality of any proof being required of the decadence of this phase of *Homo sapiens* it could be found in the present-day interest in archaeology. Most generalizations of a sweeping character such as this are unacceptable, chiefly because there is no way of putting them to proof; but the present interest of scientists in general, and of statisticians in particular, in the origins of scientific thought, is far from implying the decadence of science, whatever may be implied by an interest in the arts.

It is inviting, and at the same time profitless, to speculate why modern scientists have such an interest. The possibility of deciding priority of discovery which concerned the Victorian scientist so closely does not cause much controversy to-day, for the modern scientist would hold that to ascribe any discovery in the field of science to any single person is unrealistic. Thus, while we are taught at school, for example, that Newton and Leibnitz separately and independently 'discovered' the differential calculus, it would perhaps be more appropriate to say that Newton and Leibnitz each supplied the last link in the chain of reasoning which gave us the differential calculus—a chain which can be traced back through Pierre Fermat, Barrow, Torricelli and Galileo, and that it is surprising that there were only two mathematicians who did this.

Mathematics is essentially an expression of thought in which we build on the mental effort of our forerunners, and probability is no exception to this general rule. The real difficulty we meet with in trying to trace probability back to its origins is that it started essentially as an empirical science and developed only lately on the mathematical side. It is hard to say where in time the change came from empiricism to mathematical formalism as it appears to have taken place over hundreds of years; and the claims put forward for Pascal and Fermat as the creators of probability theory cannot entirely be substantiated.

2. When man first started to play games of chance is a time problem we shall never clearly resolve. We may place on record that it is a commonplace thing for archaeologists to find a preponderance of astragali† among the bones of animals dug up on prehistoric sites. One archaeologist stated that he had found up to seven times as many as any other bone, another put the figure at 500 (sic!), while yet a third, refusing to be drawn to a figure, stated that they were many. This fact has probably little significance. The astragalus has little marrow in it and was possibly not worth cracking for the sake of its contents as were the long bones; it is knobbly and presents no flat curves for drawing as does the shoulder

† The *astragalus* is a small bone in the ankle, immediately under the *talus* or heel-bone. See Pl. 2*a*.

blade for example. All we may do is to place on record that round about 40,000 years ago there were large numbers of the astragali of sheep, goat and deer lying about.

The astragali of animals with hooves are different from those with feet such as man, dog and cat. From the comparison in Text-fig. 1 we note how in the case of the dog the astragalus is developed on one side to allow for the support of the bones of the feet. The astragalus of the hooved animal is almost symmetrical about a longitudinal axis and it is a pleasant toy to play with. In France and Greece children still play games with them in the streets, and it is possible to buy pieces of metal fashioned into an idealized shape but still recognizable as astragali.

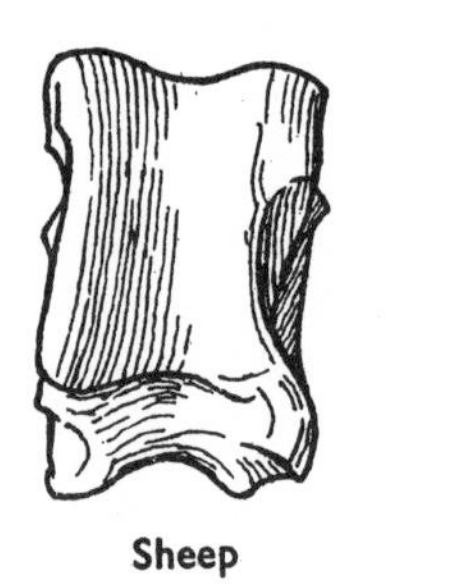

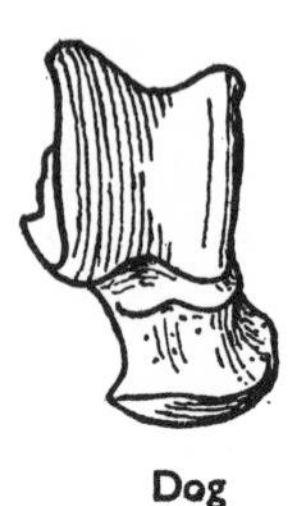

Text-fig. 1. Drawings of the astragalus in sheep and dog, natural size.

3. Some time between prehistoric man of four hundred centuries ago and the beginning of the third millennium before Christ *Homo sapiens* invented games and among these games, games of chance. We know from paintings, terra-cotta groups, etc., that the astragalus was used in Greece like the ancient quoit,* but there is no doubt from paintings on tombs in Egypt and excavated material that the use of the astragalus in games where it is desired to move counters was well established by the time of the First Dynasty. In one painting a nobleman, shown playing a game in his after-life, delicately poises an astragalus on his finger tip, a board with 'men' in front of him. A typical game of *c.* 1800 B.C. is that of 'Hounds and Jackals' illustrated in Pl. 1. The game seems similar to our present-day 'Snakes and Ladders'; the hounds and jackals were moved according to some rule by throwing the astragali found with the game and shown in the figure. Variants of this game were undoubtedly played from the time of the First Dynasty (*c.* 3500 B.C.).

It is possible but not altogether likely that these games originated in Egypt. They certainly did not originate in Greece, as has been claimed for reasons which we shall give later. However, Herodotus, the first Greek historian, like his present-day counterparts, was willing to believe that the Greeks (or allied peoples) had invented nearly everything. His claim that the Lydians introduced coinage has about as much foundation as his claim regarding games of chance. He writes (*c.* 500 B.C.) about the famine in Lydia (which was *c.* 1500 B.C.) as follows:

> The Lydians have very nearly the same customs as the Greeks. They were the first nation to introduce the use of gold and silver coins and the first to sell goods by retail. They claim also the invention of all games which are common to them with the Greeks. These they declare that they invented about the time that they colonized Tyrrhenia, an event of which they give the following account. In the days of Atys, the son of Manes, there was great scarcity through the whole land of Lydia. For some time the Lydians bore the affliction patiently, but finding that it did not pass away,

* From the name 'knucklebone' we might infer that among the early games were those in which the astragali were balanced on the bones of the knuckles and then tossed and caught again.

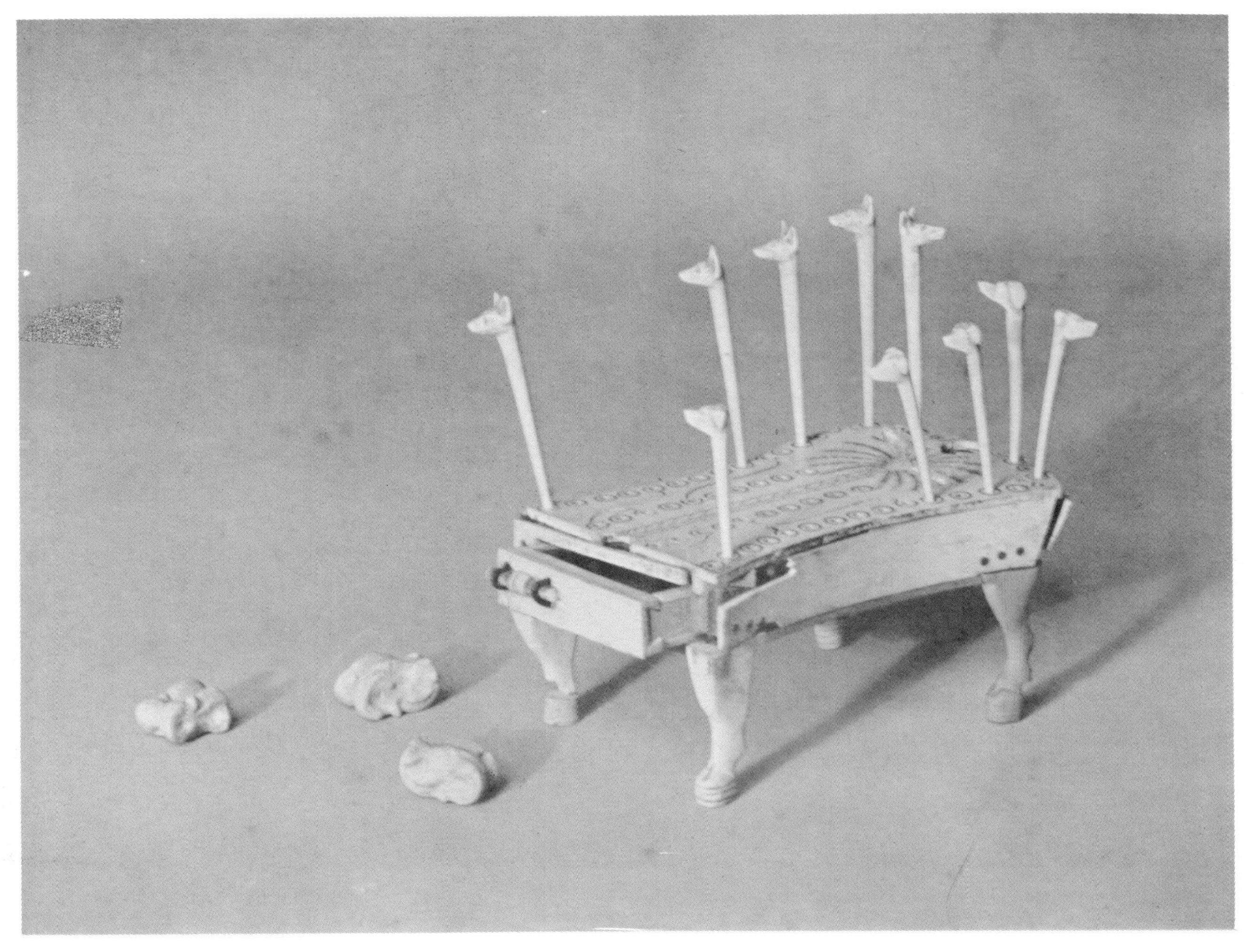

The game of Hounds and Jackals.

Plate 1

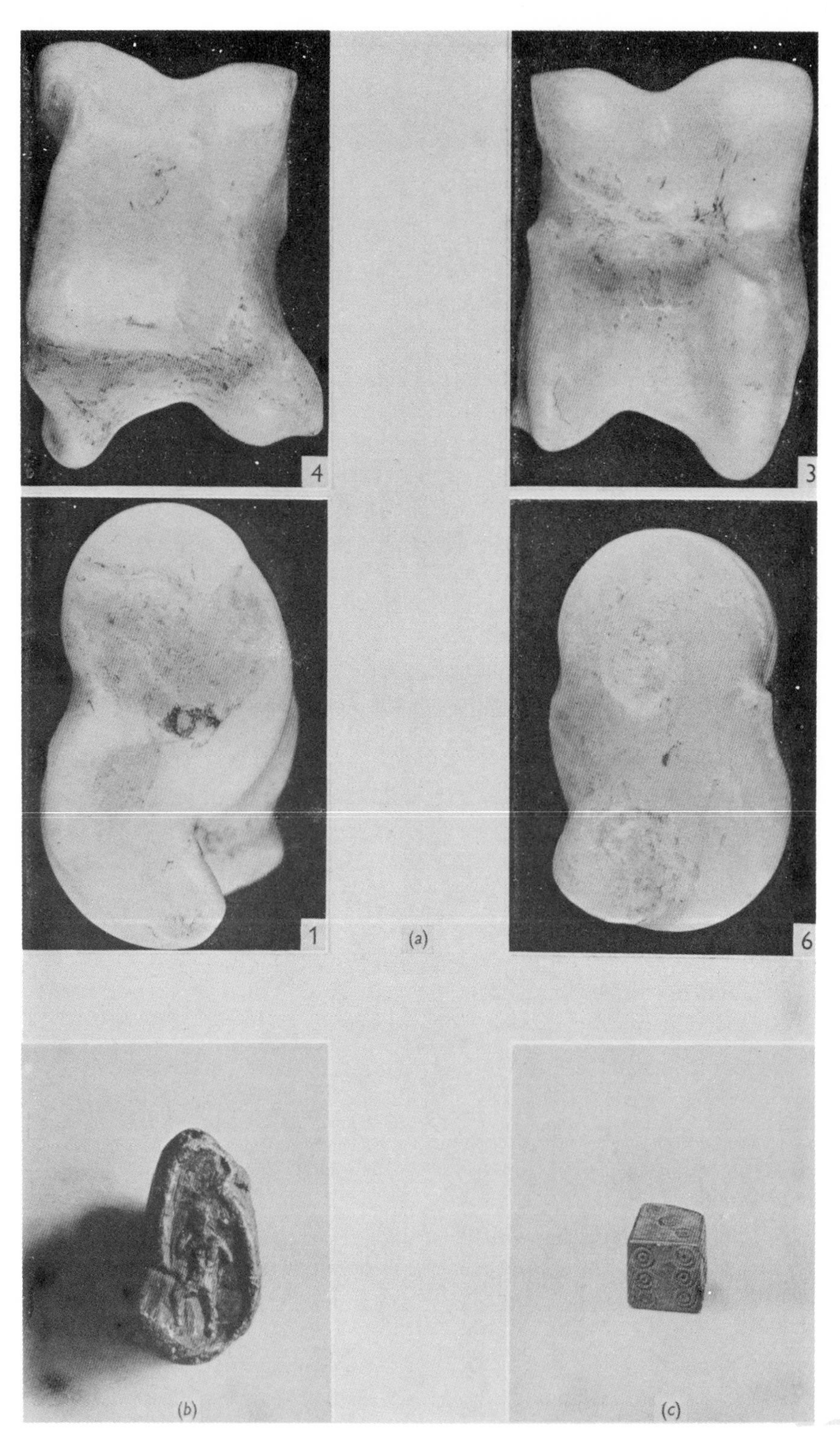

(a)

(b)

(c)

Plate 2

they set to work to devise remedies for the evil. Various expedients were discovered by various persons; dice and huckle-bones (i.e. astragali) and ball and all such games were invented, except tables,* the invention of which they do not claim as theirs. The plan adopted against famine was to engage in games on one day so entirely as not to feel any craving for food, and the next day to eat and abstain from games. In this way they passed eighteen years.

In yet another commentary we are told that games of chance were invented during the Trojan war by Palamedes. During the 10-year investment of the city of Troy various games were invented to prevent the soldiers' morale suffering from boredom.

4. The game of ball is mentioned by Homer and according to Plato was evolved in Egypt. It is not, however, a game of chance. The story of dice we shall return to, but we may first carry the story of the astragalus a little further. In the early part of the first millennium it would seem that astragali were used by both adults and children for their leisure games. Homer (*c.* 900 B.C.) tells us that when Patroclus was a small boy he became so angry with his opponent while playing a game of knucklebones that he nearly killed him. Another writer of the same period tells us that students played knucklebones everywhere, that they were acclaimed as presents and that as a prize for handwriting one student was given eighty astragali all at once! It is not difficult to imagine the small boys of that era collecting astragali as they collected marbles, much as the boys of our own era still do.

That the astragalus was used commonly in the gaming which the Greeks and later the Romans conducted with zeal and passion, the references in the literature of that time leave no room for doubt. One of the chief games may have been the simple one of throwing four astragali together and noting which sides fell uppermost. The astragalus has only four sides on which it may rest, and it has been deduced, among others by Nicolas Leonicus Thomeus (1456–1531), that a common method of enumeration was that the upper side, broad and slightly convex counted 4, the lower side broad and hollowed 3, the lateral side narrow and flat 1 and the other lateral which is slightly hollow 6. These aspects of a sheep's astragalus are shown in Pl. 2*a*. (With present-day astragali the probabilities of scoring 1 and 6 are each approximately equal to 1/10 and those of 3 and 4 approximately 4/10.) The worst throw for the Greeks with one bone was unity which they called the dog, and sometimes the vulture. The best of all throws with four knucklebones was the throw of Venus when all four sides were different which has an actual probability of about 1/26. But at different times and in different games the numbers must have been varied, for the throw of Euripides with four astragali, discussed by several fifteenth-century writers, was worth 40. How the bones fell to achieve this result is not stated, although Cardano writing in the sixteenth-century states that it was four fours. (Probability *c.* 1/39.)

5. In classical Rome the astragalus was imitated in carved stone with figures and scenes incised on the sides. A typical example is illustrated in Pl. 2*b*. Stone astragali have also been found in Egypt. At this time too we have the production of lewd figures in metal or bone varying in size from about 1 cm. to over 1 in. in height. That these figures were used for gaming may be deduced from the fact that the six possible positions in which the figure may fall are each marked with a number of dots.†

Besides the astragali it appears possible that throwing sticks were also used for games of chance, although it may be that they had a greater religious significance; we shall return

* This may have been an early form of backgammon or may have been shuffle-board.

† I have not tested these figures for bias. They are a development, I think, of dice rather than astragali.

to this point later. The throwing stick was made of wood or ivory and was often approximately 3 in. in length with cross-section when square of about 1 cm. each way. Such throwing sticks were known to the ancient Britons, to the Greeks, the Romans, the Egyptians and to the Maya Indians of the American continent. Sometimes the sticks are elliptical in cross-section with major axis of approximately 1 cm., but they are all alike in having only four numbers on them, one at each end of the upper face and one at each end of the lower parallel face. In the European throwing sticks the majority of numbers are marked by small engraved circles, but they are sometimes indicated by cuts in the wood or ivory. The Maya throwing sticks are marked by coloured scratches in ivory. The actual numbers marked vary. They are mostly 1, 2, 5, 6, but 3 and 4 have also been noticed. These throwing sticks are of little importance in gaming. They are mentioned because it is interesting to note the likelihood that gaming originated at many points, and, although this is a remark one could not defend, that it possibly was originally a debasement of a religious ceremony.

6. The six-sided die may have been obtained from the astragalus by grinding it down until it formed a rough cube. The Musée de Louvre has several astragali which have been treated in this way, but one cannot imagine they formed very satisfactory dice. The honeycombed (or cancellous) bone tissue has been exposed in several places and the crude die would clearly not have a long life. Whether the die was evolved in this way or not the evolution must have taken place some considerable time before Christ. The earliest die known was excavated in northern Iraq and is dated at the beginning of the third millennium. It is described as being of well-fired buff pottery. The dots are arranged as shown in Text-fig. 2(i), the edges at A and B being imagined folded away from the reader. It will be noted that the opposite points are in consecutive order, 2 opposite 3, 4 opposite 5 and 6 opposite 1.

A die excavated in Mohenjo-Daro (Ancient India) is also dated as third millennium, and it is also made of hard buff pottery. The order of the points is again consecutive, but this time we have 1 opposite 2, 3 opposite 4, 5 opposite 6. Few other dice have been recorded in this millennium. At the time of the XVIIIth dynasty in Egypt (*c.* 1400 B.C.) a die with the markings shown in Text-fig. 2 (ii) must have been in play. The arrangement of the five dots is unusual. Somewhere about this time, however, the arrangement of the numbers settled down to the familiar two-partitions of 7 opposite one another as shown in Text-fig. 2 (vi), which arrangement has persisted until to-day. Out of records, collected by the present writer, of some fifty dice of classical times made of crystal, ivory, sandstone, ironstone, wood and other materials, forty had the two-partition of 7 arrangement. A twelfth-century (A.D.) Greek bishop wrote that this was the way in which a die should be marked, and a sixteenth-century gambler theorizes that this arrangement was chosen to make it easy to check whether all the numbers had been marked on the die and no figure duplicated at the expense of leaving out another. One die of the first millennium is said to have 9 opposite 6, 5 opposite 3, and 4 opposite 2. It may have been especially made for a particular game; alternatively, it is possible that it may have some ceremonial significance. This is possibly also true of a die marked as in Text-fig. 2 (iii), although it might have been a die used for cheating.

7. That dice were used in Egypt is clear from the XVIIIth-dynasty specimen. It is thought, however, that dicing did not become common until the advent of the Ptolemaic dynasty (300 to 30 B.C.) which originated from Greece. Several dice are known of this

period including a beautiful specimen in hard brown limestone of side *c.* 1 in., which has the sacred symbols of Osiris, Horus, Isis, Nebhat, Hathor and Horhudet engraved on the six sides. This would almost undoubtedly have been used for some form of divination rites.

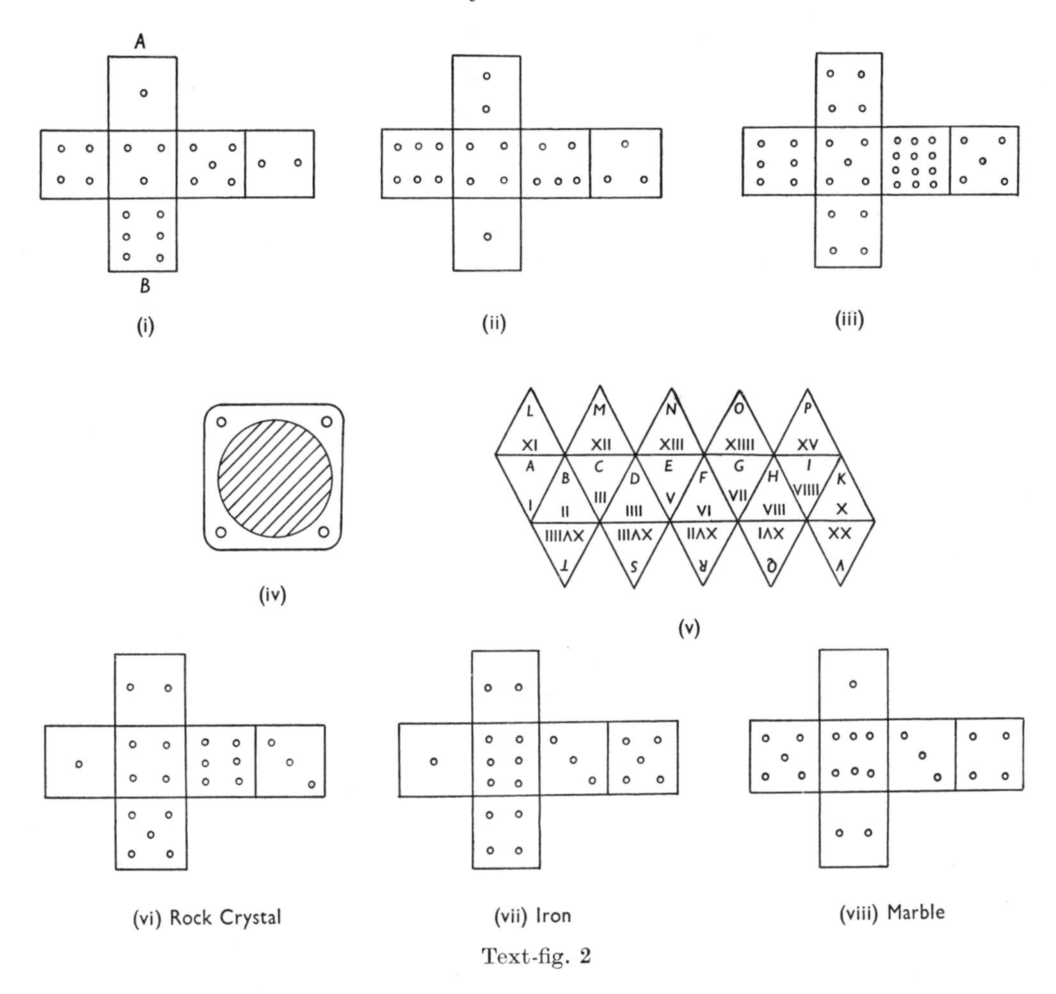

Text-fig. 2

Dice in Britain were in a very primitive state at the time of Christ. The pieces used then were formed by roughly squaring the long bone of an animal and cutting it into sections to form objects approximately cubical in shape. The marrow was taken out leaving a hollow square cylinder of which the cross-section in diagrammatic form is sketched in Text-fig. 2 (iv). (The British Museum has two of these.) These primitive dice had the two partition of 7 arrangement, 3 being opposite 4 on the hollow ends. Some dice had a 3 on each end and no 4. Several dice of this kind have been excavated in the chalk and flint country and dated late in the first millennium.

The working out of the geometry of solid figures by Greek mathematicians appears to have been followed almost immediately by the construction of polyhedral dice. A beautiful icosahedron in rock crystal now at the Musée de Louvre is the most famous of these. (In the diagram of Text-fig. 2 (v) it may be imagined the outline is folded away from the reader.) A figure with 19 faces badly cut but apparently imagined to be rectangular has a roman digit on each face from I to X, and above that the numbers rise by tens to C. The number LXXX is missing, but the number XX appears twice on one face. There was also a die of 18 faces probably formed by beating out a cubical die, a die with 14 faces and so on.

Faked dice were also not unknown. Apart from the device of leaving out one number and duplicating another it is stated that hollow dice have been found dating from Roman times. The unity on the face of the die forms a small round plate which can be lifted. It is suggested that a small ball of leather could be crammed into the hollow of the die through this hole in such a way as to cause the die to tend to fall in a predetermined manner.

8. Gaming reached such popularity with the Romans that it was found necessary to promulgate laws forbidding it except at certain seasons. What game was played by the common people we do not know, but there are many references to those played by the emperors. In Suetonius's *Life of Augustus* (Loeb's translation) we find:

> He (Augustus) did not in the least shrink from a reputation for gaming and played frankly and openly for recreation, even when he was well on in years, not only in the month of December, but on other holidays as well and on working days too. There is no question about this, for in a letter in his own handwriting he says, 'I dined, dear Tiberius, with the same company;...We gambled like old men during the meal both yesterday and to-day, for when the dice were thrown whoever turned up the "dog" or the 6, put a denarius in the pool for each one of the dice, and the whole was taken by anyone who threw the Venus'.

There are several other references to gaming in this *Life*. Whether the word talis should be translated as dice or astragali (knucklebones) is a moot point. The die as we know it is usually referred to as 'tessera'. The astragalus is often called the talus (or heel-bone), and this is the word Suetonius actually used. From the description of the play it would seem appropriate to read knucklebones for dice.

In Suetonius's *Life of Claudius* we are told that Claudius was so devoted to dicing that he wrote a book about it, and that he used to play while driving, throwing on to a board fitted especially in his carriage. From another source we learn that he played right hand against left hand.

9. These two instances are chosen to illustrate the passion for gaming which apparently possessed the Romans, and it is possible to cite many others. The question which constantly recurs to one while studying these games of the past is 'Why did not someone notice the equi-proportionality property of the fall of the die?' It is understandable that no theory was made to describe the fall of the astragalus. But the Greeks had performed the necessary abstraction of thought to make the mathematical idealization of the cube (and other solid figures); at first sight it seems curious that mathematicians did not then go on a little further and give equal weight to each side of the cube and so on. For if dicing and gaming generally were carried on by so many persons for so long that it was thought necessary to prohibit them, surely someone must have noticed that with a cube on the average any one side turned up as frequently as any other? We can only make guesses on this point, but it would seem to the writer that there are two possible explanations, the imperfections of the dice and their use in religious ceremonies.

10. *Imperfect dice.* We speak of a true or a fair die nowadays when we mean that there is no bias apparent when the die is thrown. In Roman times, and presumably earlier, it seems to have been the exception rather than the rule for the die to be true. Many dice of the classical period have been thrown by the writer and they were nearly all biased but not all in the same way. For example, three classical dice from the British Museum gave the results shown in the table from 204 throws each. The arrangement of the pips on the dice

were as in Text-fig. 2 (vi), (vii) and (viii). The rock crystal is a beautifully made die; the others are a little primitive, and the sides of the iron die are only approximately parallel. The marble and the iron dice are obviously biased, and this was true of many of the other dice examined. A photograph of a wooden die of the classical period is given in Pl. 2 (*c*). It will be noted that one of the faces shown is not square, and the impression one has is that the owner picked up a piece of wood of convenient shape, smoothed it a little and engraved the pips. It would therefore have been difficult, except over a long period, to notice any regularity.

Number of pips ...	1	2	3	4	5	6
Rock crystal	30	38	31	34	34	37
Iron	35	39	30	21	37	42
Marble	27	28	23	47	25	54

11. *Divination.* In spite of the imperfections of the dice it is probable that some theory might have been made if magic or religion or both had not been involved also. A scheme whereby the deity consulted is given an opportunity of expressing his wishes appears to be a fundamental in the development of all religions. As late as 1737 we have John Wesley deciding by the drawing of lots whether to marry or not (John Wesley's *Journal*, vol. 1, 1737, Friday, 4 March), and in the practices of present-day primitive tribes we get an echo from the classical era. At that time pebbles of diverse shapes and colours, arrows, astragali and dice were all used to probe the divine intention. In the temples there were various and varied rites attached to the process of divination by lot, but the main principle was the same. The question was posed, the lot was cast, the answer of the god was deduced. The dice (astragali, etc.) were thrown sometimes on the ground, sometimes on a consecrated table.

It was customary in classical Greece and Rome for the four astragali of the gamblers to be used in the temples. The prediction was that the throw of Venus (1, 3, 4, 6 uppermost) was favourable and the dogs unfavourable. In the temple of the oracles tablets were hung up and the priest, or possibly the suppliant, interpreted the throw of the four bones by reference to the tablets. Cases have been recorded, however, where five astragali were used. Greek inscriptions found in Asia Minor give a fairly complete record of how the throws of five were interpreted. Each throw was given the name of a god. Thus Sir James Frazer translates (commentary on Pausanias):

> 1. 3. 3. 4. 4 = 15 *The throw of Saviour Zeus*
> One one, two threes, two fours,
> The deed which thou meditatest, go, do it boldly.
> Put thy hand to it. The gods have given these favourable omens.
> Shrink not from it in thy mind. For no evil shall befall thee.

It is not clear whether the order of the numbers is important or not. If order does not matter then the probability of this throw is about 0·08.* The tesserae of the gambler were also used in divination ceremonies as well as the astragali, and it is possible that the same interpretation was given to the numbers falling uppermost, although the presence of the 2 and the 5 would make this a little awkward.

* I propose to write at greater length on 'divination probabilities' on a further occasion.

In addition to the divination carried out by the priests it was apparently a commonplace for individuals to perform acts of divination with regard to events in their daily lives. Thus Lucian, telling the story of the young man who fell madly in love with Praxiteles's Venus of Cnidos, writes:

> He threw four knucklebones on to the table and committed his hopes to the throw. If he threw well, particularly if he obtained the image of the goddess herself, no two showing the same number, he adored the goddess, and was in high hopes of gratifying his passion: if he threw badly, as usually happens, and got an unlucky combination, he called down imprecations on all Cnidos, and was as much overcome by grief as if he had suffered some personal loss.

Again we have from Propertius:

> When I was seeking Venus (i.e. good fortune) with favourable tali, the damned dogs always leaped out.

12. It is perhaps of interest here to interpolate a note on divination as reported practised by the Buddhists of present-day Tibet. According to Hastings (*Dictionary of Comparative Religions*) the simplest method is carried out by the people themselves. Many laymen are equipped with a pocket divination manual (*mô-pe*) and the augury found by casting lots. This lot-casting can either be odds and evens (the random pouring of grain, pebbles or coins from a horn, cup, etc.) or dice on a sacred board or cards on which there are magic signs, or sheets or passages of scriptures drawn from a bowl. The reincarnation prediction is, it is said by Waddell,* usually carried out by a priest. The rebirth chart seen by the writer consists of 56 2 in. squares (8×7). Each square corresponds to a future state. A six-sided die with letters on it is thrown down on to the rebirth chart, and according to the square on which it lands and the letter which falls uppermost so the priest predicts. Waddell, who visited Tibet as a member of a British Mission, obtained one of these charts and a die (*c.* 1893). He remarks: 'The dice (sic!) accompanying my board seems to have been loaded so as to show up the letter *Y*, which gives a ghostly existence, and thus necessitates the performance of many expensive rites to counteract so undesirable a fate.' Possibly a similar chicanery was practised in Roman times! It would seem a reasonable inference anyway that the mystery and awe which the religious ceremony would lend to the casting of lots for purposes of divination would prevent the thinking person from speculating too deeply about it. Any attempt to try to forecast the result of a throw could undoubtedly be interpreted as an attempt to forecast the action of the deity concerned, and such an act of impiety might be expected to bring ill luck in its train. In addition, as we have noted, a method for such forecasting could not easily be made owing to the imperfections of most of the dice. On the other hand, it is possible that probabilities were known to the priests since the ceremonial dice are well made.

13. Through the Dark Ages the Christian church appears to have carried on guerilla warfare against gaming with knucklebones and dice. The writers of the Renaissance make many references to bishops who write *de aleatoribus* or *contra aleae ludum* during the first fourteen hundred years of the Christian era. It is likely therefore that the bishops wished to get rid of the sortilege as a religious ceremony, and they succeeded to a certain extent in doing this, although divination by lot still survives to-day in the Moravian sect. What the bishops could not do was to stop men playing games of chance. There are several references in early French literature to gaming. The play of Jean Bodel, *Le Jeu de Saint Nicolas*, written *c.* A.D. 1200, has a scene where thieves are gambling in a tavern. They are playing

* L. A. Waddell, *The Buddhism of Tibet*, W. Heffer and Sons Ltd. 1934 (2nd edition).

the dice game of 'Le hasard',* the rules of which were set down clearly by Pierre-Raymond Montmort in his book some five hundred years later. (*Analys sur le jeux d'azard* 1708, p. 113). Bodel's play is interesting for the suggestion that the thieves knew how to manipulate the dice to produce a desired result.

14. With the invention of printing (*c.* 1450) and its rapid development during the latter half of the fifteenth century the references to games of chance become more numerous, but there seems to be no suggestion of the calculation of probabilities. Thus we find in the writing of François Rabelais—a man who might be expected to know the latest in games of chance as played in taverns—the following interesting passage: 'Then they studied the Art of painting or carving; or brought into use the antic play of tables, as Leonicus hath written of it or as our good friend Lascaris playeth at it'† (*Gargantua and Pantagruel*, Urquhart's translation, Book I, Chapter XXIV). *Gargantua and Pantagruel* was issued in sections at intervals between 1532 and 1552. The date of this reference will therefore be not long after 1532.

The Leonicus of the reference is Nicolas Leonicus Tomeus, a professor of Greek and Latin at Padua who was born at Venice in 1456. He was well known for his learning and philosophical bent and acted as tutor to the English Cardinal Pole when as a young man he visited Italy. According to Erasmus he was 'a man equally respectable for the purity of his morals and the profundity of his erudition'. His letters, which have been translated by Cardinal Gasquet, give an interesting picture of the life of an intellectual of that time. He died at Padua in 1531 and his collected works were printed at Basel in 1532. Rabelais is clearly referring to Leonicus's treatise *Sannutus, sive de ludo talario*, a dialogue in the manner of Plato concerning the game of knucklebones (astragali). There is, however, little relevant to the calculus of probability in this work. The discussion turns on references to the game in Roman literature and a description and argument of the value of the various types of throw.‡

A similar type of disquisition was written by Calcagnini about this time but possibly a little later than that of Leonicus. Celio Calcagnini was born at Ferrara in 1479 and died there in 1541. He was a poet, a philosopher and astronomer of repute; his treatise *Quomodo caelum stet, terra moveatur, vel de perenni motu terrae commentatio*, in which he held that the earth moved round the sun, anticipated Galileo Galilei by some years, for Galileo was not born until 1564. The dissertation of Calcagnini entitled *De talorum, tesserarum ac calculorum ludis ex more veterum* is less philosophical in tone than that of Leonicus. It is of interest to probabilists only in that it was an influence over Cardano, who, from his several references, had clearly studied it closely.

* According to the editor, F. J. Warne, of the text of the play, *Le Jeu de Saint Nicolas*, 'hasart' meant the throw of a certain number of points at dice, varying according to the game played. In present-day probability theory the meaning is of course much wider.

† Rabelais actually wrote 'en usage l'anticque jeu des tables ainsi qu'en a escript Leonicus'. Duchat in the commentary on the 1741 edition says 'Ce n'est point *tables* qu'il faut lire ici, comme dans toutes les Editions, mais *tales*'. Presumably Duchat (followed by later commentators) makes this correction because of the work of Leonicus referred to. It is just possible that Rabelais meant what he wrote and that he was referring to the ancient board game (from which the modern game of backgammon developed) in which the 'men' may have been moved by throwing astragali, the counting of the throws being that described by Leonicus.

‡ I have not been able to trace why Lascaris is mentioned by Rabelais. Andre-Jean Lascaris surnamed Rhyndaconus (1445–1535), a Greek scholar born in Phrygia, was Librarian to François I. He rescued many Greek manuscripts from the Turks. Possibly he collected references to gaming in Greek literature much as Leonicus did for Roman?

15. We arrive at the sixteenth century, then, with a well-known humanist Leonicus, and a great astronomer Calcagnini, writing on games of chance with no attempt or reference to the calculation of a probability. (This does not mean of course that some calculations had not been made in a manuscript which we do not know about.) There were, moreover, other scholars and bishops writing on the same topic about this time, so that interest in the subject was keen. As far as we know at present it was left to Gerolamo Cardano to make the step forward. Cardano, the illegitimate son of a geometer, Fazio Cardano, was born in Pavia in 1501. His illegitimacy was a bar to his professional advancement on more than one occasion, and it is possible that the bitterness engendered by this fact was responsible for his not too scrupulous regard for the attribution of other scientists' ideas. The crime of plagiarism was a common accusation among scientific workers of the sixteenth and seventeenth centuries, but of none was it raised more loudly than of Cardano who was strongly disliked by his contemporaries and despised by his successors. Until about the middle of the nineteenth century his biographers unite in regarding him as a charlatan; possibly at the present time the pendulum has swung too far the other way, and more is read into his writings than is justified. The truth would seem to lie somewhere between the extremes of charlatan and persecuted savant.

Cardano was physician, philosopher, engineer, pure and applied mathematician, astrologer, eccentric, liar and gambler, but above all a gambler. He himself owns that on one occasion he sold his furniture and his wife's possessions in order to get money to indulge his passion for gaming, and there is no doubt that this passion was one of the things which ruled him through his whole life. His chief interest professionally was medicine, but he interested himself also in the communication of spirits and the casting of horoscopes. He does not seem to have been too successful at this last, but he was not deterred from casting that of Jesus, a performance the impiety of which probably led to his imprisonment. Even allowing for the exaggerations of his biographers there seems to be no doubt that he was eccentric to the point of madness. This did not prevent him, however, from making contributions to pure mathematics, and it is to this combination of pure mathematician and gambler that we owe the *Liber de Ludo Aleae*. This treatise was found in manuscript in Cardano's papers after his death in Rome in 1576, and was first published in his collected works in 1663 at Lyon. Cardano implies that it was written *c.* 1526; the exact date is not important since no question of priority or plagiarism is involved, but it is curious that a manuscript of this kind should have survived fifty years of his remarkably variegated career.

16. The first complete translation of *de Ludo Aleae* into English is given in *Cardano, the Gambling Scholar*, by Oystein Ore, published in 1953. Ore remarks that the book is badly composed and that understanding of Cardano's work has possibly been hindered by this. There are some, however, who will not agree with his commentary on the treatise and who may feel that as much prescience is now attributed to Cardano as there was before too little. The crux of Cardano's work is to be found in the section entitled 'On the cast of one die' in Ore's translation:

> The talus has four faces and thus also four points. But the die has six; in six casts each point should turn up once; but since some will be repeated, it follows that others will not turn up. The talus is represented as having flat surfaces, on each one of which it lies on its back;...and it does not have the form of a die. One half of the total number of faces always represents equality; thus the chances are equal that a given point will turn up in three throws, for the total circuit is completed

in six, or again that one of three given points will turn up in one throw. For example, I can as easily throw one, three or five as two, four or six. The wagers are therefore laid in accordance with this equality if the die is honest....

We have therefore the necessary abstraction made; if the die is honest, i.e. if we may give equal weight to each side, then we may calculate the chances. There is no doubt, I think, that Cardano was led to this conclusion empirically and his generalization of it is partially wrong. For he goes on to discuss casts of two dice and three dice giving tables which are correct if 'the dice be honest'. When we come, however, to the section 'On play with knucklebones', it seems that he falls into error. The knucklebones (or astragali) have four sides. The different combinations of numbers which may arise in the throwing of four astragali are correctly enumerated, but the chances are calculated under the assumption that all sides are equi-probable; which they are not. Possibly Cardano had never played with astragali, for it is likely that if he had he would have noticed that to assume the sides of the astragalus had equal weight in his enumeration of alternatives was not adequate. But this fumbling suggests that he was not quite clear in his own mind about what he was proposing.

I do not think that the fact that Cardano did not quite see the mathematical abstraction clearly can detract from the fact that he did, on paper at any rate, as far as we know, calculate the first probability by theoretical argument, and in so doing he is the real begettor of modern probability theory. The claims of his biographer that he anticipated the law of large numbers, etc., may not be acceptable; it would appear that Cardano was judging from his experience rather than his algebra.

17. It would be strange if Cardano, following the mode of his age, did not communicate some of his thoughts about gaming to his pupils. Fear of being accused of plagiarism, fear of being plagiarized, may have kept him silent, but the whole tone of his treatise is a practical one; practical advice about playing, laying odds and so on make up a large portion of it. He would therefore almost certainly have discussed its contents with his friends, particularly if he thought about it over as long a period of time as he suggests. The fact that *de Ludo Aleae* did not appear in print until 1663 does not therefore seem to be a reason why Cardano's ideas should not have been common knowledge to scholars in Italy after his death, and the way in which Galileo-Galilei plunges into *his* discussion of dice playing, without much preamble, tends to lend colour to this.

Galileo-Galilei was born in Pisa in 1564, the son of Vincent Galilei, a musicographer well known in his day. He died in 1642 at Arcetri after a career as full of achievement as any that has ever been known. His contributions to science, both as astronomer and as mathematician, are striking for their originality of thought and clarity of purpose. Why this prince of scholars has never received the full recognition which is his due it is difficult to say. It is thought by some modern writers that his sensible recantation of the earth's movement, after physical torture at the hands of the Inquisition at the age of 70, has caused a revulsion to him among the scientists of later years. This is probably not so; what is more likely is that the envious fellow-scholars who delivered him to the Inquisition conspired after his death to belittle the work which he had done. In this they were possibly helped by Galileo's literary style which is noteworthy for clarity but not brevity, being in fact prolix and tedious in the extreme; no *i* is left undotted, no *t* is left uncrossed.*

* E. S. Pearson suggests to me that this prolixity was one which Galileo shared with many other Renaissance writers, and that it arose from the struggle which the early mathematicians must have had to formulate mathematical abstractions on paper. I think that this may well be so.

This being so if there was any doubt about the general method of procedure in calculating chances with a die we should have had a long disquisition on the subject. However, in *Sopra le Scoperte de i Dadi** he plunges straight away into his argument. The problem† is one already touched on by Cardano. Three dice are thrown. Although there are the same number of three partitions of 9 as there are of 10, yet the probability of achieving 9 in practice is less than that of throwing 10. Why is this? I quote a little from E. H. Thorne's translation of this note. The note begins:

> The fact that in a dice game certain numbers are more advantageous than others has a very obvious reason, i.e. that some are more easily and more frequently made than others, which depends on their being able to be made up with more variety of numbers. Thus a 3 and an 18, which are throws which can only be made in one way with 3 numbers (that is, the latter with 6, 6, 6 and the former with 1, 1, 1, and in no other way) are more difficult to make than, for example, 6 or 7, which can be made up in several ways, that is a 6 with 1, 2, 3 and with 2, 2, 2 and with 1, 1, 4 and a 7 with 1, 1, 5; 1, 2, 4; 1, 3, 3; 2, 2, 3. Again, although 9 and 12 can be made up in as many ways as 10 and 11 and therefore they are usually considered as being of equal utility to these, nevertheless it is known that long observation has made dice players consider 10 and 11 to be more advantageous than 9 and 12.

This extract serves to show how he begins the topic assuming that the calculations are known; it also serves to illustrate the prolixity of Galileo's style. After some discussion of the six 3 partitions of 9 and of 10, he goes on:

> Since a die has six faces and when thrown it can equally well fall on any one of these, only six throws can be made with it, each different from all the others. But if together with the first die we threw a second, which has also six faces, we can make 36 throws each different from all the others, since each face of the first die can be combined with each of the second....

After saying that the total number of possible throws with three dice are 216, he gives a table of the number of possible throws for a total of 10, 9, 8, 7, 6, 5, 4, 3, noting that the numbers 11–18 inclusive are symmetrical with these. Thus the number of possible throws for 10 is 27 and 25 for 9. His treatment of the problem is exactly that which we should use to-day and leaves us in no doubt that the calculation of a probability from the mathematical concept of the equi-probable sides of the die was clearly known to the sixteenth-century mathematicians of Italy. We can marvel at the person asking Galileo the question; he obviously gambled sufficiently to be able to detect a difference in empirical probabilities of 1/108.‡

18. Galileo's collected works were first published in Bologna in 1656, but this fragment on gambling was not included. It does appear in the more complete collection published at Florence in 1718. Since, however, Galileo thought the problem of little interest, for he did not pursue it, there seems to be no reason why he should have made a secret of it, and following the custom of his day he probably instructed his pupils. At any rate it is evident that the mathematical probability set was no stranger to the French mathematicians of the seventeenth century, as is witnessed by the now famous correspondence between Pascal and Fermat in 1654. The first letter of the series, from Pascal to Fermat, setting out the problem of points is missing. We have, however, Fermat's reply to it, and the subsequent

* This is Galileo's own title. *Considerazione sopra il Giuco dei Dadi*, a later title, appears first in the collected works of 1718.

† Like Pascal sometime later, Galileo wrote to answer a problem put to him by a gambler.

‡ M. G. Kendall points out to me that the problem posed by the Chevalier de Méré to Pascal concerning the problem of points involved similar small probabilities.

follow-up,* and from the way in which Fermat writes it seems clear that the actual definition of probability is assumed known. What the two savants were interested in was the application of this definition to specific problems which were concerned with dice playing between gamblers of equal skill and opportunity. The approach to the problems is similar to that of Galileo, and the generalizations which are made from the particular cases discussed are not well supported.

It is true that Galileo wrote on one problem only and fairly briefly at that, but it is difficult to see why Pascal and Fermat should be preferred as the originators of probability theory before Galileo or Cardano. It may well be that the precocity of Pascal as a mathematician led to much of his work being accepted with acclamation, and certainly without its priority being questioned. We find, for example, the famous Arithmetic Triangle in Stifel's *Arithmetic* (1543), in the *General Tratato* of Tartaglia in 1556, in the *Arithmetic of Simon Steven of Bruges* (Leiden, 1625). It is possible that Pascal may not have known of these writers. However, he certainly knew of Pierre Herigone's *Cours Mathematique* (Paris, 1634), since he makes several references to it in his own *Usage du Triangle Arithmétique pour trouver les puissances des binômes et Apotômes*. Herigone uses a table of numbers analogous to the Arithmetic Triangle to find binomal coefficients. Perhaps this same aura which dazzled Pascal's contemporaries (and at the same time caused them to overlook some of Fermat's work) still blinds us to-day.

19. If we take the origins for granted and look at developments of the theory, then by far the greatest impetus to theory during the years 1650–60 must have come from the publication of *De Ratiociniis in Aleae Ludo* by Christian Huygens. Huygens as a young man of 26 arrived in Paris in July 1655 on the equivalent of the English 'Grand Tour'. He did not meet Pascal, Fermat or Carcavi, the intimate friend of Pascal, but he did meet Roberval, professor of mathematics at the Collège Royal de France, who is mentioned by Pascal as having been also approached by the Chevalier de Méré. Huygens stayed in Paris from July to November, and after his return to Holland he began a correspondence with both Carcavi and Fermat which lasted over a period of years. The young man's imagination was obviously fired by the discussions he had in Paris, and his mathematical ambitions stimulated by the immense activity of the group which some ten years later (1665) was to found the *Académie des Sciences*. He set himself to work, and in March 1656 he wrote to Prof. van Schooten that he had prepared a manuscript about dice games. Francis Schooten was professor of mathematics at Leyden and had been Huygens's teacher. He took the young Huygens's manuscript (which was written in his native language), translated it into Latin and published it as an appendix to his *Exercitationes Mathematicae* in 1657. (A French translation of this appendix can be found in *Oeuvres de Huygens*, tome 14, on 'Calcul des Probabilités' published by *La Société Hollandaise des Sciences* in 1920.) In this *Tractatus de Ratiociniis in Aleae Ludo* Huygens sets out in a systematic manner what he must have learnt in Paris and adds some results which he may have achieved himself.

In the letter to Francis Schooten he writes

> ...quelques-uns des plus Célèbres Mathématiciens de toute la France se sont occupés de ce genre de Calcul, afin que personne ne m'attribue l'honneur de la première Invention qui ne m'appartient

* It is interesting to see Pascal fall into the same kind of trap which caused D'Alembert such controversy. In discussing the game of heads and tails and the tossing of a coin D'Alembert argued that the probability of throwing a head with two tosses of a coin was 2/3. For we may have TT, TH or H—when we stop, the second throw being immaterial, since we have achieved what we want.

pas. Mais ces savants...ont cependant cachés leurs méthodes. J'ai donc dû examiner et approfondir moi-même toute cette matière à commencer par les éléments, et il m'est impossible pour la raison que je viens de mentionner d'affirmer que nous sommes partis d'un même premier principe....

Accordingly Huygens begins by proving his basic propositions, deals at some length with the problem of points and then passes on to dice playing. His last proposition (XIV) has a familiar ring:

> If another gambler and I throw 2 dice turn and turn about with the condition that I will have won when I throw 7 points and he will have won when he throws 6, if I allow him to throw first, find my chance and his of winning.

His delineation of his fourteen propositions is admirably clear and concise, and it is no marvel that the tract was used by mathematicians as a reference book up to the time of James Bernoulli (who reprinted it) and beyond. Possibly by this crystallization of the ideas of the French mathematicians Huygens has earned the right to be regarded as the father of the probability theory.

20. After Huygens the interest of probabilists was not solely in gaming, although this interest did not die away for another hundred years or so. But with Huygens the new calculus seems fairly launched, and this is therefore a suitable point to make a break. There are many questions which one leaves unanswered. The drawings and paintings by palaeolithic man of himself are very rare, and there is probably no hope of finding pictures of his recreations. If he prized the astragalus as a toy it seemed a possibility that he might have carved or decorated it in some way, but I have not been able to find any record of this. But while we cannot pull aside the curtain from four hundred centuries the possibility does exist that the pre-historians may be able, one day, to take us back a little farther than the third millennium. The farther back one goes the more fragmentary the evidence, but the earliest dice found are described as being of 'well-fired buff pottery', and they certainly would not have been the first made.

The tantalizing period to the present writer is the period from the invention of printing to A.D. 1600. In this period we have two mathematicians only calculating probabilities, and yet this was in the immense intellectual ferment of the Italian Renaissance. It seems hardly possible that there were not other natural philosophers who attempted similar calculations, but such documents, if they exist, will only now come to light by chance.

The correspondence between the French mathematicians of the first half of the seventeenth century is almost complete, and presumably the possibility does exist here of finding further letters. They all seemed at one time or another to send letters to one friend under cover of letters to another, and such letters may conceivably still be ascribed to the wrong person. However, enough information does exist regarding the seventeenth-century mathematicians to make a coherent study, and if I appear to have done them scant justice it is because I find the period so interesting that I hope to write about it more fully on another occasion elsewhere.

Collecting information about dicing and gaming has been a hobby of mine for some time, and the list of persons who have drawn my attention to one aspect or another of it is formidable. I want to thank Prof. B. Ashmole of the British Museum who allowed me critically to examine the dice of the classical period which are in his care and M. Jean Charbonneaux of the Musée de Louvre who did me the same service. To Prof. C. M. Robertson of my own college I owe not only the privilege of tossing many dice but many stimulating discussions and useful references. A. J. Arkell allowed me to examine the dice brought by Prof. Sir Flinders Petrie from Egypt and to photograph various gaming boards not reproduced here. The breadth of knowledge and wide reading of Miss M. S. Drower have acquainted me with many Egyptian board games which provide a fascinating puzzle for those interested in deducing how they are played. Miss J. Lowe and R. Graves drew my attention to various references in classical literature. The illustration of the Hounds and Jackals game is by the courtesy of the Metropolitan Museum of Art of New York.

I want to thank Dr J. H. Willis who translated the *Sive de ludo talario* of Nicolas Leonicus for me, Dr E. H. Thorne who supplied a translation of Galileo's letter on dice, Prof. B. Woledge who drew my attention to early French plays and Miss J. Townend who drew Text-fig. 1. Miss J. Pearson and Miss J. Edmiston helped me to find many references and A. Munday and Miss A. Lodge helped with photographs. The manuscript as a whole owes much to the keen critical faculties of Prof. E. S. Pearson and Prof. M. G. Kendall. Part of this work was carried out with the aid of a grant from the Central Research Fund of the University of London.

THE BEGINNINGS OF A PROBABILITY CALCULUS

By M. G. KENDALL

1. The first article in this series by Dr F. N. David (1955)† has reviewed the development of dicing and gaming up to the time of Fermat and Pascal, who are popularly but erroneously supposed to have founded the calculus of probability. In this paper I shall try to trace the evolution of the idea of a probability calculus with especial reference to what Dr David calls the tantalizing period prior to A.D. 1600.

2. During the dark ages gaming was prevalent throughout Europe. At some unknown point of time dice finally ousted tali as instruments of play; and since cards were not introduced until about A.D. 1350 gaming must have been conducted mainly with dice for nearly a thousand years. Efforts on the part of Church and State to control the evils associated with it were as ineffectual then as they are today, and nothing is more indicative of the persistence of gambling than the continual attempts made to prevent it. The sermon of St Cyprian of Carthage *De Aleatoribus* (*c*. A.D. 240) was echoed twelve hundred years later in the more famous sermon of St Bernardino of Siena *Contra Alearum Ludos* of A.D. 1423. The gambling of the Germans referred to by Tacitus may perhaps have become more moderate but was equally prevalent in the thirteenth century when we find Friedrich II (1232) issuing a law *de aleatoribus* and Louis IX (1255) forbidding not only the play but even the manufacture of dice. A long series of edicts prohibiting the clergy from gaming (e.g. by Otto der Grosse, A.D. 952, the Councils of Trèves in A.D. 1227 and 1238, the Council of Worcester in A.D. 1240) are themselves eloquent of the failure on the part of the authorities to repress the evil.

3. We must remember, however, that all these banns and prohibitions were not really directed against games of chance as such but against the vices which accompanied them. There seems to have been nothing impious in creating a chance event or in using it for purposes of amusement. The Church was much more concerned about the drinking and swearing which accompanied gaming; and the State was more concerned about the idleness, thriftlessness and crime which were so often found among gamblers. Chaucer's Pardoner puts the official view of his day by giving an example of the blasphemy which usually accompanied a gambling game (almost certainly of hazard)*

> By Goddës precious heart and by his nails
> And by the blood of Christ that is in Hayles
> Seven is my chance, and thine is cinq and trey.
> By Goddës armës, if thou falsely play
> This dagger shall throughout thine hertë go !—
> This fruit cometh of the bitchëd bonës two:
> Forswearing, irë, falseness, homicide.

Even chess, the most innocent of all games, was classed among the major vices, at least for officials. The interdict of Louis IX referred to above says: 'They shall abstain...from

* Here and elsewhere I have modernized the spelling of the English quotations to some extent where the metre permits.

†Paper No. 1 in this volume

dice and chess, from fornication and frequenting taverns. Gaming-houses and the manufacture of dice are prohibited throughout the realm.'*

4. San Bernardino enumerates at length fifteen 'malignitates impiisimi ludi' but they are all moral evils (love of gain, idleness, corruption of youth, etc.) with the exception of blasphemy and contempt of the prohibitions of the Church. One feels that if there had been anything to say about the impiety of eliciting chance events for innocent entertainment the saint would certainly have said it. The general attitude of his time seems to have been one of toleration of the actual play but stern opposition to its associated vices. There is some positive evidence to the same effect. About A.D. 960 a certain bishop Wibold of Cambray invented a clerical version of dice to which I shall refer later; this sagacious realist evidently recognized the impossibility of stamping out the evil and hence attempted to turn it into good. Participants in the third crusade (A.D. 1190) had, in their briefing instructions, a carefully drawn up statement of the extent to which they might gamble; no person below the rank of knight was permitted to play at all for money; knights and the clergy might play but could not lose more than twenty shillings in twenty-four hours. Chaucer, in the Franklin's Tale, refers to the playing of chess and tables (backgammon) with the laudable object of distracting the heartbroken Dorigene. In 1484 Margery writes to John Paston (24 December):

> Please it you to wit that I sent your eldest son to my Lady Morley to have knowledge what sports were used in her house in Christmas next following after the decease of my lord her husband; and she said that there were none disguisings, nor harping, nor luting, nor singing nor none loud disputes; but playing at the tables (backgammon) and chess and cards; such disports she gave her folks leave to play and none other.

5. We may also notice a series of laws, beginning in the reign of Edward III, prohibiting the playing of certain games in order to promote manly sports. An act of Henry VIII added dice and cards to the list of unlawful amusements, although Henry, like many other monarchs, set a very bad example to his subjects. These laws were militaristic in origin. The common people were not to waste their leisure in playing peaceful games like bowls, ninepins, hockey and dice; their duty was to practice archery in readiness for the next war.

6. I recall these facts to establish two points. The first is that playing with dice (and later with cards) continued from Roman times to the Renaissance without interruption and was practised not only among the educated classes but among the middle classes and among the lower classes also. The second is that although the various Governments and the Church discouraged gaming to the point of prohibition, a great deal of play went on either as innocent pastime or, by popular approval, in defiance of the law.

7. One of the exasperating features of the many references to dice-playing between A.D. 1000 and 1500 is that authors invariably assume that their readers are familiar with the games they mention; and hence no rules of play are offered. We are thus very much in the dark about the exact nature of the games which were played. There are two in particular which have a long and interesting history: hazard, the ancestor of the modern American crap game; and primero, the ancestor of poker. It is instructive to consider briefly their line of development.

* 'Abstineant...a ludo etiam cum taxillis vel aleis vel scacis, et a fornicatione et tabernis. Scolas etiam deciorum prohibemus omnino....Fabrica vero deciorum prohibeatur ubique in nostro regno.' Taxillus is a diminutive of talus, but I do not know whether in this context it refers to the talus or the die.

8. The Romans played with four tali (huckelbones) but with only three tesserae (dice). At some early stage versions of dice-playing with only two dice are mentioned; for example, Bishop Eustathius, in a commentary on the Odyssey written in A.D. 1180, refers to games with two dice. The Chaucerian extract quoted above also mentioned two dice. In 1707 Montmort wrote of 'le quinquenove, le jeu de trois dez et le jeu du hazard. Les deux premiers sont les seules jeux de dez qui soient en usage en France, le dernier n'est commun qu'en Angleterre.' Both quinquenove and hazard were played with two dice; all three games are variants of the same idea.

9. Hazard, the game as distinct from its modern meaning of chance in general, was, I believe, brought back to Europe by the Third Crusaders. Godfrey of Bouillon gives a false derivation: 'À Hazait [Hazar] s'en ala ung riche mandement, et l'apiel-on Hazait pour le fait proprement que ly dés fu fais et poins premierment.' There can be little doubt that the word derives from the Arabic *al zhar*, meaning a die.* Wherever it came from, the name and the game must have spread rapidly through Europe. Jean Bodel's play *Le Jeu de Saint Nicolas*, ascribed to the year A.D. 1200, refers to *hazart*. Salimbene (the son of a crusader), writing about 1287, refers to playing 'ad azardum alias ad taxillum'. Dante's *Purgatorio*, written between A.D. 1302 and 1321, refers to *azar*; and Chaucer (about A.D. 1375) uses the word several times. The exact rules of play are not, so far as I am aware, on record. There were doubtless many variants. But the quotation from Chaucer given above suggests that the essential features of the modern game of craps were present at an early date; the addition of the numbers on two dice and the 'chances' of each player are clearly indicated.†

10. It might have been supposed that during the several thousand years of dice playing preceding, say, the year A.D. 1400, some idea of the permanence of statistical ratios and the rudiments of a frequency theory of probability would have appeared. I know of no evidence to suggest that this was so. Up to the fifteenth century we find few traces of a probability calculus and, indeed, little to suggest the emergence of the idea that a calculus of dice-falls was possible. It may be that gamblers had a rough idea of relative frequencies of occurrence—it is hard to see how they could fail to acquire such a thing; and as there is some evidence of the manufacture of false dice from Roman times onwards there was presumably a complementary notion of fair throwing. It may also be that some intelligent man worked out the elements of a theory for himself but guarded his secret on account of its cash value. But I do not really believe this. Other people tried to do the same thing later, but not with permanent success.

11. The earliest work I know which would seem to have mentioned the number of ways in which dice can fall is the game invented by Wibold, referred to above. So far as I am aware no contemporary manuscript has survived but an account of the game (a very obscure one, incidentally) was given by the chronicler Baldericus in the eleventh century, the work being first published in 1615. Wibold enumerated 56 virtues—one corresponding

* Libri (1838–41, vol. 2, p. 188) also gives a false derivation: 'Ce mot vient d'asar, qui en arabe signifie *difficile*', the difficulty in question being that of obtaining two aces or two sixes. Libri undoubtedly got this from the Dante Commentary mentioned in section 17 below.

† The 'chance' at hazard is not a probabilistic one. A player either calls or throws a 'main' (e.g. in one version, any number from 5 to 9 inclusive). He then throws again and may (*a*) 'throw out', in which case the dice pass to his opponent; (*b*) win outright by throwing a certain score; (*c*) throw another score which becomes his 'chance'. He then goes on throwing until either the 'main' or his 'chance' turns up, losing in the first and winning in the second case.

to each of the ways in which three dice can be thrown, irrespective of order. Apparently a monk threw a die three times, or threw three dice, and hence chose a virtue which he was to practice during the next 24 hours. It does not sound much of a game, but perhaps I have misunderstood Baldericus's account. The important point is that the partitional falls of dice were correctly counted. There was no attempt at assessing relative probabilities.

12. The use of dice for the purpose of choosing among a number of possibilities may well be much older than Wibold and certainly continued for long after his time. There exist several medieval poems in English, setting out the interpretations to be placed on the throws of three dice. The best-known is the *Chaunce of the Dyse* which is in rhyme royal; one verse for each of the 56 possible throws of three dice. For example, the throw 6, 5, 3 gives

Mercury that disposed eloquence
Unto your birth so highly was incline
That he gave you great part of science
Passing all folkës heartës to undermine
And other matters as well define
Thus you govern your wordës in best wise
That heart may think or any tongue suffise.

Another incomplete poem in the Sloane manuscripts also deals with the throws systematically but in a different manner; e.g. for 6, 5, 3,

Thou that has six, five and three
Thy desire to thy purpose may brought be
If desire be to thee y-thyght
Keep thee from villainy day and night.

Poems like these were doubtless used for elementary fortune-telling—one threw the dice to pick the phrase peculiar to oneself. Those mentioned come from what are probably early fifteenth-century manuscripts and have a certain interest in connexion with divination probabilities. For my present purposes the point to be noticed is that, for purely astragalomantic reasons, the different possible throws were enumerated and known without any reference to gaming or a probabilistic basis.

13. A similar idea is expressed in San Bernardino's sermon of A.D. 1423. The Saint makes a very detailed comparison of the Church of Christ and the church of Satan, represented in this instance by gaming. The Church corresponds to the gaming house; the altar to the gaming table; the sacrificial vessels to the dice box; and so on. In the middle of all this nonsense occurs a passage which does a little to compensate us for having to read it. 'The missal I compare to the die; for in flexibility, permanence and scope it is in no way inferior to the missal of Christ himself; and just as that missal is composed of a single alphabet of twenty-one letters, so in the [game of] dice there are twenty-one throws.'*

The twenty-one possible throws are undoubtedly those with two dice. This number is correct on the interpretation of the indistinguishability of partitions. One cannot help but admire the twisted ingenuity of the comparison, or speculate on what the Saint would have said had he been dealing with the gospel in Greek or Aramaic.

14. The earliest approach to the counting of the number of ways in which three dice can fall (permutations included) appears to occur in a Latin poem *De Vetula*. This remarkable work was regarded as Ovid's for some time and is included among certain medieval

* 'Missale vero taxillum, esse volo: qui quidem, et tractabilior, et durabilior, atque continentia non erit minor, quam sit missale ipsius Christi, cum in eius missali solum alphabetum, hoc est viginti una literae comprehendantur, ac totidem puncta in decio concludantur.'

editions of his poems. It is, however, supposititious and several candidates have been proposed for authorship. The one generally preferred is Richard de Fournival (1200–50), a gifted humanist of the Middle Ages and Chancellor of the cathedral of Amiens. If this is correct the poem was presumably written between A.D. 1220 and 1250. It contains a long passage dealing with sports and games, and with dicing in particular.* It is, perhaps, worth giving in full what is (if genuine) the first known calculation of the number of ways of throwing three dice. The text (taken from an edition published at Wolfenbüttel in 1662) is given in an appendix (pp. 33–34 below).

The relevant passage may be briefly and freely construed as follows:

If all three numbers are alike there are six possibilities; if two are alike and the other different there are 30 cases, because the pair can be chosen in six ways and the other in five; and if all three are different there are 20 ways, because 30 times 4 is 120 but each possibility arises in 6 ways. There are 56 possibilities.

But if all three are alike there is only one way for each number; if two are alike and one different there are three ways; and if all are different there are six ways. The accompanying figure shows the various ways.

[It follows, but is not stated, that the total number of ways is

$$(6 \times 1) + (30 \times 3) + (20 \times 6) = 216.]$$

15. The accompanying figures are shown in Plates 1 & 2, taken from Harleian MS. 5263. The figure referred to above is also given in my Fig. 1, taken from the printed edition published at Wolfenbüttel in 1662. The total of the last column is 108 which, doubled, gives us the total number of ways of throwing three dice. If this is a thirteenth-century product (the manuscript seems to be fourteenth century) it is astonishingly in advance of its time; and some of the phrases have a very modern ring (e.g. 'tria schemata surgunt', three cases arise; 'quemlibet cum dederis, reliqui duo permutant loca', if you fix one, the others permute in two ways).

16. There exists a medieval translation into French of the *De Vetula*, edited and published in 1862 by H. Cocheris, who is mainly responsible for the theory that de Fournival was the author. The translation takes considerable liberties with the original text and is not always easily matched against it. This poem is attributed to the fourteenth century. So far as I can see, the translator seems to have failed to understand the main point. He merely enumerates the 16 possible scores with three dice and points out that some of them occur more often than others. The essential step in the *De Vetula* has been lost.

17. In the sixth canto of the *Purgatorio* Dante mentions the game of hazard:

Quando si parte il giuoco della zara
Colui che perde si riman dolente
Ripetendo le volte e tristo impara:

(When a game of hazard breaks up the loser remains behind mournfully recalling the throws and learning by sad experience.)

A commentary on this passage published in 1477 says

Concerning these throws it is to be observed that the dice are square and every face turns up, so that a number which can appear in more ways [*sc.* as the sum of points on three dice] must occur more frequently, as in the following example: with three dice, three is the smallest number which can be thrown, and that only when three aces turn up; four can only happen in one way, namely as two and two aces.

* I am not competent to express an opinion about the attribution of the *De Vetula* to de Fournival, but I have considerable doubts on the point if the later printed versions correctly record what the author wrote about dice. Some of the critical passages may, however, be interpolations by later hands.

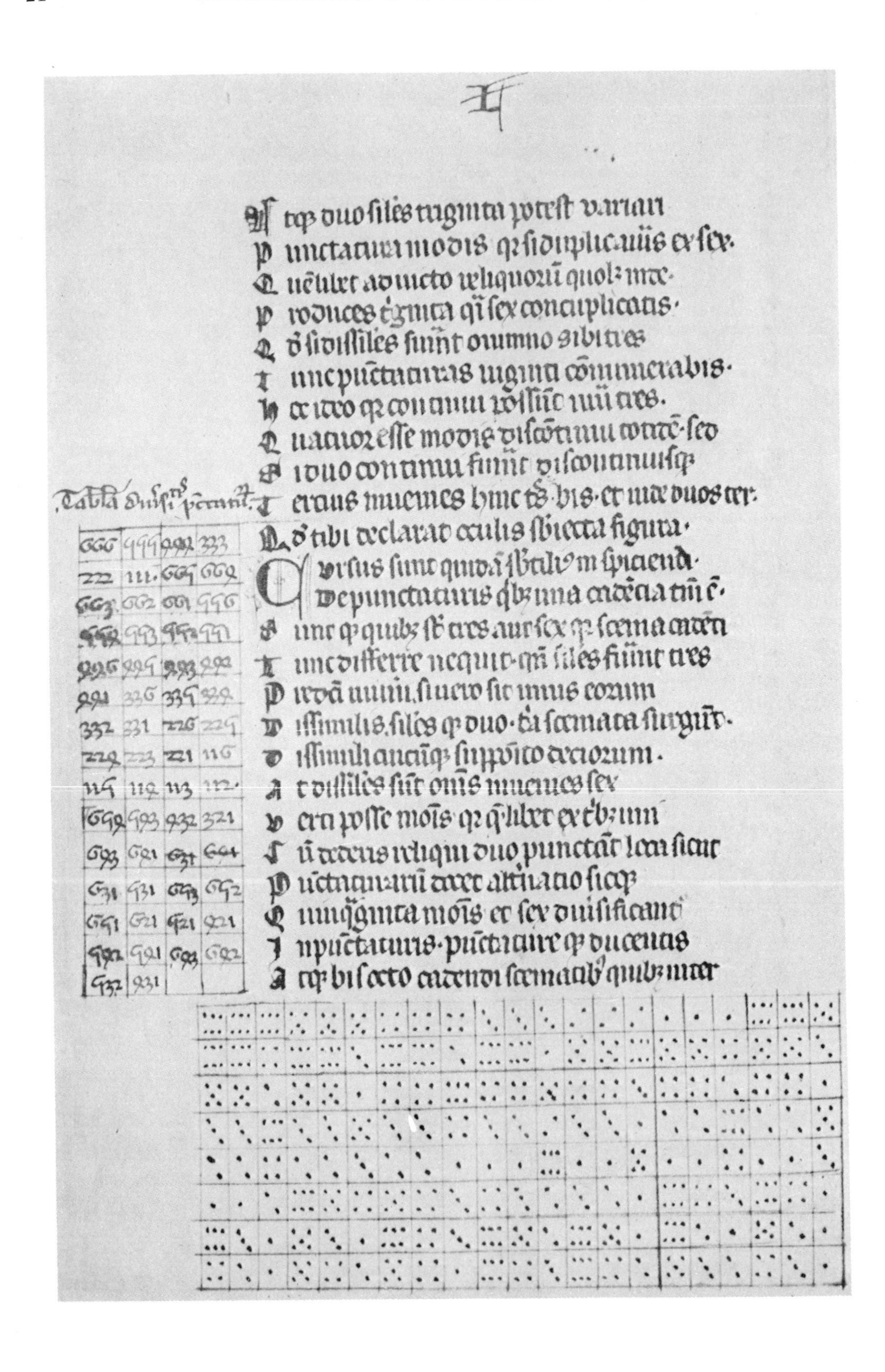

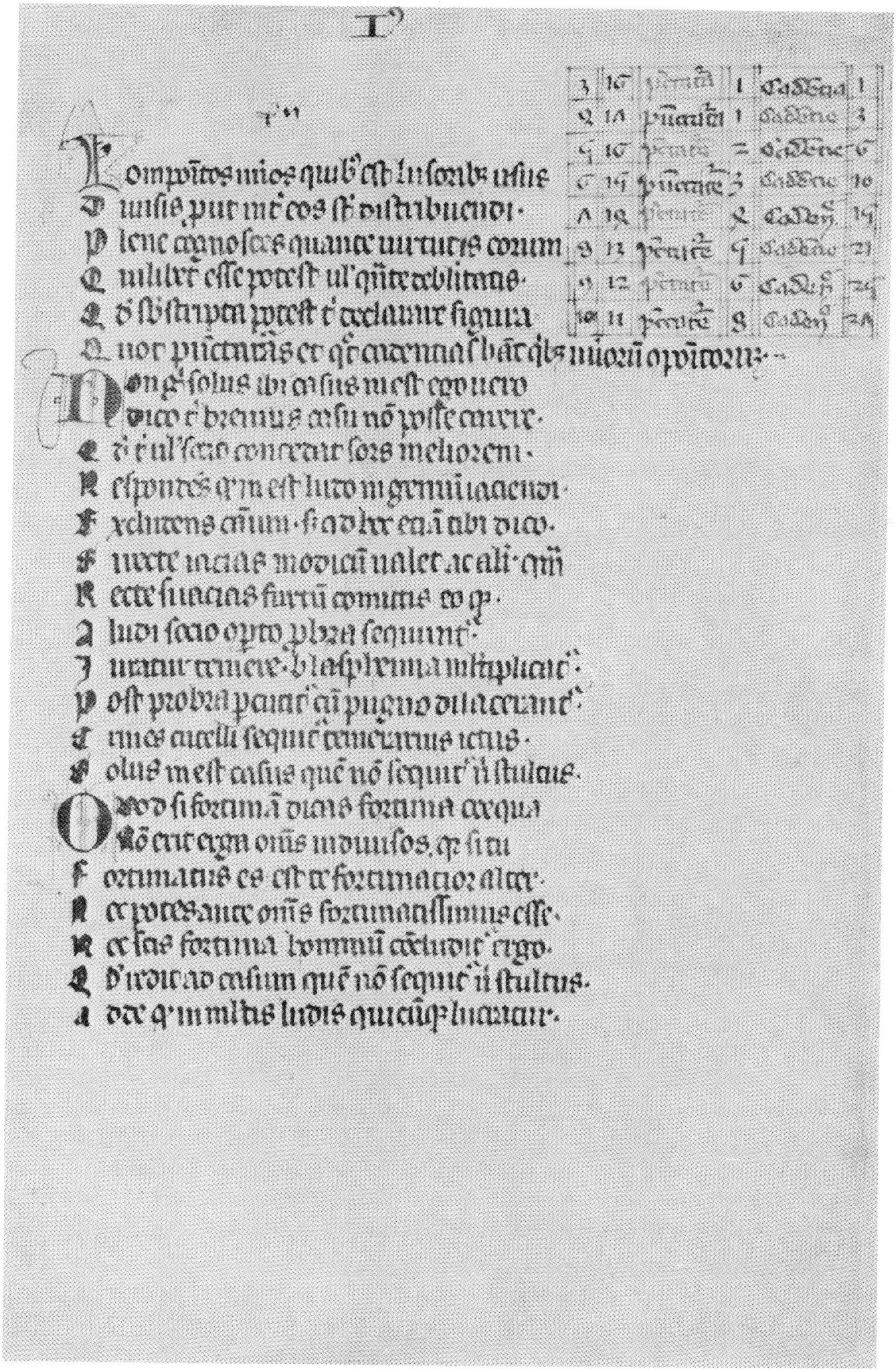

At this point the author seems to be on the verge of the usual fallacy that a three will occur equally as frequently as a four. But (more by luck than knowledge, in my view) he veers off this point and proceeds: 'and so, as these numbers can only happen in one way at each throw, in order to avoid tedium and too long a wait, they are not reckoned in the game and are called hazards. And so for 17 and 18. . . .The numbers in between can happen in more ways; the number which can happen in most ways is said to be the best throw of the set.'

19

Quinquaginta modis & ſex diverſificantur
In punctaturis, punctaturæque ducentis
Atque bis octo cadendi ſchematibus, quibus inter
Compoſitos numeros, quibus eſt luſoribus uſus,
Diviſis, prout inter eos ſunt diſtribuendi,
Plenè cognoſces, quantæ virtutis eorum
Quilibet eſſe poteſt, ſeu quantæ debilitatis:
Quod ſubſcripta poteſt tibi declarare figura.

Tabula III.

Quot Punctaturas, et quot Cadentias habeat quilibet numerorū compoſitorum.

3	18	Punctatura	1	Cadentia	1
4	17	Punctatura	1	Cadentiæ	3
5	16	Punctaturæ	2	Cadentiæ	6
6	15	Punctaturæ	3	Cadentiæ	10
7	14	Punctaturæ	4	Cadentiæ	15
8	13	Punctaturæ	5	Cadentiæ	21
9	12	Punctaturæ	6	Cadentiæ	25
10	11	Punctaturæ	6	Cadentiæ	27

Fig. 1

18. From these passages it seems clear that by the end of the fifteenth century the foundations of a doctrine of chance was being laid. The necessary conceptualization of the perfect die and the equal frequency of occurrence of each face are explicit. The idea of attaching binomial coefficients to the possibilities with two or more dice in order to calculate their relative frequency of occurrence had occurred in the *De Vetula* but seems to have been lost to sight. Not until 1556 did Tartaglia publish the scheme now known (very unjustly) as Pascal's arithmetic triangle, and then not in a probabilistic context. Nevertheless, if it is the first step which counts, that step had already been made by A.D. 1500.

19. Although the pioneer step in the *De Vetula* seems to have been overlooked by later writers in the next two centuries, the idea of enumerating the ways of obtaining given scores when permutations were taken into account must have been rediscovered by the beginning of the sixteenth century; for Cardano's *De Ludo Aleae* contains the essential ideas and is dated on internal evidence as written about 1526.* Oddly enough, however, we find the first problems in probability (so far noticed in the records) in quite a different context.

20. Fra Luca dal Borgo, or Paccioli, was an itinerant teacher of mathematics whose *Summa de Arithmetica, Geometria, Proportioni et Proportionalità*, published in 1494, was widely studied in Italy. He considers a simple version of what later became known as the problem of points: A and B, playing at a fair game (not dice, but *balla*, presumably a ball game) agree to continue until one has won six rounds; but the match has to stop when A has won five and B has won three. How should the stakes be divided?

21. Paccioli makes very heavy weather of this, but his solution amounts to saying that the stakes should be divided in the proportion 5:3. The error was noted by Tartaglia in his monumental *General Trattato* of 1556 (which date, we may remark, is thirty years after Cardano says that he was in possession of the basic principles embodied in the *De Ludo Aleae*). Tartaglia was always glad to point out errors in Paccioli with an acid superiority which foreshadows many of the modern writings on probability and statistics. He would have been more justified on this occasion if the alternative solution which he propounds had been correct, which it is not. He points out that according to Paccioli's rule, if A had won one game and B none, A would take all the stakes, which is obviously unjust. He then argues that the difference between A's score (five) and B's score (three) being two, and this being one-third of the number of games needed to win (six), A should take one third of B's share and the total stake should be divided in the ratio 2:1. Or so I interpret his rather prolix discussion. It would appear that if A has x and B y games in hand when the total number required to win is z, Tartaglia's rule requires that A takes a proportion $\frac{1}{2}+(x-y)/(2z)$ of the stake.

22. Two years after the *Trattato* there appeared a short work by G. F. Peverone, *Due Brevi e Facili Trattati, il Primo d'Arithmetica, l'Altro di Geometria.* In the first of these Peverone considers a similar problem, without reference to other writers. A has won 7 and B 9 games in a match going to 10 games. He gives two examples, which are effectively the same, and argues in this way:

> A should put 2 crowns and B 12 crowns [or, equivalently, the stake should be divided in the proportion 1:6]. For if A, like B, had one game to go each would put two crowns [or divide the stakes in equal proportions]. If A had two games to go against B's one, he should put 6 crowns against B's two, because, by winning two games he would have won four crowns,† but with the risk of losing the second after winning the first; and with three games to go he should put 12 crowns because the difficulty and risk are doubled.

* Dr David informs me that she is convinced that Cardano obtained the substance of his work on gambling from other sources. This would be in accordance with Cardano's character, for he was not an originator in spite of his extensive knowledge and peculiar gifts. On the other hand, it must remain a conjecture until those sources can be traced.

† 'Se giuocassero a 1 giuoco, bastarebbero scutti 2; et a due giuochi 6, per che vincendo solo 2 giuochi guadagnarebbe scutti 4; ma questo sta con pericolo di perdere il secondo, vinto il primo: però deve guadagnare scutti 6, et a 3 giuochi scutti 12, per che si indoppia la difficoltà e pericolo.'

23. I think this must be one of the nearest misses in mathematics. As far as the second game the argument is correct. If B has one game to go and is staking two crowns, then for A:

with one game to go he stakes 2 crowns,
with two games to go he stakes $2+4 = 6$ crowns,
with three games to go he stakes $2+4+8 = 14$ crowns,

and so on. Peverone was perfectly well acquainted with geometrical progressions and uses the word *progressione* in one exposition of his answer to this problem. Having got as far as the staking of 6 crowns by A with two games to go, if he had only stuck to his own rule and considered the conditional probabilities of gain more closely, he would have solved this simple case of the problem of points, in essence, nearly a century before Fermat and Pascal.

24. Students of the modern versions of hazard and primero have been struck by the accurate judgement of probabilities which they embody. The chance of success of the first player at craps, for example, should be 1/2 and is actually 244/493; the relative values of flush and straight at poker are correct although intuitively it is not clear what the order should be. It seems, however, that this situation has been reached empirically and not by calculation. Cardano fortunately gives us an account of primero as known to him.* From the pack of 52 the eights, nines and tens were removed, leaving 40. Four were dealt to each player two at a time. The cards had individual values, two counting 12, three 13, four 14 and five 15; six counted 18 and seven 21; an ace 16 and court cards 10. There were five combinations:

(*a*) Numerus (two or three cards of the same suit);
(*b*) Primero (all cards of different suits);
(*c*) Supremus (the three cards 7, 6, ace in the same suit);
(*d*) Fluxus (four cards of the same suit);
(*e*) Chorus (all cards of the same denomination).

These were valued in that order, a primero beating a numerus and so forth. The categories do not overlap and if two players held the same combination the elder hand won, irrespective of suit.

Now the chances of these events, or rather the number of ways in which they can occur on random drawing, are

Chorus		10
Fluxus		840
Supremus		120
Primero		8,990
Two of a suit	54,000	
Three of a suit	14,280	
Two pairs	12,150	80,430
		90,390

* In the Middle Ages many games were played in several forms. For example, there were about a dozen different versions of chess. Primero as described by Cardano does not seem to have incorporated a draw for new cards. Sir John Harington, in the reign of Elizabeth I, refers to a later version called 'prime' which does. Robert Greene, in 1591, makes a character say 'what will you play at, primero, primo visto, sant, one-and-thirty, new cut, or what shall be the game?' Shakespeare also mentions primero. I have seen it stated that the game (more properly primera) was imported from Spain on the occasion of the marriage of Mary Tudor with Philip II. The terms used in it certainly suggest a Spanish origin, but I do not see why the marriage of Henry VIII with Catharine of Aragon could not have been the occasion, or, indeed, that a specific occasion need be invoked.

In other words, the relative value of the fluxus and the supremus in Cardano's version were in the inverse order of their probabilities. At what point the present (correct) order in the modern game of poker emerged, I do not know, but it seems to have been before anyone was in a position to calculate the chances and persuade his fellows on the basis of mathematics that the orders should be reversed. In my opinion relative chances were all reached on the basis of intuition or trial and error in the games played up to the middle of the seventeenth century.

25. It seems clear that in fifteenth-century Italy the basic problems of chance in gaming had been raised and some small progress made towards solving them. A more thorough examination of the Italian mathematical books of the period may reveal further evidence on the point. One suspects that some of the simpler problems were circulated as a kind of puzzle, just as they are at the present day, without becoming of any recognized scientific importance. Galileo in his fragment *Sulla Scoperta dei Dadi*, written some time before 1642 (the date of his death), gives a complete solution of a problem in direct probability by correct enumeration of all possibilities, and he writes as if the problem were a new one, mentioning no previous authors.* Nevertheless, if Cardano's treatise is to be correctly assigned to 1526 the ideas must have been current for a century before Galileo wrote. It would appear that a calculus of probability not only was late in developing but that, once begun, it progressed exceedingly slowly.

26. Before we consider the reasons for this, something remains to be said about developments in France in the first half of the seventeenth century. The cradle of the probability calculus was undoubtedly, in my opinion, in Italy. From the fourteenth century, however, there were close connexions between France and Italy of a political as well as a geographical kind and an intellectual movement in one often generated a sympathetic movement in the other. The invasion of Italy by Charles VIII in 1494, though militarily and politically a failure, is generally regarded as a useful piece of intellectual cross-fertilization. Undoubtedly, a great many Italian works of art and ideas found their way to France with the remnant of Charles's army, although I doubt whether a copy of Paccioli's book was amongst them. In this case also a search among French books on mathematics written between A.D. 1400 and 1650 might prove to be very instructive.

27. The lack of written references to problems in the probability calculus is not necessarily indicative of a lack of contemporary interest. Knowledge of chances was so rudimentary that any capacity to gauge them accurately in play was worth a good deal of money. Huyghens, visiting France in 1657, found intense interest being taken in the doctrine of chances among mathematicians but encountered also a certain coyness about the disclosure of results. This was presumably due to fear of anticipatory publication rather than loss of income. Huyghens, being the man he was, merely worked out the theory for himself. A Latin translation of his little book, *De Ratiociniis in Ludo Aleae*, printed by van Schooten in 1657, was the first book published on the probability calculus and exercised a profound influence on James Bernoulli and Demoivre.

28. Now we come to the most interesting question of this period. Why was it that the calculus of probabilities was so long in emerging? We cannot suppose that the Greeks were incapable of making the necessary generalizations, even if they were hampered in working

* This is not a very weighty consideration. Early writers on probability, like those of the present day, often failed to mention their indebtedness to their predecessors. Laplace was notoriously bad at it.

out the details by their arithmetic and algebra. The same is true of the Arabs and of the early medieval Europeans. Dr David has suggested that imperfections in the dice may have something to do with it, but I cannot believe that this was a major reason. Some of the dice were, in fact, quite well made. The races which built the Parthenon, Trajan's Column, St Sophia and Notre Dame were quite capable of turning out a few cubes as good as any of those in current use. Nor do I think backward mathematical notation had much to do with it. We have seen that the partitional falls of dice were counted without difficulty in the tenth century. Four other possibilities are worth examining:

(*a*) the absence of a combinatorial algebra (or at any rate, of combinatorial ideas);
(*b*) the superstition of gamblers;
(*c*) the absence of a notion of chance events;
(*d*) moral or religious barriers to the development of the idea of randomness and chance.

29. Combinatorial algebra does not seem to have been cultivated by the ancients. Interest in it awoke in the sixteenth and seventeenth centuries. Leibniz published a tract *De Arte Combinatoria* in 1660 and Wallis a *De Combinationibus Alternationibus et Partibus Aliquotis Tractatus* in 1685. Doubtless the essential ideas could be traced back a good deal earlier.* Thus, when the calculus of probability was really under way, a combinatorial algebra lay ready to hand. Nevertheless, it seems to me, the absence of such an algebra cannot be held to account for the late emergence of the doctrine of chance. Cardano managed well enough without it and Galileo's enumeration of the 216 ways of throwing three dice is perfect, though apparently based only on arithmetical methods.

30. The superstition of gamblers is well known and has been remarked upon by many early writers. If men were logical and observant beings, one would have deemed it impossible for any person to engage very much at play and, at the same time, to believe that the favours of fortune were distributed unequally in the long run. But it seems quite possible for a player to believe two incompatible propositions; and with sufficient ingenuity, I suppose, it is also possible to reconcile a belief in the law of large numbers with a belief that the luck will change if one takes a different chair. One can say much on this subject. I am content to record the opinion that, although the psychology of the gambler may have done something to hinder the development of a concept of probabilistic law, it cannot have prevented the leading minds of the age from arriving at such a concept.

31. If we discount such factors as ill-made dice, indifferent mathematical expertise, superstition and so forth; and if we agree that play with dice and cards were so prevalent as to arouse general interest among intelligent people; then we seem driven to the conclusion that the late emergence of the probability calculus was due to some more fundamental factor. The very notion of chance itself, the idea of natural law, the possibility that a proposition may be true and false in fixed relative proportions, all such concepts are nowadays so much part of our common routines of thought that perhaps we forget that they were not so to our ancestors. It is in basic attitudes towards the phenomenal world, in religious and moral teachings and barriers, that I incline to seek for an explanation of the delay. Mathematics never leads thought, but only expresses it.

* The origins of combinatorial algebra would themselves make an interesting historical study. Wallis, at the age of 25, established a reputation for himself by deciphering Royalist letters intercepted during the Civil War. Bacon, in the reign of Elizabeth I, also took a keen interest in cryptography. Both men used ciphers based on combinations of symbols.

32. The Greeks and the Romans (so far as one can make summary statements about races whose members held such differing views) seem, on the whole, to have regarded the world as partly determined by chance. Gods and goddesses had influence over the course of events and, in particular, could interfere with the throwing of dice; but they were only higher beings with superhuman powers, not omnipotent entities who controlled everything. And the vaguer deities, Fortuna, the Fates and Fate itself appear to modern eyes more in the retributive role of a personified guilty conscience than as masters of the universe. The situation was radically changed by Christianity. For the early fathers of the Church the finger of God was everywhere. Some causes were overt and some were hidden, but nothing happened without cause. In that sense nothing was random and there was no chance. 'Nos eas causas', says St Augustine, 'quae dicuntur fortuitae (unde etiam fortuna nomen accepit) non dicimus nullas, sed latentes; easque tribuimus vel veri Dei, vel quorumlibet spirituum voluntati.' This view prevailed also in medieval times. Thomas Aquinas, arguing that everything is subject to the providence of God, mentions explicitly the objection that, if such were the case, hazard and luck would disappear. He replies that there are universal and particular causes; a thing can escape the order of a particular cause but not of a universal cause; and so far as it escapes it is said to be fortuitous with respect to that cause. St Thomas has an Aristotelean view of primary and secondary causes but we need not follow closely his struggles with the problems of causality, predestination and free-will. He reflected the spirit of his age, wherein God and an elaborate hierarchy of His ministers controlled and fore-ordained the minutest happening; if anything seemed to be due to chance that was our ignorance, not the nature of things.*

33. St Thomas is sometimes quoted as having expressed himself in favour of a frequency theory of probability, but, in my opinion, this rests on a source of confusion which it may be useful to remove in passing. Throughout this article I have been speaking of the doctrine of chances (which Demoivre translated as *Mensura Sortis*), not probability in the wider sense. Early writers used *probabilitas* with a different meaning, as relating to the degree of doubt with which a proposition is entertained. At the outset of our science the two things were distinct and it is a pity that they have not remained so and that our language has tended to confuse them. It seems to have been James Bernoulli who first thought of applying the doctrine of chances to the art of conjecture; and although we find applications to the assessment of the credibility of witnesses as early as 1697, it was not until Bayes' time (1763) that it was also applied to the acceptability of hypotheses. The resulting confusion, as is well known, has existed ever since and at the present time seems, if anything, to be getting worse. If any justification for the study of the history of probability and statistics were required, it would be found simply and abundantly in this, that a knowledge of the development of the subject would have rendered superfluous much of what has been written about it in the last thirty years.

34. Aquinas does not give a definition of *probabilitas*, but refers to Aristotle; 'Probabilia sunt quae videntur omnibus, aut plerisque, aut sapientibus, et his vel omnibus vel

* The same idea, of course, has come down to modern times in a line of direct descent but in a less deistic form; e.g. Spinoza, writing in 1677, says 'for a thing cannot be called contingent unless with reference to a deficiency in our knowledge'; D'Alembert, in 1750, says: 'il n'y a point de hasard à proprement parler mais il y a son équivalent: l'ignorance où nous sommes des vraies causes des événements.' More recently Paul Lévy in 1939: 'Nous pensons, quoique depuis les travaux d'Heisenberg d'éminents savants ne soient pas de cet avis, que la notion du hasard est une notion que le savant introduit parce qu'elle est commode et féconde, mais que la nature ignore.'

plurimis maxime nobilibus et probatis.' St Thomas himself regarded *probabilitas* as a quality which gave rise to an opinion. He says explicitly that it admits of degrees. Of chance (casus), he says: 'Ea quae accidunt semper vel frequenter non sunt casualia neque fortuita, sed quae accidunt in paucioribus.' And again 'Sicut in rebus naturalibus in his quae ut in pluribus agunt, gradus quidam attenditur quia quanto virtus naturae est fortior, tanto rarius deficit a suo effectu, ita et in processu rationis qui non est cum omnimoda certitudine, gradus aliquis invenitur, secundum quod magis et minus ad perfectum certitudinem acceditur.' As I understand the position, St Thomas recognized that *probabilitas* preceded certainty in the formation of knowledge; and that the frequency of events had something to do with the 'fortuitous' nature of the causality and the relative intensity of the underlying cause. But I cannot see in his writings an explicit statement that *frequentia* increased *probabilitas* or that the two were very closely related. The point really needs a deeper study than I felt worth while to give it, but it seems plain to me that the doctrine of chances was not present to his mind when *probabilitas* was under discussion.

35. A good case can be made for the thesis that the religious attitude of the times discouraged by implication the development of a study of random behaviour. Even this does not entirely satisfy me as the complete explanation, but I think it very likely that before the Reformation the feeling that every event, however trivial, happened under Divine providence may have been a severe obstacle to the development of a calculus of chances. It seems to have taken humanity several hundred years to accustom itself to a world wherein some events were without cause; or, at least, wherein large fields of events were determined by a causality so remote that they could be accurately represented by a non-causal model.* And, indeed, humanity as a whole has not accustomed itself to the idea yet. Man in his childhood is still afraid of the dark, and few prospects are darker than the future of a universe subject only to mechanistic law and to blind chance.

Whatever the reasons may be, it appears undeniable that the doctrine of chances took a remarkably long time to develop. Once launched, of course, it proceeded very rapidly; there is only a hundred years between Bernoulli's *Ars Coniectandi* and Laplace's *Traité*. But the results of that century of discovery required several thousand years of germination. Until more intensive research may have been able to lay bare the early essays and modes of thought of scientists in the fifteenth, sixteenth and seventeenth centuries the birth process of the probability calculus must remain somewhat enigmatic.

36. I have to express acknowledgements to several colleagues for help in tracing references; especially to Prof. Corrado Gini, to Father Dionisio Pacetti, O.F.M., whose authoritative knowledge of the works of St Thomas Aquinas was put freely at my disposal, to Prof. Sixtos Rios for some information about the game of primero, to Prof. W. Rose and to Mr G. Woledge for some references to early German and French gaming, and to

* As is well known, the assurance of lives was forbidden in several countries under Roman Catholic influence in the sixteenth century; e.g. in the statutes of Genoa for 1588: 'sine licentia Senatus non possunt fieri securitate...super vitam Pontificis neque super vitam...aliorum dominorum aut personarum ecclesiasticarum.' It is rather remarkable, but I suppose accidental, that nearly all the chief early writers on the probability calculus were subject to Catholic persecution. Cardano and Galileo were both victims of the Inquisition. The Bernoulli family were exiled in Switzerland, having been driven from Antwerp by Spanish persecution in the Netherlands. De Moivre lived in England because of the revocation of the Edict of Nantes. Pascal was a member of Port-Royal, though he did not live to see Jansenism driven into exile. Fermat and Montmort escaped; but Fermat lived in the provinces and published nothing on probability, and Montmort also lived a retired life.

Dr F. N. David, with whom I have had many discussions on this fascinating subject and who read this article in manuscript.

REFERENCES

DAVID, F. N. (1955).* Dicing and gaming (a note on the history of probability). *Biometrika*, **42**, 1.
LIBRI, C. D. S. (1838–41). *Histoire des Sciences Mathématiques en Italie*, 2 vols. Paris.

APPENDIX

Extract from 'De Vetula'

Forte tamen dices, quosdam praestare quibusdam
Ex numeris, quibus est lusoribus usus, eo quod
Cum decius sit sex laterum, sex & numerorum
Simplicium, tribus in deciis sunt octo decemque,
Quorum non nisi tres possunt deciis superesse.
Hi diversimode variantur, & inde bis octo
Compositi numeri nascuntur, non tamen aequae
Virtutis, quoniam majores atque minores
Ipsorum raro veniunt, mediique frequenter,
Et reliqui, quanto mediis quamvis propiores,
Tanto praestantes, & saepius advenientes.
His punctatura tantum venientibus una,
Illis sex, aliis mediocriter inter utrosque,
Sicut sint duo majores, totidemque minores,
Una quibus sit punctatura, duoque sequentes,
Hic major, minor ille, quibus sit bina duobus.
Rursum post istos sit terna, deinde quaterna,
Quinaque, sicut eis succedunt appropiando
Quattuor ad medios, quibus est punctatio sena,
Quae reddet leviora tibi subjecta tabella.

Hi sunt sex & quinquaginta modi veniendi,
Nec numerus minor esse potest, vel major, eorum.
Nam quando similes fuerint sibi tres numeri, qui
Jactum componunt quia sex componibiles sunt,
Et punctaturae sunt sex, pro quolibet una.
Sed cum dissimilis aliis est unus eorum,
Atque duo similes, triginta potest variari
Punctatura modis, quia, si duplicaveris ex se
Quemlibet, adjuncto reliquorum quolibet, inde
Producens triginta, quasi sex quintuplicabis.
Quod si dissimiles fuerint omnino sibi tres,
Tunc punctaturas viginti connumerabis.
Hoc ideo, quia continui possunt numeri tres
Quattuor esse modis; discontinui totidem: sed
Si duo continui fuerint, discontinuusque
Tertius invenies hinc tres bis, & inde duos ter:
Quod tibi declarat oculis subjecta figura.

Rursum sunt quaedam subtilius inspicienti
De punctaturis, quibus una cadentia tantum est;
Suntque; quibus sunt tres aut sex quia schema cadendi
Tunc differe nequit, quando similes fuerint tres
Praedicti numeri. Si vero sit unus eorum
Dissimilis, similisque duo, tria schemata surgunt,
Dissimili cuicunque superposito deciorum.
Sed si dissimiles sunt omnes, invenies sex
Verti posse modis, quia, quemlibet ex tribus uni

*Paper No. 1 in this volume

Cum dederis, reliqui duo permutant loca; sicut
Punctaturarum docet alternatio. Sicque
Quinquaginta modis et sex diversificantur
In punctaturis, punctaturaeque ducentis
Atque bis octo cadendi schematibus, quibus inter
Compositos numeros, quibus est lusoribus usus,
Divisis, prout inter eos sunt distribuendi
Plene cognosces, quantae virtutis eorum
Quilibet esse potest, seu quantae debilitatis:
Quod subscripta potest tibi declarare figura.

A NOTE ON PLAYING CARDS

By M. G. KENDALL

1. In an earlier article in this series (Kendall, 1956)[‡] I referred briefly to the introduction of playing cards into Europe. Subsequent correspondence arising out of that reference suggests that it may be useful to expand a little what was there said about the impact of cards on gambling.

2. Playing cards as we know them to-day in western Europe can be traced back in a clear line of descent to the beginning of the fifteenth century; but from that point backwards their history becomes more and more vague and their genealogy more and more fabulous. Where they originated is unknown. Claims have been put forward on behalf of origins in China, India, Arabia and Egypt. It is at least equally possible that they were independently invented in Europe. From the first the pictorial representations on the cards were thoroughly Western and do not suggest, to my eye at least, any trace of Eastern descent such, for example, as does the rook in chess.* Gambling with paper tickets is said to have been known in China in the twelfth century, and it is possible that the idea of playing cards drifted across to Europe along one of the early trade routes; but for the translation of the idea into practice one does not need to look further afield than fourteenth-century Italy.

3. No mention of cards has been traced in the West before A.D. 1350, and the absence of reference in authors like Chaucer and Dante, who mentioned everything, shows that they cannot have been known much before that date. They seem to have spread in the Mediterranean countries and Germany fairly rapidly. Numerals are said to have been added to the picture cards at Venice in A.D. 1377. Cards are mentioned at Nürnberg in A.D. 1380. In A.D. 1397 there appeared an ordinance in Paris prohibiting play at various games, including cards, to that part of the population who were engaged in manufacturing. (The story that cards were invented to amuse the mad Charles VI of France is false, although packs were made for him in A.D. 1393.) Specimens of such early dates have not survived. By A.D. 1423, however, the card pack appears to have evolved into its modern form. San Bernardino, whose sermon served me in good stead in the previous article (1956), refers to *charticellae in quibus variae figurae pingantur*, and goes on to mention the four suits, the Kings, Queens, Valets and Chevaliers and, as I interpret him, the trump cards of the tarot pack.†

4. The game of tarock (French *tarot*, Italian *tarocchi*) is still played in southern Europe in various versions and is probably the oldest card game. In its modern form the tarot pack consists of 56 ordinary cards divided into four suits, 21 trump cards and a wild card or joker, 78 cards in all. Fifteenth-century packs of this type exist. There are some early packs of even larger size, notably the Minchiate pack of 97 cards, 56 ordinary cards, 35 trump cards and 6 wild cards. Historians of playing cards also mention a very rare set of engravings known as the Mantegna tarot; but they are probably not by Mantegna, and I doubt very much whether they are tarot cards. They are engraved on sizeable but thin sheets of paper, and can hardly have been used in any sort of game involving shuffling and dealing; they are divided into five sets of ten, the first, for example, enumerating various social grades from the beggar to the Pope, the second giving the nine Muses and Apollo, the third giving ten branches of learning, and so on. My own opinion is that these cards were a teaching device and that the unknown inventor of the tarot pack copied some of them. Where he got the others is a mystery unless Miss Moakley is right in identifying them with Petrarch's *Trionfi*.

* Modern tarot packs are worthless evidence in this connexion. Under the influence of occultists, notably Court de Gebelin, who suggested in the eighteenth century that the trump cards incorporated the lost book of Thoth, and Eliphas Lévi, whose work on magic popularized the idea in the nineteenth century, tarot cards have acquired symbols such as sphinxes which are absent from earlier western cards.

† The early suits were clubs, coins (diamonds), cups (hearts) and swords (spades), which San Bernardino identifies with brutality, avarice, drunkenness and hatred. The court cards denote those who are outstanding in these vices. 'Presbyterorum et presbyterarum in numerabilem multitudinem esse volo: unde admittantur pusilli et magni, et feminae et masculi, periti et ignari, sapientes et stulti.' A hundred and fifty years later one John Northbrooke was inspired by this passage to coin the famous description of cards as 'the devil's picture books'.

Miss Gertrude Moakley (1956) has recently suggested that the trump cards were representations of the *Trionfi* described by Petrarch in one of his most popular poems.

‡Paper No. 2 in this volume

5. At some unknown point of time the tarot pack was simplified, or so I believe. The trump cards were dropped, and of the 56 ordinary cards one of the court cards was also dropped. (In most countries it was the Chevalier who was dropped, but in Spain it was the Queen, for reasons which it is interesting but unprofitable to speculate upon.) Thus there evolved the basic pack of 52 cards which is in general use to-day. The tarot pack survived independently, but in northern Europe is now mainly used for fortune-telling.

6. The student of the history of probability is interested in these matters only in two respects: the degree to which cards extended and encouraged gambling, and the reasons for the choice of the number of cards in the early tarot packs. From what San Bernardino says it seems that gambling began at a very early stage; presumably, as soon as a game came into existence, the adversaries began to wager on the outcome. However, extensive gambling with cards was of very slow development. Cardano mentions the game of primero, but early writers on chance confine themselves mainly to dice. The reasons, I think, were twofold: first, the permutational arithmetic required to deal with probabilities at cards was too complicated; secondly, cards were very expensive and dice much more common. Cards did not oust dice until the eighteenth century. A third reason, possibly, is that cards and backgammon involved more skill and had a higher social status. James I (1603) puts it rather well:

'As for sitting, or human pastimes—since they may at times supply the room which, being empty, would be patent to pernicious idleness—I will not therefore agree with the curiosity of some learned men of our age in forbidding cards, dice and such like games of hazard; when it is foul and stormy weather, then I say, may ye lawfully play at the cards or tables; for, as to dicing, I think it becometh best deboshed soldiers to play at on the heads of their drums, being only ruled by hazard, and subject to knavish cogging; and as for the chess, I think it over-fond because it is overwise and philosophic a folly.'

James, apparently, was not very good at chess, but his balanced broadmindedness in a Puritan age compels respect.

7. It is interesting to inquire why the early tarot packs consisted of $56+21+1$ or $56+35+6$ cards, but before dabbling in numerology let us note how treacherous a subject it is. One can make out a very good case for a connexion between the modern pack of playing cards and the calendar. The four suits correspond to the four seasons, the thirteen cards in a suit correspond to the lunar months, the 52 cards to the weeks of the year. If we score 11 for a Knave, 12 for a Queen, and 13 for a King, and add all the points for the 52 cards we get 364, which, adding one for the joker, gives us the number of days in the year.* Many a historical point has been argued on less coincidental evidence than this, but there is, in fact, no connexion between the playing pack and the calendar. The early suit cards were 56 in number.

8. Nevertheless, there are some striking resemblances between the number of the cards in the early packs and the number of ways in which dice can fall. The main number, 56, is the number of ways in which three dice can be thrown, permutations excluded, and I pointed out in my previous article (1956) that these ways were well known by the fourteenth century. The number 21, likewise, is the number of ways in which two dice can be thrown, permutations excluded. The number 35 of the larger Minchiate tarot could arise either as the number of ways of throwing three dice when sixes are ignored, or the number of ways of throwing four (four-sided) astragali. I am not prepared to lean very heavily on these coincidences; but they suggest that perhaps the constructors of the first packs, having to choose somehow, were influenced by their knowledge of dice-throwing.

9. In conclusion, I should like to correct one guess made in my earlier article. I suggested that the game of hazard was brought back to Europe by the third crusaders. It may, indeed, have been brought back in some such way, but if so, must have been imported by earlier crusaders. The word *hasart* occurs in line 10557 of Wace's *Le Roman De Brut*, dated A.D. 1155, and also in Chrétien de Troyes' *Erec et Enide*, line 356, dated A.D. 1160–70. For these references I am indebted to Prof. Brian Woledge, who remarks, incidentally, that the appearance of the initial 'h' in 'hazard' is an etymologica mystery which has never been solved.

REFERENCES

JAMES I (1603). *Basilikon Doron, or a King's Christian Duty towards God.*

KENDALL, M. G. (1956). Studies in the history of probability and statistics. II. The beginnings of a calculus of probabilities. *Biometrika*, **43**, 1.†

MOAKLEY, GERTRUDE (1956). The tarot trumps and Petrarch's *Trionfi*. *Bull. N.Y. Publ. Lib.* **60**, 55.

* Nor is this the worst. The number of letters in the sequence Ace, Two, Three, ..., Ten, Jack, Queen, King, is 52; so also in the sequence As, Deux, Trois, ..., Dix, Valet, Reine, Roi and in As, Zwei, Drei, ..., Zehn, Bube, Dame, Konig, the ch being taken as one letter. I owe this information to the firm of Thos. de la Rue and Co. Ltd.

† *Paper No. 2 in this volume*

THE BOOK OF FATE

By M. G. KENDALL

1. Divination by chance mechanism is a very ancient practice and is reported from many countries: China, Tibet, Greece, Rome and among the Germanic tribes known to the ancient world. Sticks and dice were both in use. This, for example, is Tacitus's account of the Germani:

'To divination and the drawing of lots (*auspices sortesque*) they pay as much attention as anyone. The method of drawing lots is standardized. They take a bough from some nut tree and cut it into strips, on which certain runes are written. These strips are scattered at random on a white cloth. Then the priest (if it is a public occasion) or the head of the household (if it is a family matter) prays to the gods, turns his eyes to heaven and take up three sticks, one at a time. The runes so collected are then interpreted.'

2. This is one of the earliest instances I know of divination in Europe by reference to a previously prepared script. The Romans also had their Sybilline Books. It was perhaps more common in Rome, and not uncommon among the Germanic tribes, to divine from natural phenomena such as the voices of birds or the entrails of a sacrifice. Mantic practices of all kinds flourished under the later Roman Empire until Constantine's time, when the official adoption of Christianity put a stop to the investigation of God's will by lot. Occasional attempts at revival were discouraged. Three medieval prelates who were each claiming possession of the bones of a saint decided, in what nowadays would probably be regarded as a very sporting spirit, to settle the dispute by drawing lots. But they were reprimanded for doing so and their agreement was cancelled by higher authority.

3. Vestiges of the religious practices of one age provide the superstitions of the next and the pastimes of the one after that. The Gods were transformed into demons and then into fairies; their rites become customs; divination dwindled into fortune telling. But, although consulting the oracle may have become only an entertainment it was taken seriously enough to survive every effort to suppress it. There does not exist much documentary evidence of the practice before the advent of printing but occasional references are enough to show that it never disappeared. In an earlier article (1956)* I have referred to some of the medieval poems like the 'Chaunce of the Dyse' which told fortunes or character by dice throwing. The so-called *sortes Virgilianae* seem also to have been practised over the period among those who were lucky enough to have books. A volume of Virgil was opened at random, a passage also chosen at random and then applied to the situation requiring elucidation. (Virgil in the dark ages was regarded as a notable necromancer.) Some comments of striking aptitude are reported but the only example I can quote is modern, probably apocryphal and illustrates the depths of degradation to which Sortilege has descended: when Margaret Bondfield, as Minister of Labour, was about to move a resolution requiring the renewal of unemployment assistance, known at the time as 'the dole' the *sortes* are said to have yielded the famous line from the Aeneid:

'Infandum, regina, jubes renovare dolorem.'

4. The advent of playing cards and of printing revived interest in this type of entertainment. In 1540 Francesco Marcolini of Forli published *Le Sorti*, a large and elaborately illustrated volume for consulting the oracle by using a pack of playing cards. This must have been a fairly popular work, for a second edition appeared in 1550. The scope and elaboration in this book suggests that there may somewhere be earlier prototypes. Marcolini has 50 questions, 13 for men, 13 for women and 24 for men and women. To each question he has 90 answers written in rhymed triplets, 4500 answers in all. The sortilege consists of going from a question to one of the 90 corresponding answers by an elaborate piece of mystification involving the drawing of cards at three different stages. The questions themselves are an interesting collection of the topics which exercised the minds of the gentlefolk of sixteenth-century Italy, and most of the politer ones would be found in any fortune-telling book of today. For example, question 3 for men is: whether it is better to choose a beautiful or a plain wife (*bella o brutta*). One possible answer to this (to take an example offered in illustration by Marcolini himself) is

'Una brutta e una bella ne torrai
Se vuol la legge; et se no vuol, sei matto
Se una sol moglie e brutta pigliorai.'

This, I take it, means: 'Have one of each if the law permits, but if not, you would be crazy to choose a plain one.'

**Paper No. 2 in this volume*

5. For students of combinatorial analysis and experimental design there is some mild interest in the mechanism by which Marcolini gets from question to answer. He works with a pack from which four cards of each suit, 3, 4, 5, 6 have been stripped. In point of fact, he makes no distinction between suits, so he might just as well work with a pack of nine cards of one suit. Two are chosen (effectively with replacement) so that there are 45 possible pairs (the order of a pair being immaterial). These 45 are displayed in full on a single page, one page to each of the 50 questions. The 45 possibilities, however, are divided into five groups of nine so that effectively there are only five possibilities per question. Marcolini then adds two more stages each involving choice of one card. But he again reduces the possibilities so that effectively there are only $5 \times 18 = 90$ per question. How or why these numbers were chosen is a mystery. Personally I do not feel much confidence in Marcolini's command of combinatorial analysis. There appear to me to be places where he failed to preserve the balance of his incomplete blocks and just faked them. But it is easy to be superior with the advantage of four centuries' experience, and his drawings are fascinating.

6. In *Le Sorte* we have the last stage of degeneration of the old sortilege, a deliberate obfuscation of the selective process. Henceforward the Book of Fate is of no scientific interest. It still exists. I recently bought a publication called *Napoleon's Book of Fate* part of which is evidently a lineal descendant of *Le Sorte*.

7. One curious development in the musical field may be worth recording. In 1757 a certain Johann Phillip Kirnberger published in Berlin a book which enabled Polonaises and Minuets to be composed by throwing dice. What is more, they were composed for two violins and pianoforte. This game seems to have caught on. Haydn is credited with a *Guioco Filarmonico* published in 1790 for the composition of minuets. To C.P.E. Bach has been attributed a publication of the same year for composing waltzes. In 1793, after Mozart's death, there was published (in four languages) a similar method, of composing with two dice. This work is attributed to Mozart in the Köchel–Einstein index (Anhang 294[d]). It reappeared in 1806 in England (C. Wheatstone, London) as '*Mozart's Musical Game*, fitted in an elegant box, showing by an easy system to compose an unlimited number of Waltzes, Rondos, Hornpipes and Reels.' I have not been able to trace the Wheatstone edition or to find out what was in the elegant box, but the German edition and Kirnberger's work are both in the British Musum.

These works are prefaced by fairly lengthy descriptions of how to apply the tables which are given. They give no account of the method of construction. I am certain that the general method in all these works must have been the same. We write down a simple harmonic sequence of, say, eight chords, finishing on the tonic. We can then compose a large number of bars from each of these chords by using notes of the triads or their inversions. We choose eleven of them and number them from 2 to 12 corresponding to the total points on two dice; and hence we can pick out one at random.

Likewise we can do the same for each of the eight chords in the sequence. And if we put the resulting eight bars together *in the right order* we shall, with a little luck, have an acceptable air. The same procedure can be followed with the left hand.

There only remains to put in the mystique. This is done by arranging our 88 bars in random order and then providing a table to unscramble them.

8. The student of probability and statistics will, I fear, derive little of relevance to his subject in these fortune-telling books and games. The chance mechanisms are of interest, perhaps, but there is no sign that, in any of the cases I have described, there was any appreciation of relative frequency.

References

Marcolini, F. (1540). *Le Sorte.* Venetia.

Kendall, M. G. (1956). Studies in the history of probability and statistics. II. The beginnings of a probability calculus. *Biometrika*, **43**, 1.*

Kirnberger, J. P. (1757). *Der allerzeit fertige Polonaisen und Menuettencomponist.* Berlin: G. L. Winter.

Köchel, L. von (1937). *Chronologisch-theoretisches Verzeichnis Sämtlicher Tonwerke W. A. Mozarts.* Dritte Auflage, bearbeitet von Alfred Einstein. Leipzig: Breitkopf und Härtel.

**Paper No. 2 in this volume*

RANDOM MECHANISMS IN TALMUDIC LITERATURE

By A. M. HASOVER

SUMMARY

Occurrences in Talmudic literature of probability notions are reviewed. A variety of random mechanisms were used and the ideology connected with them is discussed.

1. INTRODUCTION

In recent years several papers in Statistical Journals have been devoted to the development of probabilistic notions in the ancient world. We shall only cite the paper of David (1955),* which deals in part with probabilistic notions in the ancient world, and that of Mahalanobis (1957) on the Indian–Jaina dialectic of syādvāda. The reader is also referred to David's book (1962) for a more detailed treatment of her research in this field.

In the present paper, probabilistic notions and, in particular, random mechanisms in Talmudic literature are investigated. We shall show that in Ancient Israel a variety of random mechanisms were used in different situations, and that there was a well-defined ideology about random methods and games of chance, based partly on psychological and moral grounds.

The Rabbis were acutely aware of the relationship between the 'fairness' of a random mechanism and the symmetry of the various outcomes. They were also apparently intuitively aware of the law of large numbers in the case of equiprobable outcomes, although no mathematical analysis was developed, as far as I have been able to ascertain.

For a detailed introduction to the Talmud, see Chajes (1952).

2. THE JEWISH ATTITUDE TO CHANCE MECHANISMS

As with many other matters connected with ethics and religion, the Jewish attitude to the use of chance mechanisms in the ancient world was diametrically opposite to that of neighbouring nations. While dice gambling was in great favour among Greeks and Romans, it was forbidden among Jews, and persons indulging in it were subject to various legal disabilities. Again, astragali and dice were freely used for divination in temples devoted to idol worship, a practice strictly forbidden under penalty of death among the Jews (Deut. xviii. 9–14).

On the other hand, random mechanisms were used extensively in religious ceremonies, as well as for various legal purposes. The main technique used was that of drawing lots out of an *urn*. But other methods were also used, as we shall see in the sequel.

There were two main ideas underlying the use of lots. The first one was that the use of lots was a fair method of allocating duties or rewards among various contenders. This idea is clearly expressed in Prov. xviii. 18 'The lot causeth disputes to cease, and it decideth between the mighty'. According to Rashi's Commentary of the Bible, the Hebrew word translated by 'mighty' actually means people who have a mighty quarrel between them. Thus the meaning of the second part of the proverb is that the lot separates even those engaged in the mightiest quarrel. But we shall see later that the only methods of drawing lots which were in use had strictly equiprobable outcomes. Thus it seems that the notion of a 'fair game' was quite familiar to the Rabbis.

The second main idea underlying the use of lots is that the result of a lot undertaken by the Commandment of G'd will in fact give expression to G'd's will. This idea is most clearly expressed in the controversy between Joshua and Achan described in Sanh. 43 *b*. The background of the event is to be found in Joshua vii. 1–26.

We read:

'When the Holy One, blessed be He, said to Joshua, "Israel hath sinned", he asked Him, "Sovereign of the Universe, who hath sinned?" "Am I an informer?" He answered, "Go and cast lots". Thereupon he went and cast lots, and the lot fell upon Achan. Said he to him "Joshua, doest thou convict me by a mere lot? Thou and Eleazar the Priest are the two greatest men of the generation, yet were I to cast lots upon you, the lot might fall on one of you". "I beg thee", he replied, "cast no aspersions on [the efficacy of] lots, for Eretz Yisrael is yet to be divided by means of lots, as it is written, *The land shall be divided by lot*"' (Num. xxvi. 55) (Sanh. 43 *b*).

**Paper No. 1 in this volume*

From this passage, it appears that the concept of blind chance had been considered and consciously rejected in favour of the notion of complete control of the results of lot-drawing by Divine Providence, at least in those cases where the drawing of lots was done by command of G'd.

3. Some examples of the use of random mechanisms

We shall now describe various random mechanisms encountered in Talmudic literature. Some were used only once, like the method of division of Israel between the tribes, and others were used regularly, like the method of allocation of daily chores to the priests in the Temple.

(i) *The division of Israel*

The main passage describing this event is found in Baba Bathra, 122*a*:

'Eleazar was wearing the Urim and Tummim, while Joshua and all Israel stood before him. An urn [containing the names] of the [twelve] tribes, and an urn [containing descriptions] of the boundaries were placed before him. Animated by the Holy Spirit, he gave directions, exclaiming: "Zebulun" is coming up and the boundary lines of Acco are coming up with it. [Thereupon] he shook well the urn of the tribes and Zebulun came up in his hand. [Likewise] he shook well the urn of the boundaries and the boundary lines of Acco came up in his hand. Animated again by the Holy Spirit, he gave directions, exclaiming: "Naphtali" is coming up and the boundary lines of Gennesar are coming up with it. [Thereupon] he shook well the urn of the tribes and Naphtali came up in his hand. And [so was the procedure with] every [other] tribe.'

It is interesting to note here that the results of the drawing were announced before the drawing in order to emphasize that the lot was an expression of Divine will. Rabbi Samuel ben Meir (1085–1158) in this commentary to this section of Baba Bathra writes:

'They were asking the Urim and Tummim first, before the appointed person drew from the urn, in order that the minds of the Israelites should cool off, seeing that the lot came up as prophesied, and they would thus be convinced that the division was honest.'

Let us remark that the method of lot-drawing was not a direct method, but a more elaborate coincidence method, providing an additional safeguard against cheating. In fact, according to the Jerusalem Talmud (Yoma, Chap. 4, Sect. 1) the drawing from the two urns was performed by two different priests.

(ii) *The drawing of lots for the scapegoat on the day of atonement*

Every year, on the Day of Atonement, the High Priest used to sacrifice two he-goats in the Temple of Jerusalem, in accordance with Lev. xvi. 5–10.

'And from the congregation of the children of Israel shall he (Aaron) take two goats for a sin-offering, and one ram for a burnt offering.... And he shall take the two goats, and place them before the Lord at the door of the tabernacle of the congregation. And Aaron shall put lots upon the two goats; one lot "for the Lord" and the other lot "for Azazel". And Aaron shall bring near the goat upon which fell the lot "for the Lord", and offer him for a sin-offering. But the goat on which fell the lot "for Azazel', shall be placed alive before the Lord, to make an atonement with him, by sending him away to Azazel into the wilderness.'

The details of this ceremony are discussed at great length in the tractate Yoma (37*a*–41*a*), but the account given there is not a connected one and is intermingled with discussions on other topics. So instead of giving the original Talmudic text, we shall quote the account given in the Mishneh Torah of Maimonides (Day of Atonement, Chap. 3). Every detail of this account is of course based on the Talmudic text, even though the sources are not indicated. We shall do the same later on for the details of the lot-drawing for the daily sacrifice.

Maimonides writes:

'Concerning the two lots: on one of them was written "for the Lord", and on the other was written "for Azazel". They might be made of any material: wood, stone or metal. However, one was not to be large and the other small, or one of silver and the other of gold. Rather, both were to be alike. They were, originally, made of wood, but in the Second Temple they were made of gold. Both lots were placed in a vessel that could contain two hands, so that one might put in both his hands without reaching purposely (for a particular lot). This vessel was unhallowed. It was made of wood and was called "the urn"....'

'The High Priest shook the urn and brought up in his two hands the two lots for the two he-goats. He then put the two lots upon the two animals, the one in his right hand on the animal on the right and the one in his left hand on the animal on the left....'

Let us note that the Talmud explains that it was necessary to shake the urn, lest the high priest take one lot intentionally (Yoma 39*a*). For it was considered a happy omen when the lot for the Lord came up in the right hand, and the temptation was great to improve upon chance by dexterous manipulation.

An additional precaution was the fact that the two lots were not just inscribed, but actually engraved in order that the writing should not rub off (Jerusalem Talmud, Yoma, Chap. 4).

(iii) *The allocation of daily duties in the temple*

A completely different type of random mechanism, used for a very different purpose, namely the allocation of daily duties in the temple, is described in Yoma 22*a*. It is important to note that while in the two preceding ceremonies the drawing of lots was called ' Goral ', which means a little ball or stone (Jastrow, 1950), the allocation of daily duties is called ' Payis ', a word whose root verb means ' to pacify '. Thus the purpose of the allocation by lots seems to have clearly been the achievement of a fair and honest division. The fact that it was not carried out by the High Priest indicates that it was not primarily an appeal for G'd's decision.

We shall here also quote the description given in the Mishneh Torah (Daily Offerings, Chap. 4):

' How was the lot cast? The priests would stand in a circle and agree upon a number, say eighty, or a hundred or a thousand, or any other number upon which they would agree. The officer would say to them, " Show your fingers ", and they would thrust out their fingers, one or two. If perchance someone thrust out three fingers, it was still counted. The thumb, however, might not be thrust out in the Sanctuary because of cheaters; the thumb being short, it might easily be thrust out and bent in. If one, therefore, did thrust out a thumb it was not counted. The officer would begin to count from the designated person whose mitre he had removed at first. He would count on the fingers and go round and round until he reached the number upon which they had agreed. The person at whose finger the number was reached would come out first in the lot for the service. '

' Why was the number agreed upon counted on the thrust-out fingers and not on the persons themselves? Because it was forbidden to count Israelites except by means of some other object; for it is said: " and he numbered them by sheep " (1 Sam. xv. 4)'

There are several points to be clarified in connexion with this passage. First of all the impression is gained that the officer was counting the individual fingers. This would of course make the lot asymmetrical, and give a better chance to those who thrust more fingers than others.

It is clear that here is an acid test of whether the Rabbis understood the concept of equiprobable outcomes or not. Moreover, as the lot-drawing was repeated day after day, and taking into account the quarrelsome temperament of the priests, which is stressed many times in the Talmud, and their deep personal involvement in the outcome of the lot-drawing, it is reasonable to infer that any departure from an equitable distribution of the chores would have been noticed. Thus the law of large numbers is also involved in clarifying the exact method of counting used.

The Talmud discusses the matter and unequivocally states that *all* the fingers of one person were counted for *one*. Here is the quotation:

' *And how many did they put forth? One or two.* If they may put forth two, why is it necessary to mention that they may put forth one?—R. Hisda said: This is no difficulty: The one speaks of healthy persons, the other of sick ones. Thus it has been taught: One finger is put forth, but not two. To whom does this rule apply? To a healthy person, but a sick one may put forth even two. But the Yehidim (i.e. the scholars) put forward two and *one counts only one thereof*. But has it not been taught: One does not put forth either the third finger or the thumb because of tricksters, and if one had put forth the third finger it would be counted, but if one had put forth the thumb it would not be counted, and not alone that but the officer strikes him with a *pekia* (i.e. a whip)?—What does " it would be counted " mean? *Only one*' (Yoma 23*a*). Normally only one finger was lifted. But if there were present some older, weaker or sick priests for whom it was inconvenient to put one finger forth and hold it aloft until the count was over, the officer would require all to put forth two fingers, which is less of an effort.

The theoretical basis of the method is of course that the sum modulo m of a random variable X equidistributed on the numbers $1, \ldots, m$ and any random variable on the integers independent of X is equidistributed on $1, \ldots, m$.

The reason why putting forth the thumb was forbidden was that a trickster, foreseeing at the end of the count where it would end, might place his index-finger at some distance from the thumb, so that the officer would count his two fingers as belonging to two people, with the result that the count would be wrong and designed to serve the trickster's end.

It is worthwhile to mention that the number agreed upon for the counting had to be much larger than the number of priests present (Yoma 22*a*, Rashi's commentary). This eliminated any practical possibility of cheating, as the priests were very unlikely to carry out in their heads a division operation involving large numbers, and ensured the independence of the decisions of the priests and the counting officer.

Finally, it is interesting to note that in older times there used to be a race up the ramp of the Altar, and he that came first within four cubits of the Altar secured the task of clearing the ashes. Only if two of the participants were equal did they use lots. But it once happened that two were equal, and one of them pushed the other so that he fell and his leg was broken; and when the Court saw that they incurred danger, they ordained that they should not clear the Altar save by lots (Yoma 22*a*).

This confirms our statement that in this case the use of the lot was designed to prevent quarrels, i.e. to 'pacify' the participants in the allocation of the daily chore. Let us also note here that with respect to the burning of the Incense, a highly valued duty, sampling without replacement was used in preference to the usual sampling with replacement, so as to ensure that all new priests take a turn before a new round started.

(iv) *Lot-casting for the sacrifices on festivals*

A different technique of lot-casting was used to divide the meat of sacrifices between the priests on festivals. Details are given to the Commentary on the Mishnah by Maimonides (Shabbat 148*b*). Each priest participating in the drawing gave some object belonging to him. They then called some outside person and asked him to put one object on each portion. Each priest then took the portion on which the object belonging to him had been deposed. It is explicitly stated that the object of the lot-casting was to avoid quarrels between priests (Shabbat 149*b*). This drawing was called '*Halashim*'.

4. The Jewish attitude to dice gambling

The Talmudic word for dice is 'Kubia', from the Greek *κυβια*, itself a derivative of *κυβοσ*, a cube. Dice gambling has, however, a wider meaning in the Talmud. Thus we read in Sanhedrin 25*b*: 'Dice players include the following: Those who play with *pispasim* (i.e. polished blocks or stones), and not only with *pispasim*, but even with nut-shells and pomegranate peels.'

The Jewish attitude to dice playing revolves around the concept of *Asmakhta* (literally *reliance*). An *Asmakhta* denotes a promise to pay on fulfilment of a condition which the contracting party expects not to be fulfilled. According to some teachers, an *Asmakhta* is not a valid obligation (see Baba Bathra 168*a* and Baba Metsia 66*a*–66*b*).

We read in Sanhedrin 24*b*:

'*Mishnah:* And these are ineligible [to be witnesses or judges]: a gambler with dice, a usurer, a pigeon-trainer, and traders [in the produce] of the sabbatical year . . . R. Judah said: When is this so?—If they have no other occupation but this, but if they have other means of livelihood, they are eligible.

Gemara: What [wrong] does the dice player do? Rammi ben Hama said: [He is disqualified] because it (i.e. gambling) is an *Asmakhta*, and an *Asmakhta* is not legally binding.

R. Shesheth said: Such cases do not come under the category of *Asmakhta*; but the reason is that they (i.e. dice players) are not concerned with the general welfare (i.e. they do not contribute to the welfare of civilised society).'

The reason for the condemnation of gambling was that the gambler always expects to win, and therefore to get something for nothing. This the Rabbis considered immoral and akin to robbery. Robbery and dice-gambling are actually coupled in many disqualifications in the Talmud, e.g. Shebuoth 47*a* where it is explained that the dice gambler is disqualified only by the Rabbis, and not by the Torah, since he is a robber, but does not use violence.

On the other hand, the opinion of R. Shesheth that dice gambling is not an *Asmakhta* is explained as follows by the Tosafist Rabbenu Tam (1100–71). '[Dice gambling] is not an *Asmakhta* because, since there are two [players], each one conveys possession [of the stake] to the other, on the understanding that if he wins, the other one will convey possession [of the stake] to him; it is for the very enjoyment [of this prerogative] that he agrees to convey possession to his opponent' (Tosafot Sanhedrin 25*a*).

But even if the Rabbis admitted that at least some gambers adopted this more sober view of their chances of winning, they nevertheless condemned gambling as antisocial in as far as the Jewish society of the time was concerned. Gambling (and for that matter, even games not involving chance) was considered a waste of time, turning the Jews away from their prime duty: to study and practice the Torah. Thus we read in Kiddushin 21*b*: 'Rabbi Nahman said to Rabbi Anan: When you were at Meir's academy you wasted your time playing *Iskumandri*'. This is variously interpreted as meaning a game similar to chess or checkers, or dog racing.

That the use of chance mechanisms as such was not condemned, but only the intent to win, is shown by the fact that it was permitted to draw lots for the various portions at the Sabbath table between members of the one family (Shabbath 148*b*):

'[On the Sabbath] . . . a man may cast lots with his sons and the members of his household for the table (i.e. which portion of the food shall belong to each), provided that he does not intend to offset a large portion against a small one (i.e. all portions must be alike in size)....'

The use of lots in civil law

Only one application of lot-drawing to civil law is mentioned in the Talmud, namely in the division of an estate between brothers. We read 'It was taught: Rabbi Jose said: When brothers divide [an estate] (into equal shares), all of them acquire possession [of their respective shares] as soon as the lot for one of them is drawn. On what ground [is possession acquired]?—Rabbi Eleazar said: [Possession is acquired in the same way] as [at] the beginning of [the settlement of] the land of Israel. As [at that] beginning [the acquisition was] by lot, so here [also it is] by lot. Since then, however, [the division was made] through the urn, and the *Urim and Tummim*, [should not the division] here also [be made] through the urn and the *Urim and Tummim*?—Rabbi Ashi replied: [The lot alone suffices here] because [in return for] the benefit of mutual agreement they determine to allow each other to acquire possession [by the lot alone]' (Baba Bathra 106*b*).

There must, however, have been many more uses of lot-drawing, as is indicated by Prov. xviii. 18 quoted above.

This paper is dedicated to K. Q. Admor M. M. Schneerson Shlita, without whose inspiration it would never have been written. Thanks are also due to Rabbi I. Groner, who carefully read and criticized the manuscript, to Rabbi Dr N. Schlesinger for many stimulating discussions we had on the topic, and to Dr Thompson, of the Department of Semitic Studies of Melbourne University, for helpful comments.

References

Chajes, Z. H. (1952). *The Student's Guide Through the Talmud*. London: East and West Library.
David, F. N. (1955). Dicing and gaming. (A note on the History of Probability.) *Biometrika* **42**, 1–15.*
David, F. N. (1962). *Games, Gods and Gambling*. London: Charles Griffin and Co., Ltd.
Jastrow, M. (1950). *A Dictionary of the Targumim, the Talmud Babli and Yerushalmi, and the Midrashic Literature*. New York: Pardes Publishing House, Inc.
Mahalanobis, P. C. (1957). The foundations of statistics. *Sankhyā* **18**, 183–94.

**Paper No. 1 in this volume*

WHERE SHALL THE HISTORY OF STATISTICS BEGIN?

By M. G. KENDALL

1. A history must start somewhere, but history has no beginning. It is therefore natural, or at least fashionable, for the historian of any particular period or of any particular subject to preface his main treatment by tracing the roots of his theme back into the past as far as he possibly can. The writer on any modern idea who can claim that the Chinese thought of it first in the Shang period is usually regarded as having scored a point. The underlying assumption that historical development is continuous is sometimes justifiable and sometimes, though not justifiable, innocuous. A history of marine insurance from the foundation of Edward Lloyd's coffee house in 1712 may reasonably begin with an account of the sea-laws of Rhodes of two thousand years earlier. There is no great harm in a history of Newmarket race course glancing at Eohippus and passing lightly over the chariot races of the ancient world before lining up at the starting post of A.D. 1640.

2. But there are dangers in pursuing the roots of a subject down to its slenderest fibrils. A word, an outward form, a material expression may be continuous whereas the concepts which it embodies may drastically change. In earlier articles in this series Dr David and I have tried to emphasize that, although the instruments of gaming had existed for several thousand years, probability theory as a conceptual abstraction of the laws of chance did not come into being until the sixteenth century. In this present note I want to stress the fact that something of the same kind is true of what we call 'statistics'.

3. The temptation to begin a history of statistics with references to endeavours in the ancient world to record information about states is one which no writer has been able to resist. The numbering of the people of Israel, Augustus's balance sheets of the Roman Empire, Charlemagne's inventory of his possessions, the Doomsday book, are felt to be a natural preliminary to the emergence of descriptive statistics in the eighteenth century; as though they were early attempts undertaken in the spirit of modern statistics, hampered by the backward state of social organization, but inspired by the same concepts.

4. To believe this, in my opinion, is to fail to understand either the basis of the statistical approach or the nature of the statistical method. In this note I put forward the thesis that statistics as we now understand the term did not commence until the seventeenth century, and then not in the field of 'statistics'.

5. The early inventories, of which the Doomsday book is an outstanding example, were not, of course, called statistical. The first use of the word known to me occurs in a work by an Italian historian Girolamo Ghilini, who in 1589 refers to an account of the *civile, politica, statistica e militare scienza*. From then onwards we have a series of well-known landmarks in the history of political description: Sansovino's (1561) *Dal governo ed amministrazione dei diversi regie e republiche*; Botero's (1650) *Le relazione universale*; Seckendorf's (1656) *Teutscher Fürstenstaat*; Oldenburger's (1673) publication of Conring's lectures; Anchersen's (1741) *Statuum Cultiorum in Tabulis*; Süssmilch's (1741) *Göttliche Ordnung*; Achenwall's (1748) Public Dissertation; Büsching's (1754) *Neue Erdbeschreibung*. These works are often referred to as 'statistical' and their authors used the word to denote the method of their studies. But they are entirely concerned with the description of political states and any numerical information in them appears by accident or convenience. They are the forerunners of our geographical and political works of reference like *The Statesman*'s *Yearbook*, containing an immense amount of interesting information, but 'statistical' only in the sense of relating to what the Germans called *Staatenkunde*. The point has been well made by Westergaard (1932) in his modest and scholarly history. If it were not, he says, for the fact that the same word 'statistics' is used to denote both this early work and statistics in the modern sense, the student of the latter would probably pay no attention to the former; and Westergaard implies that he would lose little by ignoring it, a judgment with which I agree. The true ancestor of modern statistics is not seventeenth-century statistics but Political Arithmetic.

6. There is another reason, of quite a different kind, for not seeking the origins of modern statistics too curiously among the ruins of the medieval world. The early inventories of the Doomsday type were pictures of a stationary state of society. 'The feudal state of the Middle Ages', says Burckhardt (1929), 'knew of nothing more than categories of signorial rights and possessions. It looked upon production as a fixed quantity, which it approximately is, so long as we have to do with landed property only'. In those days the world was a small affair, only a few thousand years old, the fixed centre of the universe, guided under a rigid set of rules by state and church. There was no particular point in trying to keep track

of minor changes in such a system. The main motives for statistical inquiry, the estimation of military man-power and of taxation revenue, were absent. Not that soldiers and taxes were unimportant, but that the state's resources could be estimated simply on the basis of what had resulted from the last levy.

7. For the dissolution of this system we look naturally to those states which first began to found their economy in trade and manufacture; and we find, accordingly, that in fourteenth- and fifteenth-century Italy statistics in the sense of political arithmetic began to emerge. There are statistical accounts of some of the Italian city-states as early as the thirteenth century—there have survived accounts of Asti in A.D. 1250 and of Milan in A.D. 1288. But these are inventories, although greater emphasis on goods, shops and persons is already noticeable. To Venice, apparently, goes the honour of priority in a new approach. It was the Venetians who first counted the population as souls, that is to say, included all living individuals and did not, for example, confine themselves to citizens who were capable of bearing arms or paying taxes. It was at Venice on one occasion, too, that a proposal to go to war with a neighbouring state was rejected on the economic grounds that fighting the customer was bad for both sides. This general interest in numerical assessment was echoed at Florence and, a little later, among some of the despotisms of northern Italy.

8. Bearing in mind then, that fifteenth-century Italy saw striking developments in accountancy and mathematics, and hence that the necessary technical equipment was available we may conclude that the commercial requirements of the age alone necessitated the collection of a great deal of material which would nowadays be called descriptive statistics. But we are, I think, still short of a statistical approach. Counting was by complete enumeration and still tended to be a record of a situation rather than a basis for estimation or prediction in an expanding economy. In any case the Risorgimento occurred too soon for a Europe which was to spend many years fighting destructive wars at home and colonizing an unknown continent abroad. There appears to have been very little accomplished outside Italy either in descriptive statistics or in probability in the sixteenth century. Not until Europe had settled down about the time of our own restoration (A.D. 1660) did political arithmetic (including life assurance) begin in earnest.

9. Progress was then rapid, and as in many scientific developments, the basic ideas seem to arise independently at many points in Western Europe. John Graunt's famous *Observations* appeared in 1662. Hudde's book on *Annuities* appeared in 1671. Ludwig Huyghens writes to his brother Christian in 1669: 'I have just been making a table showing how long people of given age have to live....Live well! According to my calculations you will live to be about 56½ and I to 55.' At the close of the century we have, within a few years, Petty's *Political Arithmetic* (1690), Halley's *Estimate* (1693), Gregory King's *Observations* (1696), Davenant's *Discourses on the Public Revenues* (1698) and, in 1703, the first census of modern times, in Iceland. There are many accounts of this work, especially of Graunt's and Petty's. Anyone who reads Graunt or King must, in my view, feel that this is the point at which statistics really began. The whole attack on the problems can be appreciated by a statistician or a demographer nowadays. These men, one feels, thought as we think today, with most imperfect material but with enviable penetration. They *reasoned* about their data.

10. This is not to imply that statistics in the old sense came suddenly to an end, or that the inferential type of approach spread from demographic applications to other fields with any speed. The political description of states continued and flourished, becoming more numerical as additional data were collected, but remaining essentially the systematic collection of facts. During the eighteenth century we find the two subjects running side by side with very little apparent connexion between them. In the one, enumerative surveys such as Sinclair's (1793) *Statistical Account of Scotland*, and discourses by men such as Wargentin, Büsching and Schlöger, whose very names are unknown to modern statistics; in the other, contributions to demography and actuarial science by men such as Demoivre, Daniel Bernoulli, Laplace, Euler and Poisson. It was not until the middle of the nineteenth century that these two branches of knowledge showed any sign of coalescence.

11. I conclude then, that statistics in any sense akin to our own cannot be traced back before about A.D. 1660. To look backwards from that point is interesting but can be misleading, unless its purpose is, not to show that modern statistical ideas can be traced back into medieval and ancient times, but that such ideas were almost unknown prior to A.D. 1660. There *are* new things under the sun and the attitude of mind towards experimental knowledge about natural and social phenomena which flourished after the Restoration is one of them.

References

Burckhardt, J. (1929). *The Civilization of the Renaissance in Italy* (translated from the 15th edition by S. G. C. Middlemore. Harrap.

Westergaard, H. (1932). *Contributions to the History of Statistics.* London: P. S. King & Son.

MEDICAL STATISTICS FROM GRAUNT TO FARR

By MAJOR GREENWOOD

INTRODUCTION

Under the Fitzpatrick Trust, a Fellow of the Royal College of Physicians of London is chosen annually by the President and Censors to deliver two lectures in the College on 'The History of Medicine'. I had the honour of being chosen for this office in 1940 but, for obvious reasons, the lectures were not delivered, and it may be safely assumed that some years will pass before a medical audience will have time to attend to the history of a subject the modern practice of which does not make a strong appeal to physicians.

The nature of the intended audience inclined me to stress the medical rather than the purely statistical aspects of the story and I have trodden ground over which a greater man passed some years ago. I hope that Karl Pearson's studies of some or all of these old heroes will eventually be printed, and I know that my slight essays can ill sustain a comparison. But, precisely because they are slight and linger over small traits and human oddities, they may, in these times, wile away an hour or two. I have eliminated some explanations which no statistician or biometrician needs and the medical technicalities are few. Perhaps a note on the London College of Physicians as it was in the days to which these studies relate should be added.

The College was more than a century old when John Graunt was born, and the corporation consisted wholly of physicians who were Doctors of Medicine of Oxford or Cambridge; these were the *Fellows*. Physicians not Doctors of Medicine of Oxford or Cambridge were admissible only to the grade of *Licentiate*, and it was not until the nineteenth century, when Farr was a young man, that the exclusive privilege of the senior universities was abolished. It was not until Farr was a middle-aged man that the College had any direct contact with general practitioners of medicine and began to examine persons who did not seek to practise solely as physicians. In modern usage the College licence, L.R.C.P. (now only granted jointly with the membership of the Royal College of Surgeons, M.R.C.S.), is a diploma obtained by a large proportion of general medical practitioners in the South of England. Down to Farr's time, the L.R.C.P. was a 'specialist' diploma and could not have been taken by a general practitioner (the apothecary of those days) at all. The old L.R.C.P. is represented by the M.R.C.P. of our own time but with this distinction. Now, Fellows (F.R.C.P.) are normally chosen from the body of M.R.C.P.'s. In the past only Doctors of Medicine of Oxford or Cambridge could be Fellows, and before election but after examination were known as 'candidates', not licentiates. The great physician

Sydenham was never more than a licentiate. He graduated M.B. at Oxford and, for some unknown reason, never proceeded M.D. until near the end of his life, when he took the higher degree not at Oxford but at Cambridge.

I. THE LIVES OF PETTY AND GRAUNT

It is always rash to assign an absolute beginning to any form of intellectual effort, to say that this or that man was the very first to fashion some organon which has proved valuable. All we are justified in saying is that this or that man's work can be shown to have so directly influenced the thought of his contemporaries or successors that from his day the method he used has never been forgotten. It may be that the lost works of the school of the Empirics Galen despised anticipated the numerical method of Louis—some words of Celsus are consistent with the hypothesis. It may be that in the long succession of parish clerks who for more than a century transcribed the London Bills of Mortality, one or two suggested that these figures might have some other use than that of warning His Highness of the need to move into Clean Air. But we do not know. We do know that out of the casual intercourse of two Englishmen in the seventeenth century was produced a method of scientific investigation which has never ceased to be applied and has influenced for good or ill the thought of all mankind. In that sense at least we may fairly hold that John Graunt and William Petty were the pioneers not only of medical statistics and vital statistics but of the numerical method as applied to the phenomena of human society.

John Graunt and William Petty were both of Hampshire stock. Petty was of Hampshire birth, born on Monday, 26 May 1623, and was three years younger than John Graunt, who was born at the Seven Stars in Birchin Lane on 24 April 1620.

Materials for writing Petty's life are abundant; indeed a good biography of him was written nearly fifty years ago by his descendant Lord Edmond Fitzmaurice, and since then much of the material used by Lord Edmond has been printed. Sources for Graunt's biography are scanty, the most valuable John Aubrey's brief life of him.* Graunt and Petty became acquainted in or before 1650. The circumstances of that first acquaintance are interesting to those who meditate upon the perepeteia of human fate. It was the contact of client and patron.

John Graunt's early life and manhood were those of the Industrious Apprentice. His father was a city tradesman, who bred his son to the profession of haberdasher of small wares. John 'rose early in the morning to his study before shop-time' and learned Latin and French, but did not neglect his business. He was free of the Drapers' Company and went through the city offices as far as

* *Brief Lives, chiefly of Contemporaries*, set down by John Aubrey, between the years 1669 and 1696, edited by Andrew Clark, Oxford, 1896, **1**, 271 *et seq.*

common councilman; he was captain and then major of the trained bands (the ancestor of the Honourable Artillery Company). At the time of the Great Fire he is said to have been an opulent merchant. Even fifteen years earlier he—and no doubt his father (1592–1662)—had city influence. At that time a Gresham professorship was vacant and a young Dr Petty was anxious to obtain it. This young man's career had been unlike that of an industrious apprentice; it had been, even for the seventeenth century, romantic. His father was a clothier in Romsey, who 'did dye his owne cloathes' in a small way of business. When William was a child, 'his greatest delight was to be looking on the artificers—e.g. smyths, the watch-maker, carpenters, joyners etc.—and at twelve years old could have worked at any of these trades. Here he went to schoole, and learnt by 12 yeares a competent smattering of Latin, and was entred into Greek' (Aubrey, Clark's edition, **2**, 140).

But the precocious lad did not find a patron in Romsey and was shipped for a cabin boy at the age of fourteen. His short sight earned him a taste of the rope's end, and after rather less than a year at sea he broke his leg and was set ashore in Caen to shift for himself. 'Le petit matelot anglois qui parle latin et grec' attracted sympathy and obtained instruction in Caen. Caen was not a famous seat of learning like Leyden or Montpellier, but the Fellows and licentiates of the College of Physicians admitted between 1640 and 1700 include the names of four persons who studied or graduated in Caen (Nicholas Lamy, Theophilus Garencières, John Peachi and Richard Griffiths). Petty, however, was not then thinking of medicine but mathematics and navigation and came home to join the navy. In what capacity he served is unknown; he merely says (in his Will) that his knowledge of arithmetic, geometry, astronomy conducing to navigation, etc., and his having been at the University of Caen, 'preferred me to the King's Navy where at the age of 20 years, I had gotten up about three score pounds, with as much mathematics as any of my age was known to have had'. His naval career was short, for in 1643 he was again on the continent. Here he wandered in the Netherlands and France and studied medicine or at least anatomy. He frequented the company of more eminent refugees, such as Pell and Hobbes, as well as that of the French mathematician Mersen. He was very poor and told Aubrey that he once lived for a week on three pennyworth of walnuts, but on his return to England the three score pounds had increased to seventy and he had also educated his brother Anthony.

At first Petty seems to have tried to make a living out of his father's business, but he soon went to London with a patented manifold letter writer and sundry other schemes of an educational character. These occupied him between 1643 and 1649 and made him acquainted with various men of science, among others Wallis and Wilkins, but were not remunerative, and in 1649 he migrated to Oxford.

Petty was created Doctor of Medicine on 7 March 1649 by virtue of a dispensation from the delegates (no doubt the parliamentary equivalent of the Royal Mandate of later and earlier times). He was also made a Fellow of Brasenose and had already been appointed deputy to the Professor of Anatomy. He was admitted a candidate of the College of Physicians in June 1650 (he was not elected a Fellow until 1655 and was admitted on 25 June 1658). At Oxford he became something of a popular hero by resuscitating (on 14 December 1651) an inefficiently hanged criminal, who, condemned for the murder of an illegitimate child, is said to have survived to be the mother of lawfully begotten offspring.

Academically Petty rose to be full Professor of Anatomy and Vice-Principal of Brasenose. It is at this point (as usual the precise dates are dubious) that he became a candidate for a Gresham professorship and made contact with John Graunt.

Although, as I have said, the materials for a biography of Petty are abundant, all we know of his early years comes from himself or from friends of later life who knew no more than he told them. We have no independent means of judging the extent of his culture. There is good evidence that he knew more Latin than most Fellows of the College of Physicians know now; none that he was an exact scholar (indeed we have his own word, which I am not prepared to gainsay,* to the contrary). He was certainly admitted to friendship by some men, such as Wallis and Pell, who were serious mathematicians, as by others, such as Hobbes, who were not. But whether he could fairly be called a mathematician is doubtful. Of his medical knowledge we know little. He left medical manuscripts, but these are still unpublished; of his clinical experience we know nothing.

Petty told Aubrey that 'he hath read but little, that is to say, not since 25 aetat., and is of Mr. Hobbes his mind, that had he read much, as some men have, he had not known as much as he does, nor should have made such discoveries and improvements'. But it is at least certain that he made a favourable impression upon men who had read a good deal and that the young Dr Petty of 1650 was thought a promising man. Still it *had* been an odd career and one wonders what a steady business man in the city of London thought of it.

Why the anatomy professor who had resuscitated half-hanged Ann Green should be made a professor of music is not obvious, and if the Gresham appointments were jobs, why should the job be done for Petty? The modern imaginative historian might suggest various reasons. For instance, that Petty made a

* If No. 88 of *The Petty Papers* (**2**, 36) is a typical example of Petty's Latin Prose style, there is not much to be said for it. Here is an example: 'An dulcius est humanae naturae permultos suam potestatem in unum quendam et in perpetuum transferre, id est pendis amittere quam ipso puel deindem servare, vel paulatium et in breve tempus irogare, a seipsis demo reformendam et disponendam alioquin pro ut, mutato tam rerum quam animi indies suaserit?' Some of the gibberish may be due to the editor's failure to decipher the handwriting, but no emendation could twist this into unbarbaric prose.

conquest of Graunt, perhaps had Hampshire friends who were friends of the Graunt family, perhaps talked about political arithmetic. We have no evidence at all. If the Gresham Professor of Music *had* duties, Petty did not perform them; about the time of his appointment he obtained leave of absence from Brasenose and within a year (in 1652) had left for Ireland, where he was to be very busy for some time to come and to make, or found, his material fortunes.

Macaulay (chap. III) says that at the end of the Stuart period the greatest estates in the kingdom very little exceeded twenty thousand a year.

The Duke of Ormond had twenty-two thousand a year. The Duke of Buckingham, before his extravagance had impaired his great property, had nineteen thousand six hundred a year. George Monk, Duke of Albemarle, who had been rewarded for his eminent services with immense grants of crown land, and who had been notorious both for covetousness and for parsimony, left fifteen thousand a year of real estate, and sixty thousand pounds in money, which probably yielded seven per cent. These three Dukes were supposed to be three of the very richest subjects in England.

In 1685 Petty made his Will. This Will is a curiously interesting document, because it is also an autobiography. It is rich in arithmetical statements and, like much of Petty's arithmetic, the statements may be optimistic. Petty's final casting of his accounts is in this fashion: 'Whereupon I say in gross, that my reall estate or income may be £6,500 per ann. my personall estate about £45,000, my bad and desparate debts, 30 thousand pounds, and the improvements may be £4000 per ann., in all £15,000 per ann. *ut supra*.'

The details of the calculation are perplexing enough; still if the above cited dukes *were* the richest subjects of the king and if (Macaulay) 'the average income of a temporal peer was estimated by the best informed persons, at about three thousand a year', Sir William Petty, of the year 1685, had travelled as far from the young Oxford professor of 1650 as that budding physician from the little English cabin boy who spoke Latin and Greek, in Caen, in 1638. The details of the fortune-building are not our concern. The shortest account is Petty's own in his Will. He says that by the end of his Oxford career he had a stock of four hundred pounds and received an advance of one hundred more on setting out for Ireland.

Upon the tenth of September, 1652, I landed att Waterford, in Ireland, Phisitian to the army, who had suppressed the Rebellion began in the year 1641, and to the Generall of the same, and the Head Quarters, at the rate of 20s. per diem, at which I continued, till June, 1659, gaining by my practice about £400 per annum, above the said sallary. About September, 1654, I, perceiving that the admeasurement of the lands forfeited by the forementioned Rebellion, and intended to regulate the satisfaction of the soldiers who had suppressed the same, was most insufficiently and absurdly managed, I obtained a contract, dated the 11th. of December, 1654, for making the said admeasurement, and by God's blessing so performed the same as that I gained about nine thousand pounds thereby, which with the £500 above mentioned, my sallary of 20s. per diem, the benefit of my practice, together with £600 given me for directing an after survey of the adventrs lands, and £800 more for 2 years sallary as Clerk of the Councell, raised me an estate of about thirteen thousand pounds in ready and reall money, at a time, when, without art, interest, or authority, men bought as much lands for 10s, in reall money as in this year, 1685, yield 10s. per ann. rent above his *Maties* quitt rents (*The Life of Sir William Petty*, by Lord Edmond Fitzmaurice, London 1895, p. 319).

No one would willingly rake over the embers of Irish history—still glowing after nearly three hundred years. Petty believed himself to be a good man struggling against adversity and a public benefactor treated with gross injustice to the day of his death. Lecky (*History of Ireland*, vol. **1**, chap. 1, p. 111 of popular edition) took a less favourable view. Even if the subject were relevant to my undertaking, which it is not, I have not the training in historical research to justify me in writing about it. There are, however, some points of psychological interest.

Petty did not, like his contemporary Thomas Sydenham, actually take up arms against the king, but he was even more plainly a protégé of the king's enemies. Sydenham's military career was unimportant; there is no reason to believe that he ever exchanged a word with a member of the Cromwell family. Petty was the confidential adviser and close personal friend of Henry Cromwell; his services to the Commonwealth authorities were the foundation of his fortune. Like many people who have social gifts he had the gentle art of making enemies.

Pepys, Aubrey and Evelyn concur in the judgment that Petty was a most entertaining companion. Evelyn says he was a wonderful mimic. He could speak 'now like a grave orthodox divine; then falling into the Presbyterian way; then to Fanatical, to Quaker, to Monk, and to Friar and to Popish Priest'. The gift he exercised among his friends.

My Lord D. of Ormond once obtained it of him, and was almost ravished with admiration; but by and by he fell upon a serious reprimand of the faults and miscarriages of some Princes and Governors, which, though he named none, did so sensibly touch the Duke, who was then Lieutenant of Ireland, that he began to be very uneasy, and wished the spirit layed, which he had raised; for he was neither able to endure such truths, nor could but be delighted. At last he turned his discourse to a ridiculous subject, and come down from the joint-stool on which he had stood, but my lord would not have him preach any more (Evelyn).

My lord Duke was not the first or last person to fail to relish a joke against himself.

In *The Londoners* a challenged party names garden hoes as the weapons. That was Mr Robert Hichens's fun. In real life, Petty, challenged to mortal combat by a Cromwellian soldier, pleaded his myopia and demanded that the duel should take place in a cellar and the weapons be axes.

A man like this makes friends or at least admirers, also enemies. Long before the king enjoyed his own again, Petty had a host of enemies. When the king returned, one might have expected that Petty's position would be critical. According to his own account he *did* lose something, but he was knighted and the losses, such as they were, did not seem to stay the growth of his fortune. At the Restoration he was already prosperous and he died wealthy. Perhaps the explanation is that Petty was really as great a public benefactor as he thought he was. Perhaps the reason is personal. King Charles loved wits (in the old and new sense of the word) and Petty was a wit. The scanty specimens of what Petty's modern representative calls 'Rabelaisian' printed from the Petty papers would not have appealed to such a connoisseur in this genre as the king—we know from Halifax that the king liked to be the raconteur in this field and indeed repeated himself often—but he would have relished a good mimic. Still more important might have been their common virtuosity.

Charles was interested in experimental science, and although Petty certainly knew more than the king, he may not have known very much more. Neither Charles nor James would have been able to find more common ground with Isaac Newton than in a later age Bonaparte found with Laplace. But the ingenious Dr Petty, who had resuscitated half-hanged Ann Green (which would be a capital story if well told), invented an unsinkable ship, had a dozen plans for doubling the king's revenue, and knew something of everything, probably did more than Wilkins to interest the king in the new society of virtuosos (how the king must have relished the story of the planting of horns in Goa*), and he may incidentally have interested the king in his business affairs. This is all speculation; what is sure is that when Petty was back in London and able to renew personal intercourse with John Graunt, their relation was no longer that of client and patron. For a few years more, Graunt was to be a solid merchant, but before long Petty was the patron and Graunt the client.

At this point it will be convenient to conclude the biographical facts relating to Graunt. I take them mainly from Aubrey.

Graunt continued to be a prosperous city tradesman for many years after his first meeting with Petty. 'He was', says Aubrey, 'a man generally beloved; a faithful friend. Often chosen for his prudence and justice to be an arbitrator; and he was a great peace-maker. He had an excellent working head, and was facetious and fluent in his conversation.' Pepys thought as well of Graunt as did Aubrey, admiring both his conversation and his collection of prints—'the best collection of anything almost that ever I saw'.

From the Restoration for several years Graunt figures in London intellectual society (he was elected F.R.S. in 1663), but a material calamity was at hand. The Fire of 1666 no doubt caused Graunt direct financial loss; this might have been repaired. But, although brought up in Puritan ways, 'he fell', to quote Aubrey, 'to buying and reading of the best Socinian bookes, and for severall

* Sir Philiberto Vernatti, Resident in Batavia, had certain inquiries sent him by order of the Royal Society. The eighth question was: 'What ground there may be for that Relation, concerning Horns taking root, and growing about Goa?' This is Sir Philiberto's answer: 'Inquiring about this, a friend laughed, and told me it was a jeer put upon the Portuguese, because the women of Goa are counted much given to lechery' (Sprat's *History of the Royal Society of London*, 2nd ed. London 1702, p. 161).

years continued of that opinion. At least, about...he turned a Roman Catholique, of which religion he dyed a great zealot.'

Graunt's path to Rome was similar to that of young Edmund Gibbon, but the results on the career of a city tradesman in the days of Oates *triumphans* were more serious than a visit to Lausanne. Graunt became bankrupt. His name dropped out of the list of the Royal Society after 1666, and in 1674 he died. There is evidence that in these last years of worldly misfortune, when the wheel had come full circle since Graunt had secured the Gresham professorship for Petty, Petty helped Graunt. When Petty was in Ireland, Graunt acted in some sort as his London agent, and Petty conceived a plan of settling Graunt in Ireland. But (we have, of course, only Petty's word for this) Graunt was not an easy man to help; it is possible, of course, that he may have resented Petty's admonitions. 'You have done amiss in sundry particulars, which I need not mention because you yourself may easily conjecture my meanings. However we leave these things to God and be mindful of what is the sum of all religion, and of what is and ever was true religion all the world over.' This is an extract from a letter of January 1673 to Graunt (*The Petty-Southwell Correspondence*, p. xxix) printed by the late Marquis of Lansdowne. If Lord Lansdowne was right (the whole letter is not printed) in thinking this a reference to Graunt's conversion (or perversion) 'of which', says Lord Lansdowne, 'Petty seems to have disapproved on temporal rather than spiritual grounds', it might have hurt a sensitive man.

Graunt died on Easter Eve 1674 and was buried the Wednesday following in St Dunstan's church in Fleet Street. 'A great number of ingeniose persons attended him to his grave. Among others, with teares, was that ingeniose great virtuoso, Sir William Petty, his old and intimate acquaintance, who was sometime a student of Brasenose College.' Sir William outlived his friend thirteen years and lies in Romsey Abbey. Until a descendant in the nineteenth century (the third Marquis of Lansdowne) erected a monument, 'not even an inscription indicated that the founder of political economy lay in Rumsey Abbey' (Fitzmaurice, p. 315).

Graunt had a son who died in Persia and a daughter who, according to Aubrey, became a nun at Ghent. Nothing is known of descendants.

Petty's widow was raised to the peerage and her elder sons, Charles and Henry, died without issue. But the title was revived in favour of the grandson of John Fitzmaurice, the second surviving son of Thomas Fitzmaurice, Earl of Kerry, who, as the above-mentioned grandson remarked, had 'married luckily for me and mine, a very ugly woman who brought into his family whatever degree of sense may have appeared in it, or whatever wealth is likely to remain in it'. This ill-favoured woman was Petty's daughter Anne, to whom her father wrote:

> My pretty little Pusling and my daughter Ann
> That shall bee a countesse, if her pappa can.

The cynical grandson was George III's prime minister and afterwards his *bête noire*, 'The Jesuit of Berkley Square' and first Marquis of Lansdowne.

Of the two friends, one has left an intellectual monument only; descendants of the other have been famous in English history.

Of these, best known are the first and third Marquises of Lansdowne, William (1737–1805) and Henry (1780–1863). Of the first marquis, much better known as Lord Shelburne (the title created for Lady Petty), every schoolboy—not only Macaulay's schoolboy—has heard; the quarrel between Charles Fox and Shelburne, the party split, the coalition ministry and so on. Schoolboys who have reached the sixth and Lecky's *History of England in the Eighteenth Century*, know a little more. Shelburne, who had much more than a tincture of his great-grandfather's ability and applied himself to economic studies, was one of the earliest to appreciate the importance of Adam Smith and was highly thought of by two good judges of scientific ability, Benjamin Franklin and Jeremy Bentham.

As a public man, no parliamentary statesman before or since obtained so universal a dislike, a positive hatred shared by those who knew him and those who did not.

> There is certainly nothing in the actions of Shelburne to justify this extreme unpopularity. Much of it was, I believe, simply due to an artificial, overstrained, and affectedly obsequious manner, but much also to certain faults of character, which it is not difficult to detect. Most of the portraits that were drawn of him concur in representing him as a harsh, cynical, and sarcastic judge of the motives of others; extremely suspicious; jealous and reserved in his dealings with his colleagues; accustomed to pursue tenaciously ends of his own, which he did not frankly communicate, and frequently passing from a language of great superciliousness and arrogance to a strain of profuse flattery (Lecky, **5**, 136).

How far some of these characteristics may be recognized in Shelburne's ancestor, we shall inquire in due course.

The contrast between Malagrida* and his son Henry is shattering. It is *this* Marquis of Lansdowne of whom nearly everybody thinks when he sees the title in a book, and rightly so. Walter Bagehot wrote:

> You may observe that when an ancient liberal, Lord John Russell, or any of the essential sect, has done anything very queer, the last thing you would imagine anybody would dream of doing, and is attacked for it, he always answers boldly, 'Lord Lansdowne said I *might*'; or if it is a ponderous day, the eloquence runs, 'A noble friend with whom I have had the inestimable advantage of being associated from the commencement (the infantile period I might say) of my political life, and to whose advice,' etc., etc., etc.—and a very cheerful existence it must be for 'my noble friend' to be expected to justify—(for they never say it except they have done something very odd)—and dignify every aberration. Still it must be a beautiful feeling to have a man like Lord John, to have a stiff, small man

* Malagrida was an Italian Jesuit settled in Portugal who was burned in 1761. The supposed jesuitical propensities of Shelburne led to the name becoming his popular title. Hence Goldsmith's unintended *mot*: 'Do you know that I never could conceive the reason why they call you Malagrida, for Malagrida was a very good sort of man.'

bowing down before you. And a good judge (Sydney Smith) certainly suggested the conferring of this authority. 'Why do they not talk over the virtues and excellencies of Lansdowne? There is no man who performs the duties of life better, or fills a high station in a more becoming manner. He is full of knowledge, and eager for its acquisition. His remarkable politeness is the result of good nature, regulated by good sense. He looks for talents and qualities among all ranks of men, and adds them to his stock of society, as a botanist does his plants; and while other aristocrats are yawning among stars and garters, Lansdowne is refreshing his soul with the fancy and genius which he has found in odd places, and gathered to the marbles and pictures of his palace. Then he is an honest politician, a wise statesman, and has a philosophic mind', etc., etc. Here is devotion for a carping critic; and who ever heard before of *bonhomie* in an idol? (Bagehot, *Works*, **2**, 64–5).

Of the father, Atticus (an alias of 'Junius') wrote:

The Earl of Shelburne had initiated himself in business by carrying messages between the Earl of Bute and Mr. Fox, and was for some time a favourite with both. Before he was an ensign he thought himself fit to be a general, and to be a leading minister before he ever saw a public office. The life of this young man is a satire on mankind. The treachery which deserts a friend, might be a virtue compared to the fawning baseness which attaches itself to a declared enemy (*Letters of Junius*, Wade's edition, **2**, 248).

Naturally justice was no more to be expected in eighteenth-century newspaper diatribes than in the twentieth century, but a clever caricaturist does not represent Charles Fox as a living skeleton. Those who attacked the son—there were such people—took a different line, as Bagehot hints. Perhaps even in his very different character something of the ancestral Petty survives. We shall try to discover what this was.

Forty years ago Hull brought out an edition of Petty's tracts in which he included Graunt's work. In 1927 the fifth Marquis of Lansdowne printed a selection from the Petty papers and in 1928 the correspondence between Petty and his wife's cousin,* Sir Robert Southwell (*The Petty-Southwell Correspondence*, edited by the Marquis of Lansdowne, London 1928).

We shall have to examine in detail both the 'works' and the 'papers', but, as a light upon the character of Petty, the Southwell correspondence is the strongest we have. Southwell himself was some generations farther away from adventuring than Petty. He came of an 'undertaker' stock—the adventurers in Ireland of Queen Elizabeth's time—and his father was vice-admiral of Munster before him. He was born in 1635 (died in 1702), regularly educated (Queen's College, Oxford and Lincoln's Inn), knighted in 1665, for some time Clerk of the Privy Council, in the diplomatic service, held other offices, was a member of parliament and eventually settled in a country house near Bath. He was President of the Royal Society 1690–5. He might be described as a lesser William Temple; better educated and less selfish, not so able, but with the same cool, cautious judgment; a psychological antithesis of his correspondent.

* Petty married in 1667 Lady Fenton, widow of Sir Maurice Fenton and daughter of Sir Hardress Waller who, knighted in 1629, fought for the Parliament and was one of the King's judges; he was a major general in Ireland in 1650–1 and a patron of Petty there.

The correspondence covers the eleven years 1676–87. Both men were, even by modern standards, middle aged. They write one to another with complete frankness; there is a remarkable absence of the elaborate verbal formalities which in seventeenth-century and even eighteenth-century letters are so wearisome.

Petty's side of the correspondence consists roughly of domesticities 10 parts, eager accounts of his quarrels and law suits concerning money 40 parts, discussion of papers or projected papers 40 parts, add autobiographical boasting to make up the 100.

In the purely domestic part of the correspondence, Petty is seen as a kind, good-natured father interested in the doings of his relations by marriage, also as a very bad judge of others' feelings. I remember to have read an unpublished letter by the famous Edwin Chadwick, the great and very unpopular sanitarian of a century ago. It was written to a friend whose wife had just died of puerperal fever. Chadwick expressed regret in the shortest possible formula and assured his correspondent that the best solace he could have would be to assist in pushing forward a bill (which I think he enclosed) to promote some sanitary reform which would have the effect of making it less likely that other men would lose their wives in childbed. I remember thinking that, however sensible the recommendation, the man who gave it was not likely to bring much comfort to his friend.

Petty was very much like Chadwick here. Southwell lost his wife in 1681 and Petty condoled with him as follows:

> When your good father dyed, I told you that hee was full of years and ripe fruit, and that you had no reason to wish him longer in the paines of this world. But I cannot use the same Argument in this Case for your Lady is taken away somewhat within half the ordinary age of Man and soon after you have been perfectly married to her; for I cannot believe your perfect union and assimulacon was made till many years after the Ceremonies at Kinsington.
>
> What I have hitherto said tends to aggravate rather than mitigate your sorrow. But as the sun shining strongly upon burning Coles doth quench them, so perhaps the sadder Sentiments that I beget in you may extinguish those which now afflict you. The next Thing I shall say is, That when I myself married, I was scarce a year younger then you are now, and consequently do apprehend That you have a second Crop of Contentment and as much yet to come as ever I have had.

This remark, curiously enough, was not well received.

> You doe not onely condole the great loss I have sustained in a wife, but you seeme to think it reparable.... But when by 19 yeares conversation I knew the greate vertues of her mind, and discover since her death a more secrett correspondence with Heaven in Acts of Pietye and devotion (which before I knew not of), you will allow me, at least for my Children's sake, to lament that they have too early lost their guide.

Petty could not, it seems, understand that Southwell was wounded and returned to the charge in a letter which is lost. That letter provoked a reply which even Petty could not misunderstand and elicited an apology (*Correspondence*, p. 90).

Petty was quite incorrigible. A few years later Southwell had another family bereavement and is condoled with in the following terms:

That by the death of your Father, Mother and Sister, of Sir Edward Deering and your three nephews, you are the Head and Governor of both Familyes. That by the death of Rupe, Ingenious Neddy culminates; and by that of your Excellent Lady you are entitled to that million I mentioned of unmarryed teeming Ladyes.

Once again, Southwell was not comforted. 'Cousin, you doe wipe off Teares at a very strange rate, but why did nature furnish Them if there must be no Sorrow?'

Petty had a very quick perception of when and where *his* shoe pinched, but no imaginative sympathy.

Passing to Petty's financial affairs and lawsuits, the position was this. By original grants, by purchase and in various ways, Petty had widely scattered Irish interests. Questions of the validity of the original grants, of rent charges due to the crown or to other grantees, of matters of fact and matters of law were endless. Petty saw himself steadily as a great public benefactor harassed by scoundrels, and it never occurred to him even as a theoretical possibility that others had rights. Of his manner of proceeding the editor of the correspondence gives a typical example (*Correspondence*, p. 90). In 1681 Petty gave evidence before Lord Chief Baron Hen as to 'Soldier's land' which he had bought in Kerry and, it seems, the court decided against him.

Petty gave vent to his chagrin in a long and scurrilous lampoon against the offending judge, entitled: 'HENEALOGIE or the legend of Hen-Hene and Pen-Hene', in two parts. Whereof the first doth in 24 chapters of Raillery, contain the enchantements, metamorphoses and merry conceits relating to them. The second part contayning (in good earnest) the foolish, erroneous, absurd, malicious and ridiculous 'JUDGEMENTS of HEN-HENE'. Fortunately perhaps for the repute of its author, this diatribe was never made public.

Fortunately, also, for a more material reason; it would probably have led to a *second* incarceration for contempt of court.

Southwell evidently viewed his good cousin's proceedings with a mixture of gentlemanlike annoyance and practical minded contempt. He expressed these feelings more than once; the following extract from a letter of 1677 is typical; the particular suit in progress to which reference is made was a claim for £5000 in respect of a sum of £2500 actually advanced by Petty to the Farmers of Revenue.

And suffer from me this expostulation, who wish your prosperity as much as any man living; and having opportunities to see and heare what the temper of the world is towards you, I cannot but wish you well in Port, or rather upon the firm Land, and to have very little or nothing at all left to the mercy and good will of others. For there is generally imbibed such an opinion and dread of your superiority and reach over other men in the wayes of dealing, that they hate what they feare, and find wayes to make him feare that is feard. I doe the more freely open my soul to you in this matter, because tis not for the vitells that you contend, but for outward Limbs and accessions, without which you can subsist with Plenty and Honour. And therefore to throw what you have quite away, or at least to put it in dayly hazard onely to make it a little more than it is, Is what you would condemne a thousand times over in another. And you would not think the Reply sufficient that there was plain Right in the Cause and Justice of their side, for iniquities will abound and the world will never be reformed.

After all this is said, I mean not that you should relinquish the pursute of your 2500£, which is money out of your Pockett and for which you are a Debtor unto your Family. But for other pretensions, lett them goe for Heaven's Sake, as you would a hott coale out of your hand: and strive to retire to your home in this Place, where you had the respect of all, and as much quiet as could be in this life, before your medling with that pernicious business of the Farme.

There is no reason to suppose that Petty ever took such sensible advice. Yet, somehow, he kept his head well above water.

In the later part of the correspondence Petty indulges in that complacent financial retrospect which he inserted in his Will and I have, perhaps too harshly, described as autobiographical boasting. It is possible that Southwell had heard of these financial triumphs rather often; at least there is a hint of this in the following:

I will onely note that since you are soe Indulgent as to think me worthy of being your Depositary in this great Audit, and expect by the Course of Nature that I should speake when you are Silent, you must allow me liberty without blame to aske questions when you seeme defitient or Redundant.

That you are defitient may be suggested when, on the fortunate syde, I find noe Item for my Lady or of the hopefull stock she has brought you (p. 227).

The shrewd thrust of the last sentence was deadly. The subject does not recur.

I have indicated the character of the non-scientific part of the correspondence because we must examine Petty's scientific writings in greater detail. I think, however, we have enough to justify a provisional diagnosis of Petty's psychological type.

In literature and in life the perennial boy is often encountered. But while Peter Pan and Mr Reginald Fortune make far more friends than foes, that is not so true of their living counterparts. The exuberant flow of ideas and schemes, the intense and restless interest in *everything* which is characteristic of the clever child, often is extraordinarily attractive when it is associated with and controlled by the trained intelligence of a man. But the bad as well as the good points of a childlike or adolescent soul* are to be brought into the account. The

* The first Marquis of Halifax said of King Charles that 'his inclinations to love were the effects of health and a good constitution, with as little mixture of the *seraphic* part as ever man had', and Petty held that the King was typical. In *The Petty Papers* (no. 93 of vol. **2**) there is a memorandum headed 'Californian Marriages with the Reasons thereof'. 'In California', says Petty, '6 men were conjugerted to 6 women in order to beget many and well conditioned children, and for the greatest venereall pleasure, in manner following, viz.'

He then sets out the plan. One man 'excelling in strength, nimbleness, beauty, wit, courage clever child is often naïvely and intensely selfish, and so remains as the eternal boy; his quite crude and unashamed egoism, his inability to understand that others have feelings and even rights, repel as strongly as his intellectual freshness attracts. How far he is a success in life depends on which way the balance turns.

Petty seems to me a good example of this psychological type; its good points, the restless energy and exuberant flow of ideas, were sources of strength in such a time as that of the Civil War and Restoration, which, particularly the Restoration period, was in virtues and vices an age of grown-up children. Indeed his emotional adolescence may have shielded him from the deadly enmity of real men. Its bad points made him enemies, but they were children like himself. Nearly a century later, in a time of adults, these same characteristics, restless intellectual energy and vanity, exhibited by one no longer a rollicking adventurer but a great landowner, produced an unfavourable balance and we have 'Malagrida'. In Malagrida's son, one has a change; the attractive traits, the eager interest in all sorts of things is still there, but the childish hungry vanity has been softened or sublimed. The cynic may say that it was easy for a great Whig lord 150 years ago to be agreeable, to keep himself *hors concours*; perhaps it was, although the *Dropmore Papers* raise doubts. The fact, however, is certain. In the third Lord Lansdowne one sees the good and in the first the bad effects of the perennial boyishness of the ancestor. The ancestor lived in a state of society where the good points outweighted the bad points. That is why, although he made enemies and was often vexed, he was able to view his career with complacency and to bequeath a great fortune. But it is not Petty as a man but Petty as a scientific worker who is the proper object of my study.

How far does the psychological make-up which, as I think, characterized Petty conduce to scientific investigation? We might expect that it would be an immense stimulus to pioneering, that such a man would direct attention to a number of problems which deserved study, but that it would not lead to the production of any solid contribution to knowledge. Our task is to examine in some detail Petty's scientific work.

and good sense' subsequently called the Hero, is allowed four women for his sole use. One Great Rich Woman is allowed five men who are to serve her when she pleases, but another woman is allotted to the five men for use in common by the five.

It may be said this fable is only an after dinner jest—perhaps that *is* the whole explanation. But Petty does go to the trouble of financial calculations, and does seem to suggest a serious consideration. ('The encrease of children will be great and good.' 'No controversy about joynture, dower, maintenance, portion etc.') Nobody emotionally adult would be likely to make Californian Marriages a basis for practical statecraft.

II. PETTY'S SCIENTIFIC WORK

It is no part of my undertaking to survey the whole of Petty's scientific activities, but to speak only of his medical and vital statistical work.

In Hull's edition of Petty's writings, the editor discusses Petty's status as an economist and remarks that Petty's view that value depended upon labour was probably derived from Hobbes. The corn rent of agricultural lands was in Petty's view determined by the excess of their produce over the expenses of cultivation, paid in corn, and the money value of the excess will be measured by the amount of silver which a miner, working for the same time as the cultivator of the corn land, will have left after meeting his expenses with a part of the silver he secures (Hull, p. lxxiii). Why there should be any surplus, he explains by density of population.

Prof. Hull refrained from attempting to assess Petty's work in terms of modern economic theory. A mere medical statistician will naturally follow this example. More than a century ago, Mr Chainmail had learned from Mr MacQuedy that the essence of a safe and economical currency was an interminable series of broken promises and added: 'There seems to be a difference among the learned as to the way in which the promises ought to be broken; but I am not deep enough in their casuistry to enter into such nice distinctions.' Medical statisticians may well adopt Mr Chainmail's modest attitude towards the whole field of economic theory. Confining ourselves to statistics, we must consider what Petty thought should be done and what he actually did himself.

Under the first heading, praise can be unstinted. More than 150 years before the establishment of the General Register Office, Petty specifically proposed the organization of a central statistical department the scope of which was wider than that of our existing General Register Office. It was to deal not only with births, marriages, burials, houses, the ages, sexes and occupations of the people, but with statistics of revenue, education and trade (see *The Petty Papers*, **1**, 171–2). He did not confine himself to vague recommendations, but drew up an enumeration schedule to be used for each parish. On this was to be entered: The number of housekeepers and of houses; the number of hearths; the number of statute acres; the number of people by sex and in age groups, viz. under 10, between 10 and 70, over 70; for males those aged 16 to 60, and for females those between 16 and 48 and how many of these latter were married; how many persons were incurable impotents and how many lived upon alms. This, it will be noted, is a better enumeration schedule than any used in England before the census of 1821. Further in his notes (printed in *The Petty Papers*) are various suggestions for the utilization of data collected in this way.

The most striking is this: 'The numbers of people that are of every yeare old from one to 100, and the number of them that dye at every such yeare's age, do shew to how many yeare's value the life of any person of any age is equivalent and consequently makes a Par between the value of Estates for life and for years' (*The Petty Papers*, **1**, 193).

This is, I think, the most remarkable thing Petty ever wrote, for it *suggests* that he had grasped the principle of an accurate life table, viz. a survivorship table based upon a knowledge of rates or mortality in age groups. No such table was constructed from population data until the end of the eighteenth century, because until then data of the age distribution of the *living* population were not obtained. Whether Petty also realized that under certain conditions a life table could be constructed without knowledge of the ages of the living population is a controversial matter which I shall discuss later on.

Then he makes suggestions which are relevant enough to modern demographic problems.

> By the proportion between marriages and births, and of mothers to births, may be learnt what hindrance abortions and long suckling of children is to the speedier propagation of mankind; as also the difference of soyles and ayres to this foecundity of women.
>
> By the proportion between maryd and unmaryd teeming women, may be found in what number of yeeres the present stock of people may bee encreased to any number assigned answerable to the defect of the peopling of the nation for strength or trade.

There are not wanting some suggestions which imply that even if Petty's opinion of the Faculty were higher than that of Sydenham (whom we honoured *posthumously*) it was tinged with scepticism.

> Whether they [viz. fellows and licentiates of the College of Physicians] take as much medicine and remedies as the like number of any other society.
>
> Whether of 1000 patients to the best physicians, aged of any decade, there do not die as many as out of the inhabitants of places where there dwell no physicians.
>
> Whether of 100 sick of acute diseases who use physicians, as many die and in misery, as where no art is used, or only chance. (*The Petty Papers*, **2**, 169–70.)

This statistical experiment has not yet been performed and indeed might be hardly so conclusive as Petty implied.

When one passes from what Petty suggested to what he actually did himself, our praise must be qualified. As Prof. Hull said, he was 'more than once misled into fancying that his conclusions were accurate because their form was definite'.

In judging Petty it is but fair to contrast him with College contemporaries whose names are more honoured by us. Among his contemporaries in the College were Thomas Browne and Thomas Sydenham. Browne was a much older man than Petty, Sydenham almost his coeval. Of Browne's quality as a physician we know nothing; but his literary influence indirectly—through Samuel Johnson—and directly upon generations of readers has been greater than that of any other practising medical man. Browne, like Petty, had an enormous range of interests and his book learning was greater. But, as we shall see, when he tackles a problem of demography, Petty's rashest guesses seem by comparison as soberly scientific as an annual report of the Registrar-General.

Sydenham was an iconoclast in clinical practice and believed himself to be emancipated from the rule of ancient authority. No fantastic arithmetical calculations are to be found in *his* writings. In fact, with a single exception (*Observations Medicae*, **2**, i), no arithmetic at all. It never seems to have entered his mind, although his greatest work purports to give the history of the diseases in London through a generation, that the arithmetical statements of the London Bills of Mortality were of any value whatever.

Sydenham was too wise a man for us to think that he rejected the evidence because the data were compiled by illiterate old women. He would have known that the sworn searchers had the loquacity of their sex and rank and were likely to ask what 'the doctor said'. He rejected it, because counting and measuring things did not come within his purview, just as the first beginnings of pathology and medical chemistry seemed to him irrelevant.

For the most part, Petty's statistical work was severely practical, but there is one excursion into theory which is interesting. It is to be found in a section of his tract on the use of what he calls Duplicate Proportion and is reprinted by Hull (pp. 622–3).

Petty states that there are more persons living between the ages of 16 and 26 than in any other decade of life. The statement is not true for modern populations and was probably not true for the English population of Petty's time. In 1861–71 (before the fall in the birth rate and infant mortality rate) there were 5·4 millions living under 10, and 4·0 between 15 and 25). But perhaps Petty meant that there were more living in the decade 16 to 26 than in any *later* decade, in which case his statement was of course right unless the birth rate was falling.

He then asserts that the

> Roots of every number of Men's Ages under 16 (whose Root is 4) compared with the said number 4, doth show the proportion of the likelyhood of such men reaching 70 years of Age. As for example: 'Tis 4 times more likely that one of 16 years old should live to 70, than a new born Babe. 'Tis three times more likely, that one of 9 years old should attain the age of 70, than the said infant. Moreover, 'tis twice as likely, that one of 16 should reach that Age, as that one of four years old should do it; and one third more likely, than for one of nine.

We have no life table for England in 1674. Perhaps the nearest modern experience might be the Liverpool Table calculated by Farr seventy years ago. According to that table the chance of a new-born child living to be 65 was 0·0976 and the chance of a person of 15 living to 65 was 0·202, which is about double the infant's chance, not four times as large. For the Healthy Districts, the chances are 0·4246 and 0·54585; that is, in a ratio of 1·28 to 1.

Petty's statements are wildly wrong. The interesting point is how did he reach them? The only figures he had were printed by Graunt.

This 'Life Table' gives l_x as follows:

l_0	100	l_{46}	10
l_6	64	l_{56}	6
l_{16}	40	l_{66}	3
l_{26}	25	l_{76}	1
l_{36}	16		

Now if we take 2 as the survivors to 70 (it does not of course matter what the numerator is for comparative purposes), then the infant's chance of surviving to 70 is 0·02 and the person of 16 has the chance $1/20 = 0{\cdot}05$, a ratio of 2·5, not wildly different from the Liverpool Table figure and very different from 4·0.

A fortiori when Petty, having passed above age 16, asserts that 'it is five to four, that one of 26 years old will die before one of 16; and 6 to 5 that one of 36 will die before one of 26', we are in a region of pure fantasy because, even if he had had the statistical data, Petty would not have had the technical knowledge to solve the problem involved, viz. to find the probability that of two lives aged respectively x and y, the former will fall before the latter.

If we keep within the range of the simple arithmetic which Petty used, the result cannot be obtained.

He then passes to this statement:

> To prove all which I can produce the accompts of every Man, Woman, and Child, within a certain Parish of above 330 Souls; all which particular Ages being cast up, and added together, and the Sum divided by the whole number of Souls, made the Quotient between 15 and 16; which I call (if it be Constant or Uniform) the Age of that Parish, or *Numerus Index* of Longaevity there. Many of which Indexes for several times and places, would make a useful Scale of Salubrity for those places, and a better Judg of Ayers than the conjectural Notions we commonly read and talk of. And such a Scale the *King* might as easily make for all his Dominions, as I did for this one Parish.

The puzzle is to discover why Petty thought this statistical experiment proved his point and why he regarded the mean age of the population of a parish its index of longevity. The first question I cannot answer at all; about the second I can make a guess. *If* the parish population were supported solely by births and there was no migration, then, if the death rates at ages did not vary, the population would be a stationary population and both the mean age of the living and the mean age at death would be constant. The expectation of life is greater than the mean age of the living unless the rates of mortality at early ages are very high and the more favourable the rates of mortality the greater will be the difference. In Petty's day, when mortality at early ages was very high, the two constants were probably not far apart, but it is certain that both expectation of life and mean age of a life table population were greater than 16; probably of order 28 to 32.

I think we may be sure that the parish Petty counted was not stationary in the statistical sense, but had an excess of births over deaths, and that his average threw no light upon the rates of mortality.

Passing to practical statistics, it will be convenient first to note rapidly statistical observations which are incidental in treatises of primarily financial or economic interest. In the *Verbum sapienti*, which although not printed until 1691 was written as early as 1665, Petty attempts to reckon what a man is worth. Here is the method. He concludes from financial data that the annual proceed of the Stock or Wealth of the nation yields 15 millions, but that the expenses of the nation are 40 millions. So the balance of 25 millions must be derived from the labour of the people. He assumes that the population is 6 millions and that half of these can work, and earn £8. 6*s*. 8*d*. a head per annum. This would be 7*d*. a day, abating 52 Sundays and half as many other days for sickness, holidays, etc. 'Whereas the Stock of Kingdom, yielding but 15 Millions of proceed, is worth 250 Millions; then the People who yield 25, are worth 416 2/3 Millions. For although the Individuums of Mankind be reckoned at about 8 years purchase; the Species of them is worth as many as Land, being in its nature as perpetual, for ought we know.'

Why an individual's working life is worth only 8 years' purchase is not clear. One would be inclined to put it as the average number of years lived in the working period of life. Perhaps Petty took Graunt's table and worked out the average number of years of life lived between the ages of 16 and 56; it *is* nearly 8.

He then calculates the money loss due to 100,000 dying of the plague and makes it nearly 7 millions, adding that £70,000 would have been well disposed in preventing this 'centuple loss'. Perhaps this is the first printed statement of the neglected truth that public health measures pay.

Since Petty's day, others, including Farr himself, have done sums of this kind; it is a popular occupation in the United States of America.

Farr went to work more elaborately, making out a balance sheet of a man from the cradle to the grave. But the principle was much the same. We cannot say it is a *wholly* useless pastime. There is of course the difficulty that if more lives are saved the price of labour might fall. But to Petty that would have been no difficulty, because he held that wealth is *purely* relative, viz. that if the income of each person in a community is halved, everybody is as well off as before.

In the *Political Anatomy of Ireland*, Petty seeks to determine war losses in Ireland.

The number of the People being now *Anno* 1672 about 1,100,000 and *Anno* 1652 about 850 M. Because I conceive that 80 M. of them have in 20 years encreased by generation 70 M. by return of banished and expelled *English*; as also by the access of new ones, 80 M. of New *Scots*, and 20 M. of returned *Irish*, being all 250 M.

Now if it could be known what number of people were in *Ireland* Ann. 1641, then the difference between the said number, and 850, adding unto it the increase by generation in 11 years will shew the destruction of people made by the Wars, *viz*. by the Sword, Plague and Famine occasioned thereby.

I find by comparing superfluous and spare Oxen, Sheep, Butter and Beef that there was exported above 1/3 more *Ann*. 1664 than in 1641, which shews there were 1/3 more of people, *viz.* 1,466,000. Out of which Sum take what were left Ann. 1652, there will remain 616,000 destroyed by the Rebellion.

Whereas the present proportion of the *British* is as 3 to 11; But before the Wars the proportion was less, *viz.* as 2 to 11 and then it follows that the number of *British* slain in 11 years was 112 thousand Souls; of which I guess 2/3 to have perished by War, Plague and Famine. So as it follows that 37,000 were massacred in the first year of Tumults: So as those who think 154,000 were so destroyed, ought to review the grounds of their Opinions.

It follows also, that about 504 M. of the *Irish* perished, and were wasted by the Sword, Plague and Famine, Hardship and Banishment, between the 23 of *October* 1641 and the same day 1652. Wherefore those who say, That not 1/8 of them remained at the end of the Wars, must also review their opinions; there being by this Computation near 2/3 of them; which Opinion I also submit.

Assuming, which is rash, that the estimates of population in 1672 and 1652 are correct, the assumption that population varied inversely as exportation of cattle seems bold. Might it not be that shipping facilities were better in 1664 than in 1641? Had there been no exportation we could not infer the population to be infinite.

Again Petty has multiplied the estimate for 1672 by 1·333. But he needed the population of 1664, which presumably was smaller than that of 1672. If his estimate is right, the population was increasing at the rate of about 12·5 thousands per annum, so he should have multiplied 1,000,000 not 1,100,000 by 1·333 and has overestimated the 1641 population by 133,330, and therefore the number destroyed by the same amount, an overstatement of 20 %. But this is not all. If we assign the decrement of population between 1652 and 1641 wholly to sword, plague and famine, we must assume that births continued at the peacetime rate; not a likely assumption. Lastly, it seems unreasonable to assign the casualties to the two races in precise proportion to their estimated numerical strength in the population of 1641.

How it follows that 37,000 were massacred in the first year of tumults I do not know.

In a later work (*Treatise of Ireland*, pp. 610–11) Petty has another shot at this problem.

He now assumes that Graunt's deduction from a Hampshire parish register, viz. that christenings are to burials in the ratio of 5 to 4, applies to Ireland, and that the death rate is 1 in 30, i.e. about what Graunt estimated for London and much higher than his estimate for the country. He then proceeds in this way. He estimates the population of 1653 to be 900,000 and that of 1687, 1,300,000. Then taking 1/30 for the death rate and 1/24 for birth rate, he makes the population of 1652, 985,000. He does not comment on the great decrease between 1652 and 1653; but there was still war in Ireland in 1652.

He now says that the population of 1641 was greater than that of 1687, 'as appears by the Exportations, Importations, Tyths, Grist-Mills and the Judgment of Intelligent Persons'. This time he takes the population to be 1,400,000—a little less than in the earlier estimate—and by the same kind of reasoning again makes the war losses to be about 600,000. One is reminded of Hull's remark that Petty confused the accurate with the definite. Also one notes the inevitable tendency of a polemical writer—which Petty very decidedly was—to maintain his original assertion. Those of us who have *never* yielded to this temptation may cast stones at him. It is not I believe too cynical to say that *any* calculation Petty made would have made the war losses around 600,000.

Returning to the *Political Anatomy of Ireland*, we find here a distinct claim that the mean age at death (not the mean age of the living) measures longevity.

> As to Longaevity, inquiry must be made into some good old Register of (suppose) 20 persons, who were all born and buried in the same Parish, and having cast up the time which they all lived as one man, the Total divided by 20 is the life of each one with another; which compared with the like Observation in several other places, will show the difference of Longaevity, due allowance being made for extraordinary contingencies and Epidemical Diseases happening respectively within the period of each Observation (p. 172).

Apart from what we should think the absurdity of basing important conclusions upon an average of 20—and Petty only gives 20 as a figure—the mean ages at death of different populations are not comparable unless in each place the population is stationary in the sense described above. But, since so acute a man as Edwin Chadwick made the same mistake in the nineteenth century as Petty in the seventeenth century and it continues to be made in various places in the twentieth century, we need not be superior.

We now come to Petty's purely statistical work which is concerned with the growth of population; before examining this in detail, it will be convenient to consider the methods available in the seventeenth century for estimating population and notions then current on what may be called the theory of population growth.

It is hard to believe that in the ancient world nobody studied demography arithmetically. There is evidence that the Romans enumerated citizens—the word census is pure Latin—and it has been suggested that the Romans made life tables. Gouraud, cited by Todhunter (*History of the Mathematical Theory of Probability*, p. 14), refers to a passage cited from Ulpian in the *Digest* which I have discussed elsewhere.* The question was of the value of annuities and the conclusion I reached was that Ulpian had no vital statistical basis whatever for his figures, that he simply began with the capital value the law gave for *any* usufruct and then, realizing that people do die eventually, made some subtractions, ending with the absurd (vital-statistically speaking) conclusion that after the age of 60 the rate of mortality was independent of age.

There is not, I think, any reason to believe that the practical Romans had anticipated Graunt and Petty.

That is not to say that nobody studied any demographical problems arithmetically. Indeed one fellow of the College of Physicians who has had—and will

* *Journ. Roy. Stat. Soc.* **103** (1940), 246.

continue to have—a hundred readers for every one reader of Graunt and Petty, made an elaborate demographical calculation. This was Sir Thomas Browne. Sir Thomas devoted the sixth chapter of the sixth book of *Pseudodoxia* to the vulgar opinion that the earth was slenderly peopled before the Flood.

This vulgar opinion Sir Thomas found to be very wide of the mark. Indeed, far from the earth being slenderly peopled, 'we shall rather admire how the earth contained its inhabitants, than doubt its inhabitation: and might conceive the deluge not simply penall, but in some way also necessary, as many have conceived of translations, if *Adam* had not sinned, and the race of man had remained upon earth immortal'. Indeed Sir Thomas estimates that by the seventh century of the world's history its population amounted to 1,347,368,420. He reaches this result in the following way:

Having thus declared how powerfully the length of lives conduced unto populosity of those times, it will yet be easier acknowledged if we descend to particularities, and consider how many in seven hundred years might descend from one man; wherein considering the length of their dayes, we may conceive the greatest number to have been alive together. And this that no reasonable spirit may contradict, we will declare with manifest disadvantage; for whereas the duration of the world unto the flood was about 1,600 years, we will make our compute in less than half that time. Nor will we begin with the first man, but allow the earth to be provided of women fit for marriage the second or third first centuries; and will only take as granted, that they might beget children at sixty, and at an hundred years have twenty, allowing for that number forty years. Nor will we herein single out *Methuselah*, or account from the longest livers, but make choice of the shortest of any we find recorded in the Text, excepting *Enoch*: who after he had lived as many years as there be days in the year was translated at 365. And thus from one stock of seven hundred years, multiplying still by twenty, we shall find the product to be one thousand, three hundred forty seven millions, three hundred sixty eight thousand, four hundred and twenty.

	1.	20.
	2.	400.
	3.	8,000.
	4.	160,000.
Century.	5.	3,200,000.
	6.	64,000,000.
	7.	1,280,000,000.
		1,347,368,420.

Simply as a sum, there are difficulties about this result. If our 20 are equal numbers of males and females, it is not 20 which should be multiplied by 20 but 10. If they are all males, then women are left out of the reckoning. But, perhaps, as the Text does not record the ages of women, Sir Thomas esteemed them as ephemerids, sufficiently plentiful however to provide a wife for every husband. But then I think he should have said that the 20 to be begotten between 60 and 100 were all males. Anyhow the sum must be wrong because *some* of the 64,000,000 short-lived women of the sixth century should survive into the seventh. Indeed Sir Thomas uses his data a trifle capriciously.

We must surely play a game according to the rules. We are to accept the

Text word for word as it stands. But, omitting Adam, whose age at his begetting of Cain is not recorded, and Noah, who seems to have reached middle age—500 years—before becoming a father, the reproductive habits of eight fathers are recorded. Two begat males at the age of 65, one at 70, one at 90, one at 105, one at 162, one at 182 and one at 187. When this primary business was over, they are all recorded to have begotten an unspecified number of sons and daughters. So, if we are to be faithful to the Text, a very much more complicated arithmetical problem presents itself. A male begets another male at an average age of about 100, he then begets males and females at an unspecified rate for say another 600 years, required the law of increase. The Text does *not* authorize Sir Thomas to start pre-diluvian breeding at 65 or to stop it at 100. His 'manifest disadvantage' is breaking the rules of the game.

Further, the Text does not entitle him to predicate of the other males the lengths of days and procreative exploits of the recorded eight.

All this, it may be said, is breaking a butterfly upon the wheel. Nobody now takes the statistics of the Authorized Version literally. The point is that Sir Thomas Browne *did*, but used them improperly. As Lord Chesterfield said to a Garter King at Arms of his day who had not followed the rules of heraldry, 'You foolish man, you don't know your own foolish business'.

Petty did not tackle pre-diluvian demography, but he did try his hand at an estimate of the world's population after the flood, 'To justify the *Scriptures* and all other good *Histories* concerning the *Number* of the People in Ancient Time' (p. 465).

As Petty was not going to allow the population of ancient times to be greater than in the seventeenth century, but to make it increase regularly from the time of Noah's Ark, common sense saved him from fantastic figures, but not from physiological difficulties. The rules of the game obliged him to start with eight landed from the Ark, so he thought it best to make them increase and multiply very fast indeed at first and progressively more slowly. At first he doubled the population every ten years, but by the birth of Christ has brought the period up to 1000 years. But doubling every ten years (in the first century from the Flood) leads one into difficulties.

We can allow the possibility of the four pairs emerged from the Ark producing 8 offspring in ten years and so becoming 16 in year 10, without too great difficulty. But ten years later they must number 32 and this *is* a difficulty. If the fecundity of the first settlers remains the same they will contribute 8 more children, giving us a population of 24, the balance of 8 must come from the four couples of children all of whom must be under 20, and this *is* a little difficult.

But at least we may say that there is nothing wholly fantastic in Petty's procedure. Petty does belong to a different arithmetical world from that of Browne. Here we may leave purely speculative demography.

To estimate the people of an area without counting them, we must count something which has a connexion with the number of the people. We may count the tax-payers, the houses, the burials, the christenings or the acreage under corn—all or any of these items vary with the number of people.

I wish to keep separate the discussions of Petty's and Graunt's statistical researches, but in the matter now to be examined Petty used some of Graunt's methods and results, so these must be considered.

Graunt used three methods of estimation. In the first place, he surmised that the number of child-bearing women in a community might be about double the number of annual births 'forasmuch as such women, one with another, have scarce more than one child in two years'. Then he surmised that families were twice as numerous as women of child-bearing age. His reasoning was that women between 16 and 76 might be twice as numerous as women between 16 and 40 or 20 and 44 (i.e. of child-bearing age), and he thought of a family as centred round a married couple. Finally, he thought that the average family would consist of eight persons, the husband and wife, three children and three servants or lodgers. So, starting with 12,000 christenings, which he thought a fair measure of annual births, he reaches 24,000 women of fertile age, then 48,000 families and lastly 384,000 persons.

It is quite certain that Graunt's estimate of an annual fertility rate of 500 per 1000 was an enormous overstatement. In London in 1851, the ratio of legitimate births to married women aged 15–45 was 251·8 per 1000. There is no reason to believe that nuptial fertility changed appreciably between 1660 and 1860. But an error of this kind would lead him to an understatement of families. Now, however, another error saves him. We cannot be so positive that eight to the family is a great overstatement as we can that the marital fertility was not 500 per 1000, but it is much higher than any nineteenth-century finding. Using *this* multiplier saves Graunt in this sense, that his quaint rule gives almost precisely the right answer for the population of London nearly 200 years after his time.

The legitimate births registered in London in 1851 were 75,097. This, according to Graunt's rule, is to be multiplied by 32. The result is 2,403,104. The enumerated population was 2,363,236; the conjecture is only 1·7 % out. *Sic me servavit Apollo.*

Graunt's next method was experimental and very briefly described. He counted the numbers of families in certain parishes within the walls and found that '3 out of 11 Families per annum have died'. He then multiplies the burials for the year (13,000) by 11/3, and proceeds as before.

Finally, he took Newcourt's map of London and

> guessed that in 100 Yards square there might be about 54 Families, supposing every House to be 20 Foot in the front: for on two sides of the said square there will be 100 Yards of Housing in each, and in the two other sides 80 each; in all 360 Yards: that is 54 Families in each square, of which there are 220 within the Walls, making in all 11880 Families within

the Walls. But forasmuch as there die within the Walls about 3200 per *Annum*, and in the whole 13,000, it follows that the Housing within the Walls is ¼ part of the whole, and consequently, that there are 47,520 Families in and about *London*, which agrees well enough with all my former computations (p. 385).

These conjectures led Graunt to think that the rate of mortality in London was about 1 in 32. In his first essay on the growth of London (pp. 458–75) Petty bases himself upon that estimate, and in the series of papers (pp. 505–44) this remains the fundamental method, but Petty allows himself to modify the multiplier, not altogether without suspicion of bias. At a quite early stage he had satisfied himself that London was the largest city in the world and much larger than Paris. This is the kind of argument. For the three years 1682–84, the average of burials in London was 22,337 and for Paris 19,887. So if the rates of mortality were the same, London was larger than Paris.* If the rate of mortality in Paris were higher than in London then the population of London must be larger still. According to Petty (*a*) a larger proportion of the Paris population died in hospital, (*b*) the mortality in hospital was heavier in Paris than in London. So it follows that the general death rate of Paris was higher.

That at *London* the *Hospitals* are better and more desirable than those of *Paris*, for that in the best at *Paris* there die 2 out of 15, whereas at *London* there die out of the worst scarce 2 of 16, and yet but a fiftieth part of the whole die out of the *Hospitals* at *London*, and 2/5 or 20 times that proportion die out of the *Paris Hospitals* which are of the same kind; that is to say, the number of those at *London* who chuse to lie sick in *Hospitals* rather than in their own Houses, are to the like People of *Paris* as one to twenty; which shows the greater Poverty or want of Means in the People of *Paris* than those of *London*. We infer from the premisses, viz. the dying scarce 2 of 16 out of the *London Hospitals*, and about 2 of 15 in the best of *Paris* (to say nothing of l'*hostel Dieu*) that either the *Physicians* and *Chirurgeons* of London are better than those of *Paris*, or that the *Air* of *London* is more wholesome (p. 508).

These, however, are only logical deductions if the user of the hospitals in London and Paris is identical. If, as implied in the first part of the quotation, we think of hospitals in the sense which our elder contemporaries think of the old-fashioned poor law infirmaries, viz. as refuges for the sick poor, it would mean that in Paris more of the aged indigent died in institutions than in London and heavy mortality might well have nothing to do with the skill or lack of skill of the medical staff. If we think of hospitals in the modern sense, then heavy mortality might be a mere reflection of the resort to these hospitals of persons suffering from illnesses which needed special treatment. In any case, Petty can hardly have it both ways. In another essay (pp. 510–11) he contrasts the higher ratio of deaths to admissions at l'hostel Dieu of Paris with that of la Charité, argues that the excess in l'hostel Dieu is unnecessary and proceeds to calculate

* It should be remembered that the London of Petty's calculations is the whole area within the Bills. The calculations of Graunt described above did not include Westminster or the six out-parishes of Surrey and Middlesex which were within the Bills: Islington, Lambeth, Stepney, Newington, Hackney, Redriff.

what the French nation would gain by saving this excess. But he has not inquired whether the patients of the two institutions were *in pari materia*.

Here is an historical problem which might be solved by those familiar with the literature of the period. Its discussion would not be relevant here. It is, however, only just to Petty to say that, unless conditions deteriorated seriously in the following century, his strictures on l'hostel Dieu were justified. In Franklin's work (*La Vie Privée d'autrefois. L'Hygiène* (Paris 1890), pp. 177 *et seq.*) an appalling account of this hospital from the pen of the eminent surgeon Tenon, printed in 1788, is quoted. Tenon's description of the routine of this great hospital compares, unfavourably, with the story of the wounded in the Mesopotamian campaign which horrified England in the war of 1914–18. He remarks, *inter alia*, 'on ne guérissoit point de trépanés autrefois à l'Hôtel-Dieu, comme on n'en guérit pas encore aujourd'hui', and cites a court surgeon of the time of Louis XIV, i.e. a contemporary of Petty, to that effect. His account of the treatment of lying-in women is grotesquely horrible.

In another essay (pp. 533–6) Petty discusses methods of estimation more carefully than in his other papers.

He proposes to show that the population of London (within the Bills) in or about 1685 was approximately 696,000.

There are, he says, three methods: (1) From houses and families. (2) From an estimated death rate. (3) From the ratio of those who die of the plague to those who escape.

This last we may deal with at once. Petty asserts that Graunt had proved that one-fifth of the people died of the plague. But in 1665, 98,000 died of the plague; therefore the population was 490,000, and allowing an increase of one-third between 1665 and 1686 we reach 653,000.

Graunt could not have proved that one-fifth of the population died of the plague unless he knew what the population was, and he never claimed to have done so.

The other methods (which Graunt used) are rational.

To estimate houses, Petty used three methods. He says that in the Fire of 1666, 13,200 houses were burned and that deaths from these houses were one-fifth of total deaths, so he reckons the houses to have been 66,000. Then as burials in 1686 were to burials in 1666 as 4 to 3, he makes the houses of 1686, 88,000. He does not, however, say upon what basis the estimate of one-fifth of the deaths in 1666 stands.

Next, he gives an estimate of the houses in 1682 given him by those employed upon a map said to have been made in that year. This map has not been identified.

Lastly, he uses the return of hearths. In Dublin in 1685 the hearths were 29,325 and the houses 6400. In London the hearths were 388,000; so the houses on the Dublin ratio should be 87,000. In Bristol he says there were 5307 houses and 16,752 hearths, which give 123,000 houses for London; the mean of the calculations is 105,000. The Hearth Office itself, he says, certified the number to be 105,315. He must now have a multiplier. He accepts Graunt's multiplier of 8 as valid for tradesmen's families, but allows for smaller families among the poor and larger among the rich, finally choosing 6. He then allows for double families in houses by adding 10,531 to his 105,315, and multiplying the sum by 6 has 695,076 for the population.

Petty's second way was from an estimated death rate.

Petty multiplies the average of the burials in 1684 and 1685 (23,212) by 30, which makes the population 696,360.

He now essays to prove that the death rate in *London* was 1 in 30. He uses four arguments, of which only one is strictly to the point, viz. Graunt's direct observation that three deaths occur annually in eleven families—which however involves the assumption of eight persons to the families observed. Two others are relevant, viz. observations, apparently direct, that in 'healthful places' the mortality is 1 in 50 and in nine country parishes 1 in 37. The fourth partly rests upon a statement which Graunt did *not* make, viz. that one of 20 children under 10 dies annually. This fictitious value Petty averages with the statement of a M. Auzout to the effect that the rate of mortality of adults in *Rome* is 1 in 40. It will be clear that Petty has proved nothing at all. What he has done is to make it unlikely that the rate of mortality was less than 1 in 30. That, perhaps, was enough. One has a certain sympathy with his round statement: 'Till I see another round number, grounded upon many observations, nearer than 30, I hope to have done pretty well in multiplying our Burials by 30 to find the number of the People.'

With this I may conclude the analysis of Petty's statistical work. It will, I think, soon be clear enough that it is not of the calibre of Graunt's. Yet I cannot take leave of it without something of an *ave*. Careless, happy-go-lucky, tendentious; yes, all that. But anybody who has felt the exhilaration, to which Francis Galton owned, in the doing of sums concerning biological problems, feels his heart warmed by the arithmetical knight errant who had so many statistical adventures.

(*To be continued*)

MEDICAL STATISTICS FROM GRAUNT TO FARR
(Continued)

III. THE STATISTICAL WORK OF GRAUNT

JOHN GRAUNT'S contribution to our subject has always been regarded as one of the great classics of science. A few have indeed doubted whether so great a work could have been achieved by one whose material success was so modest and have sought to transfer the glory to Graunt's highly successful friend Petty. This dispute I relegate to an appendix. I assume that Graunt's published book is substantially his own original work.

The history of the material Graunt used has been written more than once and I have nothing to add to Prof. Hull's story. Graunt had, for a period of more than 60 years, arithmetical statements of the numbers of males and females christened and buried and of the causes of death (not distinguished by sex) under some sixty headings. He had no information as to the ages at death. He had no information as to the number or ages of the living population.

The first act of a scientific statistician is to assess the trustworthiness of his data, to criticize his sources. This tedious preliminary to the doing of sums was not much to Petty's taste. Petty, as we have seen, often used different data to reach some conclusion, but hardly ever discusses the reliabilities of the several data. Other Fellows of our College since Petty's day have made the same mistake. The terrible 'howler' committed by Dr William Heberden the younger, and detected, not without satisfaction, by Charles Creighton is classical.† But that was not a unique instance. Indeed, even trained statisticians sometimes confuse names with things. More than one rate of mortality has risen (or fallen) only on paper. Graunt made no such mistakes.

Graunt's *general* argument is that many causes of death are 'but matters of sense', for instance, whether a child were abortive or stillborn, and that in many cases the searchers are 'able to report the opinion of the physician, who was with the patient as they receive the same from the friends of the defunct'. But sometimes the searchers will be wrong and often enough the error will not matter.

As for consumptions, if the searchers do but truly report (as they may) whether the dead corpse were very lean and worn away, it matters not to many of our purposes whether the diseases were exactly the same, as physicians define it in their books. Moreover, in case a man of seventy-five years old died of a cough (of which had he been free, he might

† Creighton, *History of Epidemics in Britain*, 2, 747–8. Heberden supposed (erroneously) that 'Griping of the Guts' of the Bills was Dysentery and had decreased. It was Infantile Diarrhoea and had simply been transferred to the rubric 'Convulsions'.

have possibly lived to ninety) I esteem it little error (as to many of our purposes) if this person be in the table of casualties, reckoned among the aged, and not placed under the title of coughs (348).*

No doubt this brutal common sense might set on edge the teeth of some Fellows of the College of Physicians even in the seventeenth century, but it was one of the qualities which made Graunt a pioneer. Making the best the enemy of the good is a sure way to hinder any statistical progress. The scientific purist, who will wait for medical statistics until they are nosologically exact, is no wiser than Horace's rustic waiting for the river to flow away.

Graunt, however, did not accept statements which he had the means of testing. Finding in a series of years that of more than a quarter of a million deaths only 392 were assigned to the Pox, he did not infer that Syphilis had been over-rated as a cause of death.

Forasmuch as by the ordinary discourse of the world it seems a great part of men have, at one time or other, had some species of this disease, I wondering why so few died of it, especially because I could not take that to be so harmless, whereof so many complained very fiercely; upon enquiry, I found that those who died of it out of the hospitals (especially that of Kingsland, and the Lock in Southwark) were returned of ulcers and sores. And in brief, I found, that all mentioned to die of the French Pox were returned by the clerks of St Giles' and St Martin's in the Fields only, in which places I understood that most of the vilest and most miserable houses of uncleanness were: from whence I concluded, that only hated persons, and such, whose very noses were eaten off were reported by the searchers to have died of this too frequent malady (356).

In principle, the argument is still valid.

His next example of criticism is the case of Rickets, which first appeared in the Bills of Mortality in 1634 and then with 14 deaths only, but by 1659 had risen to 441. Was Rickets a 'new disease' or did an old disease receive, in the Bills, a new name?

To clear this difficulty out of the bills (for I dare venture on no deeper arguments) I enquired what other casualty before the year 1634, named in the Bills, was most like the rickets; and I found, not only by pretenders to know it, but also from other Bills, that livergrown was the nearest. For in some years I find livergrown, spleen, and rickets, put all together, by reason (as I conceive) of their likeness to each other. Hereupon I added the livergrowns of the year 1634, viz. 77, to the rickets of the same year, viz. 14, making in all 91; which total, as also the number 77 itself, I compared with the livergrowns of the precedent year 1635, viz. 82. All which showed me, that the rickets was a new disease over and above. Now, this being but a faint argument, I looked both forwards and backwards, and found that in the year 1629, when no rickets appeared there were but 94 livergrowns; and in the year 1636 there were 99 livergrowns, although there were also 50 of the rickets: only this is not to be denied, that when the rickets grew very numerous (as in the year 1660, viz. 521) then there appeared not above 15 of livergrown. In the year 1659 were 441 rickets and 8 livergrown; in the year 1658 were 476 rickets and 51 livergrown. Now though it be granted that these diseases were confounded in the judgment of the nurses, yet it is most certain that the livergrown did never but once, viz. anno 1630, exceed 100; whereas anno

* Numbers in brackets are page references to Prof. Hull's edition of *The Economic Writings of Sir William Petty together with the Observations upon the Bills of Mortality more probably by Captain John Graunt*, Cambridge, 1899.

1660, livergrown and rickets were 536. It is also to be observed, that the rickets were never more numerous than now, and that they are still increasing; for anno 1649, there were but 190, next year 260, next after that 329 and so forwards, with some little starting backwards in some years, until the year 1660, which produced the greatest of all (357–8).

This is an excellent statistical argument, and, incidentally, evidence that Graunt wrote his own book, for a physician would probably have suggested that the professional interest excited by the classical treatise of Glisson (assisted by Regimonter) which was published in 1650 might easily have increased the popularity of the diagnosis. Petty, who, with Glisson, was a founder of the Royal Society, would hardly have ignored his colleague's work.

I cannot resist the desire to mention others which, while of little statistical importance, have a medical attraction. Graunt noticed that Stopping of the Stomach first appeared in the Bills of 1636, increased from 6 to 29 by 1647, by 1655 it reached 145, in 1657, 277 and 1660, 314. First he conjectured that Stopping of the Stomach might be the Green Sickness, 'forasmuch as I find few or none to have been returned upon that account, although many be visibly stained with it'. He thought that possibly Green Sickness might not appear in the Bills 'for since the world believes that marriage cures it, it may seem indeed a shame, that any maid should die uncured, when there are more males than females, that is, an overplus of husbands to all that can be wives'. Then he wondered whether Stopping of the Stomach might not be Mother, 'forasmuch I have heard of many troubled with Mother Fits (as they call them) although few returned to have died of them'. But he was diverted by guessing 'rather the Rising of the Lights might be it'. He remembered that some women troubled with the Mother fits did complain of a choking in their throats. 'Now, as I understand, it is more conceivable that the Lights or Lungs (which I have heard called the bellows of the body) not blowing, that is, neither venting out, nor taking in breath, might rather cause such a choking, than that the Mother should rise up thither, and do it. For methinks, when a woman is with child, there is a greater rising, and yet no such fits at all' (359). He notes that Rising of the Lights increased in the Bills from 44 in 1629 to 249 in 1660.

Finally, he suggests a correlation between Stopping of the Stomach, Rising of the Lights in adults and the Livergrown, Spleen and Rickets of children. 'And that what is the Rickets in children, may be the other in more grown bodies; for surely children which recover of the Rickets, may retain somewhat to cause what I have imagined: but of this let the learned physicians consider, as I presume they have' (359).

It might be suggested that one item under Stopping of the Stomach could be surgical, viz. strangulated hernia. Rupture was a heading in the Bills, but the numbers are small and show no regular increase with the increase of population. Graunt's attraction to what used to be called hysterical stigmata is interesting. One wonders how far these passages reflect conversations with Petty. It is clear that Graunt had no belief in the peripatetic uterus; Petty would have had none. The best medical opinion of the age is, of course, that of Sydenham. Sydenham (whose pathology was traditional) had a pneumatist aetiology of Hysteria, the origin was an ataxia of the animal spirit (which was the pneuma zotikon of ancient tradition). He not only believed that Hysteria might be a serious or even mortal complication of organic disease—as we do still—but that the ataxic spirits might themselves produce humoral corruption and lead to chlorosis or ovarian dropsy (*Dissertatio epistolaris*, 92). So there is nothing repugnant to the best professional opinion of the age in admitting Hysteria to the list of causes of death. Nor is there any gross absurdity in the suggested correlation of increasing Rickets and increasing Hysteria, from the point of view of a layman. But that surmise does not imply any professional hint, it rather suggests a belief in a merely physical factor, the pressure of an enlarged organ. That passage would not have been written by a physician.

These are sufficient instances of Graunt's criticism of sources—the temptation to go on quoting examples must be resisted. I pass to his great achievement, the estimation of rates of mortality at ages when the numbers and ages of the living were not recorded. For such an estimation to be correct, we all know that the population must be stationary, viz. non-increasing, not subject to migration and having constant rates of mortality in the several age groups.

It is a nice point whether Graunt or Petty appreciated the importance of these considerations. Graunt was certainly alive to the fact that the population of London was growing and that the growth was due to immigration from the country. The arithmetical position was this. In the earlier years of his series burials and christenings were about equal in numbers, in 1605 there were 5948 burials and 6504 christenings; in 1625, 7850 burials and 7682 christenings, in 1635, 10,651 burials and 10,034 christenings. Later the burials continued to increase, but the christenings either decreased or failed to increase in the same proportion. This Graunt attributed to neglect of christening owing to religious dissidence and gave excellent reasons for his view. It is clear then that there were two factors of increase, immigration and increasing numbers of births. Most of Graunt's deductions are based upon an analysis of the deaths by causes for twenty years, 1629–36 and 1647–58, which he selected as years comparatively unaffected by plague (of his total of 229,250 deaths only 16,000 were from plague).

If we treat this total as a denominator (or one-twentieth of it) it will, from the point of view of calculating mortality ratios, be affected by two errors. The deaths of immigrants will make it too large and the increasing births will make it too small. Can it be Graunt held that the errors balanced so that, arithmetically speaking, one might behave as if one were dealing with a stationary population? An alternative explanation is that Graunt did not realize the limitations of the method.

A third possibility is that, although he knew the fallacy, he believed that the incorrect method gave an approximation to truth sufficient for his purposes. This is the solution I should be inclined to adopt were I forced to choose.

As I have pointed out above, there is at least a suggestion that Petty did have some glimmering of the conditions to be fulfilled if a summation of deaths is to give a correct view of rates of mortality. I do not believe that Graunt was less informed on any point of vital statistics than Petty. However, all this is guess-work.

Graunt did not know the ages of the dead; what he did was to pick out of the list of causes of deaths those which he thought lighted only upon children 'not more than four or five years old'. He chose Thrush, Convulsions, Rickets, Teeth and Worms, Abortives, Chrysomes, Infants, Livergrown and Overlaid. These gave him some 70,000 out of some 229,000. Then he assigned half the deaths from Small Pox, Swine Pox, Measles and Worms without Convulsions also to children under six and reaches the final conclusion that about '36 % of all quick conceptions die before six years old'.

Is this conclusion—I will not say correct, because we have no data to reach a correct result—but of a reasonable order of magnitude? The answer is that it is eminently reasonable. Two hundred years after Graunt's death, William Farr printed (in the famous Supplement to the *35th Annual Report of the Registrar-General*, p. cxxxvi) an outline Life Table for London. This was, of course, computed by an approximately correct method, using knowledge of the numbers and ages of the living population, and reflects the conditions of seventy-five years ago. Interpolating in this we find that about 32 % of 'quick conceptions died before six years old'. There is no good medical reason for holding that the conditions of *child* life in London in the middle of Victoria's reign were much better than in the seventeenth century. The old genius used a bow with a frayed string and made no allowance for windage, but his arrow hit the target not far from the white. He gave the first quantitative measure of the Herodian sacrifice in towns, a sacrifice which was to continue to be offered for more than 200 years.

Graunt then passed to the other end of life and found that 7 % of the dead were 'aged'. He conceived that the searchers would mean by 'aged' persons of 70 years or upwards, 'for no man be said to die properly of Age who is much less'. His following suggestion that the proportion living beyond 70 might be used as a measure of healthfulness is not happy. But this calculation may have led him to make, or insert, the most famous passage in his book, viz. what is, in form, the first Life Table ever published.

> Whereas we have found, that of 100 quick Conceptions about 36 of them die before they be six years old, and that perhaps but one surviveth 76; we having seven decads between six and 76, we sought six mean proportional numbers between 64, the remainder, living at six years, and the one, which survives 76, and find that the numbers following are practically near enough to the truth; for men do not die in exact proportions, nor in fractions, from whence arise this Table following (386).

Graunt's figures are 100, 64, 40, 25, 16, 10, 6, 3, 1.

The one survivor to 76 is, as Graunt implies, a guess; perhaps he conjectured that his seven survivors beyond 70 died one a year. How he calculated his mean proportional numbers is unknown. Prof. Willcox conjectured that he experimented with multipliers of 5/8 and 2/3—the former nearly reproduces the figures (see Willcox, *Revue de l'Inst. Intern. de Statistique,* **5** (1937), 327). Ptoukha (*Congrès Intern. de la Population*; Démographie historique, p. 71, Paris, 1937) ingeniously suggests that he used the multiplier (64 – 1)/100 or 0·63.

We must, I fear, conclude sorrowfully that this shot did not find the bull's eye. If Graunt's survivors are compared with those shown in Halley's table (when correctly used, *vide infra*), for 100, 64, 40, 25, 16, 10, 6, 3, 1, we should have 100, 56, 50, 45, 38, 31, 22, 14, 6. It is possible that child mortality was lower in London than in Breslau, but quite incredible that later age mortality should have been so enormously higher.

But, of course, having regard to the data, it would have been more than genius, it would have been magic, had a correct result been obtained.

Prof. Willcox, whose opinion of Graunt is almost as high as mine, regards the passage as inserted on the recommendation of Petty and as Petty's composition. He thinks that it lacks Graunt's caution and suggests the flighty ingenuity of his friend. Prof. Willcox's arguments are weighty, but I am not convinced. That Graunt did not—to use the expressive slang—*feature* his table is true. It is also true (*vide supra*) that passages in Petty's undoubted writings imply that he had some conception of a survivorship table. But—and this is my main difficulty—if this were Petty's idea, I find it difficult to believe that he would not have exploited it. Halley, whose economic scent was not so keen as Petty's, saw the epoch-making importance of an idea which was to transform the business of selling annuities. It would be odd if Petty *had* seen it that he did not comment upon it. Graunt might well have hesitated, being a cautious statistician, but surely not Petty.

However, in spite of modern practice, the writing of history wholly in terms of psychology has its pitfalls.

Let us return to simpler applications of shop arithmetic. The advantages of country life over town life from the point of view of both mortality and morality had been a commonplace of poets, particularly those Roman poets who spent much of their lives in a city, long before the seventeenth century. Graunt was the first to apply an arithmetical test of mortality; he compared the statistics of Romsey with those of London. For Romsey he had ninety years' data of marriage, christenings and burials.

His statements about the population of the parish are not quite consistent. In one sentence he says that it 'both 90 years ago, and also now, consisted of about 2700', but a few lines later says 'it neither appears by the burials, christenings, or by the built of new housing, that the said parish is more populous now, than 90 years ago, by above two or 300 souls'. A little later he says 'it is clear that the said parish is increased about 300, and it is probable that 3 or four hundred more went to London; and it is known that about 400 went to New England, the Caribe Islands and Newfoundland within these last forty years' (389). Actually, from an estimate of the number of communicants (which he assumes to be rather more than half the total population) he makes the average population between 2700 and 2800. Taking the average of burials for the whole period to be 58, this gives him a death rate of a little more than one in 50, which he contrasts with the London figure of one in 32 (apparently based on his count of 11 families with 88 persons amongst whom 3 deaths occurred in a year; but this is a rate of one in 29).

There is no doubt a certain sketchiness about this, but it was not unreasonable to infer that the Romsey rate was much lower than the London rate.

Graunt found that, unlike London, Romsey had an average excess of christenings over burials, they were in the ratio of 5 to 4. He estimates that over the period the natural increase was 1059, and, as will be seen from the quotation made, he allots about a third of this respectively to London, to the colonies and to the parish itself. He argues that supposing the population of all England to be fourteen times that of London and other parishes to send one-third of their natural increase to London, then the London burials should increase about 200 per annum 'and will answer the increase we observe'.

Here again the argument is reasonable. He goes on to an investigation which has been severely criticized. He gives a table of the greatest and least number of burials in each of the ten-year periods for which he has data. In each decade but one the maximum is more than twice the minimum. But, he remarks, in no decade in the London experience is the largest number of burials twice the smallest number (he excludes deaths from plague from his statistics). 'Which shews, that the opener and freer airs are most subject both to the good and bad impressions, and that the fumes, steams and stenches of London do so medicate and impregnate the air about it, that it becomes capable of little more, as if the said fumes rising out of London met with, opposed and jostled backwards the influences falling from above, or resisted the incursion of the country airs' (392).

Prof. Hull shook his head over this passage. 'This is an attempt to explain by physical conditions the wide range in the observed country death rate which is really due to the narrowness of the field—a single market town—under investigation. It is perhaps the gravest statistical mistake that can be charged against Graunt' (lxxvii).

I do not like to leave a hero in the lurch. I must concede that if both Romsey and London burials were samples from a Poisson universe, the fact that the Poisson parameter for London was at least a hundred times that for Romsey would make it incredible that the London range, in terms of the mean or of the standard deviation, should be so wide as that for Romsey. But Prof. Hull was wrong in supposing that the wide range in the Romsey rates was due to the narrowness of the field of observation in a statistical sense.

Taking Graunt's 58 as the 'expected' annual deaths then, as 1/58 is small, the Poisson distribution is not far from the symmetry of a normal curve, and using the results of Tippett and E. S. Pearson, we may conclude that the expected range would be $23{\cdot}45 \pm 6{\cdot}073$. The observed ranges for the successive decades are 32, 48, 78, 23, 65, 39, 121, 91, 52. All but one is greater than the expectation and six diverge by more than three times the standard error.

Something more than small numbers is involved. Still, it must be confessed that Graunt did not anticipate the reasoning of James Bernoulli, although an intuition of genius may have led him to think that something more than 'chance' had play here.

Graunt devoted special attention to the demographic influence of the plague. In the first place, he remarks that the attribution plague understated the mortality due to plague. He infers this from the fact that in plague years burials from other causes exceeded the average greatly, 'from whence we may probably suspect, that about 1/4 part more died of the plague than are returned for such'. Next he inferred that after a great outburst plague lingered for several years.

> The plague of 1636 lasted twelve years, in eight whereof there died 2000 per annum one with another, and never under 300. The which shows that the contagion of the plague depends more upon the disposition of the air than upon the effluvia from the bodies of men. Which also we prove by the sudden jumps which the plague hath made, leaping in one week from 118 to 927; and back again from 993 to 258; and from thence again the very next week to 852. The which effects must surely be rather attributed to change of the air, than of the constitutions of men's bodies, otherwise than as this depends upon that (366).

Finally, he observes that within two years the city was re-peopled; a deduction from the time taken for the number of christenings to reach again the level of a pre-plague year.

We may, if we please, smile at Graunt's epidemiological inference. But it is a reasonable inference from the facts when we remember that in Graunt's day—in spite of Fracastorius—contagium was not thought of as contagium vivum, but as a mere sympathetic vibration or passing on of something.

I have, I hope, given an adequate sample of Graunt's quality, but have not mentioned the most famous of all his deductions. Both in London and the country, on the average more males were christened than females, but more males died young or entered celibate occupations. So we reach this conclusion:

> We have hitherto said, there are more males than females; we say next that the one exceed the other by about a thirteenth part. So that although more men die violent deaths than women, that is, more are slain in wars, killed by mischance, drowned at sea and die by the hand of justice; moreover more men go to colonies and travel into foreign parts than women; and lastly, more remain unmarried than of women as fellows of colleges, and apprentices above eighteen etc. yet the said thirteenth part difference bringeth the business but to such a pass, that every woman may have an husband, without the allowance of polygamy (375).

The story of how this arithmetical justification of God's providence attracted the attention of Derham, of how Derham's book fired the enthusiasm of the Prussian Army chaplain Johann Peter Süssmilch and of how Süssmilch's book influenced Malthus has been well told by Hull. I should not myself rank this section high among Graunt's researches. From a demographic point of view neither judicial hangings nor college fellowships could have had much effect in reducing the male excess.

Even copious quotation fails to convey the spirit of a complete book. I have quoted good things, but many more remain. Graunt revealed sundry important truths and not the least important was that very imperfect data, if patiently considered, will tell us something it is good for us to know. If young medical officers going to parts of the empire where organized medical and demographical information is at no higher a level than that of seventeenth-century England—and there are many such places—were restricted to a single book on statistics, I should advise them to take not a modern scientific work, but old John Graunt's *Observations*.

Appendix

Did Graunt write the book published over his name?

John Graunt and William Petty were, as we have seen, close friends. Graunt, as the world judges success, failed and Petty succeeded. But by the judgment of scientific men in the seventeenth century, and ever since, the order of intellectual precedence was reversed. From the moment of publication a few discerning people perceived the originality and importance of the *Observations*, the same people who, while admiring Petty's verve, ingenuity and worldly success, did not take over-seriously his bright ideas.

But Graunt was a man of one book. Save a note upon the multiplication of carp and the growth of salmon, he published nothing more. Petty went on writing, scheming and talking for thirteen years after Graunt's death. That often enough in that period, the *Observations* were discussed over the wine—as they were quoted in Petty's writings—we may suppose. That Graunt discussed his work with Petty both before and after publication we may also take for certain, although we have no formal proof of it. The country statistics which Graunt first used were from Petty's native parish and even if we are not disposed—as certainly I am not—to give much weight to particular turns of phraseology, still there are sufficient verbal oddities in some pages of Graunt's book to suggest Petty's hand.

In these circumstances, it would not be very surprising if Petty's associates, particularly those who were not good judges of statistical work, were to conclude that Petty's share in the remarkable achievement of Graunt were greater than appeared. It is not even judging Petty too harshly to suppose that he himself might come to share the opinion. There is no evidence that Petty ever *did* explicitly claim the credit. In one list of his writings (one of four), found among Petty's Papers, he did include the *Observations*, which at least is evidence that he thought himself entitled to a share of the credit. We may suppose that if, in familiar intercourse, somebody had said 'Come, confess Sir William, yours was the hand that guided the pen of poor John Graunt', he might not have denied it very strenuously. I think I have produced evidence enough that Petty did not under-rate his powers and was not conspicuous for delicacy of feeling. My guess would be that long before his death he did come to believe that Graunt's intellectual success was due to his help.

Whether Petty believed this or not, it is certain that friends and associates of Petty began to believe it soon after Graunt's death, and the belief has been entertained by a few people in each generation since. These, with one conspicuous exception, have been drawn from Petty's friends or descendants or from literary critics.

In the seventeenth century, of Petty's circle, Evelyn, Southwell and Aubrey believed or said that Petty wrote or inspired Graunt's book. Two Fellows of the Royal Society, Houghton and Halley, also attributed the book to Petty. The only one of the five who was certainly a competent judge of scientific merit was Halley. Halley began the memoir which contains his Breslau table with these words:

> The contemplation of the mortality of mankind has, besides the moral, its physical and political uses, both which have some time since been most judiciously considered by the curious Sir William Petty, in his moral and political Observations upon the Bills of Mortality of London, owned by Captain John Graunt. And since in a like treatise on the Bills of Mortality of Dublin....But the deductions from those bills of mortality seemed even to their authors (*sic*) to be defective. (*Phil. Trans.* no. 196 (1693), p. 596.)

Since the seventeenth century, there has been unanimity among demographic statisticians and economists that Petty could not have written Graunt's book. Halley was quite as good a judge of scientific merit as any of them and a contemporary of the canvassed writers; if I were sure that he had read and compared Petty's acknowledged works with the *Observations* I should prefer his opinion to that of other 'experts'—including, of course, my own. Halley's direct testimony, in the sense of a court of law, would be valueless; he was only six years old when the *Observations* were published and became a Fellow of the Royal Society five years after the death of Graunt. There is no evidence that either before or after the period of writing and publishing his famous memoir, Halley worked on demography. After his memoir, but in his lifetime, a new epoch in mathematical vital statistics began. De Moivre, eleven years younger than Halley, brought out his principal works in the lifetime of Halley (1656–1742) and used Halley's table. The two men must have been well acquainted, for both were enthusiastic disciples and intimate friends of Newton, but Halley, like Graunt, made only one contribution to the literature of demography.

So it may be doubted whether Halley were sufficiently interested in demographic or economic writings to have read Petty's tracts at all. Also in the passage cited above (apart from the writing of 'authors' not 'author') the collocation of the *Observations* on the Dublin Bills with those on the London Bills is curious. There is no doubt that the *Observations* on the Dublin Bills were the work of Petty, and in the first edition of them they are stated on the title-page to be by the 'Observator on the London Bills of Mortality'. But this, as Prof. Hull pointed out (xlii), was probably a catch-penny device of the publisher, Mark Pardoe, to draw a public which had just taken a fifth edition of the London *Observations*. Actually the book did not sell, and when the publisher reissued an enlarged version, Petty's name appeared on the title-page without any reference to the London *Observations*. I conclude that Halley's evidence is less weighty than it seemed. He may well have had before him copies of Graunt's book and of the two editions of the Dublin *Observations*. Having no other knowledge of the literature he would naturally enough write as he did.

If we eliminate Halley, no other expert countenanced Petty's authorship and one, Augustus de Morgan, gave an amusing but quite cogent reason for dismissing the notion.

In speaking of the variations in the annual numbers of deaths attributed to Rickets, Graunt said:

> Now, such back-starting seem to be universal in all things: for we do not only see in the progressive motion of wheels of watches, and in the rowing of boats, that there is a little starting or jerking backwards between every step forwards, but also (if I am not much deceived) there appeared the like in the motion of the moon, which in the long telescopes at Gresham College one may sensibly discern (358).

De Morgan (*Budget of Paradoxes*, 68; *Assurance Magazine*, **8**, 167) commented on the improbability that 'that excellent machinist, Sir William Petty, who passed his day among the astronomers' would attribute to the motion of the moon in her orbit all the tremors which she gets from a shaky telescope.

Down to 1927 the matter was regarded, in scientific circles, as settled. In that year the late Marquis of Lansdowne published a copious selection of the Petty Papers with what he regarded as new evidence in favour of Petty.

The only new evidence of a direct kind was a manuscript list in Petty's hand of his writings or projected writings which included the *Observations*. There are three other lists which do not include the *Observations*, and if we are to suppose that the entry really referred to the book published under Graunt's name, then we must believe that in 1685 and in 1686 Petty had forgotten his best title to scientific immortality. The remainder of the evidence consists of parallel passages and *ad captandum* arguments to the effect that it was more probable that a physician had written on questions of medical statistics than a tradesman. This publication led to a lively controversy. Of the merits of this, I, as a party to it, am not an impartial judge. Purely literary arguments do not appeal to me when the question is of scientific method. Thus, Dr L. F. Powell attached weight to the fact that Dr Johnson in conversation had attributed to Petty an observation (not statistical) which is made not in Petty's writings but in Graunt's book.

In the discussion the word 'style' is used in different senses by the combatants. The statisticians are thinking of scientific method, the literary critics of verbal arrangement. To the former the fact that, particularly in the conclusions and the Appendix, Graunt's book has turns of phraseology which suggest Petty's hand, seems of little importance. To the latter it seems very significant.

In the article by Prof. Willcox, which I have quoted above, the controversy is reviewed, and the author concurs generally with his statistical predecessors.

Prof. Willcox does, however, differ from his predecessors in one important particular. He holds that the famous life table was supplied by Petty. He argues that this is far too conjectural to have been the work of so cautious a reasoner as Graunt:

> In attempting to reconstruct its origin I have surmised that after Graunt had estimated that 36 per cent. of the deaths were due to children's diseases, that they all occurred under the age of six, and that the seven per cent. who were reported to have died 'aged' died at over 70 years of age (at one place he says over sixty), he felt unable to go further and reported his difficulty to Petty, already perhaps speculating about a series of similar problems.
>
> Petty guessed at the number of survivors at the end of each decennial age period, 6–15, 16–25 etc. incidentally and characteristically ignoring Graunt's theory that seven per cent. survived seventy, and assuming instead, without reason, that one per cent. survived seventy-six and not one per cent. eighty-six, and that the survivors at age six decreased with each age period in a geometrical progression approximately equal to the 64 per cent. which Graunt had set for the first group (326–7).

Prof. Willcox's argument is cogent. It may be strengthened by a criticism of the late Prof. Westergaard (*Contributions to the History of Statistics* (London, 1932), p. 23). In using this table, Graunt made a serious blunder. In order to estimate the number of men of military age in London he subtracted the number alive at age 56 from the number alive at age 16. But this simply gives him the number *dying* between those ages; what he wanted was some average of the l_x's. It is evident that Graunt was not at all clear in his mind as to how to use a life table.

On the other hand, if this table were really Petty's idea, it is hard to understand why he did not exploit it. If Petty had been a Halley, the explanation would be obvious. The table is wrong; the conditions for the validity of the method were not fulfilled. There is indeed (*vide supra*) some evidence that Petty did know what data were necessary in order to construct a proper life table. One seems on the horns of a dilemma. If Petty thought the table was correct why did he make no further use of the method? If he thought it was wrong, would he have urged Graunt to insert it?

Although Prof. Willcox has certainly shaken my previous conviction, I still feel reluctant to surrender Graunt's table to Petty. However, there may be an element of sentimentality in this. At least the statisticians agree that the answer to the question which I have placed at the head of this Appendix is emphatically *yes*.

REFERENCES TO THE RECENT CONTROVERSY IN CHRONOLOGICAL ORDER

1927. LANSDOWNE, Marquis of. *The Petty Papers.*
1928. GREENWOOD, M. *J.R. Statist. Soc.* **91**, 79.
1928. LANSDOWNE, Marquis of. *The Petty Southwell Correspondence*, pp. xxiii–xxxii.
1932. LANSDOWNE, Marquis of. *The Times Literary Supplement*, 8 Sept.
1932. BRETT-JAMES, N. G. *The Times Literary Supplement*, 15 Sept.
1932. GREENWOOD, M. *The Times Literary Supplement*, 22 Sept.
1932. LANSDOWNE, Marquis of. *The Times Literary Supplement*, 13 Oct.
1932. POWELL, L. F. *The Times Literary Supplement*, 20 Oct.
1933. GREENWOOD, M. *J. R. Statist. Soc.* **96**, 76.
1937. WILLCOX, W. F. *Revue de l'Inst. Intern. de Statistique*, **5**, 321.

IV. HALLEY'S LIFE TABLE

The long and fruitful life of Edmund Halley (1656–1742) belongs to the general history of science; of him it may indeed be said *nihil quod tetigit non ornavit.* He made only one contribution to our subject, but it was of first-rate importance.

The circumstances of this undertaking are obscure; Halley would have perceived the imperfections of Graunt's life table, but it is not known whether it was he who set on foot a search for better statistical material than Graunt had had. Inquiries were however made, and made after he had become a Fellow of the Royal Society, so it is at least possible that Halley, who had travelled extensively in Europe (he was at Danzig in 1679 and in Italy in 1681), suggested that something might be found abroad. By 1691, the King's Librarian, Henry Justel, who was in touch with the Society, had been brought into communication, possibly through Leibniz, with Caspar Neumann, a scientifically minded evangelical pastor of Breslau. Neumann supplied the data which Halley used.

In 1883, J. Graetzer, a medical-statistical official of Breslau, published a little monograph* which throws light upon the work. He not only extracted from the Breslau archives all the data which were or might have been communicated to the Royal Society but had the Society's archives searched, with the result that a letter from Neumann to Justel and another from Neumann to Halley, both with statistical appendices, were discovered. Thanks to the labours of Graetzer and an essay by R. Böckh (*Bulletin de l'Inst. Intern. de Statistique*, **7** (1893), 1) we can form a reasonably clear idea of Halley's method, which was not what those who have not examined the literature suppose it to have been.

It is often stated that Halley, having found that during the five years of observation the number of births only slightly exceeded the number of burials,

* Graetzer, J., *Edmund Halley und Caspar Neumann*, 1883.

treated the population as stationary and constructed a life table by a simple summation of the deaths in the manner already explained. He was much wiser. What he tried to do was to construct a *population* table, in the following way. Suppose we know how many children were born in a calendar year, say 1690, in a town not subject to migration which maintains accurate registers of ages at death, and then we discover how many of the children born in 1690 will be alive on each successive first of January by a series of subtractions. We shall have the survivors on 1 January 1691 by subtracting those of the children born in 1690 who died in 1690. We shall have the survivors on 1 January 1692 by subtracting the deaths in 1691 occurring among the survivors to 1 January 1691, and so on. This will give a precise enumeration of the living population, and this is what Halley wanted. The figures we shall obtain will *not* be the conventional l_x's of a Life or (as German writers say) Mortality Table, but what in most modern books are represented by the capital letter L or years of life lived or "persons" living between the termini (see Appendix). *If* the population is stationary, the sum of these figures gives the population of the place under study. Now for ages between 1 (last birthday) and very advanced ages, L_x is simply l_x diminished by $\frac{1}{2}(l_x - l_{x+1})$. In the first year of life (and at advanced ages) the difference is greater. Thus in the first year of life deaths are not evenly distributed throughout the year of life, more than 70% of them occur in the first six months of life, so that instead of subtracting half the deaths we must subtract nearly three-quarters. Halley himself assigned 68% of deaths in the first year of life to the first half of the first year. The reason why Halley proceeded in this way was that he knew the population *not* to be stationary. His idea was to obtain the figures for the first few years of life accurately—indeed just as they are now obtained—and then to correct for excess of births over deaths.

His masterly plan was partly defeated by the fact that his Breslau correspondent Neumann was not so good a statistician as he was. Halley's letter to Neumann has not been preserved, we have only Neumann's answer of 1 March 1694. Probably Halley asked Neumann to send him (as a check on the calculations he had already made) the exact numbers of survivors on 1 January for five years, of births in a calendar year. Neumann did send him a table, but the table, as Graetzer pointed out, is wrong. Neumann gave the correct figures for 1 January of the first successive year, incorrect figures for the other years. Between 1 January of the year following the year the births of which are under study, and the next first of January, some of those born in the starting year will die under and some over a year of age. Neumann merely deducted the former, so he has too many survivors. To reach the right figure would have meant taking more trouble and he did not appreciate the importance of this. Böckh—whose opinion of his statistical contemporaries has a tinge of bitterness rare, of course, in other scientific pursuits—remarks that it was not strange Neumann should miss the point as it had been missed by many statisticians long after his time.

It is at least clear that Halley had realized an important truth which did not become part of even expert knowledge for more than a century.

The precise arithmetical details of Halley's work are not perhaps of much medical interest. Graetzer and Böckh have done a good deal to clear it up. The data used (an average of five years) had 1238 births and 1174 deaths and the table accounts for 1238 deaths. Halley must therefore have had a plan for increasing deaths. It is likely, from an observation he makes on mortality in Christ's Hospital, that he did not *wholly* depend on the Breslau figures. Graetzer suggests that Halley may have made two graphs, one having an ordinate of 1238 at the origin and an ordinate of 64 at the oldest age, the other an ordinate of 1174 at the origin and 0 at the oldest age and that he plotted the survivors for each graph based on recorded deaths and drew a curve passing through 1238 and 0 between these graphs. It may be so. Using the original material which Graetzer published, Böckh recalculated the table. The results do not, except at ages over 60, differ materially from Halley's. So far as concerns the mean after life time (expectation of life), Halley's table gives 27·54 years at birth, Graetzer's material 27·69. For ages under 40 the re-working gives slightly lower and for later ages higher mortality. It may be noted that Halley's table gives appreciably *higher* mortality in childhood than Graunt's, more than 43% instead of 36% are dead by the age of six years. But Graunt's method would exaggerate mortality (so would Halley's method, but, owing to his precautions, not so greatly). On the other hand, Graunt's estimate of age is only an intelligent guess. Actually, as Graetzer showed, the infant and child mortality shown by Halley's table differed little from the observed rates of mortality in the city of Breslau in 1876–80.

It has been said that Halley was not greatly interested in the medical aspects of his work. After describing methods of calculating the prices of annuities, he has the following passage*:

> It may be objected that the different salubrity of places does hinder this proposal from being universal; nor can it be denied. But by the number that die being 1,174 per annum in 34,000 it does appear that about a 30th. part die yearly as Sir William Petty has computed for London; and the number that die in infancy is a good argument that the air is but indifferently salubrious. So that by what I can learn, there cannot perhaps be one better place proposed for a standard. At least 'tis desired, that in imitation hereof the curious in other cities would attempt something of the same nature, than which nothing perhaps can be more useful.

That the mortality of childhood depends upon the atmosphere is not so foolish an hypothesis as it may seem to us. Halley lived before breast-feeding became the exception rather than the rule. The 'curious' in other cities had not the wit to follow his advice. He made no other contribution to the science of vital statistics; a gain to astronomy but a heavy loss to demography.

* I have read Halley's paper in the collection of papers, many by him, collected under the title *Miscellanea Curiosa*, printed in London in 1705; the quotation is from p. 300.

APPENDIX

Halley's table is printed in two columns, the first headed 'Age current', the second 'Persons'. Thus:

Age current	Persons
1	1000
2	855
3	798
4	760
5	732
6	710
7	692
8	680
9	670
10	661

and so on.

A mistake sometimes made is to suppose that Halley meant by age current simply the end of each year of life and that the entry against each 'age current' is the number of survivors at exact age one year less than the 'age current', viz. that of 1000 born 855 survived to the first anniversary, 798 to the second anniversary, etc. The fact that Halley uses the round number 1000 for a first entry does something to encourage the mistake among readers who have not consulted the original paper and it is sometimes made by people who should know better. It is actually a terrible 'howler', leading to a wholly false view of rates of mortality in early life. Thus if 1000 and 855 were really the first two entries of a Life Table as set out now, then, as the first two entries in English Life Table no. 7 Males (mortality 1901–10) are 1000 and 856, we might conclude that mortality in the first year of life was no lower in 1901–10 than in Breslau in the last years of the seventeenth century. But the 1000 of Halley's table is *not* the number of new-born children but the average number out of 1238 born living between the ages of 0 and 1. This is what is called the L_x of a modern table or the population living between the ages x and $x+1$. If we have a column of L_x's, which is what Halley gives us, we can deduce therefrom the more familiar l_x's provided we know the starting value and the number of deaths in the first year of life. Halley gives both items. He says that of 1238 annual births 348 die annually. So that his l_0 is not 1000 but 1238, and his l_1 is 890. He chose 1000 for L_0 by assuming that of the 348 deaths in the first year of life 238 occurred in the first six months of life, 68 %. This differs very little from the modern practice; in Life Table no. 7 quoted above 73·5 % of the deaths in the first year of life are assigned to the first six months of life. Having been given l_0 and l_1 we can deduce the other l's from the values of the L's which Halley gives because, after the end of the first year of life there is little error in supposing that the deaths between two birthdays are evenly distributed over the year; so, for instance, l_2 will be equal to L_1 less half the difference between l_1 and l_2, and proceeding in this way we put Halley's table into modern form. I attach a table calculated by Böckh.

It will be seen that, if Halley's table is properly used, the comparison is not of 1000 and 855 with 1000 and 856 but of 1000 and 719 with 1000 and 856.

Actually this is still slightly optimistic, because I am comparing 'persons' with males. The 'persons' figure for 1901–10 is 1000, 869. On the other hand, in the Breslau data still births (or some of them) are included in births, so that the mortality is slightly exaggerated. If for instance 7 % are still born, the survivors to 1 will be the same, but the l_0 should be reduced to 930. Or alternatively we should write 1000 and 773.

I attach Halley's table reduced to modern form and with the corresponding expectations of life calculated by Böckh (I have reworked some of the values from the data and agree with Böckh's figures).

Halley's Table, expressed in modern form together with the Expectations of Life at quinquennial intervals (*Böckh*)

Age	l_x	$\mathring{e}_x$	Age	l_x	$\mathring{e}_x$
0	10,000	27·54	40	3,557	22·05
5	5,816	41·47	45	3,167	19·47
10	5,307	40·25	50	2,751	17·05
15	5,049	37·19	55	2,319	14·75
20	4,806	33·93	60	1,914	12·33
25	4,552	30·69	65	1,511	9·96
30	4,257	27·64	70	1,103	7·74
35	3,921	24·78	75	670	7·50

V. GUESSING THE POPULATION

My object is to trace the growth in our country of that part of statistical science which is of interest to students of medicine or public health. In speaking of such pioneers as Graunt, Petty and Halley it was proper to construe the obligation rather freely. Both Graunt and Petty did clearly perceive the relevance of their researches to matters of public health or even clinical medicine, but much of what Petty did had a more direct bearing upon political questions than those of public health. Again, the life table is a way of expressing the facts of mortality which is valuable in some medical researches, but its importance as a statistical instrument has been much greater in non-medical than medical circles, above all of course in the financing of assurance business. The commercial importance of life tables was perceived by Halley and by other mathematicians of his and the following generations.

Looking at the position after Halley's publication it was clear that progress might be made (1) in improving the accuracy of the life table, viz. by obtaining data more relevant to the conditions of life of persons who assured their lives or bought annuities, (2) in simplifying the very laborious calculations which the determination of praemia or purchase values required. Under (1) no progress worth speaking of was made in England until the end of the eighteenth century. This was partly due, as we shall see, to a not entirely unjustified disbelief in the powers of the medical profession to change the rate of mortality, partly to ignorance. No first-rate English mathematician after Halley gave any critical attention to the theory of the Life Table before the nineteenth century. Under (2) considerable progress was made, but this progress is of little or no medical interest and to describe it would involve entering upon tedious arithmetical and algebraical detail. The primary medical-statistical quaesita are correct enumerations of deaths by sex at ages and by causes, and of the numbers living in sex and age groups. When these have been satisfied, the medical statistician can get to work.

For 150 years after Graunt's death very little was done to improve matters. Down to 1801 the population as a whole had not been counted; forty years more passed before a reasonable age distribution was secured, and it was thirty-eight years after the first denominator (populations) that the first numerator (deaths) of the fundamental fractions was obtained. Until 1801 intelligent guessing was the method and the guesses of the eighteenth century deserve a few pages, if only because they prove that statistical ability is as rare as other kinds of ability and that wishful thinking is not a modern foible.

The first estimator to mention belongs to the seventeenth century and was a younger contemporary of Graunt and Petty, Gregory King (1648–1712). He was born in Lichfield, the son of a land surveyor. At the age of fourteen he was recommended as a clerk to the famous herald Dugdale with whom he worked for several years; after Dugdale had finished his Visitation, King worked for various amateur antiquaries and was eventually invited by a lady of property in Sandon (Staffordshire) to be her steward, auditor and secretary. Here he remained until 1672 when he moved to London and, no doubt through Dugdale's recommendation, had a considerable amount of employment in both heraldic work and ordinary surveying. In 1677 he became a member of the College of Arms, in which he attained the rank of Lancaster Herald and so continued until his death, but worked for other official bodies on financial subjects.

The decorous memoir by George Chalmers, from which I have extracted these particulars, does not give us a very life-like picture of the man himself. There is a certain likeness between the early careers of Petty and King. King was not indeed shipped as a cabin boy, but Mr King (the elder) drank (if we may venture so coarse an abbreviation of Chalmers's statement that the father studied and practised his profession 'with more attention to good fellowship than mathematical studies generally allow') and King junior was a pupil teacher at eleven. If he really read Hesiod and Homer, made Greek verses and taught himself to survey land in his thirteenth year he must have had Petty's precocity. Both Petty and King had experience of practical surveying and, of course, both were interested in political arithmetic. But there the parallel ends. King was a professional surveyor and archivist and had a reasonably successful professional career. Petty was—Petty. One might, perhaps, adduce as another parallel that King made some enemies and thought himself ill-used. But the job by which Sir John Vanbrugh, a stranger to the College of Arms, was made a king-at-arms over the head of an official of twenty years' standing would have galled the meekest of mankind. One may safely conclude that King had more knowledge of the data of political arithmetic than Petty and less originality. His vital statistical work was not published until nearly a century after his death, as an appendix to the second edition of *An Estimate of the Comparative Strength of Great Britain* by George Chalmers, London, 1803. Perhaps he never intended to publish it—he communicated the substance to his contemporary Davenant—and this may explain why there are no details of how some of his results were reached. The report reads rather like a document prepared for official use by persons interested in results not methods.

The starting-point of King's attempt to estimate the population was a return from the Hearth Office of the number of houses assessed to tax on Lady Day, 1690. That was 1,319,215 which, King estimated, had increased to 1,326,000 by 1695. He deducted 30,000 for empty divided houses,* took the round figure of 1,300,000 and assigned 105,000 to the London area, 195,000 to other cities and market towns and 1,000,000 to villages and hamlets. He used a series of multipliers, 5·4 for a house within the walls of London, 4·6 for a house within the liberties, 4·4 for the out parishes in Surrey and Middlesex and 4·3 for Westminster. For other towns, his multiplier was 4·3 and for villages 4·0.

Having performed his multiplications he gives London a bonus of 10%, other towns 2% and villages 1%. Lastly he estimates homeless people to number 80,000. The final result to the nearest round number is $5\frac{1}{2}$ millions.

How King obtained his multiplier is not clear. In addition to the Hearth Office data he says he used 'the assessments on marriages, births, and burials, parish registers and other public accounts' and that from these he deduced the multipliers, but this is rather vague. He also classified the population by sex, civil state and age (under 1, under 5, under 10, under 16, above 16, above 21, above 25, above 60). How he reached these figures is not explained.

But nothing succeeds like success. As we shall see, his estimate of the total

* Prof. E. C. K. Gonner (*J.R. Statist. Soc.* **76** (1912–13), 261–97), in an interesting paper which I have largely used in writing this chapter, remarks that the 'houses' of the Hearth Office must have been really families or separate occupations as King indeed realized, and thinks that King fell into some confusion in attempting to replace families by houses. Gonner argued that the best way was to proceed on the basis that the Hearth Office unit of a family should be retained and be corrected for empty houses, blacksmiths' shops, etc. on the basis of 1801 census returns and the multiplier used should be persons per family of 1801. The result is to give a figure about a quarter of a million larger than King's. The method described in the text also leads to the conclusion that King somewhat understated the population.

population is probably very near the truth and Prof. Westergaard has remarked that, judging from Swedish observations of a few years later, King's age distribution is quite reasonable.

As a statistical prophet King was no more successful than his contemporaries (and successors). He believed that down to his time the population of England had doubled in 435 years, that the next doubling would require from 1200 to 1300 years and that in A.D. 3500–3600 the population would reach 22 millions of souls, in case, as he cautiously adds 'the world should last so long'. His estimates as a matter of arithmetical curiosity are excellently fitted by a logistic with an upper asymptote of fifteen millions and would give the present population as about eight millions.

Modern statisticians, such as Farr and Brownlee, have confirmed King's estimate of the population at the end of the seventeenth century in the following way. After 1801 the population was known by actual counting and for the first forty years of the nineteenth century baptisms and burials were still the only data of births and deaths. If one started from, say, the enumeration of 1831 and worked back to the population of 1821 by adding the numbers of burials and subtracting the number of baptisms then, if these really measured deaths and births, the result ought to agree with the census enumeration, provided immigration and emigration balanced. But the burials and baptisms understated deaths and births. One might adjust the figures by multipliers to bring the result into agreement with the census and then test against another backward run of ten years. Brownlee found that if the number of burials were multiplied by 1·2 and the number of baptisms by 1·243, the agreement was good.

This may seem a highly conjectural method, but it certainly gives quite good results. The difference between births and deaths estimated in this way for the decennium 1801–10, I find to be about 12·4 per 1000 living. If one multiplies the enumerated population of 1801 by $(1{\cdot}0124)^{10}$ we reach 10·1 millions, not a bad approximation to 10·2 millions actually counted. Assuming that before 1801 burials and baptisms had the same relation to deaths and births as between 1801 and 1841, we can work backwards to the beginning of the eighteenth century with the result that the population then was about 5·8 millions, not much more than King's estimate. In view of the following discussion it will be useful to consider the probable state of the population (as determined by these methods) in the eighteenth century. In the first sixty years of the century it grew very slowly, was about 6·1 millions in 1751 and 6·5 millions in 1761. It then began to increase faster, was 7·5 in 1781, 8·2 by 1791 and 9·2 at the census of 1801 (8·9 as enumerated, but an estimate of a deficit of 1/30th was made).

From Gregory King's time to the census of 1801 we have a series of more or less intelligent guesses.

These are well described in Prof. Gonner's paper.* Two schools of thought

* *J.R. Statist. Soc.* **76** (1912–13), 261–96.

did battle in the eighteenth century; the pessimists who held that the population was decreasing and the country going steadily to the dogs, and the optimists who believed just the contrary. Both used the same weapons. The heavy artillery was a return of houses for taxation purposes increased conjecturally by a figure for houses which escaped taxation, the sum multiplied by a conjectural average of persons per house. As light artillery one had the yield of taxes on commodities and the returns of baptisms and burials.

A pessimist put the number of untaxed houses low and the multiplier low, and an optimist raised both.

The first controversy which took place in 1754 in the proceedings of the Royal Society did not attract much notice. Brackenridge (mildly pessimistic) pointed out that the number of houses assessed to house tax had decreased between 1710 and 1754 from 729,048 to 690,000, which suggested a decrease of population (by a previous conjectural calculation based on burials and baptisms, he had reckoned a small increase, which was probably correct). Much turned on the number of houses which did not pay tax (either because the occupant was in receipt of alms, did not, owing to poverty, contribute to the church or poor rate, or through mere default). Brackenridge put the number at 200,000. His critic, Forster, argued that Brackenridge under-stated the number of untaxed houses, adducing a sample of nine country parishes with 588 houses of which only 177 were taxed and a market town with 229 taxed houses out of 448. Using these figures as a basis for conjecture Forster raises Brackenridge's 890,000 to 1,427,110. From this (with a multiplier of 6 for town houses and 5 for country houses) he reaches a population of seven and a half millions—probably a considerable over-estimate.

The next controversy was a quarter of a century later (in a period when the population was certainly increasing) and its originator was Dr Richard Price (1723–91), who has attained a posthumous celebrity reminiscent of the man whose title to distinction was that he had once been kicked by George IV. Most readers know him as the preacher of a sermon which was the text of Burke's *Reflections*, most students of economic history know him as the inventor of that theory of the virtue of a Sinking Fund which has been likened to the economic system of a community which prospered by taking in one another's washing; most vital statisticians remember him as the computor of the Northampton Life Table which gave a seriously incorrect picture of prevailing mortality and indirectly cost the country a large sum of money. Finally, in the controversy about to be described, Price was pertinaciously in the wrong on all the main issues.

The apparent inference from all this is that Price was either a fool or a knave. Gainsborough's portrait of the Rev. Richard Price, which hangs (or did hang) in the Board Room of the Equitable Assurance Society, gives no support to the hypothesis that Price was a fool; his life would be a promising field of research for a young historian with a competent knowledge of economics. His importance in statistical history is not great enough to justify me in a critical study (even if I had the necessary training in finance and economics). My guess is that Price was an able, self-confident, original-minded man, who knew a good deal about many things and had no exact knowledge of anything. He had 'a way' with him, he could *interest* people. In fact he had some of the qualities of Petty. It is easy enough to make jokes about his notion of the mysterious power of money to increase at compound interest and it is possible that William Pitt the younger (who was only a boy when he adopted Price's theory) was not a good economic reasoner. Still, even 150 years ago, there were bankers and Treasury officials, and it is possible that both they (and Price) were not so much bad theoretical reasoners as shrewd opportunists, that they were deliberately blind to the speciousness of an attractive defence of a desirable financial expedient. I have myself sometimes wondered whether, in the eighteenth century, an Assurance Society would have minded very much if a Life Table had erred on the pessimistic side.

Price did not enter on the population question with an unbiased mind. He was a keen politician and he believed that the policy of the government was bad for the country; he also believed that the wealth of a country was its people. Hence he believed that the population was declining and nothing shook that belief. Had he survived another ten years, until the first census, he would probably have disputed the accuracy of the returns.

Price began with the figures of houses in 1690, which he cited from Davenant (they were really due to King, who communicated them to Davenant), making the total 1,319,000. He then gave the figures of assessed, chargeable and cottages (cottages being houses too small to be taxed) as 678,915, 25,628 and 276,149, making a total of 980,692 in 1761. In 1777 they were 682,077, 19,396 and 251,261, a total of 952,734. On this basis he concluded that the population had declined by about one and a half millions and was actually less than five millions.

Howlett and Wales, Price's chief opponents, impugned every step in the reasoning. First, they pointed out that in the estimate for 1690 there was almost certainly a confusion between families and houses. Then they argued that many householders evaded duty (for instance by the simple plan of blocking up windows (the prayer 'Lighten our darkness, we beseech thee, Oh Pitt' is still remembered) and showed by direct enumeration in certain parishes that the returns were inaccurate. Finally, they gave reason to think that Price's multiplier was too small. On each of these points they were probably right. Indeed Price was obliged to admit the validity of some of their criticisms. But he declined to budge; sometimes he took *ad captandum* advantage of arithmetical slips by his adversaries, sometimes he declined to admit that their samples were representative, sometimes he tried to ignore the effect of corrections which he was forced to make.

These were the principal arguments. Both parties used the data of burials and baptisms as subsidiary arguments. Price seems only to have used the London Bills, which rather let him down; because although they seemed to help for some part of the century, he admits that by 1773 London was increasing and, very characteristically, uses this as in his favour: 'But it appears that, in truth, this is an event more to be dreaded than desired. The more London increases, the more the rest of the country must be deserted.' Price's adversaries went farther afield and counted burials and christenings in 162 parishes in all parts of England for two quinquennia, one beginning in 1758, the other in 1773. Baptisms increased from 47,638 to 59,567, burials from 49,553 to 53,030.

But neither party put much weight upon what we should now consider primary evidence; rightly, because of its incompleteness.

But these data were not wholly neglected by medical writers as we shall see in later sections. One may fairly say on the evidence here summarized that the eighteenth-century political arithmeticians of England made no advance whatever upon the position reached by Graunt, Petty and King. They were second-rate imitators of men of genius.

(*To be concluded*)

MEDICAL STATISTICS FROM GRAUNT TO FARR
(Concluded)

VI. SOME ENGLISH MEDICAL STATISTICIANS IN THE EIGHTEENTH CENTURY

AFTER Petty more than fifty years passed before another Fellow of the College of Physicians took an interest in statistics, and he, if less eminent in political arithmetic, was much more eminent in the art of medicine; he was the elder Heberden. Heberden published no statistical work over his name, but there seems no reason to doubt the accuracy of his son's statement that the quarto volume containing a collection of the yearly Bills of Mortality in London from 1657 to 1758 and various essays was financed by Dr William Heberden and that he wrote the preface. This preface, an essay of 15 pages which ends rather abruptly, could hardly have been written by a layman. The following passage illustrates my remark:

> The deaths imputed to the measles are very remarkably different in different years; and yet it is possible that this disease is not in reality so very irregularly epidemical or fatal, as by the bills it appears to be. The scarlet fever and malignant sore throat often occasion such appearances upon the skin, as may easily be mistaken for the measles by better judges than the mothers and nurses, who thinking themselves able to distinguish this distemper, and equal to the management of it, often call in no other assistance. This mistake is well known to have been sometimes made within these few years, during which the scarlet fever and malignant sore throat have been so common. It may perhaps have happened in every year, in which an extraordinary number of deaths are charged to the measles: and consequently those two formidable distempers, (if they are two distinct distempers, and not one and the same) being disguised under the name of the measles, may have been older, and more general than is usually imagined.

The writer's observations upon the disappearance of plague have also something of a professional air—the fact that they are decidedly confused is no argument to the contrary. Sydenham taught (*Obs. Med.* **2**, 2) that plague depended upon (*a*) a special disposition of the atmosphere, (*b*) the transmission of an infecting matter, and held (*a*) to be primary, i.e. that without the atmospheric constitution there would be no epidemic. Heberden thinks the decline of the plague was due to the rebuilding of the city and—'probably the most effective'—the great quantity of water from the Thames and the New River, 'which, for the last century, has washed the houses so plentifully, and afterwards running down into the kennels and common sewers, constantly hinders, or weakens the tendency to putrefaction'. Heberden, unlike Sydenham, who believed the secret of the atmospheric constitution beyond the wit of man, seems to have attributed it to 'putrefaction', but, like Sydenham, attributed more importance to the atmospheric than the infective factor.

For the rest, Heberden continued Graunt's criticism of the material. In particular he gives good reason for thinking that beyond the omission of dissenters' christenings and burials, an important error arises from a balance of outward burials, that more coffins are taken out of London to be buried than are brought in. From an enumeration in Westminster he concludes that the deaths within the Bills are 20 % too few. He comments on an apparent increase in certain forms of death, such as apoplexies, lethargies and palsies. 'The practice of drinking spirituous liquors must, probably, answer for some part of this: and it might be of public use, if some attention were paid to the finding out of the other causes.' Rather optimistically, he thinks abundant amends might be made for these increases by the control of smallpox through inoculation. Upon this he makes a comment which has a very modern ring. 'For while inoculation prevails only among a part of any number of people, who all have an intercourse with one another, it may occasion as many deaths by spreading the distemper in, as it is called, the natural way, as it prevents among those, on whom it is practised.'

The volume contains a reprint of Graunt's work, of one of Petty's papers, and a new essay by Corbyn Morris. This essay, which shows signs of improving statistical technique, is not of medical interest except in its collection of deaths in age groups—an operation rendered possible by the introduction, in 1727, of an age classification of all deaths. Twelve age groups were given. In Heberden's preface the importance of classifying by cause of death and age is emphasized.

From this material a life table was calculated by a fellow of the Royal Society named Postlethwaite (at the request of Heberden). This table, based upon deaths alone, is, for reasons already stated, of little value.

I doubt whether even the relative mortalities are correctly shown. The age distribution of the Bills (after 1726) for deaths was: under 2, 2–5, 5–10 and thence forward by decennia. Consequently one must distribute the deaths into single years of life upon some principle of interpolation. Neither J.P. (Postlethwaite) nor J.S. (Smart)—who made a life table for the first ten of the thirty years—states the principle on which he worked. But J.P. assigns 250 of the 363 deaths at ages under 2 to the first year of life and J.S. 290 of the 386 he had to manipulate, respectively 68·9 and 75·1 % of the total mortality under 2. Both ratios are much less than given by Halley's table, 83·3 %, which itself agrees admirably with the latest English population table (E.L. No. 10, Males) 83·5 %. If we applied the 83 % ratio it would raise the rate of infant mortality to 301 per 1000. Taking the figures simply as they stand, the survivors to age 6 years are fewer than Graunt estimated—54 % not 64 % survive.

Although, as has been said, the arithmetical values are very suspect, the indication they give may be towards the truth. Creighton gave good reasons for concluding that London after the extinction of the Plague was less, not more, healthy. These reflexions are not without importance for they help to explain a certain fatalism, a scepticism as to the possibility of reducing the death-rate, which is noticeable in both statistical and medical literature for some time to come.

The next writer to be noticed is Thomas Short. Of this industrious investigator even Sir Norman Moore could obtain few personal particulars. He may have been born in 1690 and he died in 1772. He practised in Sheffield and was a Doctor of Medicine of a Scottish university but not a licentiate of the College. His principal works, *A General Chronological History of the Air, Weather, Seasons, Meteors, etc.*, published in 1749, and *New Observations, Natural, Moral, Civil, Political and Medical on City, Town and Country Bills of Mortality*, published in 1750, are differently assessed by the greatest historian of British medicine. Creighton pronounces the former to be rubbish but gives Short a not very hard pat on the back for the latter. 'That so much statistical or arithmetical zeal and exhaustiveness (in the work of 1750) should go with so total a deficiency of the critical and historical sense (in the work of 1749) is noteworthy, and perhaps not unparalleled in modern times' (Creighton, *Hist. of Epidem. in Britain*, **1**, 405). Creighton's not very wide intellectual charity did not embrace statisticians.

It must be admitted that Short is decidedly *not* a writer to commend himself to an orderly minded, careful scholar from Aberdeen. Had he lived a century later we might have supposed that his literary model was Mrs Nickleby—he just runs on and on. A table (which must have been most troublesome to compile) of monthly christenings and weddings in various towns (in three extending over more than 150 years) leads him from arithmetical comparisons of the months most apt for procreation to a vigorous denunciation of luxury, of polygamy, of taxing common necessities, of the sale of army commissions, of novel reading, boarding schools and much else. But, although few if any would be able to read Short straight through without a rest, a good many less entertaining books might be included in a bed-side book case.

There is scarcely anything within the range of human interests upon which Dr Short has not something to say. On the whole he took a gloomy view of modern life in general and of his faculty in particular, and remarked that the 'improvements in surgery in general, have far out-stripped those in physick'. Surgeons he found to have generally less learning than physicians, but compensated for this by a closer application to the study of their own profession 'without jumbling the finite mind, and mixing studies of a different nature from their own, as of the dramatists, poets, classics, architecture, politics, history critics, logics, etc. They are also less liable to theories and false reasonings, have not that contempt of the ancients, nor of observations built on practice, improved and directed by the understanding, and raised to the pitch of truth by a long enquiry into the effects of diseases and medicines.'

Short began his book with a clear plan—long before he had finished it the plan became an inextricable confusion. He argued that a statistical measure of health could not be obtained from the data of towns in general and the capital in particular, partly owing to the inaccuracy of the data, partly owing to the fact that towns attracted newcomers and were not maintained by the balance of births and deaths within the community. So he collected material from country parishes (he also obtained data from towns but the country parishes were his prime object of study). His first set of data was a collection of transcripts of the registers of eighty-three parishes. About 60% of these parishes were from Yorkshire (mainly in the neighbourhood of Sheffield) and Derbyshire, but some from as far afield as Devonshire. He set them out in two periods, the first ending before the Restoration, the second coming down to the third decade of the eighteenth century. He had another set of eighty-three parishes for which the data covered only the second period. He classified his parishes in accordance with the nature of soil, altitude, exposure, whether wooded or bare, wet or dry. Sometimes his data covered more than a century, rarely less than 20 years. He gives the total number of baptisms, of burials (sexes distinguished) and marriages and works out the various ratios, the ratio of baptisms to burials being his chief tool.

From these data he draws a great many conclusions; for instance, respecting the salubrity or insalubrity of different soils and exposures. Most of these conclusions, it may be remarked, are now part of the common stock of lay and, perhaps, professional belief. But whether Short's data were adequate to sustain the conclusions is another question.

We may begin by taking purely arithmetical points into consideration, viz. whether, *assuming* that the parishes or groups of parishes were fairly comparable and *assuming* that the ratio of baptisms to burials is a fair measure of healthiness, Short had large enough figures for his purpose. For instance, two of his conclusions were that dry open sites of moderate elevation were healthier than a clay soil. I picked out of his list nine parishes of the former and five of the latter class. In his first (pre-Restoration) period, the parishes on dry, open sites had registered 4349 baptisms and 2644 burials, a ratio of 1·64. The five parishes on clay had 2875 baptisms and 1920 burials, a ratio of 1·497. So, as he said, the dry open sites give a higher ratio. But what is the order of magnitude of the error of sampling? We may safely hold that the standard error of the number of baptisms or burials is of the order of the square root of the observed number, or the ratio of standard error to number is of order $1/n^{\frac{1}{2}}$. From this we infer that the standard error of the ratio n_1/n_2 is given by n_1/n_2 times $(1/n_1+1/n_2-2r_{n_1n_2}.1/(n_1n_2)^{\frac{1}{2}})^{\frac{1}{2}}$. Clearly, the correlation between numerator and denominator must be large, so that the second factor lies between $(1/n_1+1/n_2)^{\frac{1}{2}}$

and $(1/n_1^{\frac{1}{2}} - 1/n_2^{\frac{1}{2}})$ and will be much nearer the second value. In a sample of 1000 Registration Districts I studied many years ago, the correlation of births and deaths was 0·73 %, in our particular case the standard error of the difference between the two ratios will be likely to be not much more than 20 % of its value.

From the purely arithmetical angle, I should conclude that Short was justified in holding that his ratios did really differ significantly, as we say, from site to site. But is it fair to assume that (1) the ratio of baptisms to burials is a good index of healthiness, (2) that the comparisons are in *pari materia*? These are much more difficult questions.

So far as concerns modern experience, it is clear that the ratio of births to deaths does not give a useful index of the rate of mortality. I made an experiment on Short's lines. I took out a sample of fifty Registration Districts for the decennium 1901–10, chosen in the following way. (1) No districts with more than 10,000 births or fewer than 1000 deaths were taken. (2) Those with many institutional deaths were excluded. For each the ratio of births to deaths was calculated and the following table formed:*

Ratio of births to deaths	No. of districts	Mean of standardized death-rates
2·0–	4	10·36
1·9–	6	10·88
1·8–	7	11·04
1·7–	6	11·60
1·6–	15	11·15
1·5–	7	10·72
1·4–	4	11·48
1·3–	1	13·01

It is true that the district with highest ratio has the lowest death-rate and the district with lowest ratio the highest death-rate, but in detail there is but little correspondence. Testing the same districts on the data of 30 years earlier, 1871–80, the same result appears. It would be very rash to conclude that because a district has a ratio of births to deaths above the average its standardized death-rate is below the average.

There are many reasons why a ratio of births to deaths may be a bad measure

* The districts selected were: Hambledon, Malling, Faversham, Romney Marsh, Uckfield, New Forest, Romsey, Hartley Wintry, Royston, Winslow, Witney, Oundle, St Ives, Caxton, Whittlesea, Lexden, Risbridge, Mildenhall, Bosmere, Plomesgate, Flegg, Cricklade, Melksham, Amesbury, Sturminster, Kingsbridge, Stratton, St Columb, Langport, Dursley, Ledbury, Wem, Mastley, Meriden, Shipston on Stour, Lutterworth, Spilsby, Hayfield, Garstang, Settle, Pateley Bridge, Gt Ouseburn, Saddleworth, Thorne, Pocklington, Skirlaugh, Easingwold, Bedale, Weardale, Brampton.

and Short knew this; for instance, the deaths might be increased by immigrants or decreased by emigrants. He was a true Englishman of Whig principles, and in speaking of *foreign* registers, remarks that 'where the births vastly exceed the burials, the country is either very healthy, or it is under an arbitrary government or both'. If the former, the marriages will bear a high proportion to the baptisms, few will die in infancy. If the latter, 'tho' more males are born, yet the funerals of females far exceed them; there is little industry amongst the people, because they want property; useless standing armies are kept up in time of peace, for the grandeur of the tyrant, maintaining his tyranny and the oppression of his people'.

Quite logically he argues that if the burials greatly exceed the baptisms, either the situation is unhealthy or the government is limited (i.e. on English lines). One diagnoses the former when epidemics are frequent and the proportion of deaths in youth and childhood large; 'the latter is known from the great resort of strangers, labourers, artificers, merchants, etc., increase of business, trade and riches; or there is a large body of people mixed with the society, of different manners and principles, whose baptisms are not registered with the rest'.

Short was, then, aware of the fallacies possible, but he held that in country parishes they were not important.

I am not confident that he was wrong so far as concerns his first period data; country villages from the middle of the sixteenth century to the middle of the seventeenth century. But when we come to the next 80 years, when England was beginning to be an industrial country, the assumptions are more hazardous, and the almost invariably pessimistic conclusions he draws (it is a little odd that the population pessimists did not make more use of Short) may well be fallacious. The cause of change for the worse was, in the first place, the deluge of profaneness which came with the Restoration, in the next place the increase of 'Luxury, pride, intemperance and debauchery' associated with the growth of industry and wealth, so fortunately associated with the 'happy Revolution' and stabilized by the 'seasonable Accession of the present Royal Family'. But Short charges this intemperance and debauchery particularly upon the towns, and it does seem a little unreasonable to suppose that even the happy Revolution or the Hanoverian accession much increased the opportunities of villagers to fall into the sins of luxury and pride.

Short next considers the succession of unhealthy years in villages and towns. He nowhere states what his criterion of a sickly year was. The arm-chair statistician might suppose that he would take a ratio of baptisms to burials which fell below some assigned percentage of the average; but, in default of any specific statement, I should guess that Short called a year sickly when the burials exceeded the baptisms; since his totals show always an excess of baptisms over burials, this would seem a simple rule. But he may have been more subtle.

He remarks: 'It may be a sickly or mortal year in a town or country parish and yet the christenings may exceed the buryings considerably, either because it happens to be a very fruitful year in that place (as often tho' not generally happens) or the year may be very sickly in that parish, if compared with other years, and yet healthy if compared with other places in much worse situations and air.' This suggests that a sudden rise in the number of burials would be his criterion, and I have found half a dozen cases in his table (of nearly 400 instances, viz. country registers and market-town registers, examined from 1541 to 1741) when the burials for sickly years did not exceed the christenings.

It is a pity Short was not more explicit; his work in this field was original and suggestive. He finds that in the country parishes the frequency of unhealthy years is never more than twice in 5 or 6 and often as rarely as once in 6 or 8 years, 'which is indeed as long, if not a longer interval, than commonly happens between one visitation of smallpox and measles and another, exclusive of all other diseases'. 'When sickly years return oftenest there is a less disproportion between christenings and burials than when they come seldomest.' Healthy places have 'their frequent eruptive and inflammatory diseases', unhealthy places 'their slow intermittents, remittents, putrid and erratic fevers'. 'It is true, some rare times the former places are visited with the latter diseases, but rarely except they are epidemics; nor are they of a great spread, duration or execution. The latter's places have also the former's diseases, but (eruptive fevers excepted) more mildly and rarely; for each country or situation is more liable to some diseases than others, and by traffic and commerce endemics became epidemics, as far as air and climate will allow.' Short passes to a general survey of epidemiological history which is not, I think, without value, but has, of course, long been superseded by Creighton's classical study.

The passages upon which I have commented are all from the first quarter of Short's treatise and typify, I think, his most valuable contributions to our subject. As his book proceeds, not only does his habit of improving the occasion grow upon him but, in commenting and performing arithmetical operations upon the London Bills, he does not show to great advantage. But anybody who will trouble to dip into the book is likely to make a friend. Creighton said of him that 'his abstract results or conclusions are colourless and unimpressive, as statistical results are apt to be for the average concrete mind'. This seems to me rather misleading. I dare say nobody ever burst into tears over or was thrown into paroxysms of mirth by a statistical table—even a table compiled by the Army Medical Department. It is quite likely that several of Short's inferences were wrong. But he *did* paint a vivid picture of the changing conditions of life as he saw it. 'Colourless' is about the last adjective I should apply to his book. Had it been studied with more attention, had he been a leading London physician instead of an obscure country practitioner, medical statistics in England would have progressed faster.

VII. SOME REPRESENTATIVE CONTINENTAL DEMOGRAPHERS OF THE EIGHTEENTH CENTURY

My object is to sketch the history of distinctively medical statistics in our own country; I have neither the knowledge nor, perhaps, the desire to cover a wider field; but it would be too insular entirely to neglect continental research contemporaneous with that described in the preceding section. I propose to discuss the work of some foreign writers which is relevant to that of the British authors mentioned in the preceding section. The most eminent contemporaries of Short were Deparcieux, Wargentin, Struyck, Kerseboom and Süssmilch, and of these Deparcieux, Struyck and Süssmilch are, I think, the most interesting, a Frenchman, a Hollander and a German. None of them was a physician. Deparcieux and Struyck were competent mathematicians. Struyck wrote on the general theory of probability, Süssmilch had no more mathematics than Graunt; but, of the three, Süssmilch is better known to posterity because he is frequently cited in books which circulate outside professional statistical circles. Deparcieux (1703–68) is the least voluminous and most attractive of the three. He published in 1746 a quarto of 132 pages (with tables) entitled *Essai sur les Probabilités de la Durée de la Vie humaine*, to which he added, 14 years later, a short appendix, and his book is a model of clear writing.

Deparcieux was fully alive to the dangers of basing a life table upon data of mortality alone, and was the first writer to construct what we should regard now (subject to a few reservations) as correct tables. Of course, like his contemporaries, he could not make bricks without straw, and no more than they could provide a general population life table. He had to use data which were not random samples of human experience and is careful to point this out. His new material was drawn from two sources, the data of tontines and the mortality experience of religious orders.

A tontine (the name is derived from that of the inventor Lorenzo Tonti, a Neapolitan banker) was a system of selling annuities on the following plan. The participants are formed into age classes, each entrant pays a capital sum and receives an annuity; as the annuitants die out the amount payable to the survivors is increased and the last survivor will enjoy an income equal to that distributed originally over all members of the age class. This was the general plan of a simple tontine (*The Wrong Box* will have made us familiar with a different application); there were various modifications, but in all an exact record of deaths at ages was essential.

Deparcieux used the data of tontines established in 1689 and 1696. He had to face many difficulties. In the first place, the tontines had a series of classes, one for those entrants under the age of 5, the next for lives from 5 to 10, and so on. What is the mean age of the members of each class? There would, as Deparcieux points out, be a bias in the first class (that of children under 5) in favour of ages beyond the mean, because parents needed no statistics to convince them that the rate of mortality in the first and second years of life is higher than in the third and fourth or fifth. In the later classes, on the other hand, the bias would be in favour of entering at an age below the mean of the class limits. He makes a rather modest allowance for these factors by taking the age at entrance in the first class as 3 years, i.e. half a year more than the mean of the class limits, and in the next (and subsequent classes) as half a year less than the mean. The next difficulty is that his observations end in 1742, consequently rates of mortality at ages are derived from persons whose dates of birth are widely separated. Thus no members of the first class of the 1689 tontine can have been exposed to the risk of dying at ages beyond 57 (actually of 202 entrants, 105 were still living at the close of the observations). So a table obtained by welding these observations ignores any secular trend of mortality. It also ignores what, in modern assurance practice, is an important factor, viz. selection. A life aged n years is less likely to end within the year of entrance than a life of the same age entered 10 years earlier. In ordinary practice there are two reasons, self-selection and medical examination. In annuitant experience only the former is involved, but this is not the *less* important of the two.

In the discussion of this subject which will be found in Elderton and Oakley's *The Mortality of Annuitants* 1900–1920 (published on behalf of the Institute of Actuaries in 1924), the conclusion is reached that when *contemporaneous* lives are in question, this selection only operates seriously on the first year of annuitant life; for that year the rate of mortality is about 63 % of that suffered by annuitants of the same age who had purchased annuities 5 years earlier (what is called ultimate mortality). If then there were no secular improvement of mortality rates—as there has been over the last 60 years—and if there were no secular change in the social or economic class of annuitants, while we should expect a lighter mortality upon recent entrants, if, as in Deparcieux's data, we are only given survivors at quinquennial intervals, we should not expect large differences. Actually one can test a particular age group, viz. 45–50 on numerically extensive material. The 1689 tontine provides ten and the 1696 tontine nine groups of persons of this age the survivors of which 5 years later are recorded. It will be seen from Table 1 that the 634 'new' entrants in the 45–50 tontine class of 1689 suffered rather heavier mortality than the 118 survivors to that age from the youngest class. This, however, is merely picking out a single pair. The correct test is to treat the data together and inquire whether the hypothesis that the whole set of deaths and survivorships might have arisen by sampling a population for which the chance of living 5 years was simply the ratio of total survivors to total exposed, viz. $5009/5394 = 0{\cdot}9286$. Applying the appropriate test, viz. that known as the χ^2 test (with 18 degrees of freedom), one reaches $P = 0{\cdot}0346$. This is not a very improbable freak of chance. Compared with modern annuitants, these tontiniers of 200 years ago had a rate of mortality some 40 % greater than the annuitants of 1900–20 between the ages of 47 and 52.

Finally, one has the class of society from which annuitants are drawn. Deparcieux was of opinion that annuitants were mainly drawn (op. cit. p. 62) from the middle class of society 'ce sont les bons Bourgeois qui tiennent un honnête milieu entre toutes ces extremités, qui se font des Rentes viagères; et ce sont ceux-là qui deviennent ordinairement vieux'. Hence he judged that the rate of mortality suffered would be less than that of the general population.

Table 1. *Deparcieux's observations*

Tontine class	Exposed to risk at age 47	Survivors to age 52		Deaths	
		Observed	Expected	Observed	Expected
(1689 tontine)					
- 5	118	109	110	9	8
-10	181	173	168	8	13
-15	211	192	196	19	15
-20	216	196	201	20	15
-25	201	189	187	12	14
-30	263	249	244	14	19
-35	526	479	488	47	38
-40	472	440	438	32	34
-45	770	723	715	47	55
-50	634	575	589	59	45
(1696 tontine)					
5-10	134	130	124	4	10
-15	131	118	122	13	9
-20	108	103	100	5	8
-25	102	92	95	10	7
-30	147	135	137	12	10
-35	211	204	196	7	15
-40	220	200	204	20	16
-45	444	415	412	29	32
-50	305	287	283	18	22
	5394	5009		385	

The next part of his investigation related to the mortality experience of members of monastic orders. These he utilized with the same good sense and care.

In Table 2 are his l_x values, to which I have added those for English Life Table No. 9 Males (general mortality of 1930–2). The reason for putting l_{20} equal to 814 is simply that in his table for tontines where the starting point is age 3, his survivors to age 20 from an initial 1000 were 814.

The column headed Benedictines (*a*) is a methodologically correct table, viz. based on entrants followed until death, Benedictines (*b*) assumes a stationary population and is not therefore so exact although it utilizes more data. Actually both tables give virtually the same results. It will be seen that to age 50 all the tables agree well; after age 50 the monks fare worse than the members of tontines and worse than the nuns. All have much worse mortality than the unselected general population of England and Wales 200 years later. Deparcieux attributes to selection the equality of tontine and monastic mortalities at younger ages and to the privations and austerities of the religious life a higher mortality at later ages.

In an investigation made by Dr S. Monckton Copeman and myself some years ago (*Report on Public Health and Medical Subjects*, no. 36, H.M.S.O. 1926) into the alleged low mortality from cancer of members of certain religious orders, we had occasion to study the general mortality experience. The result was that the mortality at ages over 25 of monks was rather more favourable than that of

Table 2. *Deparcieux's observations*

Age	Tontines	Survivors from age 20		Nuns	E.L. No. 9 Males
		Bene-dictines (*a*)	Bene-dictines (*b*)		
20	814	814	814	814	814
30	734	756	749	751	788
40	657	675	681	676	755
50	581	575	583	587	698
60	463	423	432	462	594
70	310	236	235	286	405
80	118	55	51	103	151

annuitants, that of nuns less favourable. The data were, however, scanty (monks 65 observed against 79·4 expected deaths; nuns 152 observed against 124·7 expected deaths).

Deparcieux has a few remarks on general medical-statistical questions (for instance, he urges strongly the importance of mothers nursing their infants), but nothing of much significance.

The statistical writings of Nicholas Struyck (1687–1769) are more voluminous than those of Deparcieux* and cover a wider field. Struyck was the son of an Amsterdam burgher and is said to have been in relatively easy circumstances. He enjoyed a considerable reputation as a writer on mathematical, statistical, geographical and astronomical subjects and was admitted a fellow of the Royal Society of London in 1749.

* They were collected and published in French translation at the instance of the Netherlands Assurance Society in 1912: *Les Œuvres de Nicolas Struyck, qui rapportent au calcul des chances*, etc., traduites du Hollandais par J. A. Vollgraff, Amsterdam, 1912, pp. 430.

Struyck was evidently a competent mathematician and also an industrious field worker who carried out or inspired in the Netherlands many town and village enumerations of population and vital statistical records. Like Deparcieux, he constructed life tables from annuitants' data and he certainly understood the correct arithmetical procedure. His data were, however, much fewer and he does not give sufficient details of his methods of interpolation and approximation to central ages for it to be possible to say precisely how he reached the life tables for males and females printed on p. 231 of his book. The original data were 794 males and 876 females (annuitants) observed for various periods and classified in quinquennial age groups. One has the impression that, although Struyck was a mathematician, he was not very sensitive to the dangers of basing conclusions upon small absolute numbers, and in his discussion of the vital statistics of London (op. cit. pp. 348–51) he has hardly given enough weight to the disturbing influence of migration and is perilously near the fallacy of a stationary population.

From the point of view of the medical statistician, Struyck is not a very suggestive writer. As demographer, we might rank him as technically superior to Short but medically less interesting. Like his contemporaries he can chase phantom hares in a thoroughly entertaining way. His finest example is in a section on multiple births. After a sober statistical inquiry he concludes that a case of quintuplets might reasonably be expected to occur sometimes in populations of the sizes of those of France and Germany—'it would be a very rare but not an incredible event'.

The case of the countess of Hennenberg, alleged to have brought to birth 364 or 365 infants simultaneously, does, however, strike him as 'absolutely fantastic and contrary to nature', and he carefully examines the legend. The statement was that the prolific mother produced as many children as the days of the year, and that the boys were named John and the girls Elizabeth. As Struyck justly observes it would be silly to have 182 Johns and 182 Elizabeths, and by careful research he arrives at a simple rational solution. The lady performed her feat on 26 March 1266; at that period the year began with the Feast of the Annunciation which was 25 March. So the birthday was the *second* day of the year and probably the mother had twins, one christened John the other Elizabeth. *Simplex munditiis!*

The name of Johann Peter Süssmilch (1707–67) is far better known than those of Deparcieux and Struyck although it is doubtful whether his *book* is often read. The perusal of 1201 pages of text and 207 of tables (the contents of the third edition of Süssmilch's book, published in 1765) requires a powerful appetite. If Süssmilch's literary style has less complexity than that of successors who wrote after German had become a 'literary' language, it has not much charm and few of us love propaganda. Süssmilch is a pure propagandist; the title of his book is: 'Die göttliche Ordnung in den Veränderungen des menschlichen Geschlechts, aus der Geburt, dem Tode und der Fortpflanzung desselben *erwiesen*' (italics mine), von Johann Peter Süssmilch. He sets out to reveal the divine machinery for fulfilling the command: 'Be fruitful, and multiply, and replenish the earth, and subdue it.' The reason why his book has more interest for a statistician than, say, Warburton's *Divine Legation*, is that Süssmilch conceived the notion that vital statistics might be pressed into the service of orthodox Lutheran theology, and the diligence with which he pursued his arithmetical investigations gave his book importance. It was indeed the quarry from which Malthus obtained material when the interest aroused by the first edition of his famous *Essay* led him to expand what had been not much more than a Shavian paradox into a serious treatise.

As a demographer and statistician, Süssmilch was technically inferior to either Deparcieux or Struyck and, of course, far below Halley. He had none of Graunt's originality and made no methodological advance. But he was very industrious. He assembled not only a large collection of German data, similar to but wider than those of Short, but collected foreign material—including that of Graunt, King and Short—and his tables are of real value.

The general conclusions he reached—constancy of the sex ratio, greater mortality of towns, etc.—differ in no important respect from those of his predecessors or English contemporaries. His own life table (which gave an expectation of life at birth of 28·43 years) is constructed on the incorrect principles adopted by most of his contemporaries. He was, indeed, aware that to make a life table by summing the deaths at ages occurring in an increasing population was wrong and that the population he used was increasing, but he did not know how to do better—indeed, he had no material for doing better.

His contribution to purely medical statistics is small. He has a chapter on the statistics of causes of death and compares the distribution by causes in the London Bills 1728–57 with those for Berlin in the years 1745, 1750 and 1757. When allowance is made for differences of nomenclature and misprints, the proportional distributions by causes are not very different. He makes the sensible suggestion that if the Latin names of the diseases were given by the medical attendants in official returns international comparison would be facilitated. For the rest, his medical importance is slight. To criticize or make fun of his triumphant justification of the ways of God to man would be sorry trifling. Although a dull writer, he inspires a certain affection. He was a sincere, diligent man and in polemics more courteous than most. He may, perhaps, quite contrary to his intention, have had a rather depressing influence upon enthusiastic readers, in that he had no expectation of a great reduction of mortality rates and has often anticipated ideas which we usually attribute to Malthus. He perceived that at the current rate of growth the earth must eventually be over-populated, but he argued that as the density of population increased the age of marriage would rise and consequently the fertility rate would decline. 'If, however, fertility remained the same, it would only be necessary for the rate of mortality to increase a little, so that, as in large towns, one in 25 died' (op. cit. **1**, 267).

He devotes a whole chapter to what Malthus would call positive checks upon population and clearly does not expect these to be eliminated although, for the reason just quoted, he does not think plagues and wars essential conditions. One, perhaps only one, item of the vital statistical system gave the good man some qualms. His arithmetic leads him to conclude that in cities half those born are dead by the 20th year of life, and even in the virtuous country districts half are dead before the age of 25. 'What is the reason that God permits half to die before they can be of service to God and the world? All the labour and effort of birth and rearing seem to have been in vain' (op. cit. **2**, 312). In a worldly sense there is, he confesses, no explanation. One must think of earthly life as but a preparation for the hereafter.

The trend of this reasoning is not encouraging to the social or hygienic reformer. Perhaps Süssmilch did contribute a little to the view that not much could be done to reduce the general death-rate, that, at the best, town death-rates might be slightly improved. But I doubt whether he had much influence upon medical opinion in England. Statistics are not even now a favourite study of the medical profession; 200 years ago a voluminous German writer on vital statistics would have found very few readers in the College of Physicians.

VIII. METHODOLOGICAL ADVANCES

The writers who were the subject of the last sections all flourished in the first half of the eighteenth century and all have a claim to be reckoned as pioneers. Deparcieux and Struyck made definite contributions to the mathematical or arithmetical technique of life-table construction; Süssmilch and Short followed the path blazed by Graunt, but they explored a good deal of country, and Short, at least, had novel ideas as to the utilization of local records.

In the later years of the century various medical writers, for instance, Heysham, Haygarth and Percival, made effective use of local enumerations of population in their efforts to secure sanitary improvements. The public has no passion for statistics, still a death-rate *is* more telling than a mere enumeration of deaths. But none of these writers contributed anything new to statistical methodology, and simple arithmetic, not to speak of the labour of making unofficial counts of population, is not every man's hobby. Had the proposal for making an official census in the middle of the century been accepted, no doubt interest in political or medical arithmetic would have revived, but it did not pass the House of Lords. The only official data of large dimensions were still the London Bills. These were sometimes the subject of medical statistical comment. In 1800, the younger Heberden wrote a monograph the title of which suggested competition with Short or even Graunt. But it was not a successful venture and is only remembered now (if at all) because of a statistical 'howler' which the iconoclastic Charles Creighton exposed with a satisfaction not melancholy.*

A typical example of the attitude of the better class of physicians towards statistics at the end of the eighteenth century will be found in *Observations Medical and Political on the Small-Pox...and on the Mortality of Mankind at every Age in City and Country...*, by W. Black, M.D., the second edition of which appeared in 1781. Dr Black, a medical graduate of Leyden and a licentiate of the College, who survived to 1829, reprinted the life tables of his predecessors. He was alive to the importance of the statistical method and its neglect ('In the course of many years' attendance upon medical lectures, in different universities, I never once heard the bills of mortality mentioned', op. cit. p. 119) and held that 'the detached observations of physicians or other literary individuals, confined perhaps to a small town or parish: a meagre detail of village remarks (*sic*), afford in many instances a foundation too slight to erect upon them any general or permanent conclusions' (op. cit. p. 119). He accordingly devoted most of his attention to the London Bills, which he subjected to a severe but cogent criticism, and set out in detail a sensible plan for the compilation of data in London by salaried officials with medical knowledge, which, had it been adopted, would have antedated the establishment of effective registration in London by more than 50 years.

One might explain the stagnation of medical statistical research by saying that there was not enough straw for ordinary brick makers to be employed, and no medical man of sufficient ingenuity (or temerity) to find a substitute for straw emerged. If that eminent fellow and, for a very short space of time, president of the College James Jurin had lived in the second instead of the first half of the eighteenth century, it is possible that the history of medical statistics would have been different, because, some years after his death, two famous mathematicians tackled a problem in which Jurin had taken keen interest and, as he himself was an accomplished mathematician, their method would have given him intellectual pleasure.

Jurin was an enthusiastic supporter of the practice of smallpox inoculation and wished to provide an adequate statistical proof of its value. Monk provides an eulogistic, Creighton a depreciatory account of what Jurin did. A fuller account is given by Miss Karn (M. N. Karn, *Ann. Eugen.* **4** (1931), 279 et seq.).

That Jurin proved the fatality of inoculated smallpox to be very much less than that of the natural smallpox, even Creighton admitted. But that he did much more can hardly be claimed. Jurin virtually assumed that inoculated smallpox did confer an immunity, on the basis of others' testimony and the famous experiment on six criminals, or rather on the one criminal who after

* Creighton, *History of Epidemics in Britain*, **2**, 747. Heberden made two mistakes: (1) He did not recognize that 'Griping in the Guts' of the Bills of Mortality was mainly the Diarrhœa of young children. (2) That a gradual transfer from this heading to that of 'Convulsions' had been going on.

inoculation was deliberately exposed to natural infection (see Creighton, op. cit. p. 480). Whether Jurin deserves to be sneered at because he did not do what was impossible, or whether the assumptions he made were unreasonable, are questions I shall not discuss. The mathematicians added nothing to the biological discussion, the interest of their work is purely intellectual, viz. by showing how to make the most of imperfect material. The problem proposed by Daniel Bernoulli was this.

Let us assume that inoculation completely protects against dying from smallpox and that those who are thus saved from the smallpox are neither more nor less likely to die of other causes than persons who never take smallpox, then what would be the effect on general mortality of the total eradication of smallpox? Put more picturesquely, how many years would be added to the average span of human life if smallpox were extinct?

In modern times, questions like this have often been put and answered, because we know with fair accuracy the numbers living by sex and age and the numbers dying from different causes also by sex and age. In the famous Supplement to the 35*th Annual Report of the Registrar-General*, Farr dealt with several causes. His method was simple. He subtracted from the central death-rate at any age due to all causes of death the central death-rate due to the special cause, and deduced from the resultant series of modified death-rates the appropriate life table constants. These he compared with those of the general life table. He found in this way that if phthisis were eliminated the expectation of life at birth (males) would be increased from 39·7 to 43·96 years. The elimination of the zymotic diseases would increase the mean lifetime to 46·77 years.

Farr was, of course, aware that the assumption, viz. if a particular cause of mortality was eliminated the death-rates from other diseases would not be affected, might not be justified—indeed, he had written with respect of Watt's lugubrious substitution theory, in accordance with which we gain little by eliminating one disease, its killing power will be taken by another. Farr's method is quite satisfactory as an arithmetical method but requires data not available in the eighteenth century. Bernoulli made two assumptions. The first that mortality rates from all causes were known (for his arithmetical calculations he used Halley's table although he did not quite correctly appreciate the meaning of Halley's phrase 'age current'), the second that the attack and fatality rates of smallpox were independent of age. He then reasoned thus:

Suppose there survive to age x by the life table P_x persons. Of these s, say, have not had smallpox; if $1/n$th of those who have not had smallpox were attacked within a year and $1/m$th of these die of smallpox, what is the value of s in terms of P_x, m and n? If dx is an element of time, $s\,dx/n$ will be attacked and $s\,dx/(mn)$ will die of smallpox within the element of time dx, and so there die from other diseases $-dP_x - s\,dx/(mn)$ because $-dP_x$ is the total mortality. But we are only interested in s, so the decrement through mortality $-dP_x - s\,dx/(mn)$ must be multiplied by s/P_x, and we reach the equation

$$-ds = \frac{s\,dx}{n} - \frac{s}{P_x}\left(dP_x + \frac{s\,dx}{mn}\right),$$

the solution of which is

$$s = \frac{mP_x}{(m-1)\,e^{x/n}+1}.$$

So s is known. Now let z be the number who would have survived to age x had there been no smallpox. Reasoning as before

$$-dz = -\frac{z}{P_x}\left(dP_x + \frac{s\,dx}{mn}\right).$$

The integral of which is

$$z = \frac{P_x m\, e^{x/n}}{(m-1)\,e^{x/n}+1}.$$

This is the solution. Bernoulli put $n = m = 8$ and concluded that the elimination of smallpox would, on these assumptions, add about 3 years to the mean life-time.

D'Alembert criticized Bernoulli's assumption that m and n were constant and replaced his equation by the formally simpler equation

$$dz = \frac{z}{P_x}\,dP_x + \frac{z}{P_x}\,du,$$

where du is the increment of mortality in time dx due to smallpox. The formal solution is

$$z = P_x \exp\left[\int_0^x \frac{du}{P_x}\right].$$

Isaac Todhunter commented sub-acidly on this: 'The result is not of practical use because the value of the integral is not known. D'Alembert gives several formulae which involve this or similar unfinished integrations' (*History of the Theory of Probability*, p. 268). Todhunter's comment is just so far as concerns the situation when Bernoulli and D'Alembert wrote. If, in addition to a table of general mortality, one has knowledge of the deaths at ages due to smallpox, then by means of the theorem known as the Euler-Maclaurin expansion, it *is* possible to evaluate the integral and reach a solution on D'Alembert's lines as Miss Karn (op. cit. pp. 303 et seq.) has shown. But if we do have this information, the much less laborious method of Farr is adequate.

But that does not mean that the attempt of Bernoulli and D'Alembert was futile, a mere display of mathematical fireworks. The situation in which they found themselves recurs time and again in the history of statistics, indeed of all branches of science. Often a practical man objects that a mathematician will write down equations in general terms which cannot be solved and are therefore, as the practical man urges, of no use to him. Sometimes the practical man is right, but not always; not even usually. Even when the equations cannot be solved, in the sense that certain 'constants' cannot be determined or certain integrals evaluated, methods of approximation, even inspired guesses, may lead to truth. Fifty years after Bernoulli and D'Alembert, E. E. Duvillard* published a monograph which, although seldom read, for it is scarce and 'practically' obsolete, has been rightly described by Farr as a classic of vital statistics. Duvillard set himself the same problem with the difference that vaccination was substituted for inoculation as the prophylactic, and this book, of nearly 200 quarto pages, may still be read with profit.

Duvillard lived before the days of Cauchy and mathematical rigour; no doubt much of his work would hardly satisfy the standard of a modern pure mathematician. Perhaps on that account it can be read by the amateur with comparative ease, and one may take hints of how to tackle problems for the solution of which complete statistical data are still to seek. There is no proverb the vital or medical statistician should more often repeat than the saying that the best is often the enemy of the good. It is no doubt foolish to suppose, as, according to Isaac Todhunter, Condorcet did suppose, that truth could be extracted from any data, however imperfect, provided one used formulae garnished with a sufficient number of signs of integration. It is more foolish to neglect even rough approximations to unattainable solutions. But, so far as concerns our predecessors in the College, indeed in the medical profession as a whole, the seed scattered by the foreign mathematicians fell upon stony ground. Between Short and Farr, no British physician made a contribution to statistical knowledge of much importance. I have spoken of the younger Heberden's brochure. William Woolcombe of Plymouth, in a tract on the alleged increase of tuberculosis, published in 1808, showed a better grasp of statistical method than the more famous physician.

The question Woolcombe examined was whether mortality from tuberculosis of the lungs were increasing. The statistical fact was that in well-kept registers he had examined the proportion of deaths assigned to consumption had certainly increased towards the end of the eighteenth century. Woolcombe was alive to the fact (often ignored by medical writers after his time) that the proportional mortality of a disease might increase although its absolute rate of mortality was stationary or even diminishing, and he tested his conclusions by a quite logical *ex absurdo* argument. Taking the assumption that at the beginning of the eighteenth century mortality was 1 in 36 and that the proportional mortality from phthisis was a third less than in 1801, he concluded that the general rate of mortality at the beginning of the nineteenth century must be as low as 1 in 54, unless the rate of mortality from phthisis had increased. But it was certain that in 1800 the general rate of mortality was higher than 1 in 54, at least 1 in 47. Reversing the process, viz. assuming the rate in 1801 to be known, the con-

* *Analyse et tableaux de l'influence de la petite vérole sur la mortalité à chaque âge*, Paris, 1806.

clusion was reached that the rate of mortality at the beginning of the eighteenth century must have been 1 in 27 unless the rate of mortality from phthisis had increased. This Woolcombe thought improbably high. He may have been wrong, but his method was rational. That was the best piece of medical statistical reasoning I have found in English medical literature between Short and Farr.

In 1800 the taking of a census was authorized by the legislature and not a government department, but the Speaker of the House of Commons was charged with the responsibility. Naturally, Mr Speaker passed over the actual work to one of his subordinates, and fortunately that subordinate, John Rickman, whose name is immortalized by the fact that he was a friend and correspondent of Charles Lamb, was really interested in statistics. In the report on the enumeration of 1801 comments are scanty, but they increased in subsequent volumes. Rickman was wholly responsible for the work down to the report of 1831 and, although he made no advance in statistical method, he did valuable work, particularly in calling attention to the high rate of mortality in the industrial north-west and in estimating past populations of the country. But Rickman was not professionally interested in medical questions, and before Farr no medical man utilized the new material effectively. As will be seen in the next section, the first English writer to publish a work under the title Medical Statistics was rather old fashioned in his treatment of the subject.

IX. THE END OF AN EPOCH

Almost at the end of the period I have chosen was published the first English book specifically devoted to Medical Statistics, *Elements of Medical Statistics*, by F. Bisset Hawkins, printed in 1829. It is a slender volume of 233 pages similar in format and size to the *Principles of Medical Statistics* published a little more than a century later, in 1937, by my friend and colleague Dr A. Bradford Hill.

Hawkins's book was an expansion of the Gulstonian Lectures of 1828; its author's long and useful life connects men still living with what seems a remote past. He was born in 1796, and there are still more than a dozen fellows of the College who may have sat in Comitia with him. He was admitted a fellow on 22 December 1826 and died in 1894. The copy of his book which I have read was presented by him to the Statistical (now Royal Statistical) Society in 1834 and contains corrections in his hand. Hawkins defines the province of Medical Statistics to be 'the application of numbers to illustrate the natural history of man in health and disease'. In his numerical statements he uses three indices; the ordinary crude death-rate—always expressed as one death in such or such a number—the 'probable life', i.e. the age to which half these born attain; the 'mean life', i.e. the average age at death. He was certainly aware that the age and sex constitution of a group affects the death-rate. Thus (op. cit. p. 20) he writes: 'In discussing the mortality of manufacturing towns or districts, it is just to remark that the small proportion is not always *real*; because a constant influx of *adults* is likely to render the number of deaths less considerable than that which could occur in a stationary population composed of all ages.' From the use of the term *stationary population* in this passage we may also, perhaps, infer that Hawkins knew the limitations of utility of such indices as mean age at death or *vie probable*, but I cannot fairly say that in making comparisons he calls attention to the dangers.

A modern treatise, such as that of Dr Hill, devotes a large space to methods of evaluating errors of sampling or, to speak loosely, the precautions to be taken when the observations are few in number and may not have been taken without bias. Some of the methods still employed had been invented by mathematicians before Hawkins's day, but he did not use them. On p. 32 we read: 'The annual mortality of Nice, though a small town, and enjoying a factitious reputation of salubrity, is 1 in 31; of Naples, is 1 in 28. Leghorn is more fortunate, and sinks to 1 in 35. We instance those places as being the frequent resort of invalids; but how astonishing is the superiority of England, when we compare with these even our great manufacturing towns, such as Manchester, 1 in 74; such as even Birmingham, 1 in 43; or even this overgrown metropolis, where the deaths are only 1 in 40.'

In the copy I have read, the sentence 'such as Manchester, 1 in 74' has been struck through, apparently by the author. But, even with this emendation, the comparison, to the glory of our country, is, well, tendentious.

Indeed, one must admit, however regretfully, that Hawkins's book is uncritical. He has been diligent and brought together numerical data from all parts of the world and was certainly one of the first physicians to advocate a serious study of hospital records, but one can hardly say that, as a statistician, he was better equipped or more efficient than Dr Short in 1750. But his modesty is disarming: 'I should be amply rewarded if the present humble essay should form a temporary repository of the most important of their labours; if it should become one of the early milestones on a road which is comparatively new, rugged as yet and uninviting to the distant traveller, but which gradually discloses the most interesting prospects, and will at length, if I do not deceive myself by premature anticipation, largely recompense the patient adventurer' (op. cit. p. vii).

According to Munk (*Roll*, **3**, 304) Hawkins was instrumental in obtaining the insertion in the first Registration Act of a column containing the names of the diseases or causes by which death was occasioned. 'At first the insertion was voluntary; it has since been made compulsory; and has produced important additions to medical and statistical science through the indefatigable labours of Dr W. Farr.'

So the name of Francis Bisset Hawkins deserves a place in the roll of benefactors to medical statistical science.

Eight years after the publication of Hawkins's Gulstonians there appeared, as Chapter IV of the fifth part of McCulloch's *Statistical Account of the British Empire* (**2**, 567–601, London, 1837), an article on 'Vital statistics; or the statistics of health, sickness, disease and death', the work of William Farr, then in his 30th year and still a general practitioner and free-lance medical journalist. It contains perhaps a quarter of the number of words in Hawkins's book and is not free from the quaint moralizing not always wholly relevant to the statistical theme which was characteristic of Farr, but it ranks not much below Graunt's 'Observations' as an original contribution to medical-statistical science.

Farr proposed to examine 'the mortality, the sickness, the endemics, the prevailing forms of disease, and the various ways in which, at all ages, its [The British Population's] successive generations perish'.

Slow as had been the progress of official statistics between 1662 and 1837, there had been progress. The four censuses of 1801–31 provided reasonably complete accounts of total populations. In 1821, information as to age was invited and eight-ninths of the population accepted the invitation. In 1831 the clergy were asked to return not merely totals of burials but burials classified by ages for the 18 years ending in 1830. These latter returns were incomplete, but it was possible for a lesser man than Farr to approximate to a statement of rates of mortality at ages at least for the period centring on 1821. To Farr's annoyance, the census takers of 1831 did not ask for the ages of the enumerated, contenting themselves with an enumeration of males under and over 20 years of age. The data for computing mortality rates were particularly defective for towns, but a few instances of quite good voluntary enumerations, e.g. for Carlisle and Glasgow, were available.

In handling national rates of mortality at ages, Farr's article does not display any conspicuous originality; he, quite properly, used the work of predecessors and he does not comment on the defects of the data. He does, however, call attention to particular rates of mortality, for instance, those of the troops, in an emphatic way. 'By the subjoined table of the mortality of the British army it will be seen that the soldier, in the prime of his physical powers, is rendered more liable to death every step he takes from his native climate, till at last the man of 28 years is subject, in the West Indies, to the same mortality as the man of 80 remaining in Britain.' According to his table, the average strength of British troops in Jamaica and Honduras between 1810 and 1828 was 2528; in the year of least mortality the rate 47 per 1000, the average 113 and the maximum 472! In the United Kingdom the average rate was 15 per 1000.

The most original part of Farr's essay is his treatment of sickness. Here national statistics were not available; more than 70 years were to pass before any nation-wide data were collected, and the statistics of morbidity still lag behind those of mortality. All Farr had were some data of benefit societies and returns relating to workers in the Royal Dockyards and employees of the East India Company. He begins by stating that in manhood for every death we may reckon two persons constantly sick. It is not quite clear how he reached this ratio, but probably from a comparison of the mortality rates for 1815–30 shown in a table on p. 568 of his article with some theoretical rates deduced by Edmonds for Friendly Societies (op. cit. p. 574). One has:

Age	Sickness rate per 1000	Mortality rate per 1000
20–30	17·2	10·1
30–40	23·0	11·4
40–50	31·0	14·9
50–60	45·1	23·4
60–70	93·6	45·3

Taking the general rate of mortality to be 21·3 per 1000 and the population of England and Wales to be 14,000,000, he concludes that 600,000 persons are constantly sick and that the productive power of the community is reduced by one-seventeenth part (he has made allowance for attendance on the sick). He works out from the limited data available the relation between sick-time and age and concludes that it increases in geometrical progression up to the age of 50. He asks how much sickness exists among the labourers of the country independently of those definitely incapacitated by disease. Data for the Royal Dockyards lead him to conclude that 2% are constantly kept at home by illness.

In the last section of his article, Farr considers particular diseases. An instance of his acumen is to be seen in his criticism of the view (held in 1837 as in 1937) that insanity was on the increase. He pertinently remarks that if the less barbaric treatment of lunatics diminished the mortality rate a higher proportion of enumerated lunatics would be perfectly consistent with a steady rate of morbidity.

His data for rates of mortality by causes were scanty. For London over a long period he had causes of death in age groups and, from an estimate of total mortality in age groups, could pass back to rates at ages by causes. Heysham's Carlisle data were medically and statistically more precise but limited to one not large town. The data of the Equitable Assurance Society were numerous but, as, of course, Farr knew and emphasized, related to a select class of the population.

Some of his general conclusions were as follows:

It has been shown that external agents have as great an influence on the frequency of sickness as on its fatality; the obvious corollary is, that man has as much power to prevent as to cure disease. That prevention is better than cure, is a proverb; that it is as easy, the facts we had advanced establish. Yet medical men, the guardians of public health, never have their attention called to the prevention of sickness; it forms no part of their education. To promote health is apparently contrary to their interests: the public do not seek the shield of medical art against disease, nor call the surgeon, till the arrows of death already rankle in the veins. This may be corrected by modifying the present system of medical education, and the manner of remunerating medical men.

Public health may be promoted by placing the medical institutions of the country on a liberal scientific basis; by medical societies co-operating to collect statistical observations; and by medical writers renouncing the notion that a science can be founded upon the limited experience of an individual. Practical medicine cannot be taught in books; the science of medicine cannot be acquired in the sick room. The healing art may likewise be promoted by encouraging post-mortem examinations of diseased parts; without which it is impossible to keep up in the body of the medical profession a clear knowledge of the internal change indicated by symptoms during life. The practitioner who never opens a dead body must commit innumerable, and sometimes fatal, errors (op. cit. p. 601).

Farr's article closes the epoch Graunt's book opened. The seventeenth-century pioneer did not live to see the ground he broke bear a crop. The high gods used Farr better; he lived to create the best official vital-statistics of the world. It is true that the lessons he taught were learned but slowly, either by the public or the profession. The *Annual Reports* of the Registrar-General will not be found among the frequently consulted volumes on the shelves of fellows of the College of Physicians. But something has been learned. The moral truism that human vanity is a deadly sin, now exemplified on a world-wide scale, is illustrated on the humbler scale of those topics which have been my life's work and the subject of these lectures. The distrust of 'mathematical' methods which is still general in our profession is not primarily due to the mere intellectual difficulty of learning 'mathematical' methods; much that all medical students must learn is at least as difficult.

The roots are deeper. They begin with the exaggerated claims of the iatro-mathematicians of the late seventeenth and early eighteenth centuries. The personal popularity of such men as Freind and Jurin did not conceal the fact that pathology and clinical medicine reduced to mechanical and quantitative theorems, and 'proofs' were of not much greater value in the treatment of sick men than skill in playing chess to the commander of an army. It is arguable that a talent for playing chess might, other things equal, be of advantage to a military strategist (Napoleon Bonaparte was very fond of chess and played so badly that it was difficult for his staff to avoid winning), but other things are not equal. In later times, when the intellectual prestige of mathematical science had grown enormously, it was observed that such an Admirable Crichton as our Thomas Young was inferior as a practical physician to many fellows of lesser fame. In our generation when the professional mathematicians who, 50 years ago, rather despised mere statistics, have increasingly devoted themselves to the improvement of the general theory, the complexity of statistical investigations has done little to attract the amateur, and intellectual modesty has not been the most conspicuous virtue of statistical authors. Perhaps, too, it is not easy for an experienced physician 'to renounce the notion that a science can be founded upon the limited experience of an individual'.

The moral I should draw from the history of medical statistics is that the intellectual courage of an amateur often succeeds where erudition fails. While even the purest of mathematicians would not claim that statistics is only a branch of mathematics, the hardiest contemner of algebra would admit that a training in mathematical method is an advantage to the practical statistician. The mathematician would surely agree that a knowledge of the material subjected to analysis was valuable, even if not so essential as a 'practical' man would claim.

Judged by contemporary intellectual standards, neither Graunt nor Farr was a mathematician; Graunt had no medical training, Farr's clinical experience was meagre. In respect neither of method nor subject-matter was either man an expert. But they both had intellectual curiosity and courage: one may say, if one pleases, the spurious courage of the man who is brave because he does not know what the dangers are. But, as Gilbert Chesterton once said, 'There is no real hope that has not once been a forlorn hope.' In graver matters than medical statistics and more than once in our national history salvation has been wrought by courageous amateurs who acted while professionals doubted.

Those who cannot disclaim a professional status in statistics, whether officials or professors, may learn a lesson from history. It is conveyed in the four words: *maxima debetur puero reverentia*, construing *puer* by amateur or beginner or enthusiast. It is weary work to read statistical 'proofs' of this or that aetiological theory of cancer, or proposals for this or that impossible statistical investigation. But it is treachery to science to rebuff any genuinely inquisitive person; the discovery of another Graunt in a shop or another Farr in the surgery of a general practitioner would repay the life-long boredom of all extant civil servants and professors of statistics.

THE PRINCIPLE OF THE ARITHMETIC MEAN

By R. L. PLACKETT

The history of the problem of combining a set of independent observations on the same quantity is traced from antiquity to the appearance in the eighteenth century of the arithmetic mean as a statistical concept.

The problem of estimating parameters from observational data appears first to have presented itself to the Babylonian astronomers of the last three centuries B.C. Their achievements are recorded in cuneiform script on clay tablets and have been analysed by Neugebauer (1951) who has also (1955) published a collection of the texts. The following summary is abstracted from his researches. Between about 500 and 300 B.C., the Babylonians developed a systematic mathematical theory to account for the motions of the sun, moon and planets; and they evolved simple arithmetical schemes by which the positions of these bodies could be calculated at regular intervals of time. Beyond the fact that the basic parameters in the schemes represent a compromise between observation and the needs of computation, nothing has survived to indicate how they were estimated from the original data, which are themselves almost wholly absent.

Rather more information is available concerning the methods by which the Greek astronomers analysed their observational data, for their discoveries were made possible, partly by developments of mathematical technique, and partly by the steady accumulation, since about 300 B.C., of a series of observations on the positions of stars and planets, made with graduated instruments. The *Syntaxis* of Claud Ptolemy not only presents a complete account of what was known to them, but also contains nearly everything that survives of the work of their greatest representative, Hipparchus. In what follows, we refer to the edition in two volumes translated and annotated by Karl Manitius (1913).

According to I, p. 133, Hipparchus noticed inequalities in the intervals of time between successive passages of the sun through the same solstitial point, and this suggested to him the question whether or not the length of the tropical year is constant. He considered, however, that the error in his observations and in the calculations based on them might amount to as much as $\frac{1}{4}$ day, and he concluded that any variation in the length of the year was quite insignificant. Subsequently, Hipparchus estimated the maximum variation in length as $\frac{3}{4}$ day, apparently by taking half the range of his observations (I, pp. 136–7).

> In fact, Hipparchus calculates with the help of certain eclipses of the moon, observed in the immediate neighbourhood of fixed stars, how far the star called Spica was west of the autumnal point at each eclipse, and finds some indication in this way that it shows in his time a maximum distance of $6\frac{1}{2}°$ and a minimum of $5\frac{1}{4}°$. Whence he draws the conclusion, since it is not well possible that Spica should have undergone such a considerable change of position in so short a time, that probably the sun, from whose position Hipparchus determines the positions of the fixed stars, does not accomplish its return at equal intervals.

The technique of taking the arithmetic mean of a group of comparable observations had not yet, however, made its appearance as a general principle. This is shown by Ptolemy's estimation of the amount by which the length of a year exceeds 365 days. Hipparchus had made the observations given below (I, pp. 134–5):

Autumn equinox				Spring equinox			
(1)	162 B.C.	Sept. 27	18^h	(1)	146 B.C.	March 24	6^h (11^h at Alexandria)
(2)	159 B.C.	Sept. 27	6^h				
(3)	158 B.C.	Sept. 27	12^h	(2)	135 B.C.	March 23/24	midnight
(4)	147 B.C.	Sept. 26/27	midnight	(3)	128 B.C.	March 23	18^h
(5)	146 B.C.	Sept. 27	6^h				
(6)	143 B.C.	Sept. 26	18^h				

Ptolemy gives (I, p. 142) a single observation of his own on the Autumn equinox, namely, A.D. 139 Sept. $26^d\,7^h$, and compares it with the fourth observation of Hipparchus, whence he finds that in 285 Egyptian years of 365 days, the Autumn equinox advances by $70^d\,7^h$. which he writes as $70+\frac{1}{4}+\frac{1}{20}$ days. He then gives (I, p. 143) a single observation of his own on the Spring equinox, namely, A.D. 140 March $22^d\,13^h$, and by comparing it with the first observation of Hipparchus, again arrives at an advance of $70+\frac{1}{4}+\frac{1}{20}$ days in 285 Egyptian years. A year of $365\frac{1}{4}$ days would imply an advance of $71\frac{1}{4}$ days in 285 years, and the decrement of $71\frac{1}{4}-70\frac{6}{20}=\frac{19}{20}$ day in 285 years is equivalent to 1 day in 300 years. Thus Ptolemy reaches the value of $365\frac{1}{4}-\frac{1}{300}$ days for the length of the year, and this is precisely the value which Hipparchus is quoted (I, p. 145) as having found.

A similar example of Ptolemy's veneration for Hipparchus is provided by his discussion of the precession of the equinoxes, a phenomenon discovered by Hipparchus, and caused by the motion of the pole of the equator round the pole of the ecliptic, the annual movement being about 50″. According to a quotation in II, p. 15, Hipparchus estimated the change in the position of the solstices and equinoxes to be at least $\frac{1}{100}°$ per annum. Ptolemy then gives (II, pp. 18–20) a catalogue of the declinations of 18 stars as observed by (i) Timocharis and Aristyllus, about 290 B.C., (ii) Hipparchus, and (iii) himself. He selects 6 stars from the catalogue and shows that they all lead to a precessional constant of approximately $\frac{1}{100}°$ per annum, which is thus his estimate, whereas for Hipparchus it was a lower limit. These unique data have been analysed by several commentators, beginning with Delambre (1817, pp. 254–5) who showed that the average precessional constant from all 18 stars is near the correct value, whether the changes of declination from (i) to (ii), or from (ii) to (iii), are taken. Recently Pannekoek (1955) has confirmed the accuracy of Ptolemy's arithmetic; and he suggests that Ptolemy selected the 6 stars which agreed best with the value of $\frac{1}{100}°$ per annum, but which actually each exhibit too small a change of declination.

The technique of repeating and combining observations made on the same quantity appears to have been introduced into scientific method by Tycho Brahe towards the end of the sixteenth century. According to his biographer, Dreyer (1890, p. 350):

> Each observation thus gave a value for the right ascension of α Arietis. During the following six years Tycho repeated these observations as often as an opportunity offered, and, in order to eliminate the effect of parallax and refraction, he combined the results in groups of two, so that one was founded on an observation of Venus while east of the sun, the other on an observation of Venus west of the sun; while the observations were selected so that Venus and the sun as far as possible had the same altitude, declination and distance from the earth in the two cases. From the observations of 1582 Tycho selects three single determinations, and from the years 1582–88 twelve results, each being the mean of two results found in the manner just described. The fifteen values of the right ascension of α Arietis agree wonderfully well *inter se*, the probable error of the mean being only ± 6″, but the twenty-four single results in the twelve groups show rather considerable discordances, the greatest and smallest differing by 16′ 30″.

But anyhow the final mean adopted by Tycho is an exceedingly good one, agreeing well with the best modern determinations. He adopts for the end of the year 1585 26° 0′ 30″, the modern value for the same date being 26° 0′ 45″.

The observations to which Dreyer refers are reproduced below from Tycho's collected works (**2**, 170–97):

Date	Observation	Result
1582 February 26		26° 0′ 44″
1582 March 20		26 0 32
1582 April 3		26 0 30
1582 February 27	26° 4′ 16″	26 0 20
1585 September 21	25 56 23	
1582 March 5	25 56 33	26 0 38
1585 September 14	26 4 43	
1582 March 5	25 59 15	26 0 18
1585 September 15	26 1 21	
1582 March 9	25 59 49	26 0 32
1585 September 15	26 1 16	
1586 December 26	25 54 51	26 0 42
1588 December 15	26 6 32	
1586 December 27	25 52 22	26 0 37
1588 November 29	26 8 52	
1587 January 9	26 2 5	26 0 27
1588 December 6	25 58 49	
1587 January 24	26 6 44	26 0 29
1588 October 26	25 54 13	
1587 August 17	26 5 40	26 0 14
1588 April 16	25 54 48	
1587 August 17	26 1 1	26 0 4
1588 April 16	25 59 6	
1587 August 18	25 54 35	26 0 28
1588 March 28	26 6 20	
1587 August 18	25 54 49	26 0 39
1588 April 16	26 6 30	

The process of combining the first pair is thus described by Tycho (*ibid.* p. 171).

Ab hac rursus Differentia Ascensionis vsque ad Lucidam ♈ subtracta, quae est part. 83. min. 57.//.20, prouenit Ascensio Clarae ♈, part. 25./.56.//.10, cui pro Mensibus 3 residuis addantur //.13, & obtinebimus Ascensionem Rectam Lucidae ♈ part. 25. min. 56.//.23, Anno 1585 completo correspondentem. Sed Anno 82 ex Die 27 Februarij, fuit eadem Ascensio Recta prius data part. 26. min. 4.//.16, vt sit differentia vtriusque min. 7.//.53: Dimidiata min. 3.//.56½ addita minori vel subtracta a maiore, prodit vera & limitata Ascensio Recta Lucidae ♈ part. 26./.0.//.20. Quam hac Methodo nulla habita ratione Parallaxium atque Refractionum, sed illis sese mutuo sic corrigentibus, inquirere propositum erat.

The average of the twelve determinations by means of two is 26° 0′ 27″, and the average of all fifteen is 26° 0′ 29″. How Tycho arrives at 26° 0′ 30″ is not described, but we note that the co-ordinates of the nine standard stars in his catalogue are all given at 5″ intervals, more than adequate for observational purposes. In fifteen cases out of eighteen, the co-ordinates differ from their exact values by less than 1′, and Kepler has described in a famous passage (*Astronomia Nova...*, Chap. 19; *Werke*, **3**, 178) how he was able to calculate the elements of a circular orbit for Mars, differing from Tycho's observations by 8′ or less, but rejected it because he knew that errors of 8′ could not be neglected with so diligent an observer.

We see that Tycho used the arithmetic mean to eliminate systematic errors. The calculation of the mean as a more precise value than a single measurement is not far removed and had certainly appeared about the end of the seventeenth century, as is shown by the following extract from Flamsteed's discussion of the errors produced by his mural arc on the right ascensions of stars (1725, vol. 3, p. 137):

Rectarum SOLIS Adscensionum *Differentia* inter 14um *Martii* ac 15um *Septembris* [of 1690] ex Observationibus circa *Solem* pro istis Diebus reperitur, viz.

per *Calcem* CASTORIS	μ	178° 36′ 0″		
per PROCYONEM		178 36 5		
per POLLUCEM		178 36 20		
Media inter has Differentia		178 36 8		
At hanc *Mediam* subtrahendo a SOLIS *Recta* Adscensione 15to *Septembris*, viz.				182 31 53
				178 36 8
remanet eius *vera Recta Adscensio* 14to *Martii* Meridie				3 55 45
quae *verum* dat eius *Locum*			♈	4 17 7

A third example illustrates the combination of data from different observers. During 1736–7, a French expedition under Maupertuis was sent to Lapland in order to measure the length of a degree of latitude and, by comparing it with the corresponding length in France, to decide whether the earth was flattened at the poles, as maintained, e.g. by Newton, or at the equator, as held by the Cassini family. Their method of observation, as described by Outhier, has been summarized by Clarke (1880, p. 5) as follows:

Each observer made his own observation of the angles and wrote them down apart, they then took the means of these observations for each angle: the actual readings are not given, but the mean is.

In the event, the degree proved to be longer in Lapland, and Voltaire congratulated Maupertuis on having flattened both the poles and the Cassinis.

At about this time, the calculus of discrete probability assumed an organized form, and the appearance of the differential calculus made extensions to continuous probability possible. The distribution of the arithmetic mean now began to receive the attention of mathematicians who were conversant with the new techniques, and a pioneer study by Simpson was followed by a long memoir from Lagrange.

In his paper of 1755, Simpson gives the probability that the mean of t observations is at most m/t for the following two error distributions:

(i) possible errors are $-v, \ldots, -2, -1, 0, 1, \ldots, v$ and equal probabilities are attached to them;

(ii) the same set of errors with probabilities proportional to $1, 2, \ldots, v+1, \ldots, 2, 1$, respectively.

The solution for (i), when expressed as a gaming problem, was known by 1710 and Simpson's treatment by generating functions is the same as de Moivre's (Todhunter, p. 85); since the generating function for (ii) is the square of what it is for (i), Simpson's initial contribution amounted mainly to realizing the physical interpretation of a mathematical result.

What is novel in Simpson's work appears in the four pages of additional material published in 1757. Here he extends the solution of the second problem to the limiting case where the error distribution is continuous, in the form of an isosceles triangle, and, by integration, finds the probability that the mean is nearer to zero than a single independent observation.

Simpson's debt to de Moivre is clear and the widespread respect which *The Doctrine of Chances* inspired during this period is notably attested by the following quotation from a letter written by Lagrange to Laplace on 30 December 1776.

> Il est vrai que j'ai eu autrefois l'idée de donner une traduction de l'Ouvrage de Moivre, accompagnée de notes et d'additions de ma façon, et j'avais même déjà traduit une partie de cet Ouvrage; mais j'ai depuis longtemps renoncé à ce projet, et je suis enchanté d'apprendre que vous en avez entrepris l'exécution, persuadé qu'elle répondra à la haute idée qu'on a de tout qui sort de votre plume.

In the first fifty pages of his memoir, Lagrange presents a detailed discussion of discrete error distributions, on lines essentially the same as those followed by Simpson; he again makes free use of generating functions, and again extends results from discrete to continuous distributions by appropriate limiting processes. This section also includes (problem 6) a derivation of what we would now describe as the maximum likelihood estimates of the parameters in a multinomial distribution; and purports to show (problems 4 and 5) that the mode of the distribution of sample means is the same as the population mean. The chief contribution of the memoir to the probability theory of the arithmetic mean occurs in its last twelve pages, where Lagrange gives a method of obtaining the results for continuous distributions directly. He begins by evaluating

$$\int_0^\infty \frac{x^{m-1}dx}{a^x} = \frac{(m-1)!}{(\log a)^m},$$

where a is larger than unity. He now says that the coefficient of a^{p-x} in

$$(Pa^p + Qa^{p-1} + Ra^{p-2} + \dots)/(\log a)^m, \tag{1}$$

is obtained on replacing

$$1/(\log a)^m \quad \text{by} \quad \int_0^\infty x^{m-1}a^{-x}dx/(m-1)!$$

and is thus given by

$$\{Px^{m-1} + Q(x-1)^{m-1} + R(x-2)^{m-1} + \dots\}\,dx/(m-1)!.$$

He next asserts that the probability element for the sum of n independent variables, each with density function $y(x)$, is the coefficient of a^z in $\left\{\int y\,.\,a^x dx\right\}^n$, where the term 'coefficient' is used in the sense just defined. Several examples follow, in all of which the error distribution has a finite range, so that $\int y\,.\,a^x dx$ is a sum of terms like (1), and is therefore amenable to the processes he has described. The last error distribution is given by

$$y = K\cos x \quad (-\tfrac{1}{2}\pi \leqslant x \leqslant \tfrac{1}{2}\pi),$$

and the memoir concludes with a set of ingenious manipulations involving imaginary quantities.

At this interval of time, we can recognize the last part of Lagrange's memoir as a starting point for the theory of integral transforms, although its merits were scarcely visible to Todhunter, writing in 1865. However, they were at once appreciated by Laplace, who refers to 'la belle méthode que vous donnez' in a letter written to Lagrange on

11 August 1780, and who subsequently made the technique a basic part of his attack on the problem of combining observations.

I am very grateful to Dr A. Fletcher for his invaluable suggestions and guidance on astronomical matters, and for greatly improving my translations.

REFERENCES

CLARKE, A. R. (1880). *Geodesy.* Oxford: Clarendon Press.

DELAMBRE, J. B. J. (1817). *Histoire de l'astronomie ancienne*, 2 vols. Paris: Courcier.

DREYER, J. L. E. (1890). *Tycho Brahe; a Picture of Scientific Life and Work in the XVIth Century.* Edinburgh: Black.

FLAMSTEED, J. (1725). *Historia Coelestis Britannica*, 3 vols. London: Meere.

KEPLER, J. (1609, collected works 1937). *Astronomia Nova..., Gesammelte Werke*, **3**, ed. M. Caspar. Münich: Beck.

LAGRANGE, J. L. (about 1775). Mémoire sur l'utilité de la méthode de prendre le milieu entre les résultats de plusieurs observations, dans lequel on examine les avantages de cette méthode par le calcul des probabilités, et où l'on résout différents problèmes relatifs à cette matière. *Miscellanea Taurinensia*, **5**; (1868) *Œuvres*, **2**, 173–234.

LAGRANGE, J. L. (1776). Letter to Laplace, 30 December; (1892). *Œuvres*, **14**, 66.

LAPLACE, P. S. (1780). Letter to Lagrange, 11 August; (1892). *Œuvres de Lagrange*, **14**, 95.

MANITIUS, K. (1913). *Des Claudius Ptolemäus Handbuch der Astronomie*, 2 vols. Leipzig: Teubner.

NEUGEBAUER, O. (1951). *The Exact Sciences in Antiquity.* Copenhagen: Ejnar Munksgaard.

NEUGEBAUER, O. (1955). *Astronomical Cuneiform Texts*, 3 vols. London: Lund Humphries.

PANNEKOEK, A. (1955). Ptolemy's precession. *Vistas in Astronomy*, vol. 1, pp. 60–66, ed. A. Beer. London and New York: Pergamon.

SIMPSON, T. (1755). A letter to the Right Honourable George Earl of Macclesfield, President of the Royal Society, on the Advantage of taking the mean of a number of observations in practical astronomy. *Phil. Trans.* **49**, part **1**, 82–93.

SIMPSON, T. (1757). An attempt to show the advantage arising by taking the mean of a number of observations in practical astronomy. *Miscellaneous Tracts on some curious and very interesting Subjects in Mechanics, Physical Astronomy, and Speculative Mathematics, ...*, pp. 64–75.

TODHUNTER, I. (1865). *A History of the Mathematical Theory of Probability from the Time of Pascal to that of Laplace.* London: Macmillan.

TYCHONIS BRAHE DANI (1602, collected works 1915). *Opera Omnia*, Tomus II, ed. I. L. E. Dreyer. Hauniae, in Libraria Gyldendaliana.

A NOTE ON THE EARLY SOLUTIONS OF THE PROBLEM OF THE DURATION OF PLAY

By A. R. THATCHER

It is now just 300 years since the publication by Huygens of the first result on the famous problem which became known as the Duration of Play. The aim of this note is to summarize the early development of this problem and to show how easily some of the solutions found at the beginning of the eighteenth century can be linked with modern work on sequential tests, random walks and certain storage problems.

We use throughout the following notation. Call the two players A and B, and let their chances of winning a game be p and $q = 1-p$, respectively. A starts with a counters and B starts with b counters, and after each game the loser hands one counter to the winner. It is desired to find first the probability P_a that A will eventually lose all his counters without having previously won all B's, and more generally the probability $P_{a,n}$ that this will happen within n games. P_b and $P_{b,n}$ are defined similarly. $P_{a,n}+P_{b,n}$ is the probability that the play will terminate (with the 'ruin' of one of the players) within n games. It can be shown that the play must end sooner or later, so that $P_a+P_b = 1$.

In 1657 Huygens gave without proof, in the fifth and last problem of his treatise *De ratiociniis in ludo aleae*, the numerical value for P_a in a case where $a = b = 12$ and where p and q had particular values. The general result for P_a was found by James Bernoulli, who died in 1705, but it remained in manuscript until it was published 8 years later in his *Ars Conjectandi*; Bernoulli says that the proof is laborious and leaves it to the reader. Before the *Ars Conjectandi* appeared, however, de Moivre had found a simple derivation independently and published it in his treatise *De Mensura Sortis* (1711).

De Moivre's original proof, which was later reproduced in his *Doctrine of Chances* (see 1711, pp. 227–8; 1718, pp. 23–4; 1738, pp. 45–6; 1756, pp. 52–3), is very ingenious and so much shorter than the demonstrations usually given in modern textbooks that it is worth quoting. Its essence is as follows. Imagine that each player starts with his counters before him in a pile, and that nominal values are assigned to the counters in the following manner. A's bottom counter is given the nominal value q/p; the next is given the nominal value $(q/p)^2$, and so on until his top counter which has the nominal value $(q/p)^a$. B's top counter is valued $(q/p)^{a+1}$, and so on downwards until his bottom counter which is valued $(q/p)^{a+b}$. After each game the loser's top counter is transferred to the top of the winner's pile, and it is always the top counter which is staked for the next game. Then *in terms of the nominal values* B's stake is always q/p times A's, so that at every game each player's nominal expectation is nil. This remains true throughout the play; therefore A's chance of winning all B's counters, multiplied by his nominal gain if he does so, must equal B's chance multiplied by B's nominal gain. Thus

$$P_b\left\{\left(\frac{q}{p}\right)^{a+1}+\left(\frac{q}{p}\right)^{a+2}+\dots+\left(\frac{q}{p}\right)^{a+b}\right\} = P_a\left\{\left(\frac{q}{p}\right)+\left(\frac{q}{p}\right)^{2}+\dots+\left(\frac{q}{p}\right)^{a}\right\}.$$

The use of $P_a+P_b = 1$ now gives immediately

$$P_b = \frac{(q/p)^a-1}{(q/p)^{a+b}-1}, \tag{1}$$

and this is the probability of the 'gambler's ruin'.

In terms of the counters, A's total expected gain is bP_b-aP_a, while his expectation per game is $p-q$. These obvious facts are indeed only special cases of a more general result given by de Moivre (1718, pp. 135–6; 1738, pp. 48–9; 1756, pp. 55–6). De Moivre does not actually divide one expression by the other, but, since the total expectation equals the expectation per game times the expected number of games, this division is all that is required in order to get the expected number of games

$$E(N) = \frac{bP_b-aP_a}{p-q}. \tag{2}$$

De Moivre was also the first to discover and publish a general method for calculating $P_{a,n}+P_{b,n}$, thus finding the chance that the play would terminate within n games. For the case where a is infinite (so that $P_{a,n} = 0$) and $n-b$ is odd, he found

$$P_{b,n} = \text{first } \tfrac{1}{2}(n-b+1) \text{ terms of } (p+q)^n + \text{first } \tfrac{1}{2}(n-b+1) \text{ terms of } (p/q)^b(q+p)^n. \tag{3}$$

This solution, with a similar one for the case where $n-b$ is even, was given without proof in his *De Mensura Sortis* and later in *The Doctrine of Chances* (1711, p. 262; 1718, pp. 119–20; 1738, p. 179; 1756, pp. 208–9). Fieller (1931) has drawn attention to this result and also provided a simple and elegant proof.

De Moivre's first solution of the general problem of calculating $P_{a,n}+P_{b,n}$ when both a and b are finite (1711, p. 261; 1718, pp. 113–14; 1738, stated incorrectly on pp. 173–4; 1756, p. 203) called for the performance of $n-1$ multiplications and the rejection of certain terms during the process. For moderate n the calculation is not so tedious as appears at first sight, and it has the advantage of giving the answer reduced to the smallest number of terms; as de Moivre later pointed out, the rejected terms can also be used to obtain $P_{a,n}$ and $P_{b,n}$ separately.

However, a few months before de Moivre's method actually appeared (for the *Philosophical Transactions* for 1711 were delayed in the press), a different solution giving $P_{a,n}$ and $P_{b,n}$ separately had been found and was soon published by de Montmort (1713). This result is of particular interest because it provides one of the easiest solutions of the problem, since the series which can be derived from it by using modern tables is rapidly convergent over the range of values of n where the play is likely to terminate.

In 1710 de Montmort found a method for calculating $P_{a,n}$ and $P_{b,n}$ for the case $p=q$. He sent some numerical results to John Bernoulli, who passed the letter to his nephew Nicholas. In a reply dated 26 February 1711, published by de Montmort (1713, p. 308 et seq.), Nicholas Bernoulli gave without proof the general solution for the case $p \neq q$; in modern notation it can be written as follows:

$$P_{b\,n} = \sum_t \left\{ p^{ts+b} q^{ts} \sum_i \binom{n}{i} (p^{n-b-2ts-i}q^i + q^{n-b-2ts-i}p^i) \right\}$$
$$- \sum_t \left\{ p^{ts+s} q^{ts+a} \sum_i \binom{n}{i} (p^{n-b-2ts-2a-i}q^i + q^{n-b-2ts-2a-i}p^i) \right\}. \quad (4)$$

In this formula $s=a+b$; the summation over $i \geqslant 0$ continues until the terms in the series in each curly bracket, re-arranged in descending powers of p, meet in the middle (the middle term counting only once if $n-b$ is even); and the summation over t covers all values $\geqslant 0$ which leave non-negative exponents within the summation over i on the line concerned. Bernoulli stated the result for $n-b$ even, but in fact (4) is also valid if $n-b$ is odd.

Not content with this, Nicholas Bernoulli confirmed that the limit of (4) as $n \to \infty$ gives the correct value for P_b. He does not give his method but it is not difficult to guess; if for example $p>q$ it is only necessary to re-write the two lines of (4) as

$$\sum_t p^{-ts}q^{ts}[p^n + np^{n-1}q + \ldots + p^{2ts+b}q^{n-2st-b}]$$
$$- \sum_t p^{-ts-a}q^{ts+a}[p^n + np^{n-1}q + \ldots + p^{2ts+2a+b}q^{n-2ts-2a-b}]. \quad (5)$$

As $n \to \infty$ the sums in each square bracket tend to 1; this follows from (James) Bernoulli's Theorem, which at the time had not been published but which was known to Nicholas. The expression thus reduces to two geometric series, and is immediately seen to agree with (1) above. In passing, it may be noted that as $a \to \infty$ the expression (4) reduces to de Moivre's expression (3).

When de Montmort saw this extraordinary solution he admitted that he could not follow it (this was partly because Bernoulli had inadvertently used one symbol in two senses), and remarked: 'votre formule m'étonne pour sa generalité' (1713, p. 316). Later, in comparing it with his own, he said: 'je n'ai eu en vûe que la supposition des hazards égaux pour l'un et pour l'autre Joueur, au lieu que vous les supposés dans un rapport quelconque' (1713, p. 345). De Montmort's solution, which he then describes briefly, consisted of a method of picking out the binomial coefficients in (4) from Pascal's triangle; this was of course sufficient when $p=q$, and was in itself a remarkable result to have found. Nevertheless, it seems clear that the solution (4) of the general case $p \neq q$, though often described as de Montmort's, was in fact found first by Nicholas Bernoulli.

De Montmort reproduced (4) in the body of his book, gave an example and added a most interesting though far from rigorous demonstration (1713, pp. 268–72). De Moivre at first called the result 'very handsom' (1718, p. 122), but later criticized de Montmort's statement of it (which indeed is not entirely correct) and seems to hint that he had found the same method of solution before the year 1711 (see 1738, pp. 181–2; 1756, pp. 210–11). This is certainly possible, though it may be doubted whether de Moivre had carried the investigation of (4) as far as Bernoulli; perhaps he used it in particular cases, but did not pursue the matter because his own result gave $P_{a\,n}+P_{b,n}$ in a smaller number of terms.

De Moivre later solved the Duration of Play problem in two further ways, and in the course of his work made an extensive investigation of recurring series (which he was the first to explore). His results included the partial fraction expansion of a generating function (1738, pp. 197–99; 1756, pp. 224–7); he found the probability of runs of successes (1738, pp. 243–8; 1756, pp. 254–9), and of course made the original derivation of the normal distribution (1738, pp. 235–43; 1756, pp. 243–50). On the Duration of Play problem itself he expressed $P_{b,n}$ as a recurring series with fewer terms than (4); and finally he discovered the first results on the trigonometrical solution (see Feller, 1950, p. 292, equation 5·7), including the asymptotic form for $P_{b,n}$ when $a = b$ and $p = q$. For fuller details of his work, and of its subsequent development by Laplace and many others, the reader is referred to Todhunter (1865) and Fieller (1931).

It remains to show the link between these early solutions and modern work. This stems from the well-known fact that the Duration of Play situation can be regarded as a linear random walk with two absorbing barriers, such that the movement of the particle at each jump has a distribution with mean $\mu = p-q$ and variance $\sigma^2 = 4pq$. To complete the comparison a simple approximation is required, namely

$$(p/q)^\lambda \simeq \exp(2\lambda\mu/\sigma^2), \tag{6}$$

which can be shown to apply with sufficient accuracy in the cases for which it will be required.

If then in equations (1) and (2) we make the substitutions (6) and $p-q = \mu$ we shall obtain approximations for the probability of absorption at a given barrier, and for the expected number of steps before absorption at either barrier, in the corresponding random walk; and under the conditions of the central limit theorem these will be valid for all walks with given finite μ and σ, provided that the number of steps is sufficiently large. It can be seen by inspection that the transformed version of equations (1) and (2) are in fact the same as Wald's approximations for the operating characteristic and average sample number of a sequential test, in the form quoted by Page (1954, equations 5, 7).

We can similarly transform (3), making the normal approximation to the binomial expressions; it will be found that the result agrees with that given by Bartlett (1946, equation 8), obtained as the solution of a differential equation for the diffusion process. It is of interest to note that the same result can also be used to find a quick approximate solution of a storage problem considered in a recent paper by Anis (1956). This concerns a reservoir, of unlimited capacity, which has initial water level x; this level varies each year by an amount distributed with zero mean and unit variance. When n and x are sufficiently large we can ignore the end-effects and assume that the probability that the reservoir will run dry within n years is approximately the same as the probability that B will lose $b = x$ counters within n trials (where a is infinite and $p = q = \frac{1}{2}$). By de Moivre's result (3) this probability will be twice the sum of the first $\frac{1}{2}(n-x+1)$ terms of $(\frac{1}{2}+\frac{1}{2})^n$. Hence, for large n and x the probability that the reservoir will not run dry within n years can be expressed approximately as $2\int_0^{x/\sqrt{n}} e^{-\frac{1}{2}t^2}/\sqrt{(2\pi)}\,dt$, and it is easy to verify that this distribution has the same moment ratios as the limiting values found by Anis.

Finally, we come to Nicholas Bernoulli's general solution of the Duration of Play. If for any value of t either line of (4) is arranged in descending powers of p, it will be found to be the sum of multiples of two binomial expressions in the same way as (3)—see also Fieller (1931, equation 10.1), who proceeds to obtain the exact solution of the problem in a convenient form as a series of multiples of incomplete beta-functions, and also provides a rigorous proof.

The application of (6) and the normal approximation to the binomial puts the solution in the simple approximate form $\Sigma A_i \int_{a_i}^{\infty} \frac{1}{\sqrt{(2\pi)}} e^{-\frac{1}{2}x^2}\,dx$; this series agrees with the (exact) result given by Bartlett (1946, equation 17) for the diffusion process. In view of the usefulness of this series it is worth repeating here for completeness

$$P_{b,n} \simeq F(b) - w(-a)\,F(b+2a) + w(-a-b)\,F(3b+2a) - w(-2a-b)\,F(3b+4a) + \ldots, \tag{7}$$

where

$$F(\lambda) \equiv Q\left(\frac{\lambda}{\sigma\sqrt{n}} - \frac{\mu\sqrt{n}}{\sigma}\right) + w(\lambda)\,Q\left(\frac{\lambda}{\sigma\sqrt{n}} + \frac{\mu\sqrt{n}}{\sigma}\right),$$

$$Q(\lambda) \equiv \int_\lambda^\infty \frac{1}{\sqrt{(2\pi)}} e^{-\frac{1}{2}x^2}\,dx,$$

$$w(\lambda) \equiv \exp(2\lambda\mu/\sigma^2).$$

The corresponding series for $P_{a,n}$ is found by interchanging a with b and changing the sign of μ in the definitions of F and w.

It will be found that (7) converges rapidly over the range of n where the process is likely to terminate, and so (as suggested by Bartlett) provides a rapid approximation for the probability that a particle starting at the origin, with a jump distribution having mean μ and variance σ^2, will reach $x = b$ (without having previously been absorbed at $x = -a$) within n jumps. It can similarly be used to find the chance that a linear sequential test will end within n trials, or that a finite reservoir with random net input will either dry up or overflow within a given time.

REFERENCES

ANIS, A. A. (1956). *Biometrika*, **43**, 79.
BARTLETT, M. S. (1946). *Proc. Camb. Phil. Soc.* **42**, 239.
DE MOIVRE, A. (1711). De Mensura Sortis. *Phil. Trans.* **27**, 213.
DE MOIVRE, A. (1718). *The Doctrine of Chances*, 1st ed. London.
DE MOIVRE, A. (1738). *The Doctrine of Chances*, 2nd ed. London.
DE MOIVRE, A. (1756). *The Doctrine of Chances*, 3rd ed. London.
DE MONTMORT, P. R. (1713). *Essai d'Analyse sur les Jeux de Hazard*, 2nd ed. Paris.
FELLER, W. (1950). *An Introduction to Probability Theory and its Applications.* New York: Wiley.
FIELLER, E. C. (1931). *Biometrika*, **22**, 377.
PAGE, E. S. (1954). *J. R. Statist. Soc.* B, **16**, 136.
TODHUNTER, I. (1865). *History of the Theory of Probability.* Cambridge and London; Macmillan.

THOMAS BAYES'S ESSAY TOWARDS SOLVING A PROBLEM IN THE DOCTRINE OF CHANCES

THOMAS BAYES—A BIOGRAPHICAL NOTE

BY G. A. BARNARD

Bayes's paper, reproduced in the following pages, must rank as one of the most famous memoirs in the history of science and the problem it discusses is still the subject of keen controversy. The intellectual stature of Bayes himself is measured by the fact that it is still of scientific as well as historical interest to know what Bayes had to say on the questions he raised. And yet such are the vagaries of historical records, that almost nothing is known about the personal history of the man. *The Dictionary of National Biography*, compiled at the end of the last century, when the whole theory of probability was in temporary eclipse in England, has an entry devoted to Bayes's father, Joshua Bayes, F.R.S., one of the first six Nonconformist ministers to be publicly ordained as such in England, but it has nothing on his much more distinguished son. Indeed, the note on Thomas Bayes which is to appear in the forthcoming new edition of the *Encyclopedia Britannica* will apparently be the first biographical note on Bayes to appear in a work of general reference since the *Imperial Dictionary of Universal Biography* was published in Glasgow in 1865. And in treatises on the history of mathematics, such as that of Loria (1933) and Cantor (1908), notice is taken of his contributions to probability theory and to mathematical analysis, but biographical details are lacking.

The Reverend Thomas Bayes, F.R.S., author of the first expression in precise, quantitative form of one of the modes of inductive inference, was born in 1702, the eldest son of Ann Bayes and Joshua Bayes, F.R.S. He was educated privately, as was usual with Nonconformists at that time, and from the fact that when Thomas was 12 Bernoulli wrote to Leibniz that 'poor de Moivre' was having to earn a living in London by teaching mathematics, we are tempted to speculate that Bayes may have learned mathematics from one of the founders of the theory of probability. Eventually Thomas was ordained, and began his ministry by helping his father, who was at the time stated, minister of the Presbyterian meeting house in Leather Lane, off Holborn. Later the son went to minister in Tunbridge Wells at the Presbyterian Chapel on Little Mount Sion which had been opened on 1 August 1720. It is not known when Bayes went to Tunbridge Wells, but he was not the first to minister on Little Mount Sion, and he was certainly there in 1731, when he produced a tract entitled 'Divine Benevolence, or an attempt to prove that the Principle End of the Divine

Providence and Government is the happiness of His Creatures'. The tract was published by John Noon and copies are in Dr Williams's library and the British Museum. The following is a quotation:

[p. 22]: I don't find (I am sorry to say it) any necessary connection between mere intelligence, though ever so great, and the love or approbation of kind and beneficent actions.

Bayes argued that the principal end of the Deity was the happiness of His creatures, in opposition to Balguy and Grove who had, respectively, maintained that the first spring of action of the Deity was Rectitude, and Wisdom.

In 1736 John Noon published a tract entitled 'An Introduction to the Doctrine of Fluxions, and a Defence of the Mathematicians against the objections of the Author of the Analyst'. De Morgan (1860) says: 'This very acute tract is anonymous, but it was always attributed to Bayes by the contemporaries who write in the names of the authors as I have seen in various copies, and it bears his name in other places.' The ascription to Bayes is accepted also in the British Museum catalogue.

From the copy in Dr Williams's library we quote:

[p. 9]: It is not the business of the Mathematician to dispute whether quantities do in fact ever vary in the manner that is supposed, but only whether the notion of their doing so be intelligible; which being allowed, he has a right to take it for granted, and then see what deductions he can make from that supposition. It is not the business of a Mathematician to show that a strait line or circle can be drawn, but he tells you what he means by these; and if you understand him, you may proceed further with him; and it would not be to the purpose to object that there is no such thing in nature as a true strait line or perfect circle, for this is none of his concern: he is not inquiring how things are in matter of fact, but supposing things to be in a certain way, what are the consequences to be deduced from them; and all that is to be demanded of him is, that his suppositions be intelligible, and his inferences just from the suppositions he makes.

[p. 48]: He [i.e. the Analyst = Bishop Berkeley] represents the disputes and controversies among mathematicians as disparaging the evidence of their methods: and, Query 51, he represents Logics and Metaphysics as proper to open their eyes, and extricate them from their difficulties. Now were ever two things thus put together? If the disputes of the professors of any science disparage the science itself, Logics and Metaphysics are much more to be disparaged than Mathematics; why, therefore, if I am half blind, must I take for my guide one that can't see at all?

[p. 50]: So far as Mathematics do not tend to make men more sober and rational thinkers, wiser and better men, they are only to be considered as an amusement, which ought not to take us off from serious business.

This tract may have had something to do with Bayes's election, in 1742, to Fellowship of the Royal Society, for which his sponsors were Earl Stanhope, Martin Folkes, James Burrow, Cromwell Mortimer, and John Eames.

William Whiston, Newton's successor in the Lucasian Chair at Cambridge, who was expelled from the University for Arianism, notes in his Memoirs (p. 390) that 'on August the 24th this year 1746, being Lord's Day, and St. Bartholomew's Day, I breakfasted at Mr Bay's, a dissenting Minister at Tunbridge Wells, and a Successor, though not immediate, to Mr Humphrey Ditton, and like him a very good mathematician also'. Whiston goes on to relate what he said to Bayes, but he gives no indication that Bayes made reply.

According to Strange (1949) Bayes wished to retire from his ministry as early as 1749, when he allowed a group of Independents to bring ministers from London to take services in his chapel week by week, except for Easter, 1750, when he refused his pulpit to one of these preachers; and in 1752 he was succeeded in his ministry by the Rev. William Johnston, A.M., who inherited Bayes's valuable library. Bayes continued to live in Tunbridge Wells until his death on 17 April 1761. His body was taken to be buried, with that of his father, mother, brothers and sisters, in the Bayes and Cotton family vault in Bunhill Fields, the Nonconformist burial ground by Moorgate. This cemetery also contains the grave of Bayes's friend, the Unitarian Rev. Richard Price, author of the *Northampton Life Table* and object of Burke's oratory and invective in *Reflections on the French Revolution*, and the graves of John Bunyan, Samuel Watts, Daniel Defoe, and many other famous men.

Bayes's will, executed on 12 December 1760, shows him to have been a man of substance. The bulk of his estate was divided among his brothers, sisters, nephews and cousins, but he left £200 equally between 'John Boyl late preacher at Newington and now at Norwich, and Richard Price now I suppose preacher at Newington Green'. He also left 'To Sarah Jeffery daughter of John Jeffery, living with her father at the corner of Fountains Lane near Tonbridge Wells, £500, and my watch made by Elliott and all my linen and wearing apparell and household stuff.'

Apart from the tracts already noted, and the celebrated Essay reproduced here, Bayes wrote a letter on Asymptotic Series to John Canton, published in the *Philosophical Transactions of the Royal Society* (1763, pp. 269–271). His mathematical work, though small in quantity, is of the very highest quality; both his tract on fluxions and his paper on asymptotic series contain thoughts which did not receive as clear expression again until almost a century had elapsed.

Since copies of the volume in which Bayes's essay first appeared are not rare, and copies of a photographic reprint issued by the Department of Agriculture, Washington, D.C., U.S.A., are fairly widely dispersed, the view has been taken that in preparing Bayes's paper for publication here some editing is permissible. In particular, the notation has been modernized, some of the archaisms have been removed and what seem to be obvious printer's errors have been corrected. Sometimes, when a word has been omitted in the original, a suggestion has been supplied, enclosed in square brackets. Otherwise, however, nothing has been changed, and we hope that while the present text should in no sense be regarded as definitive, it will be easier to read on that account. All the work of preparing the text for the printer was most painstakingly and expertly carried out by Mr M. Gilbert, B.Sc., A.R.C.S. Thanks are also due to the Royal Society for permission to reproduce the Essay in its present form.

In writing the biographical notes the present author has had the friendly help of many persons, including especially Dr A. Fletcher and Mr R. L. Plackett, of the University of Liverpool, Mr J. F. C. Willder, of the Department of Pathology, Guy's Hospital Medical School, and Mr M. E. Ogborn, F.I.A., of the Equitable Life Assurance Society. He would also like to thank Sir Ronald Fisher, for some initial prodding which set him moving, and Prof. E. S. Pearson, for patient encouragement to see the matter through to completion.

REFERENCES

Anderson J. G. (1941). *Mathematical Gazette*, **25**, 160–2.
Cantor, M. (1908). *Geschichte der Mathematik*, vol. iv. (Article by Netto.)
De Morgan, A. (1860). *Notes and Queries*, 7 Jan. 1860.
Loria, G. (1933). *Storia delle Matematiche*, vol. iii. Turin.
Mackenzie, M. (Ed.) (1865). *Imperial Dictionary of Universal Biography*, 3 vols. Glasgow.
Strange, C. H. (1949). *Nonconformity in Tunbridge Wells*. Tunbridge Wells.
The Gentleman's Magazine (1761). **31**, 188.
Notes and Queries (1941). 19 April.

[Since this biographical note was written, Mr O. B. Sheynin has suggested that reference should be made to a second contribution from Price, "Supplement to the Essay on a Problem in the Doctrine of Chances" (*Phil. Trans.* 1765, **54**, 296–335). This is concerned with improving approximations made in the main Essay. Ed.]

AN ESSAY TOWARDS SOLVING A PROBLEM IN THE DOCTRINE OF CHANCES

By the late Rev. Mr BAYES, F.R.S.

Communicated by Mr Price, in a Letter to John Canton, A.M., F.R.S.

Read 23 *December* 1763

Dear Sir,

I now send you an essay which I have found among the papers of our deceased friend Mr Bayes, and which, in my opinion, has great merit, and well deserves to be preserved. Experimental philosophy, you will find, is nearly interested in the subject of it; and on this account there seems to be particular reason for thinking that a communication of it to the Royal Society cannot be improper.

He had, you know, the honour of being a member of that illustrious Society, and was much esteemed by many in it as a very able mathematician. In an introduction which he has writ to this Essay, he says, that his design at first in thinking on the subject of it was, to find out a method by which we might judge concerning the probability that an event has to happen, in given circumstances, upon supposition that we know nothing concerning it but that, under the same circumstances, it has happened a certain number of times, and failed a certain other number of times. He adds, that he soon perceived that it would not be very difficult to do this, provided some rule could be found according to which we ought to estimate the chance that the probability for the happening of an event perfectly unknown, should lie between any two named degrees of probability, antecedently to any experiments made about it; and that it appeared to him that the rule must be to suppose the chance the same that it should lie between any two equidifferent degrees; which, if it were allowed, all the rest might be easily calculated in the common method of proceeding in the doctrine of chances. Accordingly, I find among his papers a very ingenious solution of this problem in this way. But he afterwards considered, that the *postulate* on which he had argued might not perhaps be looked upon by all as reasonable; and therefore he chose to lay down in another form the proposition in which he thought the solution of the problem is contained, and in a *scholium* to subjoin the reasons why he thought so, rather than to take into his mathematical reasoning any thing that might admit dispute. This, you will observe, is the method which he has pursued in this essay.

Every judicious person will be sensible that the problem now mentioned is by no means merely a curious speculation in the doctrine of chances, but necessary to be solved in order to [provide] a sure foundation for all our reasonings concerning past facts, and what is likely to be hereafter. Common sense is indeed sufficient to shew us that, from the observation of what has in former instances been the consequence of a certain cause or action, one may make a judgment what is likely to be the consequence of it another time, and that the larger [the] number of experiments we have to support a conclusion, so much the more reason we have to take it for granted. But it is certain that we cannot determine, at least not to any nicety, in what degree repeated experiments confirm a conclusion, without the particular discussion of the beforementioned problem; which, therefore, is necessary to be considered by any one who would give a clear account of the strength of *analogical* or *inductive reasoning*; concerning, which at present, we seem to know little more than that it does sometimes in fact convince us, and at other times not; and that, as it is the means of [a]cquainting us with many truths, of which otherwise we must have been ignorant; so it is, in all probability, the source of many errors, which perhaps might in some measure be avoided, if the force that this sort of reasoning ought to have with us were more distinctly and clearly understood.

These observations prove that the problem enquired after in this essay is no less important than it is curious. It may be safely added, I fancy, that it is also a problem that has never before been solved. Mr De Moivre, indeed, the great improver of this part of mathematics, has in his *Laws of Chance*,* after Bernoulli, and to a greater degree of exactness, given rules to find the probability there is, that if a very great number of trials be made concerning any event, the proportion of the number of times it will happen, to the number of times it will fail in those trials, should differ less than by small assigned limits from the proportion of the probability of its happening to the probability of its failing in one single trial. But I know of no person who has shewn how to deduce the solution of the converse problem to this; namely, 'the number of times an unknown event has happened and failed being given, to find the chance that the probability of its happening should lie somewhere between any two named degrees of probability.' What Mr De Moivre has done therefore cannot be thought sufficient to make the consideration of this point unnecessary: especially, as the rules he has given are not pretended to be rigorously exact, except on supposition that the number of trials made are infinite; from whence it is not obvious how large the number of trials must be in order to make them exact enough to be depended on in practice.

Mr De Moivre calls the problem he has thus solved, the hardest that can be proposed on the subject of chance. His solution he has applied to a very important purpose, and thereby shewn that those are much mistaken who have insinuated that the Doctrine of Chances in mathematics is of trivial consequence, and cannot have a place in any serious enquiry.† The purpose I mean is, to shew what reason we have for believing that there are in the constitution of things fixt laws according to which events happen, and that, therefore, the frame of the world must be the effect of the wisdom and power of an intelligent cause; and thus to confirm the argument taken from final causes for the existence of the Deity. It will be easy to see that the converse problem solved in this essay is more directly applicable to this purpose; for it shews us, with distinctness and precision, in every case of any particular order or recurrency of events, what reason there is to think that such recurrency or order is derived from stable causes or regulations in nature, and not from any of the irregularities of chance.

The two last rules in this essay are given without the deductions of them. I have chosen to do this because these deductions, taking up a good deal of room, would swell the essay too much; and also because these rules, though of considerable use, do not answer the purpose for which they are given as perfectly as could be wished. They are however ready to be produced, if a communication of them should be thought proper. I have in some places writ short notes, and to the whole I have added an application of the rules in the essay to some

* See Mr De Moivre's *Doctrine of Chances*, p. 243, etc. He has omitted the demonstrations of his rules, but these have been since supplied by Mr Simpson at the conclusion of his treatise on *The Nature and Laws of Chance*.

† See his *Doctrine of Chances*, p. 252, etc.

particular cases, in order to convey a clearer idea of the nature of the problem, and to shew how far the solution of it has been carried.

I am sensible that your time is so much taken up that I cannot reasonably expect that you should minutely examine every part of what I now send you. Some of the calculations, particularly in the Appendix, no one can make without a good deal of labour. I have taken so much care about them, that I believe there can be no material error in any of them; but should there be any such errors, I am the only person who ought to be considered as answerable for them.

Mr Bayes has thought fit to begin his work with a brief demonstration of the general laws of chance. His reason for doing this, as he says in his introduction, was not merely that his reader might not have the trouble of searching elsewhere for the principles on which he has argued, but because he did not know whither to refer him for a clear demonstration of them. He has also made an apology for the peculiar definition he has given of the word *chance* or *probability*. His design herein was to cut off all dispute about the meaning of the word, which in common language is used in different senses by persons of different opinions, and according as it is applied to *past* or *future* facts. But whatever different senses it may have, all (he observes) will allow that an expectation depending on the truth of any *past* fact, or the happening of any *future* event, ought to be estimated so much the more valuable as the fact is more likely to be true, or the event more likely to happen. Instead therefore, of the proper sense of the word *probability*, he has given that which all will allow to be its proper measure in every case where the word is used. But it is time to conclude this letter. Experimental philosophy is indebted to you for several discoveries and improvements; and, therefore, I cannot help thinking that there is a peculiar propriety in directing to you the following essay and appendix. That your enquiries may be rewarded with many further successes, and that you may enjoy every valuable blessing, is the sincere wish of, Sir,

your very humble servant,

Richard Price

Newington-Green,
10 *November* 1763

PROBLEM

Given the number of times in which an unknown event has happened and failed: *Required* the chance that the probability of its happening in a single trial lies somewhere between any two degrees of probability that can be named.

SECTION I

DEFINITION 1. Several events are *inconsistent*, when if one of them happens, none of the rest can.

2. Two events are *contrary* when one, or other of them must; and both together cannot happen.

3. An event is said to *fail*, when it cannot happen; or, which comes to the same thing, when its contrary has happened.

4. An event is said to be determined when it has either happened or failed.

5. The *probability of any event* is the ratio between the value at which an expectation depending on the happening of the event ought to be computed, and the value of the thing expected upon it's happening.

6. By *chance* I mean the same as probability.

7. Events are independent when the happening of any one of them does neither increase nor abate the probability of the rest.

Prop. 1

When several events are inconsistent the probability of the happening of one or other of them is the sum of the probabilities of each of them.

Suppose there be three such events, and whichever of them happens I am to receive N, and that the probability of the 1st, 2nd, and 3rd are respectively a/N, b/N, c/N. Then (by the definition of probability) the value of my expectation from the 1st will be a, from the 2nd b, and from the 3rd c. Wherefore the value of my expectations from all three will be $a+b+c$. But the sum of my expectations from all three is in this case an expectation of receiving N upon the happening of one or other of them. Wherefore (by definition 5) the probability of one or other of them is $(a+b+c)/N$ or $a/N+b/N+c/N$. The sum of the probabilities of each of them.

COROLLARY. If it be certain that one or other of the three events must happen, then $a+b+c = N$. For in this case all the expectations together amounting to a certain expectation of receiving N, their values together must be equal to N. And from hence it is plain that the probability of an event added to the probability of its failure (or of its contrary) is the ratio of equality. For these are two inconsistent events, one of which necessarily happens. Wherefore if the probability of an event is P/N that of it's failure will be $(N-P)/N$.

Prop. 2

If a person has an expectation depending on the happening of an event, the probability of the event is to the probability of its failure as his loss if it fails to his gain if it happens.

Suppose a person has an expectation of receiving N, depending on an event the probability of which is P/N. Then (by definition 5) the value of his expectation is P, and therefore if the event fail, he loses that which in value is P; and if it happens he receives N, but his expectation ceases. His gain therefore is $N-P$. Likewise since the probability of the event is P/N, that of its failure (by corollary prop. 1) is $(N-P)/N$. But P/N is to $(N-P)/N$ as P is to $N-P$, i.e. the probability of the event is to the probability of it's failure, as his loss if it fails to his gain if it happens.

Prop. 3

The probability that two subsequent events will both happen is a ratio compounded of the probability of the 1st, and the probability of the 2nd on supposition the 1st happens.

Suppose that, if both events happen, I am to receive N, that the probability both will happen is P/N, that the 1st will is a/N (and consequently that the 1st will not is $(N-a)/N$) and that the 2nd will happen upon supposition the 1st does is b/N. Then (by definition 5) P will be the value of my expectation, which will become b if the 1st happens. Consequently if the 1st happens, my gain by it is $b-P$, and if it fails my loss is P. Wherefore, by the foregoing proposition, a/N is to $(N-a)/N$, i.e. a is to $N-a$ as P is to $b-P$. Wherefore (*componendo inverse*) a is to N as P is to b. But the ratio of P to N is compounded of the ratio of P to b, and that of b to N. Wherefore the same ratio of P to N is compounded of the ratio of a to N and that of b to N, i.e. the probability that the two subsequent events will both happen is compounded of the probability of the 1st and the probability of the 2nd on supposition the 1st happens.

Corollary. Hence if of two subsequent events the probability of the 1st be a/N, and the probability of both together be P/N, then the probability of the 2nd on supposition the 1st happens is P/a.

Prop. 4

If there be two subsequent events to be determined every day, and each day the probability of the 2nd is b/N and the probability of both P/N, and I am to receive N if both the events happen the first day on which the 2nd does; I say, according to these conditions, the probability of my obtaining N is P/b. For if not, let the probability of my obtaining N be x/N and let y be to x as $N-b$ to N. Then since x/N is the probability of my obtaining N (by definition 1) x is the value of my expectation. And again, because according to the foregoing conditions the first day I have an expectation of obtaining N depending on the happening of both the events together, the probability of which is P/N, the value of this expectation is P. Likewise, if this coincident should not happen I have an expectation of being reinstated in my former circumstances, i.e. of receiving that which in value is x depending on the failure of the 2nd event the probability of which (by cor. prop. 1) is $(N-b)/N$ or y/x, because y is to x as $N-b$ to N. Wherefore since x is the thing expected and y/x the probability of obtaining it, the value of this expectation is y. But these two last expectations together are evidently the same with my original expectation, the value of which is x, and therefore $P+y = x$. But y is to x as $N-b$ is to N. Wherefore x is to P as N is to b, and x/N (the probability of my obtaining N) is P/b.

Cor. Suppose after the expectation given me in the foregoing proposition, and before it is at all known whether the 1st event has happened or not, I should find that the 2nd event has happened; from hence I can only infer that the event is determined on which my expectation depended, and have no reason to esteem the value of my expectation either greater or less than it was before. For if I have reason to think it less, it would be reasonable for me to give something to be reinstated in my former circumstances, and this over and over again as often as I should be informed that the 2nd event had happened, which is evidently absurd. And the like absurdity plainly follows if you say I ought to set a greater value on my expectation than before, for then it would be reasonable for me to refuse something if offered me upon condition I would relinquish it, and be reinstated in my former circumstances; and this likewise over and over again as often as (nothing being known concerning the 1st event) it should appear that the 2nd had happened. Notwithstanding therefore this discovery that the 2nd event has happened, my expectation ought to be esteemed the same in value as before, i.e. x, and consequently the probability of my obtaining N is (by definition 5) still x/N or P/b.* But after this discovery the probability of my obtaining N is the probability that the 1st of two subsequent events has happened upon the supposition that the 2nd has, whose probabilities were as before specified. But the probability that an event has happened is the same as the probability I have to guess right if I guess it has happened. Wherefore the following proposition is evident.

* What is here said may perhaps be a little illustrated by considering that all that can be lost by the happening of the 2nd event is the chance I should have had of being reinstated in my former circumstances, if the event on which my expectation depended had been determined in the manner expressed in the proposition. But this chance is always as much *against* me as it is *for* me. If the 1st event happens, it is *against* me, and equal to the chance for the 2nd event's failing. If the 1st event does not happen, it is *for* me, and equal also to the chance for the 2nd event's failing. The loss of it, therefore, can be no disadvantage.

Prop. 5

If there be two subsequent events, the probability of the 2nd b/N and the probability of both together P/N, and it being first discovered that the 2nd event has happened, from hence I guess that the 1st event has also happened, the probability I am in the right is P/b.*

Prop. 6

The probability that several independent events shall all happen is a ratio compounded of the probabilities of each.

For from the nature of independent events, the probability that any one happens is not altered by the happening or failing of any of the rest, and consequently the probability that the 2nd event happens on supposition the 1st does is the same with its original probability; but the probability that any two events happen is a ratio compounded of the probability of the 1st event, and the probability of the 2nd on supposition the 1st happens by prop. 3. Wherefore the probability that any two independent events both happen is a ratio compounded of the probability of the 1st and the probability of the 2nd. And in like manner considering the 1st and 2nd events together as one event; the probability that three independent events all happen is a ratio compounded of the probability that the two 1st both happen and the probability of the 3rd. And thus you may proceed if there be ever so many such events; from whence the proposition is manifest.

Cor. 1. If there be several independent events, the probability that the 1st happens the 2nd fails, the 3rd fails and the 4th happens, etc. is a ratio compounded of the probability of the 1st, and the probability of the failure of the 2nd, and the probability of the failure of the 3rd, and the probability of the 4th, etc. For the failure of an event may always be considered as the happening of its contrary.

Cor. 2. If there be several independent events, and the probability of each one be a, and that of its failing be b, the probability that the 1st happens and the 2nd fails, and the 3rd fails and the 4th happens, etc. will be $abba$, etc. For, according to the algebraic way of notation, if a denote any ratio and b another, $abba$ denotes the ratio compounded of the ratios a, b, b, a. This corollary therefore is only a particular case of the foregoing.

Definition. If in consequence of certain data there arises a probability that a certain event should happen, its happening or failing, in consequence of these data, I call it's happening or failing in the 1st trial. And if the same data be again repeated, the happening or failing of the event in consequence of them I call its happening or failing in the 2nd trial; and so on as often as the same data are repeated. And hence it is manifest that the happening or failing of the same event in so many diffe[rent] trials, is in reality the happening or failing of so many distinct independent events exactly similar to each other.

* What is proved by Mr Bayes in this and the preceding proposition is the same with the answer to the following question. What is the probability that a certain event, when it happens, will be accompanied with another to be determined at the same time? In this case, as one of the events is given, nothing can be due for the expectation of it; and, consequently, the value of an expectation depending on the happening of both events must be the same with the value of an expectation depending on the happening of one of them. In other words; the probability that, when one of two events happens, the other will, is the same with the probability of this other. Call x then the probability of this other, and if b/N be the probability of the given event, and p/N the probability of both, because $p/N = (b/N) \times x$, $x = p/b$ = the probability mentioned in these propositions.

Prop. 7

If the probability of an event be a, and that of its failure be b in each single trial, the probability of its happening p times, and failing q times in $p+q$ trials is Ea^pb^q if E be the coefficient of the term in which occurs a^pb^q when the binomial $(a+b)^{p+q}$ is expanded.

For the happening or failing of an event in different trials are so many independent events. Wherefore (by cor. 2 prop. 6) the probability that the event happens the 1st trial, fails the 2nd and 3rd, and happens the 4th, fails the 5th, etc. (thus happening and failing till the number of times it happens be p and the number it fails be q) is $abbab$ etc. till the number of a's be p and the number of b's be q, that is; 'tis a^pb^q. In like manner if you consider the event as happening p times and failing q times in any other particular order, the probability for it is a^pb^q; but the number of different orders according to which an event may happen or fail, so as in all to happen p times and fail q, in $p+q$ trials is equal to the number of permutations that $aaaa$ bbb admit of when the number of a's is p, and the number of b's is q. And this number is equal to E, the coefficient of the term in which occurs a^pb^q when $(a+b)^{p+q}$ is expanded. The event therefore may happen p times and fail q in $p+q$ trials E different ways and no more, and its happening and failing these several different ways are so many inconsistent events, the probability for each of which is a^pb^q, and therefore by prop. 1 the probability that some way or other it happens p times and fails q times in $p+q$ trials is Ea^pb^q.

Section II

Postulate. 1. I suppose the square table or plane $ABCD$ to be so made and levelled, that if either of the balls o or W be thrown upon it, there shall be the same probability that it rests upon any one equal part of the plane as another, and that it must necessarily rest somewhere upon it.

2. I suppose that the ball W shall be first thrown, and through the point where it rests a line os shall be drawn parallel to AD, and meeting CD and AB in s and o; and that afterwards the ball O shall be thrown $p+q$ or n times, and that its resting between AD and os after a single throw be called the happening of the event M in a single trial. These things supposed:

Lem. 1. The probability that the point o will fall between any two points in the line AB is the ratio of the distance between the two points to the whole line AB.

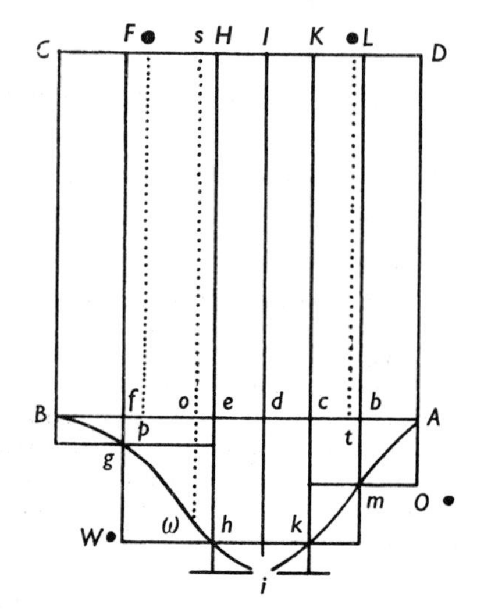

Let any two points be named, as f and b in the line AB, and through them parallel to AD draw fF, bL meeting CD in F and L. Then if the rectangles Cf, Fb, LA are commensurable to each other, they may each be divided into the same equal parts, which being done, and the ball W thrown, the probability it will rest somewhere upon any number of these equal parts will be the sum of the probabilities it has to rest upon each one of them, because its resting upon any different parts of the plane AC are so many inconsistent events; and this sum, because the probability it should rest upon any one equal part as another is the same, is the probability it should rest upon any one equal part multiplied by the number of parts. Consequently, the probability there is that the ball W should rest somewhere upon Fb is the probability it has to rest upon one equal part multiplied by the number of equal parts in Fb; and the probability it rests somewhere upon Cf or LA, i.e. that it does not rest upon Fb (because it must rest somewhere upon AC) is the probability it rests upon one equal part multiplied by the number of equal parts in Cf, LA taken together. Wherefore, the probability it rests upon Fb is to the probability it does not as the number of equal parts in Fb is to the number of equal parts in Cf, LA together, or as Fb to Cf, LA together, or as fb to Bf, Ab together. Wherefore the probability it rests upon Fb is to the probability it does not as fb to Bf, Ab together. And (*componendo inverse*) the probability it rests upon Fb is to the probability it rests upon Fb added to the probability it does not, as fb to AB, or as the ratio of fb to AB to the ratio of AB to AB. But the probability of any event added to the probability of its failure is the ratio of equality; wherefore, the probability it rests upon Fb is to the ratio of equality as the ratio of fb to AB to the ratio of AB to AB, or the ratio of equality; and therefore the probability it rests upon Fb is the ratio of fb to AB. But *ex hypothesi* according as the ball W falls upon Fb or not the point o will lie between f and b or not, and therefore the probability the point o will lie between f and b is the ratio of fb to AB.

Again; if the rectangles Cf, Fb, LA are not commensurable, yet the last mentioned probability can be neither greater nor less than the ratio of fb to AB; for, if it be less, let it be the ratio of fc to AB, and upon the line fb take the points p and t, so that pt shall be greater than fc, and the three lines Bp, pt, tA commensurable (which it is evident may be always done by dividing AB into equal parts less than half cb, and taking p and t the nearest points of division to f and c that lie upon fb). Then because Bp, pt, tA are commensurable, so are the rectangles Cp, Dt, and that upon pt compleating the square AB. Wherefore, by what has been said, the probability that the point o will lie between p and t is the ratio of pt to AB. But if it lies between p and t it must lie between f and b. Wherefore, the probability it should lie between f and b cannot be less than the ratio of pt to AB, and therefore must be greater than the ratio of fc to AB (since pt is greater than fc). And after the same manner you may prove that the forementioned probability cannot be greater than the ratio of fb to AB, it must therefore be the same.

LEM. 2. The ball W having been thrown, and the line os drawn, the probability of the event M in a single trial is the ratio of Ao to AB.

For, in the same manner as in the foregoing lemma, the probability that the ball o being thrown shall rest somewhere upon Do or between AD and so is the ratio of Ao to AB. But the resting of the ball o between AD and so after a single throw is the happening of the event M in a single trial. Wherefore the lemma is manifest.

Prop. 8

If upon BA you erect the figure $BghikmA$ whose property is this, that (the base BA being divided into any two parts, as Ab, and Bb and at the point of division b a perpendicular being erected and terminated by the figure in m; and y, x, r representing respectively the ratio of bm, Ab, and Bb to AB, and E being the coefficient of the term in which occurs a^pb^q when the binomial $(a+b)^{p+q}$ is expanded) $y = Ex^pr^q$. I say that before the ball W is thrown, the probability the point o should fall between f and b, any two points named in the line AB, and withall that the event M should happen p times and fail q in $p+q$ trials, is the ratio of $fghikmb$, the part of the figure $BghikmA$ intercepted between the perpendiculars fg, bm raised upon the line AB, to CA the square upon AB.

DEMONSTRATION

For if not; first let it be the ratio of D a figure greater than $fghikmb$ to CA, and through the points e, d, c draw perpendiculars to fb meeting the curve $AmigB$ in h, i, k; the point d being so placed that di shall be the longest of the perpendiculars terminated by the line fb, and the curve $AmigB$; and the points e, d, c being so many and so placed that the rectangles, bk, ci, ei, fh taken together shall differ less from $fghikmb$ than D does; all which may be easily done by the help of the equation of the curve, and the difference between D and the figure $fghikmb$ given. Then since di is the longest of the perpendicular ordinates that insist upon fb, the rest will gradually decrease as they are farther and farther from it on each side, as appears from the construction of the figure, and consequently eh is greater than gf or any other ordinate that insists upon ef.

Now if Ao were equal to Ae, then by lem. 2 the probability of the event M in a single trial would be the ratio of Ae to AB, and consequently by cor. Prop. 1 the probability of it's failure would be the ratio of Be to AB. Wherefore, if x and r be the two forementioned ratios respectively, by Prop. 7 the probability of the event M happening p times and failing q in $p+q$ trials would be Ex^pr^q. But x and r being respectively the ratios of Ae to AB and Be to AB, if y is the ratio of eh to AB, then, by construction of the figure AiB, $y = Ex^pr^q$. Wherefore, if Ao were equal to Ae the probability of the event M happening p times and failing q in $p+q$ trials would be y, or the ratio of eh to AB. And if Ao were equal to Af, or were any mean between Ae and Af, the last mentioned probability for the same reasons would be the ratio of fg or some other of the ordinates insisting upon ef, to AB. But eh is the greatest of all the ordinates that insist upon ef. Wherefore, upon supposition the point should lie anywhere between f and e, the probability that the event M happens p times and fails q in $p+q$ trials cannot be greater than the ratio of eh to AB. There then being these two subsequent events, the 1st that the point o will lie between e and f, the 2nd that the event M will happen p times and fail q in $p+q$ trials, and the probability of the first (by lemma 1) is the ratio of ef to AB, and upon supposition the 1st happens, by what has been now proved, the probability of the 2nd cannot be greater than the ratio of eh to AB, it evidently follows (from Prop. 3) that the probability both together will happen cannot be greater than the ratio compounded of that of ef to AB and that of eh to AB, which compound ratio is the ratio of fh to CA. Wherefore, the probability that the point o will lie between f and e, and the event M happen p times and fail q, is not greater than the ratio of fh to CA. And in like manner the probability the point o will lie between e and d, and the event M happen and fail as before, cannot be greater than the ratio of ei to CA. And again, the probability the point o will lie between d and c, and the event M happen and fail as before, cannot be greater than the ratio of ci to CA. And lastly, the probability that the point o will lie between c and b, and the event M happen and fail as before, cannot be greater than the ratio of bk to CA. Add now all these several probabilities together, and their sum (by Prop. 1) will be the probability that the point will lie somewhere between f and b, and the event M happen p times and fail q in $p+q$ trials. Add likewise the correspondent ratios together, and their sum will be the ratio of the sum of the antecedents to their common consequent, i.e. the ratio of fh, ei, ci, bk together to CA; which ratio is less than that of D to CA, because D is greater than fh, ei, ci, bk together. And therefore, the probability that the point o will lie between f and b, and withal that the event M will happen p times and fail q in $p+q$ trials, is less than the ratio of D to CA; but it was supposed the same which is absurd. And in like manner, by inscribing rectangles within the figure, as eg, dh, dk, cm, you may prove that the last mentioned probability is *greater* than the ratio of any figure less than $fghikmb$ to CA.

Wherefore, that probability must be the ratio of $fghikmb$ to CA.

Cor. Before the ball W is thrown the probability that the point o will lie somewhere between A and B, or somewhere upon the line AB, and withal that the event M will happen p times, and fail q in $p+q$ trials is the ratio of the whole figure AiB to CA. But it is certain that the point o will lie somewhere upon AB. Wherefore, before the ball W is thrown the probability the event M will happen p times and fail q in $p+q$ trials is the ratio of AiB to CA.

Prop. 9

If before anything is discovered concerning the place of the point o, it should appear that the event M had happened p times and failed q in $p+q$ trials, and from hence I guess that the point o lies between any two points in the line AB, as f and b, and consequently that the probability of the event M in a single trial was somewhere between the ratio of Ab to AB and that of Af to AB: the probability I am in the right is the ratio of that part of the figure AiB described as before which is intercepted between perpendiculars erected upon AB at the points f and b, to the whole figure AiB.

For, there being these two subsequent events, the first that the point o will lie between f and b; the second that the event M should happen p times and fail q in $p+q$ trials; and (by cor. prop. 8) the original probability of the second is the ratio of AiB to CA, and (by prop. 8) the probability of both is the ratio of $fghimb$ to CA; wherefore (by prop. 5) it being first discovered that the second has happened, and from hence I guess that the first has happened also, the probability I am in the right is the ratio of $fghimb$ to AiB, the point which was to be proved.

Cor. The same things supposed, if I guess that the probability of the event M lies somewhere between 0 and the ratio of Ab to AB, my chance to be in the right is the ratio of Abm to AiB.

Scholium

From the preceding proposition it is plain, that in the case of such an event as I there call M, from the number of times it happens and fails in a certain number of trials, without knowing anything more concerning it, one may give a guess whereabouts it's probability is, and, by the usual methods computing the magnitudes of the areas there mentioned, see the chance that the guess is right. And that the same rule is the proper one to be used in the case of an event concerning the probability of which we absolutely know nothing antecedently to any trials made concerning it, seems to appear from the following consideration; viz. that concerning such an event I have no reason to think that, in a certain number of trials, it should rather happen any one possible number of times than another. For, on this account, I may justly reason concerning it as if its probability had been at first unfixed, and then determined in such a manner as to give me no reason to think that, in a certain number of trials, it should rather happen any one possible number of times than another. But this is exactly the case of the event M. For before the ball W is thrown, which determines it's probability in a single trial (by cor. prop. 8), the probability it has to happen p times and fail q in $p+q$ or n trials is the ratio of AiB to CA, which ratio is the same when $p+q$ or n is given, whatever number p is; as will appear by computing the magnitude of AiB by the method of fluxions.* And consequently before the place of the point o is discovered or the number of times the event M has happened in n trials, I can have no reason to think it should rather happen one possible number of times than another.

In what follows therefore I shall take for granted that the rule given concerning the event M in prop. 9 is also the rule to be used in relation to any event concerning the probability of which nothing at all is known antecedently to any trials made or observed concerning it. And such an event I shall call an unknown event.

Cor. Hence, by supposing the ordinates in the figure AiB to be contracted in the ratio of E to one, which makes no alteration in the proportion of the parts of the figure intercepted between them, and applying what is said of the event M to an unknown event, we have the following proposition, which gives the rules for finding the probability of an event from the number of times it actually happens and fails.

Prop. 10

If a figure be described upon any base AH (Vid. Fig.) having for it's equation $y = x^p r^q$; where y, x, r are respectively the ratios of an ordinate of the figure insisting on the base at right angles, of the segment of the base intercepted between the ordinate and A the beginning of the base, and of the other segment of the base lying between the ordinate and the point H, to the base as their common consequent. I say then that if an unknown event has happened p times and failed q in $p+q$ trials, and in the base AH taking any two points as f and t you erect the ordinates fC, tF at right angles with it, the chance that the probability of the event lies somewhere between the ratio of Af to AH and that of At to AH, is the ratio of $tFCf$, that part of the before-described figure which is intercepted between the two ordinates, to $ACFH$ the whole figure insisting on the base AH.

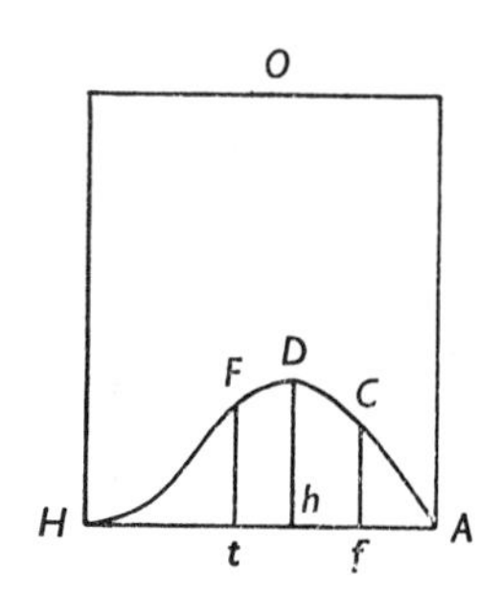

This is evident from prop. 9 and the remarks made in the foregoing scholium and corollary.

Now, in order to reduce the foregoing rule to practice, we must find the value of the area of the figure described and the several parts of it separated, by ordinates perpendicular to its base. For which purpose, suppose $AH = 1$ and HO the square upon AH likewise $= 1$, and Cf will be $= y$, and $Af = x$, and $Hf = r$, because y, x and r denote the ratios of Cf, Af, and Hf respectively to AH. And by the equation of the curve $y = x^p r^q$ and (because $Af+fH = AH$) $r+x = 1$. Wherefore

$$y = x^p(1-x)^q$$
$$= x^p - qx^{p+1} + \frac{q(q-1)\,x^{p+2}}{2} - \frac{q(q-1)\,(q-2)\,x^{p+3}}{2.3} + \text{etc.}$$

Now the abscisse being x and the ordinate x^p the correspondent area is $x^{p+1}/(p+1)$ (by prop. 10, cas. 1, Quadrat. Newt.)† and the ordinate being qx^{p+1} the area is $qx^{p+2}/(p+2)$; and

* It will be proved presently in art. 4 by computing in the method here mentioned that AiB contracted in the ratio of E to 1 is to CA as 1 to $(n+1)E$: from whence it plainly follows that, antecedently to this contraction, AiB must be to CA in the ratio of 1 to $n+1$, which is a constant ratio when n is given, whatever p is.

† 'Tis very evident here, without having recourse to Sir Isaac Newton, that the fluxion of the area ACf being

$$y\dot{x} = x^p\dot{x} - qx^{p+1}\dot{x} + \frac{q(q-1)}{2}x^{p+2}\dot{x} - \text{etc.}$$

the fluent or area itself is

$$\frac{x^{p+1}}{p+1} - \frac{qx^{p+2}}{p+2} + \frac{q(q-1)\,x^{p+3}}{2(p+3)} - \text{etc.}$$

in like manner of the rest. Wherefore, the abscisse being x and the ordinate y or $x^p - qx^{p+1} + \text{etc.}$ the correspondent area is

$$\frac{x^{p+1}}{p+1} - \frac{qx^{p+2}}{p+2} + \frac{q(q-1)\,x^{p+3}}{2(p+3)} - \frac{q(q-1)\,(q-2)\,x^{p+4}}{2.3(p+4)} + \text{etc.}$$

Wherefore, if $x = Af = Af/(AH)$, and $y = Cf = Cf/(AH)$, then

$$ACf = \frac{ACf}{HO} = \frac{x^{p+1}}{p+1} - \frac{qx^{p+2}}{p+2} + \frac{q(q-1)\,x^{p+3}}{2(p+3)} - \text{etc.}$$

From which equation, if q be a small number, it is easy to find the value of the ratio of ACf to HO and in like manner as that was found out, it will appear that the ratio of HCf to HO is

$$\frac{r^{q+1}}{q+1} - \frac{pr^{q+2}}{q+2} + \frac{p(p-1)\,r^{q+3}}{2(q+3)} - \frac{p(p-1)\,(p-2)\,r^{q+4}}{2.3(q+4)} + \text{etc.}$$

which series will consist of few terms and therefore is to be used when p is small.

2. The same things supposed as before, the ratio of ACf to HO is

$$\frac{x^{p+1}r^q}{p+1} + \frac{qx^{p+2}r^{q-1}}{(p+1)\,(p+2)} + \frac{q(q-1)\,x^{p+3}r^{q-2}}{(p+1)\,(p+2)\,(p+3)} + \frac{q(q-1)\,(q-2)\,x^{p+4}r^{q-3}}{(p+1)\,(p+2)\,(p+3)\,(p+4)}$$
$$+ \text{etc.} + \frac{x^{n+1}q(q-1)\ldots 1}{(n+1)\,(p+1)\,(p+2)\ldots n},$$

where $n = p+q$. For this series is the same with $x^{p+1}/(p+1) - qx^{p+2}/(p+2) + \text{etc.}$ set down in Art. 1st as the value of the ratio of ACf to HO; as will easily be seen by putting in the former instead of r its value $1-x$, and expanding the terms and ordering them according to the powers of x. Or, more readily, by comparing the fluxions of the two series, and in the former instead of $\dot{r}$ substituting $-\dot{x}$.*

3. In like manner, the ratio of HCf to HO is

$$\frac{r^{q+1}x^p}{q+1} + \frac{pr^{q+2}x^{p-1}}{(q+1)\,(q+2)} + \frac{p(p-1)\,r^{q+3}x^{p-2}}{(q+1)\,(q+2)\,(q+3)} + \text{etc.}$$

* The fluxion of the first series is

$$x^pr^q\dot{x} + \frac{qx^{p+1}r^{q-1}\dot{r}}{p+1} + \frac{qx^{p+1}r^{q-1}\dot{x}}{p+1} + \frac{q(q-1)\,x^{p+2}r^{q-2}\dot{r}}{(p+1)\,(p+2)} + \frac{q(q-1)\,x^{p+2}r^{q-2}\dot{x}}{(p+1)\,(p+2)} + \frac{q(q-1)\,(q-2)\,x^{p+3}r^{q-3}\dot{r}}{(p+1)\,(p+2)\,(p+3)} + \text{etc.}$$

or, substituting $-\dot{x}$ for $\dot{r}$,

$$x^pr^q\dot{x} - \frac{qx^{p+1}r^{q-1}\dot{x}}{p+1} + \frac{qx^{p+1}r^{q-1}\dot{x}}{p+1} - \frac{q(q-1)\,x^{p+2}r^{q-2}\dot{x}}{(p+1)\,(p+2)} + \frac{q(q-1)\,x^{p+2}r^{q-2}\dot{x}}{(p+1)\,(p+2)} - \text{etc.}$$

which, as all the terms after the first destroy one another, is equal to

$$x^pr^q\dot{x} = x^p(1-x)^q\,\dot{x} = x^p\dot{x}\left[1 - qx + q\frac{(q-1)}{2}x^2 - \text{etc.}\right]$$
$$= x^p\dot{x} - qx^{p+1}\dot{x} + \frac{q\,(q-1)\,x^{p+2}}{2}\dot{x} - \text{etc.}$$
$$= \text{the fluxion of the latter series, or of } \frac{x^{p+1}}{p+1} - \frac{qx^{p+2}}{p+2} + \text{etc.}$$

The two series therefore are the same.

4. If E be the coefficient of that term of the binomial $(a+b)^{p+q}$ expanded in which occurs $a^p b^q$, the ratio of the whole figure $ACFH$ to HO is $\{(n+1)E\}^{-1}$, n being $= p+q$. For, when $Af = AH, x = 1, r = 0$. Wherefore, all the terms of the series set down in Art. 2 as expressing the ratio of ACf to HO will vanish except the last, and that becomes

$$\frac{q(q-1)\ldots 1}{(n+1)(p+1)(p+2)\ldots n}.$$

But E being the coefficient of that term in the binomial $(a+b)^n$ expanded in which occurs $a^p b^q$ is equal to

$$\frac{(p+1)(p+2)\ldots n}{q(q-1)\ldots 1}.$$

And, because Af is supposed to become $= AH$, $ACf = ACH$. From whence this article is plain.

5. The ratio of ACf to the whole figure $ACFH$ is (by Art. 1 and 4)

$$(n+1)E\left[\frac{x^{p+1}}{p+1}-\frac{qx^{p+2}}{p+2}+\frac{q(q-1)x^{p+3}}{2(p+3)}-\text{etc.}\right]$$

and if, as x expresses the ratio of Af to AH, X should express the ratio of At to AH; the ratio of AFt to $ACFH$ would be

$$(n+1)E\left[\frac{X^{p+1}}{p+1}-\frac{qX^{p+2}}{p+2}+\frac{q(q-1)X^{p+3}}{2(p+3)}-\text{etc.}\right]$$

and consequently the ratio of $tFCf$ to $ACFH$ is $(n+1)E$ multiplied into the difference between the two series. Compare this with prop. 10 and we shall have the following practical rule.

Rule 1

If nothing is known concerning an event but that it has happened p times and failed q in $p+q$ or n trials, and from hence I guess that the probability of its happening in a single trial lies somewhere between any two degrees of probability as X and x, the chance I am in the right in my guess is $(n+1)E$ multiplied into the difference between the series

$$\frac{X^{p+1}}{p+1}-\frac{qX^{p+2}}{p+2}+\frac{q(q-1)X^{p+3}}{2(p+3)}-\text{etc.}$$

and the series

$$\frac{x^{p+1}}{p+1}-\frac{qx^{p+2}}{p+2}+\frac{q(q-1)x^{p+3}}{2(p+3)}-\text{etc.}$$

E being the coefficient of $a^p b^q$ when $(a+b)^n$ is expanded.

This is the proper rule to be used when q is a small number; but if q is large and p small, change everywhere in the series here set down p into q and q into p and x into r or $(1-x)$, and X into $R = (1-X)$; which will not make any alteration in the difference between the two series.

Thus far Mr Bayes's essay.

With respect to the rule here given, it is further to be observed, that when both p and q are very large numbers, it will not be possible to apply it to practice on account of the multitude of terms which the series in it will contain. Mr Bayes, therefore, by an investigation which it would be too tedious to give here, has deduced from this rule another, which is as follows.

Rule 2

If nothing is known concerning an event but that it has happened p times and failed q in $p+q$ or n trials, and from hence I guess that the probability of its happening in a single trial lies between $(p/n)+z$ and $(p/n)-z$; if $m^2 = n^3/(pq)$,‡ $a = p/n$, $b = q/n$, E the coefficient of the term in which occurs $a^p b^q$ when $(a+b)^n$ is expanded, and

$$\Sigma = \frac{(n+1)\sqrt{(2pq)}}{n\sqrt{n}} Ea^p b^q$$

multiplied by the series

$$mz - \frac{m^3z^3}{3} + \frac{(n-2)\,m^5z^5}{2n\,.\,5} - \frac{(n-2)\,(n-4)\,m^7z^7}{2n\,.\,3n\,.\,7} + \frac{(n-2)\,(n-4)\,(n-6)\,m^9z^9}{2n\,.\,3n\,.\,4n\,.\,9} - \text{etc.}$$

my chance to be in the right is greater than

$$\frac{2\Sigma}{1+2Ea^pb^q+2Ea^pb^q/n} \; *$$

and less than

$$\frac{2\Sigma}{1-2Ea^pb^q-2Ea^pb^q/n},$$

and if $p = q$ my chance is 2Σ exactly.

In order to render this rule fit for use in all cases it is only necessary to know how to find within sufficient nearness the value of Ea^pb^q and also of the series $mz - \frac{1}{3}m^3z^3 + \text{etc.}$ With respect to the former Mr Bayes has proved that, supposing K to signify the ratio of the quadrantal arc to its radius, Ea^pb^q will be equal to $\frac{1}{2}\sqrt{n}/\sqrt{(Kpq)}$ multiplied by the *ratio*, [h], whose *hyberbolic* logarithm is

$$\frac{1}{12}\left[\frac{1}{n}-\frac{1}{p}-\frac{1}{q}\right] - \frac{1}{360}\left[\frac{1}{n^3}-\frac{1}{p^3}-\frac{1}{q^3}\right] + \frac{1}{1260}\left[\frac{1}{n^5}-\frac{1}{p^5}-\frac{1}{q^5}\right] - \frac{1}{1680}\left[\frac{1}{n^7}-\frac{1}{p^7}-\frac{1}{q^7}\right] + \frac{1}{1188}\left[\frac{1}{n^9}-\frac{1}{p^9}-\frac{1}{q^9}\right] - \text{etc.}\dagger$$

* In Mr Bayes's manuscript this chance is made to be greater than $2\Sigma/(1+2Ea^pb^q)$ and less than $2\Sigma/(1-2Ea^pb^q)$. The third term in the two divisors, as I have given them, being omitted. But this being evidently owing to a small oversight in the deduction of this rule, which I have reason to think Mr Bayes had himself discovered, I have ventured to correct his copy, and to give the rule as I am satisfied it ought to be given.

† A very few terms of this series will generally give the hyperbolic logarithm to a sufficient degree of exactness. A similar series has been given by Mr DeMoivre, Mr Simpson and other eminent mathematicians in an expression for the sum of the logarithms of the numbers 1, 2, 3, 4, 5, to x, which sum they have asserted to be equal to

$$\tfrac{1}{2}\log c + (x+\tfrac{1}{2})\log x - x + \frac{1}{12x} - \frac{1}{360x^3} + \frac{1}{1260x^5} - \text{etc.}$$

c denoting the circumference of a circle whose radius is unity. But Mr Bayes, in a preceding paper in this volume, has demonstrated that, though this expression will very nearly approach to the value of this sum when only a proper number of the first terms is taken, the whole series cannot express any quantity at all, because, let x be what it will, there will always be a part of the series where it will begin to diverge. This observation, though it does not much affect the use of this series, seems well worth the notice of mathematicians.

‡ [This equation is corrected to $m^2 = n^3/(2pq)$ by Price in his 1765 *Supplement*, see p. 133. Ed.]

where the numeral coefficients may be found in the following manner. Call them A, B, C, D, E etc. Then

$$A = \frac{1}{2.2.3} = \frac{1}{3.4}, \quad B = \frac{1}{2.4.5} - \frac{A}{3}, \quad C = \frac{1}{2.6.7} - \frac{10B + A}{5},$$

$$D = \frac{1}{2.8.9} - \frac{35C + 21B + A}{7}, \quad E = \frac{1}{2.10.11} - \frac{126C + 84D + 36B + A}{9},$$

$$F = \frac{1}{2.12.13} - \frac{462D + 330C + 165E + 55B + A}{11} \text{ etc.}$$

where the coefficients of B, C, D, E, F, etc. in the values of D, E, F, etc. are the 2, 3, 4, etc. highest coefficients in $(a+b)^7$, $(a+b)^9$, $(a+b)^{11}$, etc. expanded; affixing in every particular value the least of these coefficients to B, the next in magnitude to the furthest letter from B, the next to C, the next to the furthest but one, the next to D, the next to the furthest but two, and so on.*

With respect to the value of the series

$$mz - \tfrac{1}{3}m^3z^3 + \frac{(n-2)m^5z^5}{2n.5} \text{ etc.}$$

he has observed that it may be calculated directly when mz is less than 1, or even not greater than $\sqrt{3}$: but when mz is much larger it becomes impracticable to do this; in which case he shews a way of easily finding two values of it very nearly equal between which its true value must lie.

The theorem he gives for this purpose is as follows.

Let K, as before, stand for the ratio of the quadrantal arc to its radius, and H for the ratio whose hyperbolic logarithm is

$$\frac{2^2-1}{2n} - \frac{2^4-1}{360n^3} + \frac{2^6-1}{1260n^5} - \frac{2^8-1}{1680n^7} + \text{etc.}$$

Then the series $mz - \frac{1}{3}m^3z^3 + \text{etc.}$ will be greater or less than the series

$$\frac{Hn\sqrt{K}}{(n+1)\sqrt{2}} - \frac{n\left(1-\dfrac{2m^2z^2}{n}\right)^{\frac{1}{2}n+1}}{(n+2)\,2mz} + \frac{n^2\left(1-\dfrac{2m^2z^2}{n}\right)^{\frac{1}{2}n+2}}{(n+2)(n+4)\,4m^3z^3}$$

$$- \frac{3n^3\left(1-\dfrac{2m^2z^2}{n}\right)^{\frac{1}{2}n+3}}{(n+2)(n+4)(n+6)\,8m^5z^5} + \frac{3.5.n^4\left(1-\dfrac{2m^2z^2}{n}\right)^{\frac{1}{2}n+4}}{(n+2)(n+4)(n+6)(n+8)\,16m^7z^7} - \text{etc.}$$

continued to any number of terms, according as the last term has a positive or a negative sign before it.

From substituting these values of Ea^pb^q and

$$mz - \frac{m^3z^3}{3} + \frac{(n-2)}{2n}\frac{m^5z^5}{5} \text{etc.}$$

in the second rule arises a third rule, which is the rule to be used when mz is of some considerable magnitude.

* This method of finding these coefficients I have deduced from the demonstration of the third lemma at the end of Mr Simpson's *Treatise on the Nature and Laws of Chance.*

Rule 3

If nothing is known of an event but that it has happened p times and failed q in $p+q$ or n trials, and from hence I judge that the probability of its happening in a single trial lies between $p/n+z$ and $p/n-z$ my chance to be right is *greater* than

$$\frac{\frac{1}{2}\sqrt{(Kpq)}\,h}{\sqrt{(Kpq)}+hn^{\frac{1}{2}}+hn^{-\frac{1}{2}}}\left\{2H-\frac{\sqrt{2}\,(n+1)\,(1-2m^2z^2/n)^{\frac{1}{2}n+1}}{\sqrt{K}\,(n+2)\,mz}\right\}$$

and *less* than

$$\frac{\frac{1}{2}\sqrt{(Kpq)}\,h}{\sqrt{(Kpq)}-hn^{\frac{1}{2}}-hn^{-\frac{1}{2}}}\left\{2H-\frac{\sqrt{2}\,(n+1)\,(1-2m^2z^2/n)^{\frac{1}{2}n+1}}{\sqrt{K}\,(n+2)\,mz}+\frac{\sqrt{2}\,n\,(n+1)\,(1-2m^2z^2/n)^{\frac{1}{2}n+2}}{\sqrt{K}\,(n+2)\,(n+4)\,2m^3z^3}\right\}$$

where m^2, K, h and H stand for the quantities already explained.

AN APPENDIX

Containing an application of the foregoing Rules to some particular Cases

The first rule gives a direct and perfect solution in all cases; and the two following rules are only particular methods of approximating to the solution given in the first rule, when the labour of applying it becomes too great.

The first rule may be used in all cases where either p or q are nothing or not large. The second rule may be used in all cases where mz is less than $\sqrt{3}$; and the third in all cases where m^2z^2 is greater than 1 and less than $\frac{1}{2}n$, if n is an even number and very large. If n is not large this last rule cannot be much wanted, because, m decreasing continually as n is diminished, the value of z may in this case be taken large, (and therefore a considerable interval had between $p/n-z$ and $p/n+z$), and yet the operation be carried on by the second rule; or mz not exceed $\sqrt{3}$.

But in order to shew distinctly and fully the nature of the present problem, and how far Mr Bayes has carried the solution of it; I shall give the result of this solution in a few cases, beginning with the lowest and most simple.

Let us then first suppose, of such an event as that called M in the essay, or an event about the probability of which, antecedently to trials, we know nothing, that it has happened *once*, and that it is enquired what conclusion we may draw from hence with respect to the probability of it's happening on a *second* trial.

The answer is that there would be an odds of three to one for somewhat more than an even chance that it would happen on a second trial.

For in this case, and in all others where q is nothing, the expression

$$(n+1)\left\{\frac{X^{p+1}}{p+1}-\frac{x^{p+1}}{p+1}\right\} \quad \text{or} \quad X^{p+1}-x^{p+1}$$

gives the solution, as will appear from considering the first rule. Put therefore in this expression $p+1=2$, $X=1$ and $x=\frac{1}{2}$ and it will be $1-(\frac{1}{2})^2$ or $\frac{3}{4}$; which shews the chance there is that the probability of an event that has happened once lies somewhere between 1 and $\frac{1}{2}$; or (which is the same) the odds that it is somewhat more than an even chance that it will happen on a second trial.*

In the same manner it will appear that if the event has happened twice, the odds now mentioned will be seven to one; if thrice, fifteen to one; and in general, if the event has happened p times, there will be an odds of $2^{p+1}-1$ to one, for *more* than an equal chance that it will happen on further trials.

Again, suppose all I know of an event to be that it has happened ten times without failing, and the enquiry to be what reason we shall have to think we are right if we guess that the probability of it's happening in a single trial lies somewhere between $\frac{16}{17}$ and $\frac{2}{3}$, or that the ratio of the causes of it's happening to those of it's failure is some ratio between that of sixteen to one and two to one.

Here $p+1=11$, $X=\frac{16}{17}$ and $x=\frac{2}{3}$ and $X^{p+1}-x^{p+1}=(\frac{16}{17})^{11}-(\frac{2}{3})^{11}=0{\cdot}5013$ etc. The answer therefore is, that we shall have very nearly an equal chance for being right.

* There can, I suppose, be no reason for observing that on this subject unity is always made to stand for certainty, and $\frac{1}{2}$ for an even chance.

In this manner we may determine in any case what conclusion we ought to draw from a given number of experiments which are unopposed by contrary experiments. Every one sees in general that there is reason to expect an event with more or less confidence according to the greater or less number of times in which, under given circumstances, it has happened without failing; but we here see exactly what this reason is, on what principles it is founded, and how we ought to regulate our expectations.

But it will be proper to dwell longer on this head.

Suppose a solid or die of whose number of sides and constitution we know nothing; and that we are to judge of these from experiments made in throwing it.

In this case, it should be observed, that it would be in the highest degree improbable that the solid should, in the first trial, turn any one side which could be assigned beforehand; because it would be known that some side it must turn, and that there was an infinity of other sides, or sides otherwise marked, which it was equally likely that it should turn. The first throw only shews that *it has* the side then thrown, without giving any reason to think that it has it any one number of times rather than any other. It will appear, therefore, that *after* the first throw and not before, we should be in the circumstances required by the conditions of the present problem, and that the whole effect of this throw would be to bring us into these circumstances. That is: the turning the side first thrown in any subsequent single trial would be an event about the probability or improbability of which we could form no judgment, and of which we should know no more than that it lay somewhere between nothing and certainty. With the second trial then our calculations must begin; and if in that trial the supposed solid turns again the same side, there will arise the probability of three to one that it has more of that sort of sides than of *all* others; or (which comes to the same) that there is somewhat in its constitution disposing it to turn that side oftenest: And this probability will increase, in the manner already explained, with the number of times in which that side has been thrown without failing. It should not, however, be imagined that any number of such experiments can give sufficient reason for thinking that it would *never* turn any other side. For, suppose it has turned the same side in every trial a million of times. In these circumstances there would be an improbability that it has *less* than 1,400,000 more of these sides than all others; but there would also be an improbability that it had *above* 1,600,000 times more. The chance for the latter is expressed by 1,600,000/1,600,001 raised to the millioneth power subtracted from unity, which is equal to 0·4647 etc and the chance for the former is equal to 1,400,000/1,400,001 raised to the same power, or to 0·4895; which, being both less than an equal chance, proves what I have said. But though it would be thus improbable that it had *above* 1,600,000 times more or *less* than 1,400,000 times *more* of these sides than of all others, it by no means follows that we have any reason for judging that the true proportion in this case lies somewhere between that of 1,600,000 to one and 1,400,000 to one. For he that will take the pains to make the calculation will find that there is nearly the probability expressed by 0·527, or but little more than an equal chance, that it lies somewhere between that of 600,000 to one and three millions to one. It may deserve to be added, that it is more probable that this proportion lies somewhere between that of 900,000 to 1 and 1,900,000 to 1 than between any other two proportions whose antecedents are to one another as 900,000 to 1,900,000, and consequents unity.

I have made these observations chiefly because they are all strictly applicable to the events and appearances of nature. Antecedently to all experience, it would be improbable as infinite to one, that any particular event, beforehand imagined, should follow the application of any one natural object to another; because there would be an equal chance for any one of an infinity of other events. But if we had once seen any particular effects, as the burning of wood on putting it into fire, or the falling of a stone on detaching it from all contiguous objects, then the conclusions to be drawn from any number of subsequent events of the same kind would be to be determined in the same manner with the conclusions just mentioned relating to the constitution of the solid I have supposed. In other words. The first experiment supposed to be ever made on any natural object would only inform us of one event that may follow a particular change in the circumstances of those objects; but it would not suggest to us any ideas of uniformity in nature, or give us the least reason to apprehend that it was, in that instance or in any other, regular rather than irregular in its operations. But if the same event has followed without interruption in any one or more subsequent experiments, then some degree of uniformity will be observed; reason will be given to expect the same success in further experiments, and the calculations directed by the solution of this problem may be made.

One example here it will not be amiss to give.

Let us imagine to ourselves the case of a person just brought forth into this world, and left to collect from his observation of the order and course of events what powers and causes take place in it. The Sun would, probably, be the first object that would engage his attention; but after losing it the first night he would be entirely ignorant whether he should ever see it again. He would therefore be in the condition of a person making a first experiment about an event entirely unknown to him. But let him see a second appearance or one *return* of the Sun, and an expectation would be raised in him of a second return, and he might know that there was an odds of 3 to 1 for *some* probability of this. This odds would increase, as before represented, with the number of returns to which he was witness. But no finite number of returns would be sufficient to produce absolute or physical certainty. For let it be supposed that he has seen it return at regular and stated intervals a million of times. The conclusions this would warrant would be such as follow. There would be the odds of the millioneth power of 2, to one, that it was likely that it would return again at the end of the usual interval. There would be the probability expressed by 0·5352, that the odds for this was not *greater* than 1,600,000 to 1; and the probability expressed by 0·5105, that it was not less than 1,400,000 to 1.

It should be carefully remembered that these deductions suppose a previous total ignorance of nature. After having observed for some time the course of events it would be found that the operations of nature are in general regular, and that the powers and laws which prevail in it are stable and permanent. The consideration of this will cause one or a few experiments often to produce a much stronger expectation of success in further experiments than would otherwise have been reasonable; just as the frequent observation that things of a sort are disposed together in any place would lead us to conclude, upon discovering there any object of a particular sort, that there are laid up with it many others of the same sort. It is obvious that this, so far from contradicting the foregoing deductions, is only one particular case to which they are to be applied.

What has been said seems sufficient to shew us what conclusions to draw from *uniform* experience. It demonstrates, particularly, that instead of proving that events will *always* happen agreeably to it, there will be always reason against this conclusion. In other words, where the course of nature has been the most constant, we can have only reason to reckon upon a recurrency of events proportioned to the degree of this constancy; but we can have no reason for thinking that there are no causes in nature which will *ever* interfere with the operations of the causes from which this constancy is derived, or no circumstances of the world in which it will fail. And if this is true, supposing our only *data* derived from experience, we shall find additional reason for thinking thus if we apply other principles, or have recourse to such considerations as reason, independently of experience, can suggest.

But I have gone further than I intended here; and it is time to turn our thoughts to another branch of this subject: I mean, to cases where an experiment has sometimes succeeded and sometimes failed.

Here, again, in order to be as plain and explicit as possible, it will be proper to put the following case, which is the easiest and simplest I can think of.

Let us then imagine a person present at the drawing of a lottery, who knows nothing of its scheme or of the proportion of *Blanks* to *Prizes* in it. Let it further be supposed, that he is obliged to infer this from the number of *blanks* he hears drawn compared with the number of *prizes*; and that it is enquired what conclusions in these circumstances he may reasonably make.

Let him first hear *ten* blanks drawn and *one* prize, and let it be enquired what chance he will have for being right if he guesses that the proportion of *blanks* to *prizes* in the lottery lies somewhere between the proportions of 9 to 1 and 11 to 1.

Here taking $X = \frac{11}{12}$, $x = \frac{9}{10}$, $p = 10$, $q = 1$, $n = 11$, $E = 11$, the required chance, according to the first rule, is $(n+1)E$ multiplied by the difference between

$$\left\{\frac{X^{p+1}}{p+1} - \frac{qX^{p+2}}{p+2}\right\} \text{ and } \left\{\frac{x^{p+1}}{p+1} - \frac{qx^{p+2}}{p+2}\right\} = 12.11.\left\{\left[\frac{(\frac{11}{12})^{11}}{11} - \frac{(\frac{11}{12})^{12}}{12}\right] - \left[\frac{(\frac{9}{10})^{11}}{11} - \frac{(\frac{9}{10})^{12}}{12}\right]\right\} = 0{\cdot}07699 \text{ etc.}$$

There would therefore be an odds of about 923 to 76, or nearly 12 to 1 *against* his being right. Had he guessed only in general that there were less than 9 blanks to a prize, there would have been a probability of his being right equal to 0·6589, or the odds of 65 to 34.

Again, suppose that he has heard 20 *blanks* drawn and 2 *prizes*; what chance will he have for being right if he makes the same guess?

Here X and x being the same, we have $n = 22$, $p = 20$, $q = 2$, $E = 231$, and the required chance equal to

$$(n+1)E\left\{\left[\frac{X^{p+1}}{p+1} - \frac{qX^{p+2}}{p+2} + \frac{q(q-1)X^{p+3}}{2(p+3)}\right] - \left[\frac{x^{p+1}}{p+1} - \frac{qx^{p+2}}{p+2} + \frac{q(q-1)x^{p+3}}{2(p+3)}\right]\right\} = 0{\cdot}10843 \text{ etc.}$$

He will, therefore, have a better chance for being right than in the former instance, the odds against him now being 892 to 108 or about 9 to 1. But should he only guess in general, as before, that there were less than 9 blanks to a prize, his chance for being right will be worse; for instead of 0·6589 or an odds of near two to one, it will be 0·584, or an odds of 584 to 415.

Suppose, further, that he has heard 40 *blanks* drawn and 4 *prizes*; what will the before-mentioned chances be?

The answer here is 0·1525, for the former of these chances; and 0·527, for the latter. There will, therefore, now be an odds of only $5\frac{1}{2}$ to 1 against the proportion of blanks to prizes lying between 9 to 1 and 11 to 1; and but little more than an equal chance that it is less than 9 to 1.

Once more. Suppose he has heard 100 *blanks* drawn and 10 *prizes*.

The answer here may still be found by the first rule; and the chance for a proportion of blanks to prizes *less* than 9 to 1 will be 0·44109, and for a proportion *greater* than 11 to 1, 0·3082. It would therefore be likely that there were not *fewer* than 9 or *more* than 11 blanks to a prize. But at the same time it will remain unlikely* that the true proportion should lie between 9 to 1 and 11 to 1, the chance for this being 0·2506 etc. There will therefore be still an odds of near 3 to 1 against this.

From these calculations it appears that, in the circumstances I have supposed, the chance for being right in guessing the proportion of *blanks* to *prizes* to be nearly the same with that of the number of *blanks* drawn in a given time to the number of prizes drawn, is continually increasing as these numbers increase; and that therefore, when they are considerably large, this conclusion may be looked upon as morally certain. By parity of reason, it follows universally, with respect to every event about which a great number of experiments has been made, that the causes of its happening bear the same proportion to the causes of its failing, with the number of happenings to the number of failures; and that, if an event whose causes are supposed to be known, happens oftener or seldomer than is agreeable to this conclusion, there will be reason to believe that there are some unknown causes which disturb the operations of the known ones. With respect, therefore, particularly to the course of events in nature, it appears, that there is demonstrative evidence to prove that they are derived from permanent causes, or laws originally established in the constitution of nature in order to produce that order of events which we observe, and not from any of the powers of chance.† This is just as evident as it would be, in the case I have insisted on, that the reason of drawing 10 times more *blanks* than *prizes* in millions of trials, was, that there were in the wheel about so many more *blanks* than *prizes*.

But to proceed a little further in the demonstration of this point.

We have seen that supposing a person, ignorant of the whole scheme of a lottery, should be led to conjecture, from hearing 100 *blanks* and 10 prizes drawn, that the proportion of *blanks* to *prizes* in the lottery was somewhere between 9 to 1 and 11 to 1, the chance for his being right would be 0·2506 etc. Let [us] now enquire what this chance would be in some higher cases.

Let it be supposed that *blanks* have been drawn 1000 times, and prizes 100 times in 1100 trials.

In this case the powers of X and x rise so high, and the number of terms in the two series

$$\frac{X^{p+1}}{p+1}-\frac{qX^{p+2}}{p+2}\text{etc.}\quad\text{and}\quad\frac{x^{p+1}}{p+1}-\frac{qx^{p+2}}{p+2}\text{etc.}$$

become so numerous that it would require immense labour to obtain the answer by the first rule. 'Tis necessary, therefore, to have recourse to the second rule. But in order to make use of it, the interval between X and x must be a little altered. $\frac{10}{11}-\frac{9}{10}$ is $\frac{1}{110}$, and therefore the interval between $\frac{10}{11}-\frac{1}{110}$ and $\frac{10}{11}+\frac{1}{110}$ will be nearly the same with the interval between $\frac{9}{10}$ and $\frac{11}{12}$, only somewhat larger. If then we make the question to be; what chance there would be (supposing no more known than that blanks have been drawn 1000 times and prizes 100 times in 1100 trials) that the probability of drawing a blank in a single trial would lie somewhere between $\frac{10}{11}-\frac{1}{110}$ and $\frac{10}{11}+\frac{1}{110}$ we shall have a question of the same kind with the preceding questions, and deviate but little from the limits assigned in them.

The answer, according to the second rule, is that this chance is greater than

$$\frac{2\Sigma}{1+2Ea^pb^q+\dfrac{2Ea^pb^q}{n}}$$

and less than

$$\frac{2\Sigma}{1-2Ea^pb^q-2E\dfrac{a^pb^q}{n}}$$

Σ being

$$\frac{(n+1)\sqrt{(2pq)}}{n\sqrt{n}}Ea^pb^q\left\{mz-\frac{m^3z^3}{3}+\frac{(n-2)\,m^5z^5}{2n.5}-\text{etc.}\right\}$$

* I suppose no attentive person will find any difficulty in this. It is only saying that, supposing the interval between nothing and certainty divided into a hundred equal chances, there will be 44 of them for a less proportion of blanks to prizes than 9 to 1, 31 for a greater than 11 to 1, and 25 for some proportion between 9 to 1 and 11 to 1; in which it is obvious that, though one of these suppositions must be true, yet, having each of them more chances against them than for them, they are all separately unlikely.

† See Mr De Moivre's *Doctrine of Chances*, page 250.

By making here $1000 = p$, $100 = q$, $1100 = n$, $\frac{1}{110} = z$,

$$mz = z\sqrt{\left(\frac{n^3}{pq}\right)} = 1{\cdot}048808, \quad Ea^pb^q = \tfrac{1}{2}h\frac{\sqrt{n}}{\sqrt{(Kpq)}},$$

h being the ratio whose hyperbolic logarithm is

$$\frac{1}{12}\left[\frac{1}{n}-\frac{1}{p}-\frac{1}{q}\right]-\frac{1}{360}\left[\frac{1}{n^3}-\frac{1}{p^3}-\frac{1}{q^3}\right]+\frac{1}{1260}\left[\frac{1}{n^5}-\frac{1}{p^5}-\frac{1}{q^5}\right]-\text{etc.}$$

and K the ratio of the quadrantal arc to radius; the former of these expressions will be found to be 0·7953, and the latter 0·9405 etc. The chance enquired after, therefore, is greater than 0·7953, and less than 0·9405. That is; there will be an odds for being right in guessing that the proportion of blanks to prizes lies *nearly* between 9 to 1 and 11 to 1, (or *exactly* between 9 to 1 and 1111 to 99), which is greater than 4 to 1, and less than 16 to 1.

Suppose, again, that no more is known than that *blanks* have been drawn 10,000 times and *prizes* 1000 times in 11,000 trials; what will the chance now mentioned be?

Here the second as well as the first rule becomes useless, the value of mz being so great as to render it scarcely possible to calculate directly the series

$$\left\{mz-\frac{m^3z^3}{3}+\frac{(n-2)\,m^5z^5}{2n.5}-\text{etc.}\right\}$$

The third rule, therefore, must be used; and the information it gives us is, that the required chance is greater than 0·97421, or more than an odds of 40 to 1.

By calculations similar to these may be determined universally, what expectations are warranted by any experiments, according to the different number of times in which they have succeeded and failed; or what should be thought of the probability that any particular cause in nature, with which we have any acquaintance, will or will not, in any single trial, produce an effect that has been conjoined with it.

Most persons, probably, might expect that the chances in the specimen I have given would have been greater than I have found them. But this only shews how liable we are to error when we judge on this subject independently of calculation. One thing, however, should be remembered here; and that is, the narrowness of the interval between $\frac{9}{10}$ and $\frac{11}{12}$, or between $\frac{10}{11}+\frac{1}{110}$ and $\frac{10}{11}-\frac{1}{110}$. Had this interval been taken a little larger, there would have been a considerable difference in the results of the calculations. Thus had it been taken double, or $z = \frac{1}{55}$, it would have been found in the fourth instance that instead of odds against there were odds for being right in judging that the probability of drawing a blank in a single trial lies between $\frac{10}{11}+\frac{1}{55}$ and $\frac{10}{11}-\frac{1}{55}$.

The foregoing calculations further shew us the uses and defects of the rules laid down in the essay. 'Tis evident that the two last rules do not give us the required chances within such narrow limits as could be wished. But here again it should be considered, that these limits become narrower and narrower as q is taken larger in respect of p; and when p and q are equal, the exact solution is given in all cases by the second rule. These two rules therefore afford a direction to our judgment that may be of considerable use till some person shall discover a better approximation to the value of the two series in the first rule.*

But what most of all recommends the solution in this *Essay* is, that it is compleat in those cases where information is most wanted, and where Mr De Moivre's solution of the inverse problem can give little or no direction; I mean, in all cases where either p or q are of no considerable magnitude. In other cases, or when both p and q are very considerable, it is not difficult to perceive the truth of what has been here demonstrated, or that there is reason to believe in general that the chances for the happening of an event are to the chances for its failure in the same *ratio* with that of p to q. But we shall be greatly deceived if we judge in this manner when either p or q are small. And tho' in such cases the *Data* are not sufficient to discover the exact probability of an event, yet it is very agreeable to be able to find the limits between which it is reasonable to think it must lie, and also to be able to determine the precise degree of assent which is due to any conclusions or assertions relating to them.

* Since this was written I have found out a method of considerably improving the approximation in the second and third rules by demonstrating that the expression $2\Sigma/\{1+2Ea^pb^q+2Ea^pb^q/n\}$ comes, in most cases, almost as near to the true value wanted as there is reason to desire, only always somewhat less. It seems necessary to hint this here; though the proof of it cannot be given.

DANIEL BERNOULLI ON MAXIMUM LIKELIHOOD

By M. G. KENDALL

1. Almost as soon as the calculus of probabilities began to take a definite shape mathematicians were concerned with the use of probabilistic ideas in reconciling discrepant observations. James Bernoulli's *Ars Coniectandi* was published in 1713. Within 9 years we find Roger Cotes (1722), in a work on the estimation of errors in trigonometrical mensuration, discussing what would nowadays be described as an estimation problem in a plane. Let p, q, r, s be four different determinations of a point o, with weights P, Q, R, S which are inversely proportional to distance from o (*pondera reciproce proportionalia spatiis evagationum*). Put weights P at p, etc., and find their centre of gravity z. This, says Cotes, is the most probable site of o. (*Dico punctum z fore locum obiecti maxime probabilem, qui pro vero eius loco tutissime haberi potest.*) Cotes does not say why he thinks this is the most probable position or how he arrived at the rule.

2. According to Laplace this result of Cotes was not applied until Euler (1749) used it in some work on the irregularities in the motion of Saturn and Jupiter. Further attacks on the problem of a somewhat similar kind were employed by Mayer (1750) in a study of lunar libration and by Boscovich (1755) in measurements on the mean ellipticity of the earth. There was evidently a good deal of interest being taken in the combination of observations about the middle of the eighteenth century. The ideas, as was only natural, were often intuitive and sometimes obscurely expressed, but the fundamental questions seem to have been asked at quite an early stage. For example, Simpson (1757) refers to a current opinion that one good observation was as accurate as the arithmetic mean of a set, and although from that point onwards a series of writers argued for the arithmetic mean, Laplace (1774), in his first great memoir, was clearly aware that for some distributions of error there were better estimators such as the median.

3. Simpson (1756, 1757) was the first to introduce the concept of distribution of error and to consider continuous distributions. But like most of his contemporaries he regarded it as inevitable to impose two conditions: first, the distributions must be symmetrical; secondly, they must be finite in range. Lagrange reproduced Simpson's work without acknowledgement in a memoir published between 1770 and 1773, but Lagrange's contributions are more of analytical than of probabilistic interest.

4. Daniel Bernoulli was born in 1700 and lived to be 82. Throughout his productive life he made contributions to the theory of probability and although his mathematical methods are not now of much importance, the originality of his thinking on such matters as moral expectation entitles him to a permanent place among the founders of the subject. In particular, the memoir on maximum likelihood reproduced in the following pages is astonishingly in advance of its time. The author was 78 when it was published and it appears that he excogitated the basic ideas for himself without reference to previous writings. The memoir may, in actual fact, have been written rather earlier. Laplace's

article of 1774 refers to *manuscripts* of Bernoulli and Lagrange which he had heard of but not seen. An announcement of their existence, says Laplace sublimely, *reawakened* his interest in the subject. Laplace was 25 at the time.

5. I am much indebted to my colleague Mr C. G. Allen for the translations of the articles by Bernoulli and Euler which follow. They are, I felt, of sufficient interest to justify the publication of an English version, especially Bernoulli's. The reasoning is so clear that I can leave Daniel to tell his own story, but perhaps I may direct attention to two points:

(*a*) Influenced by the belief that an error distribution must have a finite range, Bernoulli runs into trouble with the parameter determining that range. He assumes a semi-circular distribution and lays down the peculiar condition that any distribution must be abrupt at its terminals. Once this is done, however, his formulation of maximum likelihood is clear and explicit and he derives what would nowadays be called the *ML* equations by differentiating the likelihood of the sample.

(*b*) In § 16 he is right on the verge of a principle of minimal variance. In comparing two methods of estimation he points out that one (the *ML* method) gives samples which are closer to the true value than the other.

6. The commentary by Euler seems to me of less value. He points out, correctly in my opinion, that the *ML* principle is arbitrary in the sense that there is no logical reason to believe that observations come from a generating system which gives them the greatest probability. (Bernoulli admits that his reasoning on this point is metaphysical, but at least he does reason about it.) Euler then goes on to propound principles which seem to me to be much more open to doubt than the one he is trying to replace. His examples at the end, in which he has to manoeuvre his error-range to avoid imaginary solutions, ends rather lamely with the conclusion that it doesn't matter much anyway. However, it is always of interest to read what a great mind has to offer on a subject. Nor should we forget, perhaps, that at the time of publication Euler himself was 71 and had been blind for 10 years.

References

Boscovich, R. G. (1755). (In Maire, C. and Boscovich.) *De litteraria expeditione per Pontificiam ditionem ad dimetiendos duos meridiani gradus.* Romae.

Cotes, R. (1722). Aestimatio errorum in mixta mathesi, per variationes partium trianguli plani et spherici. *Opera Miscellanea*, Cantabrigiae.

Euler, L. (1749). *Pièce qui a remporté le prix de l'Académie Royale des Sciences en* 1748, *sur les inégalités du mouvement de Saturne et de Jupiter.* Paris.

Lagrange, J-L. (1770–3). Mémoire sur l'utilité de la méthode de prendre le milieu des résultats de plusieurs observations etc. *Miscellanea Taurinensia*, **5**, 167.

Laplace, P. S. (1774). Déterminer le milieu que l'on doit prendre entre trois observations données d'un même phénomène. *Mém. Acad. Paris* (*par divers savants*), **4**, 634.

Mayer, T. (1750). Abhandlung über die Umwälzung des Mondes um seine Axe. *Kosmographische Nachrichten und Sammlung.*

Simpson, T. (1756). A letter...on the advantage of taking the mean of a number of observations in practical astronomy. *Phil. Trans.* **44**, 82.

Simpson, T. (1757). An attempt to show the advantage arising by taking the mean of a number of observations in practical astronomy. *Miscellaneous Tracts*, London.

The most probable choice between several discrepant observations and the formation therefrom of the most likely induction

By DANIEL BERNOULLI†

Translated by C. G. Allen

British Library of Political and Economic Science, London School of Economics and Political Science

1. Astronomers as a class are men of the most scrupulous sagacity; it is to them therefore that I choose to propound those doubts that I have sometimes entertained about the universally accepted rule for handling several slightly discrepant observations of the same event. By this rule the observations are added together and the sum divided by the number of observations; the quotient is then accepted as the true value of the required quantity, until better and more certain information is obtained. In this way, if the several observations can be considered as having, as it were, the same weight, the centre of gravity is accepted as the true position of the objects under investigation. This rule agrees with that used in the theory of probability when all errors of observation are considered equally likely.

2. But is it right to hold that the several observations are of the same weight or moment, or equally prone to any and every error? Are errors of some degrees as easy to make as others of as many minutes? Is there everywhere the same probability? Such an assertion would be quite absurd, which is undoubtedly the reason why astronomers prefer to reject completely observations which they judge to be too wide of the truth, while retaining the rest and, indeed, assigning to them the same reliability. This practice makes it more than clear that they are far from assigning the same validity to each of the observations they have made, for they reject some in their entirety, while in the case of others they not only retain them all but, moreover, treat them alike. I see no way of drawing a dividing line between those that are to be utterly rejected and those that are to be wholly retained; it may even happen that the rejected observation is the one that would have supplied the best correction to the others. Nevertheless, I do not condemn in every case the principle of rejecting one or other of the observations, indeed I approve it, whenever in the course of observation an accident occurs which in itself raises an immediate scruple in the mind of the observer, before he has considered the event and compared it with the other observations. If there is no such reason for dissatisfaction I think each and every observation should be admitted whatever its quality, as long as the observer is conscious that he has taken every care.

3. Let us compare the observer with an archer aiming his arrows at a set mark with all the care that he can muster. Let his mark be a continuous vertical line so that only deviations in a horizontal direction are taken into account; let the line be supposed to be drawn in the middle of a vertical plane erected perpendicular to the axis of vision, and let the whole of the plane on either side be divided into narrow vertical bands of equal width. Now if the arrow be loosed several times, and for each shot the point of impact be examined and its distance from the vertical mark noted on a sheet, though the outcome cannot in the least be exactly predicted, yet there are many assumptions that can reasonably be made and

† This memoir and the following commentary by Euler appeared in Latin in the memoirs of the Academy of St Petersburg, *Acta Acad. Petrop.* (1777), pp. 3–33. A photostat copy has been deposited in the library of the Royal Statistical Society.

which can be useful to our inquiry, provided all the errors are such as may easily be in one direction as the other, and their outcome is quite uncertain, being decided only as it were by unavoidable chance. In astronomy, likewise, anything which admits of correction *a priori* is not reckoned as an error. When all those corrections have been made which theory enjoins, any further correction which is necessary in order to reconcile the several slightly discrepant observations which differ slightly from each other is a matter solely for the theory of probability. What in particular happens in the course of observation, *ex hypothesi* we scarcely know, but this very ignorance will be the refuge to which we are forced to flee when we take our stand on what is not truest but most likely, not certain but most probable (*non verissimum sed verisimillimum, non certum sed probabilissimum*), as the theory of probability teaches. Whether that is always and everywhere identical with the usually accepted arithmetical mean may reasonably be doubted.

4. Errors, which are unavoidable in observation, may indeed affect individual observations; nevertheless, any given observation has its own rights and could not be impugned if it were the only one that had been made. Any observation must therefore be in itself sound and good, and no-one ought to assign any other value than that ascertained thereby; but since they are mutually contradictory, a value has to be assigned to the whole complex of observations without touching the parts. In this way a definite error is attributed to the individual observations; but I think that of all the innumerable ways of dealing with errors of observation one should choose the one that has the highest degree of probability for the complex of observations as a whole.

The rule which I here propound will be accepted by all, provided that the degree of probability in respect of a given observation can be defined in terms of a point which is assumed to be true. I freely admit that this last condition has not been definitely met; at the same time I am convinced that all things are not equally uncertain and that better results can be got than can be expected from the commonly accepted rule. Let us see if certain assumptions should not properly be made in this argument which contribute something to a higher probability. I will begin the examination with some general considerations.

5. If the archer whom I mentioned in § 3 makes innumerable shots, all with the utmost possible care, the arrows will strike sometimes the first band next to the mark, sometimes the second, sometimes the third and so on, and this is to be understood equally of either side whether left or right. Now is it not self-evident that the hits must be assumed to be thicker and more numerous on any given band the nearer this is to the mark? If all the places on the vertical plane, whatever their distance from the mark, were equally liable to be hit, the most skilful shot would have no advantage over a blind man. That, however, is the tacit assertion of those who use the common rule in estimating the value of various discrepant observations, when they treat them all indiscriminately. In this way, therefore, the degree of probability of any given deviation could be determined to some extent *a posteriori*, since there is no doubt that, for a large number of shots, the probability is proportional to the number of shots which hit a band situated at a given distance from the mark.

Moreover, there is no doubt that the greatest deviation has its limits which are never exceeded and which indeed are narrowed by the experience and skill of the observer. Beyond these limits all probability is zero; from the limits towards the mark in the centre the probability increases and will be greatest at the mark itself.

6. The foregoing give some idea of a scale of probabilities for all deviations, such as each observer should form for himself. It will not be absolutely exact, but it will suit the nature

of the inquiry well enough. The mark set up is, as it were, the centre of forces to which the observers are drawn; but these efforts are opposed by innumerable imperfections and other tiny hidden obstacles which may produce in the observations small chance errors. Some of these will be in the same direction and will be cumulative, others will cancel out, according as the observer is more or less lucky. From this it may be understood that there is some relation between the errors which occur and the actual true position of the centre of forces; for another position of the mark the outcome of chance would be estimated differently. So we arrive at the particular problem of determining the most probable position of the mark from a knowledge of the positions of some of the hits. It follows from what we have adduced that one should think above all of a scale (*scala*) between the various distances from the centre of forces and the corresponding probabilities. Vague as is the determination of this scale, it seems to be subject to various axioms which we have only to satisfy to be in a better case than if we suppose every deviation, whatever its magnitude, to occur with equal ease and therefore to have equal probability. Let us suppose a straight line in which there are disposed various points, which indicate of course the results of different observations. Let there be marked on this line some intermediate point which is taken as the true position to be determined. Let perpendiculars expressing the probability appropriate to a given point be erected. If now a curve is drawn through the ends of the several perpendiculars this will be the scale of the probabilities of which we are speaking.

7. If this is accepted, I think the following assumptions about the scale of probabilities can hardly be denied.

(*a*) Inasmuch as deviations from the true intermediate point are equally easy in both directions, the scale will have two perfectly similar and equal branches.

(*b*) Observations will certainly be more numerous and indeed more probable near to the centre of forces; at the same time they will be less numerous in proportion to their distance from that centre. The scale therefore on both sides approaches the straight line on which we supposed the observed points to be placed.

(*c*) The degree of probability will be greatest in the middle where we suppose the centre of forces to be located, and the tangent to the scale for this point will be parallel to the aforesaid straight line.

(*d*) If it is true, as I suppose, that even the least-favoured observations have their limits, best fixed by the observer himself, it follows that the scale, if correctly arranged, will meet the line of the observations at the limits themselves. For at both extremes all probability vanishes and a greater error is impossible.

(*e*) Finally, the maximum deviations on either side are reckoned to be a sort of boundary between what can happen and what cannot. The last part, therefore, of the scale, on either side, should approach steeply the line on which the observations are sited, and the tangents at the extreme points will be almost perpendicular to that line. The scale itself will thus indicate that it is scarcely possible to pass beyond the supposed limits. Not that this condition should be applied in all its rigour if, that is, one does not fix the limits of error over-dogmatically.

8. If we now construct a semi-ellipse of any parameter on the line representing the whole field of possible deviations as its axis, this will certainly satisfy the foregoing conditions quite well. The parameter of the ellipse is arbitrary, since we are concerned only with the proportion between the probabilities of any given deviation. However elongated or compressed the ellipse may be, provided it is constructed on the same axis, it will perform the

same function; which shows that we have no reason to be anxious about an accurate description of the scale. In fact we can even use a circle, not because it is proved to be the true scale by mathematical reasoning, but because it is nearer the truth than an infinite straight line parallel to the axis, which supposes that the several observations are of equal weight and probability, however distant from the true position. This circular scale also lends itself best to numerical calculations; meanwhile it is worth observing in advance that both hypotheses come to the same whenever the several observations are considered to be infinitely small. They also agree if the radius of the auxiliary circle is supposed to be infinitely large, as if no limits were set to the deviations. Thus if the deviation of an observation from the true position is thought of as the sine of a circular arc, the probability of that observation will be the cosine of the same arc. Let the auxiliary semicircle, which I have just described, be called *the controlling semicircle* (*moderator*). Where the centre of this semicircle is located, the true position, which fits the observations best, is to be fixed. Admittedly our hypothesis is, to some extent, precarious, but it is certainly to be preferred to the common one, and will not be hazardous to those who understand it, since the result that they will arrive at will always have a higher probability than if they had adhered to the common method. When by the nature of the case a certain decision cannot be reached, there is no other course than to prefer the more probable to the less probable.

9. I will illustrate this line of argument by a trivial example. The particular problem is the reconciliation of discrepant observations; it is therefore a question of difference of observations. Now if a dice-thrower makes three throws with one die so that the second exceeds the first by one and the third exceeds the second by two, the throws may arise in three ways, viz. 1, 2, 4 or 2, 3, 5 or 3, 4, 6. None of these throws is to be preferred to the other two, for each is in itself equally probable. If you prefer the one in the middle, viz. 2, 3, 5, the preference is illogical. The same sort of thing happens if you choose to consider observations which, so far as you are concerned, are accidental, whether they are astronomical or of some other kind, as equally probable. Now suppose the thrower produces the same result by throwing a pair of dice three times. There will then be eight different ways in which he would obtain this result, viz. 2, 3, 5; 3, 4, 6; 4, 5, 7; 5, 6, 8; 6, 7, 9; 7, 8, 10; 8, 9, 11 and 9, 10, 12. But they are far from being all equally probable. It is well known that the respective probabilities are proportional to the numbers 8, 30, 72, 100, 120, 80, 40 and 12. From this known scale I have better right to conclude that the fifth set has happened than that any other has, because it has the highest probability; and so the three throws of a pair of dice will have been 6, 7 and 9. No-one, however, will deny that the first set 2, 3 and 5 might possibly have happened, even though it has only a fifteenth part of the probability corresponding to the fifth set. Forced to choose, I simply choose what is most probable. Although this example does not quite square with our argument, it makes clear what contribution the investigation of probabilities can make to the determination of cases. Now I will come more to grips with the actual problem.

10. First of all, I would have every observer ponder thoroughly in his own mind and judge what is the greatest error which he is morally certain (though he should call down the wrath of heaven) he will never exceed however often he repeats the observation. He must be his own judge of his dexterity and not err on the side of severity or indulgence. Not that it matters very much whether the judgement he passes in this matter is fitting or somewhat flighty. Then let him make the radius of the *controlling circle* equal to the aforementioned greatest error; let this radius be r and hence the width of the whole doubtful field $= 2r$.

If you desire a rule on this matter common to all observers, I recommend you to suit your judgement to the actual observations that you have made: if you double the distance between the two extreme observations, you can use it, I think, safely enough as the diameter of the controlling circle, or, what comes to the same thing, if you make the radius equal to the difference between the two extreme observations. Indeed, it will be sufficient to increase this difference by half to form the diameter of the circle if several observations have been made; my own practice is to double it for three or four observations, and to increase it by half for more. Lest this uncertainty offend any one, it is as well to note that if we were to make our controlling semicircle infinite we should then coincide with the generally accepted rule of the arithmetical mean; but if we were to diminish the circle as much as possible without contradiction, we should obtain the mean between the two extreme observations, which as a rule for several observations I have found to be less often wrong than I thought before I investigated the matter.

11. After all these preliminaries it remains to determine the position of the controlling circle, since it is at the centre of this circle that the several observations should be deemed to be, as it were, concentrated. The aforesaid position is deduced from the fact that the whole complex of observations would occur more easily, and therefore more probably, for this location than for any other position of the circle. We shall have the true degree of probability for the whole complex of observations if we note the probability corresponding to the several observations that have been carried out and multiply all the probabilities by each other, just as we did in § 9. Then the product of the multiplication is to be differentiated and the differential put $= 0$. In this way we shall obtain an equation whose root will give the distance of the centre from any given point.

Put the radius of the controlling circle $= r$; the smallest observation $= A$; the second $A+a$; the third $A+b$; the fourth $A+c$, and so on; the distance of the centre of the controlling semicircle from the smallest observation $= x$, so that $A+x$ will denote the quantity which is most probably to be assumed on the basis of all the observations. By our hypothesis the probability for the first observation alone is to be expressed by $\sqrt{\{r^2-x^2\}}$; for the second observation by $\sqrt{\{r^2-(x-a)^2\}}$; for the third by $\sqrt{\{r^2-(x-b)^2\}}$; for the fourth by $\sqrt{\{r^2-(x-c)^2\}}$ and so on. Then I would have the several probabilities multiplied together according to the rules of the theory of probability, which gives

$$\sqrt{\{r^2-x^2\}} \times \sqrt{\{r^2-(x-a)^2\}} \times \sqrt{\{r^2-(x-b)^2\}} \times \sqrt{\{r^2-(x-c)^2\}} \times \ldots.$$

Finally, if the differential of this product is put $= 0$, the equation, by virtue of our hypotheses, gives the required value x as having the highest probability. As, however, the aforesaid quantity is to be brought to its maximum value, it is obvious that its square will simultaneously be brought to the same state. So we can use, for ease of calculation, a formula which is composed entirely of rational terms, viz.

$$(r^2-x^2) \times \{r^2-(x-a)^2\} \times \{r^2-(x-b)^2\} \times \{r^2-(x-c)^2\} \times \ldots$$

and the differential is once more put $= 0$. For the rest, as many factors are to be taken as there were observations.

12. If a single observation was made, we must accept the observation as true. Now this is shown by our hypothesis. If only the first factor r^2-x^2 is taken, we shall have $-2x\,dx = 0$ or $x = 0$ and consequently $A+x = A$. So in this case our hypothesis agrees with the common one.

If two observations have been made, A and $A+a$, two factors are to be taken, namely

$$\{r^2-x^2\}\times\{r^2-(x-a)^2\} \quad \text{or} \quad r^4-2r^2x^2+x^4+2ar^2x-a^2r^2+2ax^3\times a^2x^2,$$

the differential of which

$$=-4r^2x\,dx+4x^3\,dx+2ar^2\,dx-6ax^2\,dx+2a^2x\,dx=0 \quad \text{or} \quad 2x^3-3ax^2-2r^2x+a^2x+ar^2=0.$$

The only useful root which this equation gives is $x=\frac{1}{2}a$, and $A+x=A+\frac{1}{2}a$. This also is the teaching of the common hypothesis. This agreement holds whatever be the radius of the controlling circle, a fact which shows clearly enough, in the case of several observations, that the size of our controlling circle in an enterprise of this sort need not be strictly exact, and one should not expect it to be. What is awkward—and I do not conceal it—is that for several observations a very long calculation is required, and so I hardly dare propose more than general discussions of these cases. Let me at least expound the theory of three observations, which is of the highest importance.

13. When we have three observations to deal with, viz. A; $A+a$ and $A+b$, we shall have three factors

$$\{r^2-x^2\}\times\{r^2-(x-a)^2\}\times\{r^2-(x-b)^2\},$$

for which we have to find the maximum value. If now these factors are actually multiplied together we shall obtain

$$\begin{gathered}
r^6+2ar^4x-3r^4x^2-4ar^2x^3+3r^2x^4+2ax^5-x^6\\
-a^2r^4-2ab^2r^2x+2b^2r^2x^2+2ab^2x^3-b^2x^4+2bx^5\\
-b^2r^4+2br^4x-a^2b^2x^2-4br^2x^3-4abx^4\\
+a^2b^2r^2-2a^2br^2x+4abr^2x^2+2a^2bx^3-a^2x^4\\
+2a^2r^2x^2.
\end{gathered}$$

If this expression is differentiated, and then after division by dx is put $=0$ to obtain the maximum value, the following general equation for any three observations whatsoever will result

$$\begin{gathered}
2ar^4-6r^4x-12ar^2x^2+12r^2x^3+10ax^4-6x^5\\
-2ab^2r^2+4b^2r^2x+6ab^2x^2-4b^2x^3+10bx^4\\
+2br^4-2a^2b^2x-12br^2x^2-16abx^3\\
-2a^2br^2+8abr^2x+6a^2bx^2-4a^2x^3\\
+4a^2r^2x=0.
\end{gathered}$$

The root of this equation, which is indeed of the fifth degree and consists of twenty terms, gives the distance of the centre of the controlling circle from the first observation, and the quantity $A+x$ gives the value which is most probably to be deduced from the three observations which have been made.

14. Unless the force of our fundamental arguments has been most attentively weighed there will be few perhaps who will see any relation whatever between the enormous equation and what seems to be a very simple question; for the common answer is $x=\frac{1}{3}(a+b)$. Nevertheless, our equation corresponds well enough to notions which crop up elsewhere, some of which I will now expound.

(*a*) If the radius of the controlling circle is supposed to be infinite compared with a and b, all terms are to be rejected except those in which r rises to the highest power, in which case our equation is reduced to this very simple one $2ar^4+2br^4-6r^4x = 0$ or $x = \frac{1}{3}(a+b)$. So the common rule is contained in our equation. If, however, our definition set out in § 10 is considered, it will be obvious how unfitting is the hypothesis of an infinite radius and how manifestly some more suitable one could be substituted for it.

(b) If we put $b = 2a$, it is obvious that $x = a$ whatever value is given to the radius r, and that too will be common to both theories. Let us see therefore what our equation shows for this case. Substituting for b the equation becomes

$$\begin{aligned}6ar^4-6r^4x-36ar^2x^2+12r^2x^3+30ax^4-6x^5\\-12a^3r^2+36a^2r^2x+36a^3x^2-52a^2x^3\\-8a^4x = 0.\end{aligned}$$

Now this equation, whatever be the value of r, is satisfied by $x = a$, which the nature of the case demands.

(c) If $b = -a$, x must equal 0 whatever be the value of r. This too is beautifully shown by our equation, which now becomes

$$\begin{aligned}-6r^4x+12r^2x^3-6x^5\\-2a^4x+8a^2x^3 = 0.\end{aligned}$$

A glance will show that the useful root is $x = 0$.

15. This and other similar corollaries sufficiently confirm the real connexion of our fundamental arguments with the question under discussion, however enormous the equation we have found may seem in so simple an inquiry. I proceed to examples in which the radius of the controlling circle is neither infinite nor indifferent, which is where practically all cases belong. In these examples our new theory always produces a different result from the common one; and the more the intermediate observation approaches either extreme, the greater the difference. It is on the discussion of these cases that the matter hinges, so we must have recourse to purely numerical examples.

Example 1. Let us assume three observations

$$A;\quad A+0{\cdot}2\quad\text{and}\quad A+1,$$

so that

$$a = 0{\cdot}2\quad\text{and}\quad b = 1$$

and let the value to be assumed as most likely from these three observations be $A+x$. The common rule gives $x = 0{\cdot}4$. Let us see the new one which to my mind is more probable, and let us put $r = 1$ (cf. § 10). The following purely numerical equation results

$$1{\cdot}92-0{\cdot}32x-12{\cdot}96x^2+4{\cdot}64x^3+12x^4-6x^5 = 0.$$

the solution of which is approximately $x = 0{\cdot}4427$, which exceeds the commonly accepted value by more than a tenth. This marked excess is due to the fact that the middle observation is much nearer to the first than to the third. From this it is easily deduced that the excess will be changed to a defect if the middle observation is nearer to the third than to the first, and that the nearer the middle observation is to the mean between the two extreme observations, the smaller will be this defect. To test this conjecture I retain the other values and change only the middle observation, as follows.

Example 2. Let a now $= 0{\cdot}56$, and as before $r = b = 1$. By the commonly accepted rule we shall have $x = 0{\cdot}52$. Let us see what happens with ours. The equation of § 13 gives the following numerical equation

$$1{\cdot}3728+3{\cdot}1072x-13{\cdot}4784x^2-2{\cdot}2144x^3+15{\cdot}6x^4-6x^5 = 0$$

which is approximately satisfied by $x = 0{\cdot}5128$. In accordance with our principles, the value of x is less than the arithmetical mean which is usually accepted, but the difference between the two is now quite small, viz. 0·0072, exactly as I had anticipated would be the case. Hence it can also be seen that the greatest difference between the two estimates occurs when it so happens that two observations exactly coincide and only the third diverges. There are two cases, viz. when $a = 0$ and when $a = b$. I will expound the result in each case.

Example 3. Put $a = 0$, leaving the remaining denominations unaltered. Dividing by $2b-2x$ we have the following numerical equation

$$1-6x^2-2x^3+3x^4 = 0,$$

which is approximately satisfied by $x = 0{\cdot}3977$, whereas the value of x obtained from the common rule is $x = 0{\cdot}3333$. The former exceeds the latter by 0·0644. If, however, we put $a = b$ and divide by $2x$, the following equation results

$$4-6x-6x^2-10x^3-3x^4 = 0.$$

This is approximately satisfied by $x = 0{\cdot}6022$, while the common value is 0·6666. So the difference between the two is once more 0·0644, but this time our new value is less than the common one, whereas previously it was greater. It is clear from this that our method takes better aim at a certain intermediate point than does the common method. Evidence of this sort does much to commend the method that I propose, and I will go a little more closely into this consideration, if so be that an *argumentum ad hominem* may be accepted in a matter which does not admit of mathematical demonstration.

16. If we combine the two cases in example 3, and suppose that six observations have been made, viz. $A, A, A+b$ and $A+b, A+b, A$, it is obvious that three observations support the value A and the same number the value $A+b$. We see by § 12 that in this case both methods give the required mean value as $A+\frac{1}{2}b$, or for example 3, $A+0{\cdot}5$; or, omitting the constant quantity A, simply 0·5. This value, derived from the six observations combined, will not be doubted by anyone. Now let us divide these six observations into two other triads, namely $A, A, A+1$ and $A+1, A+1, A$. In this case, rejecting once more the quantity A, the commonly accepted rule gives for the first triad $0{\cdot}\dot{3}$ and for the second $0{\cdot}\dot{6}$, both differing, the first by defect and the second by excess, by $0{\cdot}1\dot{6}$ from the mean 0·5. So for either triad of observations taken separately the common theory involves an error of $0{\cdot}1\dot{6}$, while ours involves an error of 0·1022, which is notably smaller. A great deal more evidence of this kind could be adduced to give further support to our fundamental argument; but I am afraid I should appear immoderate if I went on extending something which cannot be settled with certainty and absolute perfection. We have no higher aim than to be able to distinguish what is more probable from what is less.

17. Such further perfection as we may reasonably expect will consist in a stricter and more accurate determination of the controlling scale and its width. I will add a few further comments on this topic. It is obvious from the foregoing considerations that our estimates are not so very different from the commonly accepted rule: so it is a question of a certain correction which this rule appears to allow. This correction is provided by the actual divergences of the observations from the required true point, since they can be so arranged, for any given width of the controlling scale, as to make the most probable fit with this point. But for my part I can see no way of strictly determining the width of the aforesaid scale except that which I mentioned in § 10. If an observer, through undue mistrust of his own powers, enlarges the

dimensions of the controlling semicircle excessively, it will not give all the help it might, but what it gives will be more certain; if on the other hand he contracts the scale unduly, other things being equal he will arrive at a correction which is a little greater and somewhat less probable. Prudence seems to be as necessary here as sharp-sightedness. Should you wish to use the observations that have actually been made as a basis for an *a posteriori* estimate of the width of the controlling scale to be applied, it will be prudent to weigh in your own mind whether one should consider the observations to have turned out luckily or not. The more you assign to good luck, the less you can attribute to the skill applied in observing, and the larger accordingly will be the controlling circle which you will apply. In § 13 I assumed $r = b$; in other words the radius of the controlling circle equalled the distance between the two extreme observations. I admit, however, on better reflexion that this size of radius seems to me to argue somewhat excessive confidence; it would be safer certainly in future to put $r = \frac{3}{2}b$ or even $r = 2b$. If so, the correction would come out notably smaller but all the more certain and trustworthy.

18. If there is any validity in our principles, though they are metaphysical rather than mathematical, we may justly conclude therefrom that one should seldom if ever reject an observation, and never without the utmost circumspection. I have already given my opinion on this subject in § 2. The whole complex of observations is simply a chance event modified and confined within certain limits by the skill of the observer. It may well happen, though very rarely, that of three observations two are miraculously identical, while the third by ill luck is very wide of the other two. But if this happens to me and I am certain that I have not unduly contracted the limits of maximum possible error or shown undue confidence in my skill, I should not hesitate to refer the examination of the whole case to our principles and form my estimate from them. Only the observer must give the same attention to each of the observations. I should like them all treated equally.

19. The only remaining caution refers to the controlling scale which I have applied. We have taken a semicircle as answering sufficiently the conditions set out in § 7 and at the same time most suited to the calculations that have to be carried out. Meanwhile it is worthy of note that there are other infinite curves which undoubtedly lead to the same equation as I set out in § 13. In § 11 we made the probabilities, for a circular scale, proportional respectively to the perpendiculars

$$\sqrt{(r^2-x^2)};\quad \sqrt{\{r^2-(x-a)^2\}};\quad \sqrt{\{r^2-(x-b)^2\}}.$$

Now if instead of a semicircle we suppose a parabola (*arcum parabolicum*) constructed on the line $2r$, with its axis passing perpendicularly through the middle of this line, then keeping the same notation, we shall have perpendiculars, or the corresponding probabilities expressed by them,

$$\frac{\rho}{r^2}(r^2-x^2),\quad \frac{\rho}{r^2}\{r^2-(a-x)^2\},\quad \frac{\rho}{r^2}\{r^2-(b-x)^2\},\quad \text{etc.,}$$

where the new letter ρ denotes the longest perpendicular at the abscissa $x = 0$. Now since the factor ρ/r^2 is common to all the terms we can simply substitute unity for this factor when we have brought the product of all the several probabilities to a maximum. It follows from this that the parameter of the parabola is always arbitrary. I also pointed out in the aforementioned § 11 that if this product has been brought to its maximum all its powers will at the same time be maximized or minimized. It is obvious from this that both scales, the parabolic and the circular, lead to the same required value of x. Furthermore, it is evident that innumerable other scales fulfil the same function; they will all have this property, that from their peak they approach in either direction the line $2r$, on which the several observations are necessarily supposed to lie, and intersect it. Therefore all scales of this sort achieve our aim, and we need not be too pedantic in this matter, since we are content to strive for something better if not for the best.

20. Finally, as regards the awkward, not to say monstrous, form of our fundamental equation set out in § 13, we can mend the awkwardness somewhat; for I express the useful root as approximately

$$x = \frac{a+b}{3} + \frac{2a^3 - 3a^2b - 3ab^2 + 2b^3}{27r^2}.$$

The first term is none other than the common arithmetical mean for three observations, the second indicates approximately the further correction required by our principles. This root indeed will agree all the more accurately with the equation of § 13, the greater is assumed to be the width of the controlling scale indicated by $2r$. Far be it from us, however, to increase the value of the letter r unnecessarily merely to make calculation easier, for every useless increase takes away a little from the amount of our correction. Nor would it be less dangerous to attribute too much to one's powers of observation and so shorten the radius r unjustifiably. 'There are fixed bounds, outside of which justice cannot exist': cf. § 10. Our principles themselves show that it is impossible for r to be less than $\frac{1}{2}b$, since this involves the manifest contradiction of positing as impossible something which is supposed to have actually happened. I have not concealed, however, the somewhat free assumptions that have been made in the course of our argument; but I should not have thought that all our methods of judging the observations that have been made ought to be rejected on that account. Of this at least I am convinced, that *the common rule for three observations gives somewhat too small a result when* $a < \frac{1}{2}b$ *and too large a result if* $a > \frac{1}{2}b$, *and cannot ever be applied with greater certainty than when the intermediate observation is approximately equidistant from the two extremes.* Secondly, I think it probable that our equation in § 13 gives a safer and better determination of the position to be selected, provided the radius of the controlling circle is not rashly diminished beyond the limits which the powers of the observer permit: cf. § 17. The question that I have dealt with is properly this: given three or more shots of an arrow marked on a straight line, to determine the most probable position of the point at which the archer was aiming. But any and every observer who understands these things will form for himself criteria which will answer his purpose, according to the nature of the material (*argumento*) which he has to hand, provided he makes cautious use of the rules derived from the theory of combinations.

Recapitulation. By its very nature our problem is indeterminate, inasmuch as it depends on the practice, experience, and skill of the observer, on the precision of the instruments, on the keenness of the senses, in short on countless circumstances which may be more or less favourable. Account will be taken of all these things in assuming the width of the field of possible deviations; on this subject I have given my opinion, with all circumspection. Secondly, one has to examine the casual working of chance in favour of any given deviation (lit. the working of the casual chance which favours any deviation), since it is advantageous if any given deviation is assigned the probability which from the nature of the case fits it.† To be sure, this scale of probabilities remains in its turn uncertain and undetermined, should

† Reading *cuiuis aberrationi* for the *cuiuis aberratione* of the text.

an accurate one be desired, but displays, nevertheless, by the very nature of the case, several properties; and if these are satisfied, it may be considered to be sufficiently known, as I learned from several experiments. So a method comes to light of expressing in accordance with the proven precepts of the theory of probability the absolute probability appropriate to any given system of observations for any assumed location of that system. It only remains then to select that location of the system in question which enjoys the highest probability. It certainly seems extraordinary to me that the algebraic equation defining this location, which is so far-fetched and rises to the fifth degree for only three observations, which is expressed in a very large number of terms and is deduced from principles never used before, nevertheless, from whatever point it is examined, gives rise to nothing which is in the least displeasing, still less leads to any absurd result. The upshot of the calculations in any example is little different from that which is indicated by the common method, provided one does not recklessly jump at the precepts which I have laid down. Where the comparison of three given observations shows that the middle one is approximately equidistant from the extremes, we shall adhere without hesitation to the common rule; but if the two intervals are notably unequal, I think it is better to have recourse to our theory, provided one follows the precepts I have set out and exercises the greatest prudence in fixing just bounds to the field of possible deviations. All this I should wish to have weighed in the balance of metaphysics rather than mathematics. Those who are most shocked by our principles will have nothing further to contradict if only they make the field of possible deviations as large as possible.

Observations on the foregoing dissertation of Bernoulli

By L. EULER

1. The question which our distinguished friend Bernoulli handles here is one of no little moment, namely, how an unknown quantity should be derived from several observations which vary slightly from each other. To make the nature of the question easier to discern clearly, let us suppose that the elevation of the pole star at some place or other has to be discovered and that the observations made to this end have the following different values:

$$\Pi+a, \quad \Pi+b, \quad \Pi+c, \quad \Pi+d, \quad \text{etc.},$$

where the letters a, b, c, d, etc., are taken to be expressed in seconds. From these the true elevation of the pole at this place, $\Pi+x$, is to be deduced. Generally this quantity x is obtained by taking the arithmetic mean of all the quantities a, b, c, d, etc. Hence if the number of observations $= n$, $x = (a+b+c+d+\text{etc.})/n$.

2. In this rule it is obviously assumed that all observations are of the same degree of goodness. For if some were more exact than others, account ought to be taken of this distinction in the computation. Now although there is no apparent reason in the circumstances why one of these observations should be accorded a greater value than the rest, nevertheless, the learned author observes that these observations ought to be awarded a higher degree of goodness the nearer they approach to the truth, just as that class of observations which is thought to depart too far from the truth is usually completely rejected. The whole business therefore amounts to this: to show how the degree of goodness appropriate to the several observations is to be estimated.

3. According to the view of the distinguished author, it will be convenient to consider the deviation of each observation from the truth as already known. This will be $x-a$ for the first observation, $x-b$ for the second, $x-c$ for the third, etc., but the defect of each observation should be estimated not so much from these differences as from their squares, since the defect itself is to be reckoned as the same whether the observation errs by excess or defect. Hence if some observation agrees perfectly with the truth, its defect will be zero. If therefore the degree of goodness of this observation is indicated by r^2, it is obvious that the degree of goodness of the first observation must be indicated by $r^2-(x-a)^2$, that of the second by $r^2-(x-b)^2$, of the third by $r^2-(x-c)^2$ and so on, the value of r being such that for an observation which is to be all but rejected the degree of goodness vanishes. If we assume that this happens in the observation which gives $\Pi+u$, then since the degree of goodness of this would be $r^2-(x-u)^2$, it must be laid down in all cases that $r^2=(x-u)^2$.

4. Having established these conclusions concerning the degree of goodness of each observation, the distinguished author appeals to the following principle, for which indeed he gives no reason: that the product of all the formulae expressing the degrees of goodness of the several observations should be allotted a maximum value. On this principle therefore he bids one differentiate this product and equate the differential with nought, since this equation will then give the true value of x. This he illustrates with some examples based on sets of three observations, deriving therefrom values of x which seem to be quite in conformity with the truth.

5. This principle for only three observations led to an equation of the fifth degree, whose root x had to be found; and anyone who wished to apply the principle to four observations would arrive at an equation of the seventh degree. Five observations would lead to one of the ninth degree and so on. It is thus abundantly evident that this method cannot possibly be used where there are several observations, and this is in fact candidly conceded by the distinguished author, who presents the whole dissertation as a 'purely metaphysical speculation.

6. As, however, the distinguished author has not supported this principle of the maximum by any proof, he will not take it amiss if I propound certain doubts about it. If we assume that among the observations in question there is one that should be almost rejected, whose degree of goodness would accordingly be as small as possible, it is evident that the product of all the formulae mentioned would in fact be reduced to nothing, so that it could not possibly be considered as a maximum, no matter how great it might be, were that observation omitted. Now the principles of the theory of probability make it abundantly clear that the value of the unknown quantity x should come out the same whether an observation such as this, which has no goodness at all, is introduced into the calculation or totally rejected.

7. I do not think that it is necessary in this question to have recourse to the principle of the maximum, since the undoubted precepts of the theory of probability are quite sufficient to resolve all questions of this kind. If the first observation, which gave $\Pi+a$, is assigned the amount or degree of goodness (*pretium seu gradum bonitatis*) α, the second β, the third γ, the unknown quantity x is given by the rules of this theory thus:

$$x=\frac{a\alpha+b\beta+c\gamma+d\delta+\text{etc.}}{\alpha+\beta+\gamma+\delta+\text{etc.}}.$$

Hence
$$\alpha(x-a)+\beta(x-b)+\gamma(x-c)+\delta(x-d)+\text{etc.}=0.$$

Now it is clear that if all the grades of goodness were equal and the number of observations were n, we should have $x = (a+b+c+d+\text{etc.})/n$, as required by the common rule. From which it follows that different values may emerge for the unknown quantity x to the extent that the degrees of goodness differ.

8. Since therefore, as the distinguished author himself states, the grades of goodness indicated by the letters α, β, γ, δ are

$$\alpha = r^2-(x-a)^2, \quad \beta = r^2-(x-b)^2,$$
$$\gamma = r^2-(x-c)^2, \quad \delta = r^2-(x-d)^2, \quad \text{etc.},$$

the equation we have found becomes

$$r^2(x-a)+r^2(x-b)+r^2(x-c)$$
$$-(x-a)^3-(x-b)^3-(x-c)^3 \text{ etc.} = 0.$$

Hence if the number of observations $= n$ and we put for brevity's sake

$$a+b+c+d+\text{etc.} = A,$$
$$a^2+b^2+c^2+d^2+\text{etc.} = B,$$
$$a^3+b^3+c^3+d^3+\text{etc.} = C,$$

that equation is reduced to the following fairly simple form

$$nr^2x-Ar^2-nx^3+3Ax^2-3Bx+C = 0.$$

Thus we arrive at a cubic equation, from which the unknown x can easily be found, whatever the number of observations n.

9. If we regard the quantity r as infinite, which is the case when all the observations are assigned the same degree of goodness, then we may neglect all the other terms and directly deduce from this equation the following

$$x = \frac{A}{n} = \frac{a+b+c+d+\text{etc.}}{n}$$

just as is required by the rule which is commonly adopted. If we designate this value by the letter p, and substitute Π for $\Pi+p$ in the observations themselves, we shall have to diminish the several numbers a, b, c, d, etc., by the same quantity p, and thus the sum of them all, for which we put A, will equal 0. To avoid, however, the introduction of new letters into the calculation at this point we can from the beginning so constitute the quantity Π that if the values of the several observations are given as $\Pi+a$, $\Pi+b$, $\Pi+c$, $\Pi+d$, etc., the sum of the letters $a+b+c+d+\text{etc.} = 0$. Then to discover the quantity x we shall have the following much simpler equation

$$nx^3-nr^2x+3Bx-C = 0,$$

from which would follow, if r were infinite, $x = 0$. It is clear from this that if this equation has several real roots, the smallest should be taken as x, so that the required true value will be $\Pi+x$.

10. This same question can, however, be referred even to a quadratic equation by introducing the sort of observation which after weighing all the circumstances we decide should be totally rejected. Let such an observation be $\Pi+u$, and since *ex hypothesi* its degree of

goodness $r^2-(x-u)^2 = 0$, $r^2 = (x-u)^2$. The introduction of this value in the last equation that we found produces the following form

$$2nux^2-nu^2x+3Bx-C = 0.$$

It will be convenient to regard the term $-nu^2x$ in this equation as the greatest, so that the equation can be expressed as follows

$$x(nu^2-3B-2nux) = -C$$

from which follows

$$x = \frac{-C}{nu^2-3B-2nux},$$

where by substituting for x the value just obtained we get the following continued fraction

$$x = \frac{-C}{nu^2-3B+}\,\frac{2nuC}{nu^2-3B+}\,\frac{2nuC}{nu^2-3B+}, \quad \text{etc.,}$$

a form which will soon give the true value of x itself.

11. Since the distinguished author has founded his solution on the principle of the maximum, it will not now be difficult to produce an analytical formula of this sort which, when made equal to its maximum, yields the true value of x. Let us use for this purpose the form first discovered

$$r^2(x-a)+r^2(x-b)+r^2(x-c)+\text{etc.}$$
$$-(x-a)^3-(x-b)^3-(x-c)^3-\text{etc.} = 0,$$

which may be regarded as the differential of some formula which is to be raised to its maximum. This formula itself will emerge, if this expression is put in the form of a differential and integrated. Multiplying by $4dx$ and integrating we obtain

$$2r^2(x-a)^2+2r^2(x-b)^2+2r^2(x-c)^2+\text{etc.}$$
$$-(x-a)^4-(x-b)^4-(x-c)^4-\text{etc.}+\text{constant.}$$

If we assume $-nr^4$ as the constant, there being n observations, by change of sign the following formula results

$$\{r^2-(x-a)^2\}^2+\{r^2-(x-b)^2\}^2+\{r^2-(x-c)^2\}^2+\text{etc.}$$

12. In place therefore of the formula which our distinguished friend Bernoulli thought should be made equal to its maximum we have now arrived at another formula very well suited to the nature of the question, which when brought to its maximum gives the true value of x, since this formula is obtained by adding together the squares of all the degrees of goodness.

13. To furnish an example of our method, let us consider the observations by which the longitude of the observatory of St Petersburg is deduced from the difference between the meridians of the observatories of Paris and St Petersburg. These are reported as follows:

I	1° 51′ 50″	IV	1° 51′ 50″
II	1° 51′ 52″	V	1° 51′ 50″
III	1° 51′ 39″	VI	1° 51′ 50″

Taking the arithmetic mean of these in the usual way we obtain 1° 51′ 48½″.

14. Now let us apply our formulae to this case, taking $\Pi = 1°\,51'\,48\tfrac{1}{2}''$. The values of our six letters a, b, c, d, e, f will be

$$a = 1\tfrac{1}{2},\quad b = 3\tfrac{1}{2},\quad c = -9\tfrac{1}{2},\quad d = 1\tfrac{1}{2},\quad e = 1\tfrac{1}{2},\quad f = 1\tfrac{1}{2}.$$

Their sum $A = 0$; the sum of the squares B is found to be $\frac{1}{2}(223)$; the sum of the cubes $= -801$. Hence our equation for $n = 6$ will be

$$12ux^2 - 6u^2x + 801$$
$$+334\tfrac{1}{2}x = 0.$$

15. Now let us *define* the number u from a case which the author of the observations thinks should be rejected, such as $1^\circ 52' 20''$, which gives $u = 31\frac{1}{2}$. Let us suppose that $u = 30$, making our quadratic equation

$$360x^2 - 5065\tfrac{1}{2}x + 801 = 0$$

instead of which we may write in round figures

$$36x^2 = 500x - 80.$$

From this
$$x = \frac{250 \pm \sqrt{59{,}620}}{36},$$

that is either
$$x = \frac{250+244}{36} = 14 \quad \text{or} \quad x = \frac{250-244}{36} = \frac{1}{6}.$$

The latter value only can be considered, and might have been obtained immediately by neglecting the first term in the equation: the value of x would then have been $\frac{50}{8}$ or approximately $\frac{1}{6}$. The required difference of the meridians will therefore be $1^\circ 51' 48\frac{2}{3}''$.

16. Again, suppose that observation had been rejected which gave $1^\circ 51' 0''$: we then have $u = -48\frac{1}{2}$. Let us take $u = -48$, giving the equation

$$-576x^2 - 13{,}489\tfrac{1}{2}x + 801 = 0.$$

Neglecting the first term we obtain $x = \frac{8}{135} = \frac{1}{17}$. Now since this observation would have deserved to be rejected, if u had been in the neighbourhood of -300, hence, carrying out the calculation as before, x would have come out as about $\frac{1}{6}$. It is clear from this that in this case we could have been content with the common rule, since not even a second's difference is involved.

17. Since, however, the third of these observations differs so much from the others, it will perhaps be convenient to set the limit not far from it. If we were to do this for the case $1^\circ 51' 33\frac{1}{2}''$, $u = -15''$, our equation would accordingly be

$$-180x^2 - 1000x + 80 = 0,$$

the smaller root of which equation will be $\frac{12}{18} = \frac{2}{3}$. Hence the difference of the meridians would have come out as $1^\circ 51' 49\frac{1}{6}''$. It is once more clear from this case that no notable error is to be feared, unless we make a quite monstrous mistake in assuming a value for u. In this matter it will suffice to note that nu^2 must always be much larger than $3B$.

18. In particular this method deserves to be applied to those observations from which the learned Lexell not long ago determined the parallax of the sun. From these we take, purely by way of example, the following four conclusions drawn from the observations, namely (I) 8·52; (II) 8·43; (III) 8·86; (IV) 8·28. Taking the arithmetic mean of these we get 8·52. If therefore we put $\Pi = 8{\cdot}52$ the values of the four letters a, b, c, d can be fixed as follows:
$$a = 1, \quad b = 9, \quad c = -34, \quad d = +24$$

so that the sum comes out as $A = 0$.† All these numbers of course denote hundredths of a second. The sum of the squares $B = 1814$, the sum of the cubes $C = -24{,}750$.

† According to his original usage all these signs should be reversed. Presumably a has been rounded up to unity to make the sum of deviations zero.

19. If we now assume as the term where the degree of goodness vanishes $u = 40$, our equation emerges as

$$320x^2 - 948x + 24{,}750 = 0.$$

From this the value of x itself comes out as imaginary. Let us accordingly assume $u = 50$; the equation will then become

$$400x^2 - 10{,}000x + 24{,}750$$
$$+ 5442x = 0$$

and we still arrive at an imaginary result. If, however, we take $u = 60$ the smaller value of x will be $3\frac{5}{12}$, which might seem to be too large. If we admit it the parallax of the sun would be 8·555. But let us note that larger values of u give smaller values for x. Since the application of this method is so vague, we may well doubt whether in this fashion we can arrive any closer to the truth, and perhaps it will suffice to have learnt at any rate, whether the value of x will come out positive or negative.

20. In this case, to be sure, we have seen that the value of x is certainly positive, since we have found a negative number for C. Hence we may profitably observe in general that whenever C comes out positive, x becomes negative, while if C is negative the value of x will be positive. In either case it must of necessity be so small that the result will hardly differ from the common rule. This at any rate can be added, that the larger the number C, the greater must necessarily be the value of x. For if the sum of the cubes C actually vanished, then x would always $= 0$, whatever value is accepted for u, just as the common rule requires.

21. Thus, notwithstanding the uncertainty produced by the number u, it seems that something reasonably probable can be laid down even if we cannot reach certainty, if we pay attention to the following points. First, it is certain that whenever the sum of the cubes $C = 0$, x will always $= 0$. Secondly, the larger the quantity C, the larger will be the value of x itself, with the opposite sign. Thirdly, it is clear enough that the quantity nu^2 must be very much greater than the quantity $3B$. In view of this we can lay it down with reasonable probability that $x = -C/\lambda nB$, where the number λ, it is true, is left to our judgement. However, it will meet all cases and depart hardly at all from the truth, if we put $\lambda = 2$ or at most $\lambda = 3$. The resulting difference will usually be so unimportant that we hardly need consider it. For the case where the greatest error is to be feared would undoubtedly be that in which several observations, i in number, agree entirely in each giving the value a, while the one remaining observation gives $-ia$, so that the sum of them all $A = 0$. The sum of the squares $B = ia^2 + i^2a^2 = i(i+1)a^2$; the sum of the cubes $= ia^3 - i^3a^3 = -i(i^2-1)a^3$. For $n = i+1$ our formula gives

$$x = +\frac{i(i^2-1)a}{\lambda i(i+1)^2} = \frac{(i-1)a}{\lambda(i+1)}.$$

If therefore i is very large and we take $\lambda = 2$ the result is $x = \frac{1}{2}a$. In the earlier example where $n = b$, $B = 11\frac{1}{2}$ and $C = -801$, $x = +801/(12 \times 111\frac{1}{2}) = \frac{3}{5}$ approximately. In the second, where $n = 4$, $B = 1814$, $C = -24{,}750$, $x = 24{,}750/(8 \times 1814) = \frac{8}{5}$ approximately. These values do not appear to involve anything absurd.

If, however, anyone thinks that it would be more reasonable to take $\lambda = 3$, I hardly think the difference is worth arguing about, since the very nature of the observations does not admit of a greater degree of precision.

A NOTE ON THE HISTORY OF THE GRAPHICAL PRESENTATION OF DATA

By ERICA ROYSTON

1. The cartoon of the business man despondently watching his production curve disappearing off the chart on the wall and down through the floor into the office below appears to most people to be a joke that has worn slightly thin. Nevertheless, the very fact that graphical representation can become the subject of a cartoon shows how completely it has come into everyday usage. The graph is now generally accepted as one of the clearest and most effective ways of presenting data, whether for consumption by the layman or the specialist. Many volumes have been published on the art of graphical presentation, on the appropriate method to be used for any specific purpose and on the advantages and disadvantages of various approaches to the subject. It seems strange therefore that so little curiosity has been displayed regarding the historical origins of the graphical representation of data.* In fact, these historical origins are not at all clear.

2. As in most other historical studies of this nature it is virtually impossible to state categorically that a given method was introduced by any one person or at any one moment of time. The most that one may hope to do is to show that such a method was being used at a certain time and to investigate the author's claims, if any, to originality. Apart from this, some of the underlying concepts can be traced back to their possible origins.

3. For the purpose of this note the graphical representation of data will be taken to mean the geometric and graphical presentation of factual data as distinct from the graphical plotting of mathematical functions. Although the development of the technique of using Cartesian co-ordinates with one axis representing time seems to date back only a few centuries, the spatial representation of time originated much earlier. There exist instances of the movement of stars, or, more precisely, the inclinations of the planetary orbits being plotted as a function of time, as early as the tenth century (Funkhouser, 1936), and the emergence of written music represents one of the earliest instances of the use of time-series.†

4. The basic idea of using co-ordinates to determine the location of a point in space dates back to the Greeks at least, although it was not until the time of Descartes that mathematicians systematically developed the idea. For the purpose of tracing the origins of graphical representation the emergence of the concept of plotting mathematical functions

* While holding the part-time appointment of Professor of Geometry at Gresham College (1890–4) Karl Pearson gave a series of twelve lectures under the heading 'The Geometry of Statistics'. The synopsis which survives refers to Playfair as the father of the subject but the lectures themselves appear to have been lost. See E. S. Pearson (1938, p. 142.)

† Neumes (νευματα), the basis of present-day musical notes, were at first only approximately measured in terms of time. The duration of notes was not fixed precisely until the twelfth and thirteenth centuries, when the so-called Franconian reform (after Franco of Cologne's *Ars Cantus Mensurabilis*) took place. Together with the gradual introduction of the bar and a fixed beat in the bar, music became what may be termed a true time-series, where notes were graphically presented, using the tone as ordinate against the time as abscissa.

of the type $y=f(x)$ is important. The most usual form of time-series is, in fact, little more than this, except that usually some form of frequency is plotted as an observed, not necessarily strictly mathematical, function of time. Cartesian co-ordinates are therefore one of the foundations on which the modern graph of data is based. It is interesting to note in passing that William Playfair (1801) justifies plotting money against time as follows:

> This method has struck several persons as being fallacious because geometrical measurement has not any relation to money or to time yet here it is made to represent both. The most familiar and simple answer to this objection is that if the money received by a single man in trade were all guineas and every evening he made a single pile of all the guineas received during the day, its height would be proportioned to the receipts of that day, so that by this plain operation time, proportion and amount would be physically combined.

As will be shown, however, this picturesque description is merely an explanation to the layman, for Playfair himself almost certainly approached the subject as an extension of the use of rectangular co-ordinates, which by his time were familiar in mathematics.

5. One of the early users of graphical representation in statistics was A. F. W. Crome. He was born in Germany in 1753, the third of twenty children. His father was a clergyman and, following in his footsteps, Crome studied theology. During his studies he acted as tutor to the children of General von Holzendorf and later to those of Karl Alexander von Bismarck. He passed his examinations in 1771 and in 1783 became lecturer in geography and history at Dessau. In 1783 he became a tutor to the 16-year-old Prince of Dessau, and finally in 1786 he was made ordinary Professor of Statistics and Public Finance at Giessen University, where he remained till he retired in 1831. He died two years later.

6. With an academic background such as this it is hardly surprising to learn that Crome first evolved his system of geometric representation as an aid to teaching. Crome was, in modern terminology, an economic geographer rather than a statistician. His works are descriptive rather than analytical. Although the *Allgemeine Deutsche Biografie* calls him a 'pioneer in statistics', it is probably using the word in its original meaning, i.e. as data referring to states. Thus the adjective 'statistical', which appears in nearly all the titles of his works, e.g. *Geographisch-statistische Darstellung der Staatskräfte*, merely seems to indicate that some numerical data are given. Thus, for example, in *Über die Grösse und Bevölkerung der sämtlichen Europäischen Staaten* (1785), Crome gives a detailed description of the geographical 'vital data' of most European states, chiefly population figures and areas. To make the importance of such data clearer he devised his *Grössen Karte* which was of much the same type as that illustrated in Fig. 1. (The latter is taken from a similar treatise dealing with the German states.) He justifies the use of this geometrical representation as follows (1785):

> If therefore one does not wish to limit one's knowledge of geography to knowing the names of the cities, provinces, rivers and monarchs, but also wishes to include an overall view of the condition and might of the various empires and states in order to get a good grasp of their might, size and culture, and to compare these among themselves, one must get the largest possible overall view of the area and population of the European states—this is the purpose of the attached chart.
>
> On the one hand the eye can take in some idea of the relative size of the European states by just looking at a general map of this part of the world. But on the other hand the outline of the states is such that they cannot easily be compared.
>
> ...The proportions of the different sizes can however be more easily seen and grasped if they are brought before the eye in the form of a drawing, because the imagination is thus stimulated, than if these merely appeared in the form of numbers, especially when these consist of many digits as is often the case with areas of states....

7. The method Crome employed in these charts was simple. He expressed the area of each of the states with which he was dealing as a square proportional to that area. He then drew these squares inside each other in such a manner as to keep the vertical sides of the squares, which are of course proportional to the square root of the area represented, in scale. When the area of a state was relatively much larger than the others he broke the scale in much the same manner as is done on modern graphs.

8. The idea of representing data in this way seems to have been thought completely new, for Crome went into great detail in describing the uses and misuses of his charts. For example, he warns his readers that there is no justification for assuming that, because the squares were drawn one inside the other, the outside square, or rather the country represented by it, was larger than all those within it added together. It appears that Crome was led to the idea by the comparison of geographical areas on maps; in fact, he used spatial representation of magnitude, not the linear magnitudes moving through time of the Cartesian approach.

There is, however, one example of graphical representation being used by one of Crome's contemporaries and compatriots, namely, G. R. von Gothe's *Höhen Tableau* (Altitude Table) (1813). This is an imaginary panoramic scene, complete with trees and animals, where most of the highest known mountains of Europe and America appear in proportion, their altitudes being given by reference to a scale on the vertical axis.

9. In 1817 Crome wrote:

> I have used this method with great success in my 35 year career as Docent. To this end I published in 1782 my first *Produkten Karte* of Europe, which was followed in 1785 by the *Grössen Karte*. This was issued in an improved form in 1792 as the *Verhältniss Karte* of Europe.

It seems that the *Produkten Karte* referred to differed little from the later ones,* and it is a reissue of the *Verhältniss Karte* that is illustrated here as Fig. 1. It appears as part of a much larger chart which also gives lists of data referring to military strength, etc.

10. On the base of this chart, published in 1820, is a diagram that Crome seems to have used here for the first time. An extract of it is shown as Fig. 2. It consists basically of a series of circles each representing the density of population in the state to which it refers, the density being inversely proportionate to the area of the circle. The various half-tangents and extended radii denote total population and national income, the lines being drawn in different colours to distinguish them from one another. Thus total population is denoted as follows:

The tangent on the right-hand side of several circles indicates the population in millions as measured on the scale, the radii leaning to the right indicate 100 thousands, and the thin black lines also leaning to the right thousands. (N.B. The length of these slanting lines is misleading, as the quantity is indicated by the number of sections of the vertical scale crossed.) These quantities are then added together to give the total population. In fact, each of these lines represents a group of digits from the original figure.

Thus, for example, Luxemburg (third from left) has a population density of 1981 per German square mile (from the text below the circle). Its population is indicated by the

* This chart is missing in the copies of Crome's works available both in the British Museum and the British Library of Political Science.

extended radius sloping to the right which shows 2·5 units, the units in this case being 100,000 (see above). Total population is thus 250,000. The national income is shown as follows:

The tangent on the left indicates 1 unit, i.e. 1 million Gulden, the extended radius sloping to the left indicates 5 units, i.e. $5 \times 100{,}000$ Gulden. Total national income is thus $1{,}000{,}000 + 500{,}000 = 1{,}500{,}000$ Gulden. National income per head, as indicated by the radius extended downwards is 6 Gulden.

11. Although a little involved and misleading and on the whole giving only a very poor overall picture, this method is perhaps more precise than if the population and national income figures had all been plotted to scale. Basically this could be taken as a rather crude attempt at a bar chart, the circles being merely a generalization of the squares Crome used earlier.

12. Crome nowhere explicitly claims that he invented the methods he used, although implicitly he does so several times, as, for example, in the last passage quoted above. Nevertheless, this last method of circles, although it could merely have been a generalization of the squares, seem similar enough to that used by Playfair (1801) a few years earlier (Fig. 3) to warrant closer examination of the possibility that Crome did get some of his ideas from the latter.

13. William Playfair was born in 1759 near Dundee, and was also the son of a clergyman. His father died when William was 13, his brother John, who later became famous as a result of his work in the field of mathematics and geology, taking charge of the family. William Playfair was apprenticed to Andrew Meikle, inventor of the threshing machine. In 1780 he became a draughtsman for Boulton and Watt, and while with them he patented two inventions, the 'Eldorado sash' and a machine for making fretwork on silver tea-trays. Backed by these inventions Playfair opened a shop in London, which soon failed. In 1789 he succeeded Joel Barlow as agent of the Scioto (Ohio) Land Company, a rather unsavoury concern which sold shares in land which may or may not have existed. While in Paris Playfair probably helped to capture the Bastille; he is known also to have saved Duval from the mob in 1791. In the next year allegations concerning his 'mismanagement' of the Ohio Company, coupled with some rather plain speaking against the Revolutionaries, forced Playfair to leave Paris for Frankfurt.

14. Some of Playfair's doings appear a little odd. For example, in 1793, while in Frankfurt, a French emigré gave him details of a system of signalling by means of the semaphore system. He quickly made models of the apparatus—a large clock-like dial, several of which were to be erected across the country to form a chain whereby messages could be relayed—and sent them to the Duke of York. Thereafter he claimed to have introduced the system to England. In actual fact, R. L. Edgeworth (1797) had already used it in 1767 to relay the results of a race at Newmarket to his home some miles away. On his return to London, Playfair opened a 'Security Bank' to facilitate small loans by subdividing large securities. This very soon collapsed, and from 1795 Playfair lived only by writing. Among other things he claimed to have warned the government of Napoleon's intention to escape from Elba. After Waterloo he returned to Paris to edit *Galigni's Messenger*. While there he got involved in a case of libel, and to escape the sentence of three months' imprisonment he fled to London. The remaining years before his death in 1823 he spent in London writing pamphlets.

Figs. 1 and 2. Extracts from Crome's *Geographisch-statistische Darstellung*, 1820.

Plate 1

Royston: *Studies in the History of Probability and Statistics. III*

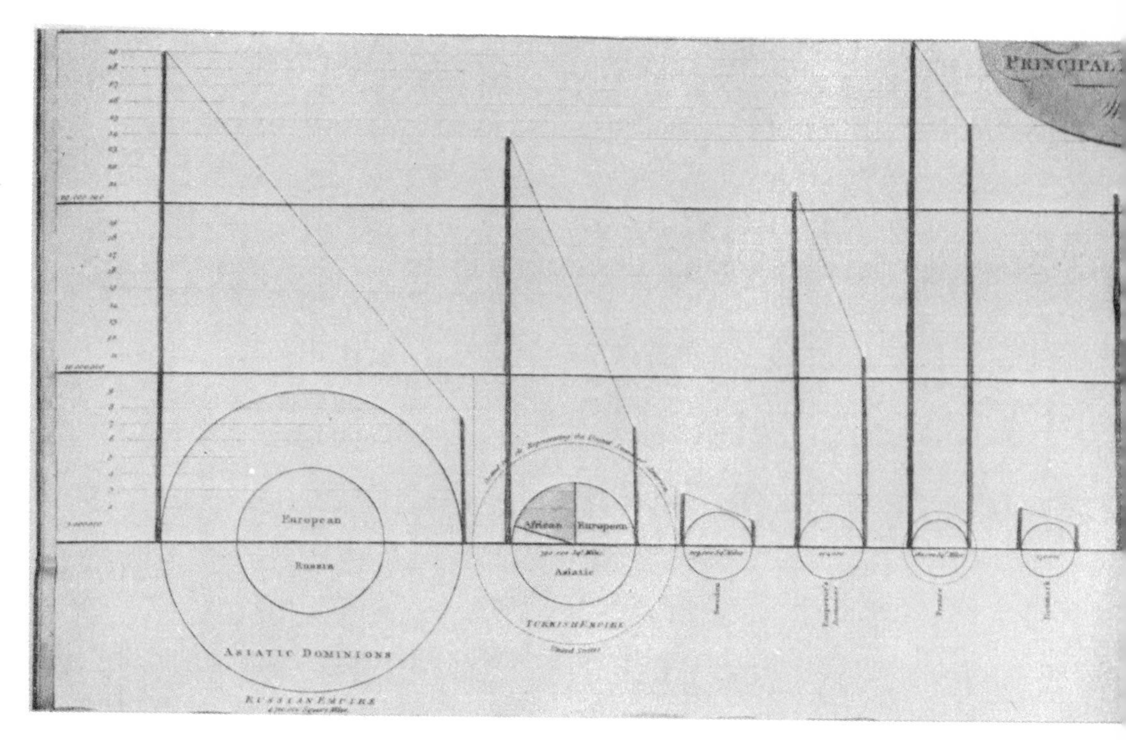

Fig. 3. Extract from Playfair's *Statistical Breviary*, 1801.

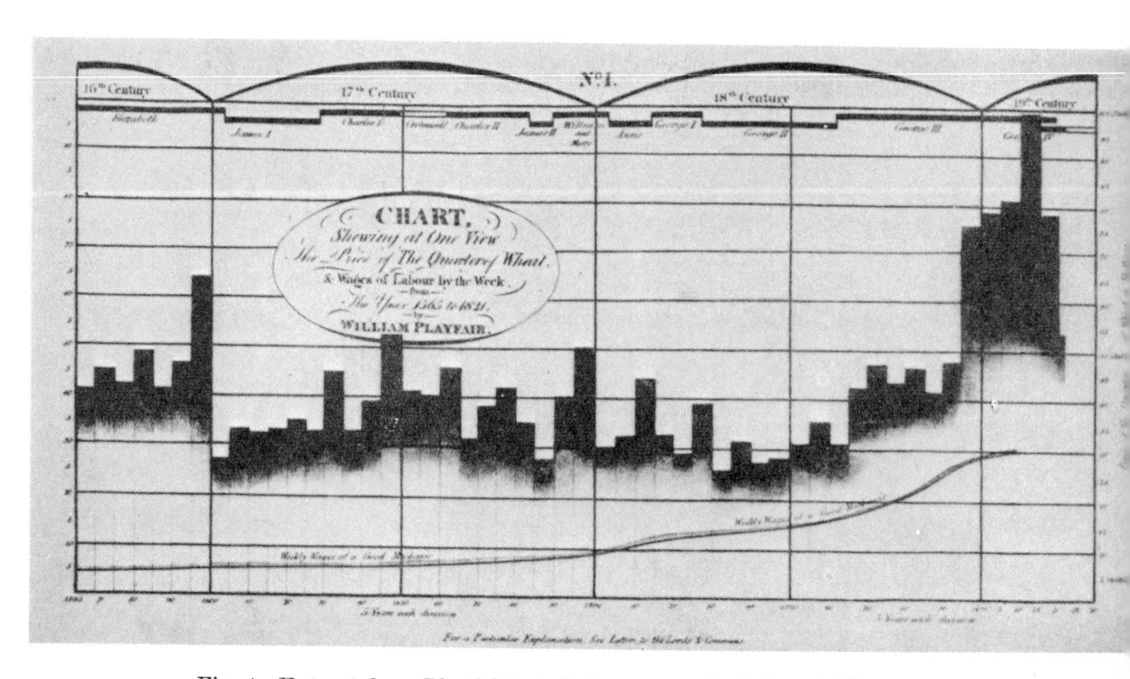

Fig. 4. Extract from Playfair's *A Letter on our Agricultural Distress*, 1821.

Plate 2

15. There is nothing of the studious or academic background of Crome here, and it is therefore rather surprising that what Playfair wrote was good sound economics and, moreover, that his works were illustrated with some extremely good graphs, histograms and pie diagrams. He wrote mainly works on general descriptive economics, one of his favourite subjects being the international balance of trade. Most of the graphs he used were plotted time-series, chiefly export and import figures expressed in millions of pounds; but he also used circles to present spatial magnitudes after the manner of Crome. His earliest work containing such graphs seems to be his *Commercial and Political Atlas*, published in 1786. This consists of graphs of exports and imports represented by coloured lines, the gap between them being termed the balance in favour of England and usually also coloured. He produced such graphs for trade to and from most of Britain's foreign markets. In the same book there are also some very good bar diagrams showing the amount of exports to and imports from each country plotted against time.

16. Playfair's professed aim in publishing his charts was to make statistics, presumably again in the sense of data relating to state-craft, a little more palatable. Statistics, one may gather from the following extract from Playfair's *Statistical Breviary* (1801), were not thought of very highly.

...for no study is less alluring or more dry and tedious than statistics, unless the mind and imagination are set to work or that the person studying is particularly interested in the subject; which is seldom the case with young men in any rank in life.

17. Most of Playfair's later diagrams were merely improvements of his earlier work. Fig. 4 shows one such graph where two series have been superimposed upon one another, one as a graph and the other as a histogram (the two series here being weekly wages and the price of a quarter of wheat (1821)). He illustrated his *British Family Antiquity* with several beautifully executed bar diagrams, which differed from the true frequency graphs used elsewhere by the fact that only the presence or absence of a given factor is plotted against time. This type of diagram, Playfair conceded, had long been used in chronology.

18. Fig. 3 shows an extract from an earlier example of Playfair's work. Here the areas of European states are expressed as circles and in some cases subdivided as pie diagrams. The line on the left-hand side of the circles indicates population in millions and that on the right-hand side national revenue in millions of pounds. The slope of the line joining these two is intended to indicate the approximate ratio of one to the other. Playfair nowhere states this and is presumably aware of the fact that the differing diameters of the circles make any exact comparison impossible. This particular figure has been mentioned earlier as closely resembling Crome's work (Fig. 2) and therefore warranting the examination of the possibility of there being some connexion between the two writers. Certainly this was possible, for Crome and Playfair were contemporaries and, moreover, both wrote in the same field. Playfair's diagram was, however, published before Crome's, and it seems unlikely that, having had sufficient flair to produce perfect time-series graphs, he should have copied Crome. Conversely, however, it seems equally unlikely, even were it possible, that Crome should have taken a basically clear and simple diagram of Playfair's and made such a complicated diagram as Fig. 2 out of it—unless, of course, he was working under the impression that he was improving on Playfair's work. On the face of it, with no direct evidence either way, it seems that if there was any connexion between the two it was Crome who was copying from Playfair and not vice versa. But it is at least equally possible that the two worked entirely independently.

19. There is nothing crude or clumsy about Playfair's work and it seems surprising that one man should have developed it to such a high pitch—for, unlike Crome, Playfair makes definite claims to originality, calling himself the 'inventor of linear arithmetic' as he termed graphical representation. In 1796, for instance, he writes:

> I confess I was long anxious to find out whether I was actually the first who applied the principles of geometry to matters of finance as it has long been applied to chronology with great success. I am now satisfied, upon due enquiry, that I was the first, for during the 11 years I have never been able to learn that anything of a familiar nature had ever before been thought of...

and later (1805):

> The impression is not only simple, but it is as lasting in retaining as it is easy in receiving. Such are the advantages claimed for the invention 20 years ago, when it first appeared. The claim has been allowed and not objected to so far as the inventor knows, either in this or in any other country.

And as an explanation of how he came to think of such graphical representation (1805);

> I think it well to embrace this opportunity, the best I have had, and perhaps the last I ever shall have, of making some return (as far as acknowledgement is a return) for an obligation, of a nature to be repaid by acknowledging publicly, that to the best and most affectionate of brothers I owe the invention of these charts.
>
> At a very early period of my life, my brother, who in a most exemplary manner maintained and educated the family his father left, made me keep a register of a thermometer, expressing the variation by lines on a divided scale. He taught me to know that whatever can be expressed in numbers may be represented by lines. The chart of the thermometer was on the same principle with those given here, the application only is different. The brother to whom I owe this now fills the Natural Philosophical Chair at the University of Edinburgh.

20. Were it not for the somewhat doubtful background revealed by his biographers, including among other matters the bogus claim concerning the semaphore, this sounds honest enough. Certainly there did not seem to be any counterclaims made by his contemporaries, for Playfair's work was acclaimed both in this country and in France, where even the Academy of Sciences 'testified its approbation of this application of geometry to accounts' (Playfair, 1798).

21. Apart from the influence exerted on Playfair by his brother, there is, however, another clue to be found in his biography. In 1780 Playfair worked as a draughtsman to Boulton and Watt. Now Watt is well known as the first engineer who illustrated in diagrammatic form the work done in a steam-engine cylinder by graphing pressure against volume, and one of his draughtsmen would undoubtedly have come into contact with graphical methods of presenting many kinds of motion. It seems very possible that Playfair's real originality lay, not in the kind of diagram he used, but in its application to descriptive statistics and economics.

22. On the whole, therefore, it seems quite possible that William Playfair, if not the 'inventor' of graphical representation as we know it to-day, was the first to introduce it into statistics; and that Crome independently had much the same ideas so far as concerns spatial presentation, but did not carry them out quite as well. The only reasons for suspecting Playfair's authenticity are his occasional dubious activities in other fields, and in the circumstances it is perhaps advisable to accord him the benefit of the doubt.

23. Little that Playfair did would seem to have been beyond the capabilities or scope of men like Graunt, Petty or Halley a century earlier. It seems somewhat surprising that the

graphical presentation of statistics should have had to wait until the end of the eighteenth century. The explanation may well be that until the middle of that century there was very little to present. To quote Playfair himself (1801):

Statistical knowledge, though in some degree searched after in the most early ages of the world, has not till within these last 50 years become a regular object of study.

REFERENCES

Crome, A. F. W. (1785). *Über die Grösse und Bevölkerung der sämtlichen Europäischen Staaten.* Leipzig.
Crome, A. F. W. (1820). *Geographisch-statistische Darstellung der Staatskräfte.* Leipzig.
Edgeworth, R. L. (1797). *A Letter to the Rt. Hon. Earl of Charlemont on the Tellograph.* Dublin.
Funkhouser, H. G. (1936). A note on a 10th century graph. *Osiris,* **1.** Bruges.
Pearson, E. S. (1938). *Karl Pearson.* Cambridge University Press.
Playfair, W. (1786). *The Commercial and Political Atlas.* London.
Playfair, W. (1796). *For the Use of the Enemies of England.* London.
Playfair, W. (1798). *Linear Arithmetic.* London.
Playfair, W. (1801). *Statistical Breviary.* London.
Playfair, W. (1805). *An Enquiry into the Permanent Causes of the Decline and Fall of Powerful Nations.* London.
Playfair, W. (1808). *British Family Antiquity.* London.
Playfair, W. (1821). *A Letter on our Agricultural Distress.* London.
von Gothe, G. R. (1813). Höhen Tableau. *Allgemeine Geographische Ephemeriden,* **41.** Weimar.

THOMAS YOUNG ON COINCIDENCES

By M. G. KENDALL

SUMMARY

Thomas Young considered the *problème des rencontres* in 1819 and arrived at the Poisson distribution with unit parameter as a limiting case. He seems to have been the first to apply the theory to linguistic problems.

Thomas Young, M.D., F.R.S. (1773–1829) is well known for his decipherment of Egyptian hieroglyphics, his theories of the undulatory nature of light and the physiology of colour-vision, and his researches on elasticity and capillarity. Dr Alan Mackay of Birkbeck College has recently drawn to my notice some comments by this remarkable polymath on coincidences, with particular reference to linguistics and history. They occur in a paper (1819) whose title would not lead one to expect them.

The paper is in the form of a letter to Captain Henry Kater, F.R.S. and is in three parts, the second and third of which are not relevant to this note, being concerned with the mean density of the earth and the irregularities of the earth's surface. The first part is entitled 'On the estimation of the advantage of multiplied observations'.

Young first of all deals with the central limit effect of the average of repeated observations, and in particular derives the mean deviation of a binominal with probability $\frac{1}{2}$ and its probable error, which in his terminology is 0·85 (about six-sevenths) of the mean deviation divided by $\sqrt{n}$, where n is the number of observations. He then makes a rather mysterious comment about sampling variance:

'We might obtain a conclusion nearly similar by considering the sum of the squares of the errors, amounting always to $n2^n$: but besides the greater labour of computing the sum of the squares of the errors of any series of observations, the method, strictly speaking, is somewhat less accurate, since the amount of this sum is affected in a slight degree by any error which may remain in the mean, while the simple sum of the errors is wholly exempted from this uncertainty.'

He proceeds with observations which show that he appreciated the difference between sampling errors and bias. 'It may therefore be inferred from these calculations, first, that the original conditions of the probability of different errors, though they materially affect the observations themselves, do not very greatly modify the nature of the conclusions respecting the accuracy of the mean result, because their effect is comprehended in the magnitude of the mean error from which these conclusions are deduced: and secondly, that the error of the mean, on account of this limitation, is never likely to be greater than six-sevenths of the mean of all the errors, divided by the square root of the number of observations. But though it is perfectly true, that the probable error of the mean is always somewhat less than the mean error divided by the square root of the number of observations, provided that no constant causes of error have existed; it is still very seldom safe to rely on the total absence of such causes; especially as our means of detecting them must be limited by the accuracy of our observations, not assisted, in all instances, by the tendency to equal errors on either side of the truth: and when we are comparing a series of observations made with any one instrument, or even by one observer, we can place so little reliance on the absence of some constant error, much greater than the probable result of the accidental causes, that it would in general be deceiving ourselves even to enter into the calculation upon the principles here explained: and it is much to be apprehended, that for want of considering this necessary condition, the results of many elegant and refined investigations, relating to the probabilities of error, may in the end be found perfectly nugatory.'

He then proceeds:

'There are cases in which some little assistance may be derived from the doctrine of chances with respect to matters of literature and history: but even here it would be extremely easy to pervert this application in such a manner, as to make it subservient to the purpose of clothing fallacious reasoning in the garb of demonstrative evidence. Thus if we were investigating the relations of two languages to each other, with a view of determining how far they indicated a common origin from an older language, or an occasional intercourse between the two nations speaking them, it would be important to inquire, upon the supposition that the possible varieties of monosyllabic or very simple words must be limited by the extent of the alphabet to a certain number; and that these names were to be given promiscuously to the same number of things, what would be the chance that, 1, 2, 3 or more of the names would be applied to the same things in two independent instances.'

Young then derives (apparently in ignorance of de Moivre's work) expressions for coincidences of a familiar kind: if n names are assigned at random to n objects, what are the probabilities that no name, one name, ..., n names are correctly assigned? He considers the case of large n, and effectively arrives at the Poisson distribution with unit parameter

No coincidence	0·3678794	Four only	0·0153283
One only	0·3678794	Five only	0·0030657
Two only	0·1839397	Six only	0·0005109
Three only	0·0613132	Seven only	0·0000730

He concludes:

'It appears therefore that nothing whatever could be inferred with respect to the relation of two languages from the coincidence of the sense of any single word in both of them; and that the odds would only be 3 to 1 against the agreement of two words: but if three words appeared to be identical, it would be more than 10 to 1 that they must be derived in both cases from some parent language, or introduced in some other manner; six words would give near 1700 chances to 1, and 8 near 100,000: so that in these last cases the evidence would be little short of absolute certainty.

'In the Biscayan, for example, or the ancient language of Spain, we find in the vocabulary accompanying the elegant essay of Baron W. Von Humbolt, the words *berria*, new; *ora*, a dog; *guchi*, little; *oguia*, bread; *otsoa*, a wolf, whence the Spanish *onza*; and *zazpi*, or, as Lacroze writes it, *shashpi*, seven. Now in the ancient Egyptian, new is Beri; a dog, Uhor; little, Kudchi; bread, Oik; a wolf, Uonsh; and seven, Shashf; and if we consider these words as sufficiently identical to admit of our calculating upon them, the chances will be more than a thousand to one, that, at some very remote period, an Egyptian colony established itself in Spain: for none of the languages of the neighbouring nations retain any traces of having been the medium through which these words have been conveyed.

'On the other hand, if we adopted the opinions of a late learned antiquary, the probability would be still incomparably greater that Ireland was originally peopled from the same mother country: since he has collected more than 100 words which are certainly Egyptian, and which he considers as bearing the same sense in Irish; but the relation, which he considers he has magnified into identity, appears in general to be that of a very faint resemblance: and this is precisely an instance of a case, in which it would be deceiving ourselves to attempt to reduce the matter to a calculation.

'The mention of a single number, which is found to be indisputably correct, may sometimes afford a very strong evidence of the accuracy and veracity of a historian. If the number were indefinitely large, the probability that it could not have been suggested by accident would amount to an absolute certainty: but where it must naturally have been confined within certain moderate limits, the confirmation, though somewhat less absolute, may still be very strong. For example, if the subject were the number of persons collected together for transacting business, it would be a fair assumption that it must be between 2 or 3 and 100, and the chances must be about 100 to 1 that a person reporting it truly must have some good information; especially if it were not an integral number of tens or dozens, which may be considered as a species of units. Now it happens that there is a manuscript of Diodorus Siculus, which, in describing the funerals of the Egyptians, gives 42 for the number of persons who had to sit in judgement on the merits of the deceased: and in a multitude of ancient rolls of papyrus, lately found for Egypt, it may be observed, that 42 personages are delineated, and enumerated, as the judges assisting Osiris in a similar ceremony. It is therefore perfectly fair to conclude from this undeniable coincidence, that we might venture to bet 100 to 1, that the manuscript in question is in general more accurate than others which have been collated; that Diodorus Siculus was a well informed and faithful historian; that the graphical representations and inscriptions in question do relate to some kind of judgement; and lastly, that the hieroglyphical numbers, found in the rolls of papyrus, have been truly interpreted.'

Reference

Remarks on the probabilities of error in physical observations, and on the density of the earth, considered, especially with regard to the reduction of experiments on the pendulum. In a letter to Capt. Henry Kater, F.R.S. By Thomas Young, M.D. *For. Sec.* R.S.
Phil. Trans. 1819, 70–95.

NOTES ON THE HISTORY OF CORRELATION.

Being a paper read to the Society of Biometricians and Mathematical Statisticians, June 14, 1920.

By KARL PEARSON

(1) As I have often stated, Laplace anticipated Gauss by some 40 years. In his memoir of 1783, *Histoire de l'Académie*, pp. 423—467, he gives the expression for the probability integral

$$\frac{1}{\sqrt{2\pi}}\int_x^\infty e^{-\frac{1}{2}x^2}dx$$

and suggests (p. 433) its tabulation as a useful task. It is clear that to do this is to recognise the existence of the probability-curve

$$y=\frac{1}{\sqrt{2\pi}}e^{-\frac{1}{2}x^2},$$

or in its doubly projected form

$$y=\frac{N}{\sqrt{2\pi}\sigma}e^{-\frac{1}{2}\frac{x^2}{\sigma^2}}.$$

Laplace's investigation while not proceeding from the very simple axioms of Gauss, which lead directly to the above equation, is more satisfactory than Gauss' because we see better the nature of the approximations by which the curve is reached and get hints of how to generalise it. Many years ago I called the Laplace-Gaussian curve the *normal* curve, which name, while it avoids an international question of priority, has the disadvantage of leading people to believe that all other distributions of frequency are in one sense or another 'abnormal.' That belief is, of course, not justifiable. It has led many writers to try and force all frequency by aid of one or another process of distortion into a 'normal' curve.

Gauss starting with a normal curve as the law of distribution of errors reached at once the method of least squares. To understand the origin of the correlational calculus we must really go back to Gauss' fundamental memoirs on least squares, namely the *Theoria combinationis observationum erroribus minimis obnoxiae* of 1823 and the *Supplementum* of 1826.

We observe or measure *directly* a certain number of quantities $a, b, c, d, \ldots$. Each of these quantities is supposed by Gauss to be independent and to follow the normal law. The combined probability of the system* is accordingly

$$P \propto e^{-\frac{1}{2}\left\{\left(\frac{a-\bar{a}}{\sigma_a}\right)^2+\left(\frac{b-\bar{b}}{\sigma_b}\right)^2+\left(\frac{c-\bar{c}}{\sigma_c}\right)^2+\ldots\right\}},$$

or the product of the independent probabilities, where $\sigma_a, \sigma_b, \sigma_c, \ldots$ are the variability in errors of $a, b, c, \ldots$ and $\bar{a}, \bar{b}, \bar{c}, \ldots$ the means. This probability will be a maximum when

$$u^2_{a, b, c, \ldots} = S\left(\frac{a-\bar{a}}{\sigma_a}\right)^2$$

is a minimum. This is really the principle of weighted least squares. Its validity depends upon the normal law of distribution of error. Without this law holding it may be a utile method, but we have no means of proving it the 'best.'

The investigator in Gauss' case is, however, not interested in the quantities observed, but in certain indirectly ascertained quantities $x_1, x_2, \ldots x_n$ which are functions of them. Thus

$$\begin{aligned} x_1 &= f_1(a, b, c, \ldots), \\ x_2 &= f_2(a, b, c, \ldots), \\ \ldots &= \ldots\ldots\ldots\ldots\ldots, \end{aligned}$$

where $f_1, f_2, \ldots$ are known functions. Now Gauss cannot as a rule express from these general equations $a, b, c, \ldots$ in terms of $x_1, x_2, \ldots x_n$.

He assumes that all of them differ slightly from their mean or 'true' values and accordingly expands by Taylor's theorem and reaches the result†

$$\begin{aligned} x_1 - \bar{x}_1 &= \alpha_1(a-\bar{a}) + \beta_1(b-\bar{b}) + \gamma_1(c-\bar{c}) + \ldots, \\ x_2 - \bar{x}_2 &= \alpha_2(a-\bar{a}) + \beta_2(b-\bar{b}) + \gamma_2(c-\bar{c}) + \ldots, \end{aligned}$$

where the $\alpha, \beta, \gamma, \ldots$ are $\frac{df}{da}, \frac{df}{db}, \frac{df}{dc}, \ldots$ and can be ascertained *a priori*. Clearly Gauss supposes that a linear relationship is adequate, in other words he replaces statistical differentials by mathematical differentials, a step he does not really justify.

From these linear equations we can find the $a-\bar{a}, b-\bar{b}, c-\bar{c}, \ldots$ in terms of the indirectly observed variables $x_1-\bar{x}_1, x_2-\bar{x}_2, x_3-\bar{x}_3, \ldots$ by solution in determinantal form, say

$$\begin{aligned} a-\bar{a} &= A_1(x_1-\bar{x}_1) + B_1(x_2-\bar{x}_2) + C_1(x_3-\bar{x}_3)\ldots, \\ b-\bar{b} &= A_2(x_1-\bar{x}_1) + B_2(x_2-\bar{x}_2) + C_2(x_3-\bar{x}_3)\ldots. \end{aligned}$$

Substituting in u^2 we find

$$u^2_{x_1, x_2, \ldots} = S\left(\frac{A_1^2}{\sigma_a^2}\right)(x_1-\bar{x}_1)^2 + S\left(\frac{B_1^2}{\sigma_a^2}\right)(x_2-\bar{x}_2)^2 + 2S\left(\frac{A_1B_1}{\sigma_a^2}\right)(x_1-\bar{x}_1)(x_2-\bar{x}_2) + \ldots.$$

Hence the probability of $x_1, x_2, \ldots$ occurring is

$$P \propto e^{-\frac{1}{2}u^2_{x_1, x_2, \ldots}}.$$

* I use throughout notation which I assume now-a-days to be more familiar than that of Gauss.

† $\bar{a}, \bar{b}, \bar{c}, \ldots$ are actually in Gauss' method approximate or *guessed* solutions not means, but this does not affect the general nature of the discussion.

This is a normal surface which contains *the product terms*. As we now interpret it we say that the x's are *correlated* variates. And in this sense Gauss in 1823 reached the normal surface of n correlated variates. But he does not seek to express all his relations in terms of the S.D.'s $\sigma_{x_1}, \sigma_{x_2}, \sigma_{x_3}, \ldots$ and the correlations $r_{12}, r_{23}, \ldots$ of these variates. These x-variates are not for Gauss, nor for those who immediately followed him, the *directly* observed quantities. What he is seeking is the expression for σ_x, or the probable error of an indirectly observed variate in terms of

$$S\left(\frac{A_1^2}{\sigma_a^2}\right), \quad S\left(\frac{B_1^2}{\sigma_a^2}\right), \quad S\left(\frac{A_1 B_1}{\sigma_a^2}\right), \ldots.$$

In this case A, B, C are ratios of minors and determinants of the $\alpha, \beta, \gamma, \ldots$ which are Gauss' known quantities. His object therefore is to express σ_x not from direct observations but in terms of $\alpha, \beta, \gamma, \ldots$ through the sums of determinantal terms.

Writers on Least Squares and Adjustment of Observations then take w any function of $x_1, x_2, \ldots x_n$, i.e.

$$w = F(x_1, x_2, \ldots x_n),$$

express the relation in a linear form, i.e.

$$w - \bar{w} = \lambda_1 (x_1 - \bar{x}_1) + \lambda_2 (x_2 - \bar{x}_2) + \ldots,$$

and then, to find σ_w^2, go through lengthy analysis to determine

$$\text{Mean } (x_1 - \bar{x}_1)^2, \quad \text{Mean } (x_2 - \bar{x}_2)^2, \quad \text{Mean } (x_1 - \bar{x}_1)(x_2 - \bar{x}_2), \text{ etc.}$$

in terms of the original $\alpha, \beta, \gamma, \ldots$. There is not a word in their innumerable treatises that what is really being sought are the mutual correlations of a system of correlated variables. The mere using of the notation of the correlational calculus throws a flood of light into the mazes of the theory of errors of observation. There is much more in the theory of least squares than I have stated; there are equations of conditions—the angle and side equations of geodesy, etc.—these only complicate the matter. The point is this: that the Gaussian treatment leads (i) to a non-correlated surface for the directly observed variates, (ii) to a correlation surface for the *indirectly* observed variates. This occurrence of product terms arises from the geometrical relations between the two classes of variates, and not from an organic relation between the indirectly observed variates appearing on our direct measurement of them.

It will be seen that Gauss' treatment is almost the inverse of our modern conceptions of correlation. For him the *observed* variables are independent, for us the observed variables are associated or correlated. For him the non-observed variables are correlated owing to their known geometrical relations with observed variables; for us the unobservable variables may be supposed to be uncorrelated causes, and to be connected by unknown functional relations with the correlated variables. In short there is no trace in Gauss' work of observed physical variables being—apart from equations of condition—associated organically which is the fundamental conception of correlation.

(2) The next important work to be considered is that of August Bravais. It is entitled "Sur les probabilités des erreurs de situation d'un point." It was published in the *Mémoires présentés par divers savants à l'Académie royale des Sciences de l'Institut de France*, T. IX. Paris, 1846, pp. 256—332. It appears, however, to have been reported favourably upon in 1838*. Bravais was in many respects a remarkable man. Essentially a geologist he wrote also on astronomy, physics, meteorology and the theory of probabilities. He made a voyage to Lapland for geodesic purposes and took the opportunity of measuring a number of Lapp skulls! He had a width of action most sympathetic to the biometrician.

Writing in 1895 of the history of correlation I said:

"The fundamental theorems of correlation were for the first time and almost exhaustively discussed by Bravais [Title as above of his memoir] nearly half a century ago. He deals completely with the correlation of two and three variables." Then speaking of Galton's coefficient of correlation I say: "This indeed appears in Bravais' work, but a single symbol is not used for it. It will be found of great value in the present discussion. In 1892 Professor Edgeworth, also unconscious of Bravais' memoir, dealt in a paper on 'Correlated Averages' with correlation for three variables (*Phil. Mag.* Vol. XXXIV. 1892, pp. 194—204). He obtained results identical with Bravais', although expressed in terms of 'Galton's functions'" [i.e. coefficients of correlation].

Again later, p. 287, in giving the fundamental equation for the correlation of three variates I wrote: "This agrees with Bravais' result, except that he writes for r_1, r_2, r_3 the values $\Sigma(yz)/(n\sigma_2\sigma_3)$ etc., which we have shown to be the best values (see *loc. cit.* p. 267)." Again on p. 301 I write before proving the general theorem of multiple correlation: "*Edgeworth's Theorem.* We may stay for a moment over the results above to deduce Professor Edgeworth's Theorem," with the footnote, 'Briefly stated with some rather disturbing printer's errors in the 'Phil. Mag.' Vol. XXXIV. p. 201, 1892."

Now all these statements if they were correct would indicate that Bravais discovered correlation before Galton and that Edgeworth first published the form of the multiple correlation surface. They have been accepted by later writers, notably Mr Yule in his manual of statistics, who writes (p. 188):

"Bravais introduced the product-sum, but not a single symbol for a coefficient of correlation. Sir Francis Galton developed the practical method, determining his coefficient (Galton's function as it was termed at first) graphically. Edgeworth developed the theoretical side further and Pearson introduced the product-sum formula."

Now I regret to say that nearly the whole of the above statements are hopelessly incorrect. Bravais has no claim, whatever, to supplant Francis Galton as the discoverer of the correlational calculus. For the most part he is simply taking a very special case of the Gaussian analysis, and nowhere on p. 267 of his memoir can I now find that he has used the expressions for the correlation symbols without

* *Comptes rendus*, T. VII. p. 77.

their names. Again Edgeworth did *not* obtain results identical with Bravais', he went on a route of his own to find the true multiple correlation surface and gave as I said in 1895 only doubtful results. But I fear they were not all due to printer's errors. On re-examining his memoir 25 years later I think he harnessed imperfect mathematical analysis to a jolting car and drove it into an Irish bog on his road, and that it was doubtful analysis not errors of printing which led to his obscure conclusions. I was scarcely justified in 1895 in calling the multiple regression result Edgeworth's Theorem. He had tried in 1892 to solve the problem, and he can hardly be said to have succeeded properly. It is very difficult to explain now how my errors of ascription came about, still less possible is it to understand why later writers have not corrected my false history, but merely repeated it.

As far as I can remember what happened at all, it was as follows. I know that I was immensely excited by Galton's book of 1889—*Natural Inheritance*—and that I read a paper on it in the year of its appearance. In 1891—2 I lectured popularly on probability at Gresham College, taking skew whist contours as illustrations of correlation. In 1892 I lectured on variation, in 1893 on correlation to research students at University College, the material being afterwards published as the first four of my *Phil. Trans.* memoirs on evolution. At this time I dealt with correlation and worked out the general theory for three*, four and ultimately n variables. The field was very wide and I was far too excited to stop to investigate properly what other people had done. I wanted to reach new results and apply them. Accordingly I did not examine carefully either Bravais or Edgeworth, and when I came to put my lecture notes on correlation into written form, probably asked somebody who attended the lectures to examine the papers and say what was in them. Only when I now come back to the papers of Bravais and Edgeworth do I realise not only that I did grave injustice to others, but made most misleading statements which have been spread broadcast by the text-book writers.

(3) Let us now examine Bravais' memoir. He commences by stating that he is going to measure the errors of the determination of the coordinates x, y, z of a point in space. These coordinates are not measured directly but are functions of the observed elements $a, b, c, \ldots$, and he puts

$$x = \phi(a, b, c, \ldots),$$
$$y = \psi(a, b, c, \ldots),$$
$$z = \chi(a, b, c, \ldots).$$

He then expands x, y, z linearly in terms of a, b, c assuming that mathematical differentials may be used for errors; thus he writes

$$\delta x = A\delta a + B\delta b + C\delta c + \ldots,$$
$$\delta y = A'\delta a + B'\delta b + C'\delta c + \ldots,$$
$$\delta z = A''\delta a + B''\delta b + C''\delta c + \ldots.$$

He tells us that the A, B, C are differential coefficients, i.e. of the known functions ϕ, ψ, χ, and that to justify the neglect of higher powers and products we must get

* Published in the *R. S. Proc.* Vol. LVIII. p. 241, 1895.

rid of constant sources of error which arise chiefly from vices of method of observation, ignorance of physical laws, etc. That they can be removed by increasing the number of our observations, and in surveying—which he has essentially in mind—by using the repeating circle, which destroys the majority of constant errors and lessens the influence of variable causes by the fact itself of repeating the observed angles. It is clear that he is thinking solely of theodolite work, and that his x, y, z are Gauss' indirectly observed quantities, his directly observed quantities being angles and bases $a, b, c, \ldots$.

He now changes his notation; he uses x, y, z for the errors δx, δy, δz, and m, n, p for δa, δb, δc, and takes equations

$$x = Am + Bn + Cp + \ldots$$

and calls x, y, z the *dependent* variables, m, n, p the *independent* variables. He says that Laplace has shown that a variation of x between x and $x + \delta x$ will be of the form

$$\sqrt{\frac{h_x}{\pi}}\, e^{-h_x x^2}\, dx,$$

where h_x is given by

$$\frac{1}{h_x} = \frac{A^2}{h_m} + \frac{B^2}{h_n} + \frac{C^2}{h_p} + \ldots.$$

It is therefore clear that he supposes that his *observed* quantities $m, n, p, \ldots$ are *uncorrelated* in our sense of the word. In fact he gives for two and three variates the expressions

$$\sqrt{\frac{h_m h_n}{\pi \cdot \pi}}\, e^{-(h_m m^2 + h_n n^2)}\, dm\, dn \quad \text{(p. 261)},$$

$$\sqrt{\frac{h_m h_n h_p}{\pi \cdot \pi \cdot \pi}}\, e^{-(h_m m^2 + h_n n^2 + h_p p^2)}\, dm\, dn\, dp \quad \text{(p. 264)}.$$

There is obviously not a single step, not a line in this, which does not occur in Gauss, except that Gauss would use

$$u^2 = h_m m^2 + h_n n^2 + h_p p^2$$

and not trouble to state that the probability was given by the exponential.

Now Gauss' problem was to express the variability of x in terms of the variability of the observed quantities $a, b, c, \ldots$ or m, n, p, and of the differential coefficients A, B, C. This is absolutely the same as Bravais' problem, and Bravais' treatment goes very little further than Gauss'—indeed it is essentially narrower as while Gauss neither limits the number of his variables nor their nature, Bravais treats only of *position* in space.

I will now give the value of the expression Bravais reaches for his surface of two dimensions, expressing by d^2w the briquette of frequency on $dxdy$:

$$\frac{d^2 w}{dx\, dy} = \frac{1}{\pi} \frac{1}{\left\{\Sigma \dfrac{(AB' - A'B)^2}{h_m h_n}\right\}^{\frac{1}{2}}}\, e^{-\dfrac{\left\{x^2 \Sigma \frac{A'^2}{h_m} - 2xy\, \Sigma \frac{AA'}{h_m} + y^2 \Sigma \frac{A^2}{h_m}\right\}}{\Sigma\left(\frac{(AB' - A'B)^2}{h_m h_n}\right)}} \quad \text{(p. 272)}.$$

Now, if you take

$$x = Am + Bn + \dots,$$
$$y = A'm + B'n + \dots,$$
$$\sigma_x^2 = A^2\sigma_m^2 + B^2\sigma_n^2 + \dots,$$
$$\sigma_y^2 = A'^2\sigma_m^2 + B'^2\sigma_n^2 + \dots,$$
$$\text{Mean}\,(\delta x\,\delta y) = r_{xy}\sigma_x\sigma_y = AA'\sigma_m^2 + BB'\sigma_n^2 + \dots.$$

Whence

$$\sigma_x^2\sigma_y^2(1 - r_{xy}^2) = \Sigma\,(A^2B'^2 + A'^2B^2 - 2AA'BB')\,\sigma_n^2\sigma_m^2$$
$$= \Sigma\,(AB' - A'B)^2\,\sigma_n^2\sigma_m^2,$$

whence, remembering $\quad h_m = \dfrac{1}{2\sigma_m^2}, \quad h_n = \dfrac{1}{2\sigma_n^2},$

we easily deduce the

$$z = \frac{1}{2\pi\sigma_x\sigma_y\sqrt{1 - r_{xy}^2}}\,e^{-\frac{1}{2}\left(\frac{x^2}{\sigma_x^2} - \frac{2r_{xy}xy}{\sigma_x\sigma_y} + \frac{y^2}{\sigma_y^2}\right)\frac{1}{1 - r_{xy}^2}}$$

of our familiar notation.

But this is precisely what Bravais does *not* do, and for the simple reason that his x, y, z are not variables which he has directly determined and for which he can directly find σ_x, σ_y and r_{xy}. He is merely seeking to express the variability of x and y in terms of the directly determined constants and certain differential coefficients. This is one of the fundamental problems of the Method of Least Squares and had already been solved by Gauss. Bravais adds so far *nothing whatever* to Gauss' solution of 20 years earlier. If Bravais discovered correlation, then Gauss had done so previously.

As a matter of fact while the above expression shows how a hasty examination of Bravais' memoir might lead one to believe he had reached the correlation surface, he was in fact occupied with an entirely different problem, one which was really only a particular case of Gauss' earlier and more comprehensive work.

We cannot pass over, however, the really valuable portion of Bravais' memoir. It lies in this: Having got his coefficients of x and y in terms of the differential coefficients A, B, C, ... he writes the surface

$$\frac{K}{\pi}e^{-K^2(\alpha_1x^2 - 2\beta_0xy + \alpha_2y^2)},$$

and then discusses the properties of a surface of which the contours are

$$ax^2 + 2cxy + by^2 = D,$$

i.e. the familiar ellipses of our normal surface. He gets the conjugate of x-axes as the locus of maximum y's and determines the probability of points lying in certain areas—bounded by similar ellipses or in angular sectors. He gets the line $x = -\dfrac{c}{a}y$, which corresponds to Galton's regression-line. But this is not a result of observing x and y and determining their association, but of the fact that x and y

are functions of certain independent and directly observed quantities. When he thinks of c and a at all, it is not in terms of observations on x and y but of the differential coefficients A, B, C of the geometrical relations between position in space and the angles by which that position is found.

Next we come to his surface of three variates and the treatment is identical. He writes

$$z = \frac{G}{\pi^{\frac{3}{2}}} e^{-(ax^2+by^2+cz^2+2ezy+2fxz+2gxy)} \quad \text{(p. 296)},$$

and his primary object is to determine a, b, c, e, f, g in terms of the differential coefficients A, B, C and the variabilities of the observed independent variates.

Thus he gives

$$a = G^2 \Sigma \frac{1}{h_m h_n} \{(A'B'')\}^2, \quad e = G^2 \Sigma \frac{1}{h_m h_n} \{(A''B)(A''B')\}, \quad \frac{1}{G^2} = \Sigma \frac{\{(AB'C'')\}^2}{h_m h_n h_p}.$$

There is throughout merely the standpoint of the Gaussian method of treating errors of observation, and if we are to attribute any discovery of the idea of correlation to Bravais we must with the same confidence assert that Gauss was the primary originator of the whole idea. To my mind this is absurd*. In the case of both these distinguished men the quantities they were observing were *absolutely independent*; they neither of them had the least idea of correlation between observed quantities. The product terms in their expressions—never analysed in the sense of correlation—arise solely not from organic relationships, but from the *geometrical* relationships which exist between their observed quantities and the indirectly observed quantities they deduce from them. Bravais himself (p. 331) says that the application of his results are narrowly circumscribed by the nature of his assumptions—astronomy and the great geodesic surveys alone provide sufficiently accurate material. As far as Gauss and Bravais are concerned we must, I think, hold that they contributed nothing of real importance to the problem of correlation, and that my statement of 1895 was a totally erroneous one.

The same criticism applies to all the treatment of the normal surfaces by later writers, which are described at very considerable length by Czuber in his *Theorie der Beobachtungfehler*, Leipzig, 1891. In all cases the variables are *indirectly* observed quantities and the product terms arise because they are mathematically supposed to be linear functions of the directly observed, but quite independent variables. That the directly measured quantities might themselves be correlated does not seem to have occurred to the many writers on the theory of observations.

As far as I am aware there is nothing to record on our subject beyond the work of the writers on the theory of observations referred to above until we reach Francis Galton himself. His first statement of his ideas was in a lecture at the

* I feel quite certain that if any one had told either Gauss or Bravais that $\Sigma(ab)$ for their *observed* measurements need not be zero, they would have been laughed out of court, as the astronomers now laugh at us, when we assert that their measurements of different stellar magnitudes are very probably correlated!

Royal Institution, Feb. 9, 1877. He had found it very difficult to collect human material for two generations and after careful consideration selected sweet pea seeds. These seeds were both measured and weighed and actually observations were taken on foliage and length of pod although as far as I am aware Galton never published the reductions of the latter. As he himself writes in 1885: "It was anthropological evidence that I desired, caring only for the seeds as means of throwing light on heredity in man. I tried in vain for a long and weary time to obtain it in sufficient abundance, and my failure was a cogent motive, together with others, in inducing me to make an offer of prizes for family records, which was largely responded to, and furnished me last year with what I wanted*."

The title of Galton's R. I. lecture was *Typical Laws of Heredity in Man.* Here for the first time appears a numerical measure r of what is termed 'reversion' and which Galton later termed 'regression.' This r is the source of our symbol for the correlation coefficient, which was really the first letter of 'reversion' not of 'regression.' The main results are given in a mathematical appendix†. Galton works with the modulus—i.e. our $\sqrt{2}\sigma$—probably because the tables of the probability integral were then given in the modulus as argument. But we can at once convert into more customary notation. Thus we find the now familiar result

$$v = \sqrt{1 - r^2}\, c_1,$$

or, translating his symbols:

$$\text{Variability of family} = \sqrt{1 - r^2} \times \text{variability of general population.}$$

Galton had already reached the idea of homoscedasticity in the arrays of offspring. "I was certainly astonished to find the variability of the produce of the little seeds to be equal to that of the big ones; but so it was and I thankfully accept the fact, for had it been otherwise, I cannot imagine, from theoretical considerations, how the typical problem could be solved" (p. 10).

Next Galton supposes the mean taken of both parents and notes that the "variability of the parentage," what he would have called later the mid-parentage, $= \frac{1}{\sqrt{2}}$ variability of either parent. He has not yet reached the idea of reducing one sex to the standard of the other, and the result is only true, if we have to deal with characters not sexually differentiated.

Now we come to the test point‡:

"Reversion"—Galton tells us, p. 10—"is the tendency of the ideal mean filial type to depart from the parental type, reverting to what may be roughly and perhaps fairly described as the average ancestral type. If family variability had been the only process in simple descent that affected the characteristics of a sample the dispersion of the race from its mean ideal type would indefinitely increase with

* Address to Anthropological Section, *B. A. Report*, 1885, p. 1207.

† Royal Institution of Great Britain, Friday, February 9, 1877.

‡ Let the reader remember that these words were spoken just 40 years ago, and that they waited 12 to bring forth fruit!

the number of generations, but reversion checks this increase, and brings it to a standstill."

Galton's proof assuming homoscedasticity is of a very simple nature. Let the reversion be λx, where x is the parental character. Then the mean variability of the offspring generation

$$= \sigma^2(1-r^2) + \lambda^2 (\text{mean } x^2)$$
$$= \sigma^2(1-r^2) + \lambda^2\sigma^2.$$

Therefore unless $\lambda = r$ the population cannot remain stable. Or without any hypothesis as to normality, only on the basis of linearity of reversion, homoscedasticity and stability the Galton coefficient r of reversion must be equal to the r which gives the reduction of the 'family variability.' Thus the lecture of 1877, while it contains points which later work was to clear up, still in the main lines gives on the data for size in sweet pea seeds the fundamental properties of the regression line. I have worked out Galton's data for sweet peas and show you a diagram of the result which Miss A. Davin has prepared for me. The parent seed was of course selected seed, and Galton took 100 of each parental grade and determined the mean of the offspring, which of course were non-selected seed, i.e. not seedsman's seed. Galton fixes the regression in round numbers at $\frac{1}{3}$, I make it slightly larger. In any case the regression coefficient is small, if we consider the sweet pea, as Galton did, as self-fertilising. It has been so proclaimed in several botanical investigations on heredity in the sweet pea. But in 1907 I watched a row of sweet peas and observed *Megachile Willughbiela*, the leaf-cutting bee, in quite considerable numbers visiting the flowers. The Superintendent of the R. H. S.'s garden at Wisley also replied to an inquiry that he had no doubt some English insect cross-fertilised sweet peas because in trying new sorts the gardeners had to place the rows in different parts of the garden to reduce the risk of cross-fertilisation. Darwin's statement* that "in this country it"—the sweet pea—"seems invariably to fertilise itself," appears open to question. Galton's coefficient may therefore, although low, be not so low as it appears on the assumption of self-fertilisation.

The next few years Galton was occupied in collecting material for further investigation of regression and heredity. He had established his Anthropometric Laboratory at South Kensington and by offering prizes obtained his Records of Family Faculties. The first-fruits of these data are to be found in his Presidential Address to the Anthropological Section of the British Association at Aberdeen in 1885. The part of this Address dealing with regression was considerably extended in a paper read to the Anthropological Institution in the same year. Galton now deals with the inheritance of stature and transmutes female to male stature before determining his mid-parentages. He does this, not as we should do now by multiplying by the ratio of paternal and maternal standard deviations, but by the multiplying factor of mean statures 1·08. This is roughly permissible if the coefficients of variation for the two sexes are the same as they very nearly are for stature. In this paper we have the first published diagram of the *two* regression

* *Cross and Self-fertilisation of Plants*, 1878, p. 153.

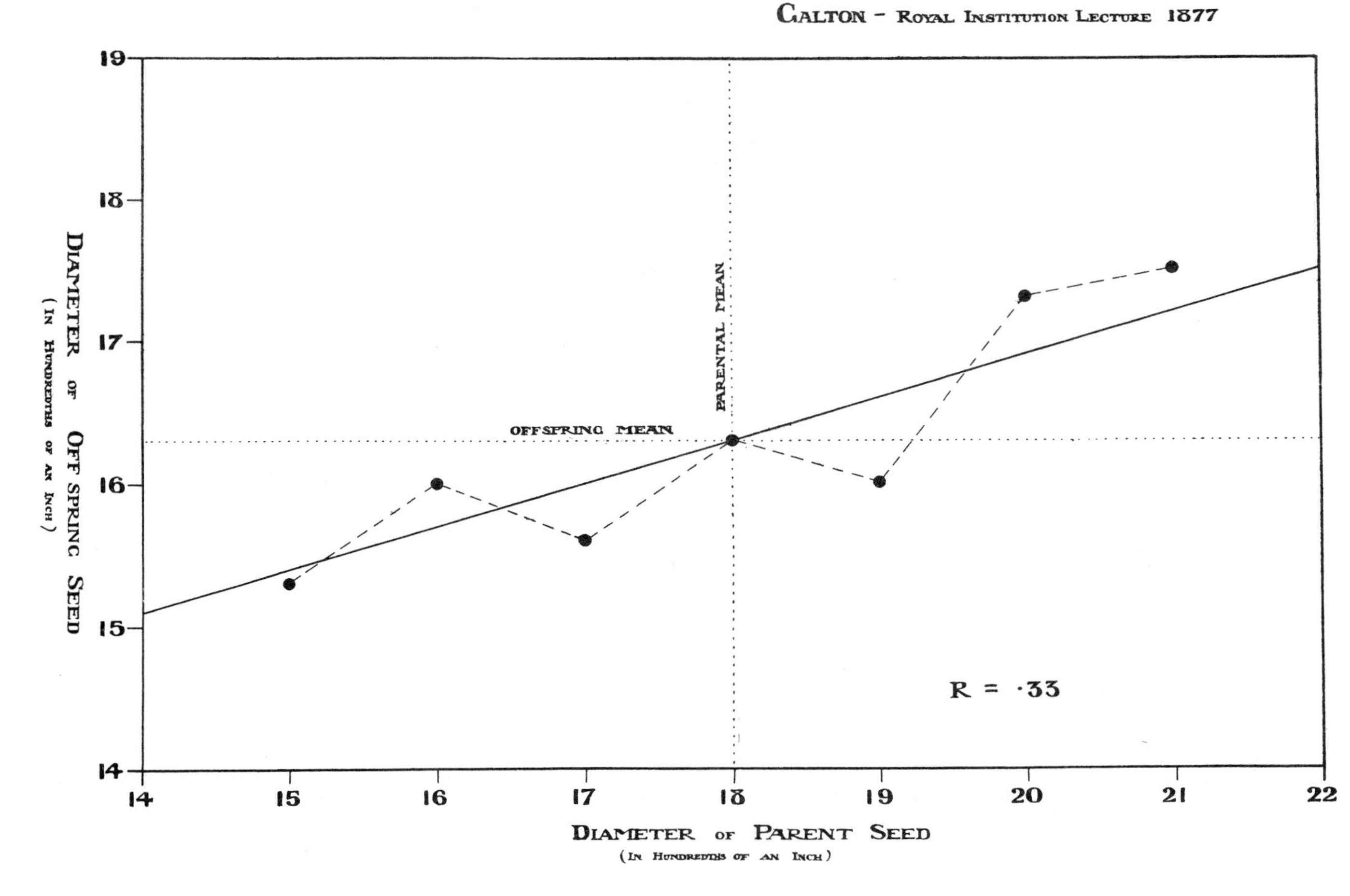
INHERITANCE IN SIZE OF SWEET PEA SEEDS.
GALTON - ROYAL INSTITUTION LECTURE 1877
19 18 17 16 15 14
DIAMETER OF OFF SPRING SEED
(IN HUNDREDTHS OF AN INCH)
OFFSPRING MEAN
PARENTAL MEAN
R = ·33
14 15 16 17 18 19 20 21 22
DIAMETER OF PARENT SEED
(IN HUNDREDTHS OF AN INCH)

lines and the first correlation table (of stature in parentage and offspring) as we should now call it.

Also Galton gives the diagram* which indicates how he discovered observationally the form of the normal frequency surface. He proceeded to smooth his correlation

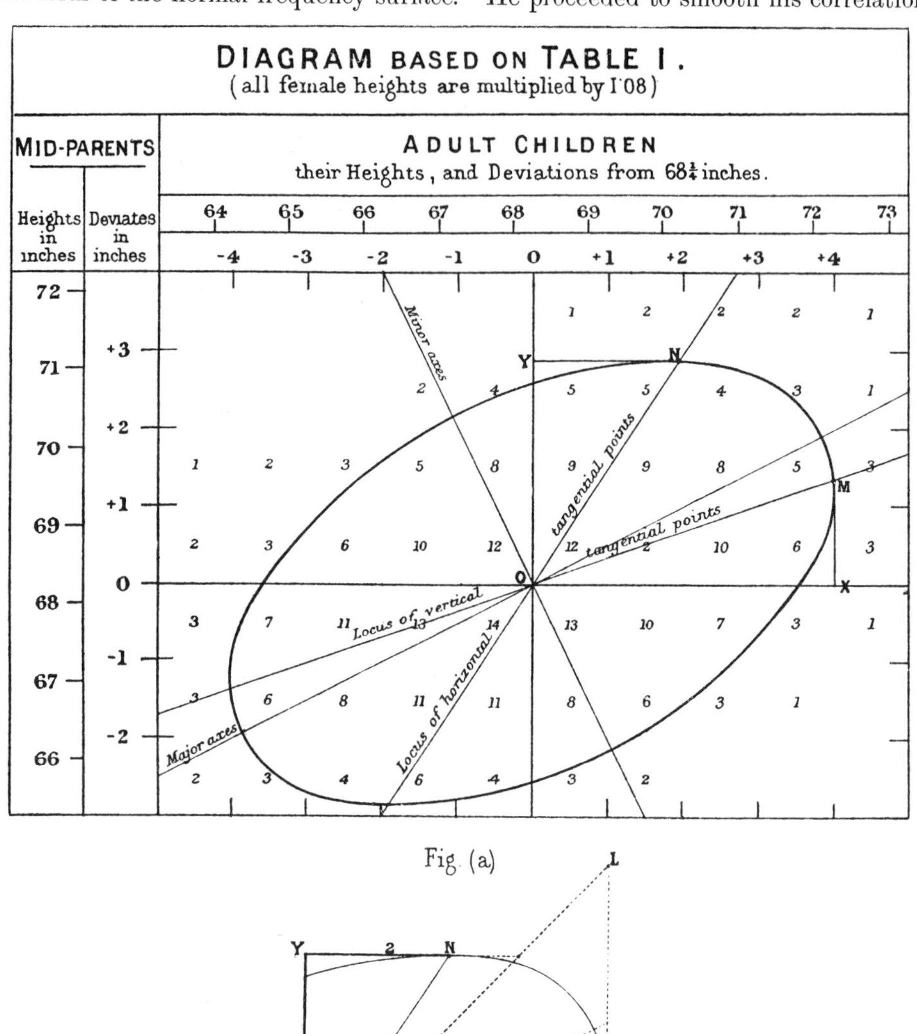

Fig. (a)

table by taking the mean of four adjacent cells, and then drawing contour lines through points of the same frequency. He found such contour lines were a system of concentric similar and similarly placed ellipsoids and that the regression lines

* Reproduced here by permission of the Royal Anthropological Institute.

were what the mathematician terms the conjugate diameters of the variate axes. He discovered that the sections parallel to the variate axes were 'apparently' normal curves of equal S.D. but that this S.D. was reduced and bore a constant ratio to the S.D. of the general population. He knew 8 years earlier the relation of $\sigma\sqrt{1-r^2}$ to the 'reversion' coefficient r. That Galton should have evolved all this from his observations is to my mind one of the most noteworthy scientific discoveries arising from pure analysis of observations.

Why Galton did not at once write down the equation to his surface as

$$z \propto e^{-\frac{1}{2}\frac{y^2}{\sigma_y^2}} \times e^{-\frac{1}{2}\frac{1}{\sigma_x^2(1-r^2)}\left(x - r\frac{\sigma_x}{\sigma_y}y\right)^2}$$

has always been a puzzle to me. Actually he carried the problem, stated in the language of probability, to Mr J. D. Hamilton Dickson, a mathematician of Peterhouse, Cambridge, who after stating the wording of Galton's problem, wrote down the answer substantially as above in the fourth line of his memoir*! The fact is that Galton's statement of his problem, involving as it did the assumption of normal distribution, homoscedasticity and linear regression, provided the answer the moment his results were read in symbols. The explanation of Galton's action possibly lies in the fact that Galton was very modest and throughout his life underrated his own mathematical powers.

Thus in 1885 Galton had completed the theory of bi-variate normal correlation. The next stage in the theory of correlation, multi-variate correlation, was directly indicated by the general problem of ancestry. As is now well known the fundamental regression equation is

$$\frac{x_0 - \bar{x}_0}{\sigma_0} = -\Sigma\left(\frac{R_{0s}}{R_{00}}\frac{x_s - \bar{x}_s}{\sigma_s}\right) \quad \text{.........................(i)},$$

where R_{pq} is the p, q minor of the determinant

$$R = \begin{vmatrix} 1 & r_{01} & r_{02} & \dots & r_{0n} \\ r_{10} & 1 & r_{12} & \dots & r_{1n} \\ \dots & \dots & \dots & \dots & \dots \\ r_{n0} & r_{n1} & r_{n2} & \dots & 1 \end{vmatrix}$$

and the variability of the array is

$$\sigma_0\sqrt{\frac{R}{R_{00}}} \quad \text{...................................(ii)}.$$

Galton endeavoured to reach this by a short cut, and thus evolved his law of ancestral heredity. This was a brilliant and suggestive step, but he was not able to state the conditions under which it is theoretically correct or bring forward data at that time to confirm its observational accuracy.

* *R. S. Proc.* Vol. XL. p. 63, 1886. Galton himself writes (*B. A. Report*, 1885, p. 1211), "I may be permitted to say that I never felt such a glow of loyalty and respect towards the sovereignty and magnificent sway of mathematical analysis as when his answer reached me, confirming, by purely mathematical reasoning, my various and laborious statistical conclusions with far more minuteness than I had dared to hope, for the original data ran somewhat roughly, and I had to smooth them with tender caution."

Another feature of Galton's work at this time must be noted. He worked with the median instead of the mean, and he used probable errors or quartile values instead of standard deviations. Further, to obtain r, he somewhat laboriously expressed both variates in terms of their quartile deviations: thus r became the slope of his regression line. It was then determined by graphically fitting a good line, or from certain chosen arrays. Thus he worked with somewhat primitive statistical tools, and the wonder is that he achieved as much by their aid as he did.

Given A and B with regression r_{ab}, B and C with regression r_{bc}, Galton assumed $r_{ac}=r_{ab}\times r_{bc}$ to obtain his kinship relations. A nephew is the son of a brother. Hence

$$r \text{ for uncle and nephew} = r \text{ for brothers} \times r \text{ for father and son.}$$

This of course is incorrect; it implies the vanishing of the corresponding partial correlation coefficient. Again, I think, his mid-parental correlation is not theoretically consonant with his parental correlations.

Another noteworthy point of the 1877 R. I. and the 1885 A. I. papers is the ample provision of mechanical models to illustrate by dropping shot or seeds the properties of bi-variate frequency. One wonders whether these elaborate quincunxes have been preserved and if so where they are at the present time. I reproduce one of them by permission from the *Journal of the Royal Institution*.

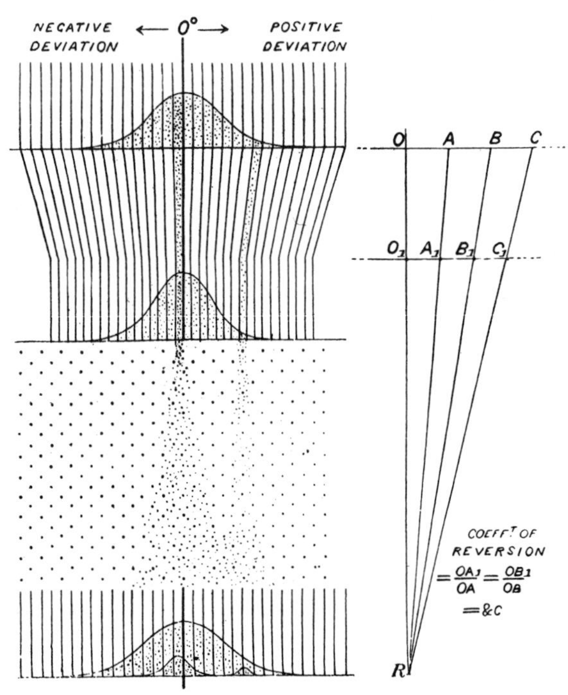

In 1886 Galton published a paper in the *Royal Society Proceedings** on "Family Likeness in Stature." This contains Hamilton Dickson's note and further data from Galton's *Family Records*.

* Vol. XL. pp. 42—66.

He gives under the headings "Mean regression w," and "Quartile of individual variability" the coefficients of correlation of various pairs of relations: Midparent and Offspring, Brothers, Fathers and Sons, Uncles and Nephews, Grandparents and Grandsons, but he does not realise that on the theory of multiple regression there are certain inconsistencies in his values. I do not think that there is much additional contribution to theory in this paper.

In 1888, however, Galton took a great step forward. He recognised that the whole statistical apparatus he had evolved for the treatment of the problem of heredity had a vastly wider significance. In a paper read to the Royal Society on December 5, 1888*, entitled "Correlations and their Measurement chiefly from Anthropometric Data," the term correlation first appears in our subject. Thus Galton's opening lines run:

"Co-relation or correlation of structure" is a phrase much used in biology, and not least in that branch of it which refers to heredity, and the idea is even more frequent than the phrase; but I am not aware of any previous attempt to define it clearly, to trace its mode of action, or to show how to measure its degree.

Two variable organs are said to be correlated when the variation of the one is accompanied on the average by more or less variation of the other, and in the same direction (p. 135).

The last words seem to us now out of place, but Galton had not yet reached the idea of negative correlation. Also the balance is still swinging between 'co-relation' and 'correlation' although it has ultimately fallen to the more weighty word. How clearly Galton grasped the essence of correlation may be shown by the following sentences which might have saved many ingenious later investigators thinking they had made an important discovery. "It is easy to see that co-relation must be the consequence of the variations of the two organs being partly due to common causes. If they were wholly due to common causes, the co-relation would be perfect, as is approximately the case with the symmetrically disposed parts of the body. If they were in no respect due to common causes, the co-relation would be *nil*. Between these two extremes are an endless number of intermediate cases, and it will be shown how the closeness of co-relation in any particular case admits of being expressed by a single number" (p. 135). This single number it is needless to say is our present coefficient of correlation. Galton drops now the w of his 1886 work and returns to the r of his 1877 lecture, and the symbol r has remained to the present day.

Galton's process is the same as in the heredity problem. He used median and quartile and reduces the deviations to their respective quartiles as unit. He then smooths his means of arrays, draws a line to represent them and reads off its slope as r. He thus determines seven correlations which he here terms "indices of correlation†." They are between Stature and Cubit, Stature and Head Length, Stature and Middle Finger Length, Cubit and Middle Finger Length, Head Length and Head Breadth, Stature and Height of Knee, Cubit and Height of Knee. He fully

* *R. S. Proc.* Vol. XLV. pp. 135—145.

† On p. 143 r, the index of co-relation, is identified with the 'regression' or 'reversion' of Galton's earlier papers.

realises (1) that r is the same when obtained from either variate as 'relative,' (2) that r is always less than unity, (3) that r measures the closeness of co-relation, and (4) provides the regression line (p. 145).

On p. 144 the term "partial co-relation" is used but hardly in our modern sense although Galton is feeling his way towards multiple correlation. One problem he gives on p. 144 perhaps deserves mention, namely, if the n variates be expressed in terms of their quartiles then the quartile variability of their sum is $\sqrt{n}$ if they are independent and n if they be "rigidly and perfectly co-related." "The actual value would be almost always somewhere intermediate between these extremes, and would give the information that is wanted."

What Galton needs is the "multiple correlation coefficient," i.e.

$$\sqrt{1-\frac{R}{R_{00}}},$$

but he is not yet on the right track for it.

In 1889 appeared Galton's book *Natural Inheritance* embodying most of the work we have discussed in the earlier memoirs of 1877 to 1888. Beyond this Galton did not carry the subject of correlation. He, in my opinion to-day, created it; there is nothing in the memoirs of Gauss or Bravais that really antedates his discoveries. They were dealing with the relatively narrow problem of determining the probable errors of indirectly observed quantities deduced from independent or uncorrelated directly observed quantities. The product-terms that arise in their investigations were expressed in terms of differential coefficients; they were not treated as a means of determining organic relationships between directly measured variates. Galton, starting from the organic relationship between parent and offspring, passed to the idea of a coefficient measuring the correlation of all pairs of organs, and thence to the 'organic' relationship of all sorts of factors. If you think Galton did not appreciate the width of his new methods you must turn to the last paragraph of his *Introduction* to the *Natural Inheritance.*

"The conclusions cannot, however, be intelligently presented in an introductory chapter. They depend on ideas that must first be well comprehended, and which are now novel to the large majority of readers and unfamiliar to all. But those who care to brace themselves to a sustained effort, need not feel much regret that the road to be travelled over is indirect, and does not admit of being mapped beforehand in a way they can clearly understand. It is full of interest of its own. It familiarizes us with the measurement of variability, and with curious laws of chance *that apply to a vast diversity of social subjects.* This part of the inquiry may be said to run along a road on a high level that affords wide views in unexpected directions, and from which easy descents may be made to totally different goals to those we have now to reach. I have a great subject to write upon..." (p. 2).

Galton realised as fully as any of us now the width of application that would open up to the new calculus of correlation, and what easy descents there would be from the "high level road" to strange goals. His notebooks of this period show that he was applying correlation and the regression line in a variety of ways thus to the relation between wing and tail length in birds, to fertility and to disease. His advance was chiefly hampered by the restriction of his data and the need for organised observers and computers.

The publication of *Natural Inheritance* provided Francis Galton with at least three recruits for the field of correlation: Weldon, Edgeworth and myself.

Weldon started in 1889 measuring the organs of shrimps at Plymouth and he was able to announce early in 1890—the letter is now in the glass case in our library here—the first correlation coefficients, or as he termed them "Galton Functions," between organs in shrimps. This was rapidly followed by his work on crabs, and the attempt to show that Galton functions were the same for all local races of the same species. In his first paper on shrimps Weldon writes*:

"In making this investigation I have had the great privilege of being constantly advised and helped in every possible way by Mr Galton. My ignorance of statistical methods was so great that without Mr Galton's constant help, given by letter at the expenditure of a very great amount of time and trouble, this paper would never have been written."

The pupil, however, was soon to outdistance the master in the width of his theoretical knowledge. A second paper on the shrimp followed in 1892†, and this deals more closely with the correlations. Weldon now replaces medians by means in both marginal totals and arrays. He still uses probable errors or quartiles, and goes through the laborious process of reducing each deviation to the probable error. He uses r "in accordance with Mr Galton's notation" to represent the constant which measures the "degree of correlation" between organs. I think, but it is not quite clear, that he determined his probable error from the mean error, not from the quartile. He then determined r from each individual array and took the mean value of these r's as the true r. He accordingly introduced a greatly increased accuracy into the computing of correlation. He dealt with five local races of shrimps and found correlations for 22 pairs of organs. His regression diagram, p. 8, is still an admirable sample of this type of work. The correlations between post-spinous portion and total carapace length may be cited as illustrations of what Weldon and Galton were testing:

Plymouth (1000)	$r = 0{\cdot}81$
Southport (800)	$r = 0{\cdot}85$
Roscoff (500)	$r = 0{\cdot}80$
Sheerness (380)	$r = 0{\cdot}85$
Helder (300)	$r = 0{\cdot}83$

The suggestion that r has the same value for all races of the same species was supposed to be confirmed by these results. We now realise that without a knowledge of the probable error of r, such a statement is illusory. But it was this very series of values which led to the investigation of the probable error of r and so to the extension of the correlational calculus.

* *R. S. Proc.* Vol. XLVII. p. 445, 1890.

† *R. S. Proc.* Vol. XLI. p. 2.

In this paper Weldon also published for the first time with due appreciation of their meaning *negative* correlation coefficients. In conclusion Weldon remarks: "A large series of such specific constants would give an altogether new kind of knowledge of the physiological connexion between the various organs of animals; while a study of those relations which remain constant through large groups of species would give an idea, attainable at present in no other way, of the functional correlations between various organs which have led to the establishment of the great subdivisions of the animal kingdom" (p. 11). In these lines we can read the starting-point of biometry as applied to other types of life than man.

I will not keep you longer over Weldon's contributions than to say that in 1893 appeared his third statistical paper* on "Correlated Variations in Naples and Plymouth shore Crabs." Weldon dealt with 23 pairs of organs in both Naples and Plymouth races. He proposes to call r "Galton's function"†. The paper shows that the 23 values of r at Plymouth and Naples are fairly close, but was again inconclusive because the significance of the differences could not be ascertained without a knowledge of the probable error of r.

We may next turn to Edgeworth, whose fundamental paper is that on "Correlated Averages" which appeared in the *Philosophical Magazine* of August, 1892, pp. 190—204. Edgeworth starts by referring to Galton's memoir of 1888 and Weldon's of 1892 on shrimps. He assumes for the probability that any particular values $x_1, x_2, \ldots$ shall occur

$$\Pi = Je^{-R} dx_1 dx_2 dx_3 \ldots,$$

where R is

$$= p_1 (x_1 - \bar{x}_1)^2 + p_2 (x_2 - \bar{x}_2)^2 + \ldots + 2q_{12} (x_1 - \bar{x}_1)(x_2 - \bar{x}_2) + \ldots.$$

He does not justify this assumption but hopes to do so in a subsequent paper. He states that Galton by the happy device of measuring each deviation by the corresponding quartile had reduced in the case of two variates

$$R = \frac{x_1^2}{1-\rho^2} - \frac{2\rho x_1 x_2}{1-\rho^2} + \frac{x_2^2}{1-\rho^2}$$

to the discovery of a single constant ρ. This is hardly accurate; to reduce the expression R to the above it would be needful to measure not in terms of the quartile but of $\sqrt{2}$ S.D., which is I think sometimes termed the 'modulus‡. Edgeworth replaces Galton's "Index of Co-relation" and Weldon's "Galton's Function" by the term "coefficient of correlation." He then proceeds to weaken down Weldon's process of finding a mean r by suggesting that it will be adequate to find it by taking some of the ratios of 'subject' and mean 'relative' instead of the whole series. I look upon this suggestion as a distinctly retrogressive step.

* *R. S. Proc.* Vol. LIV. pp. 318—329.

† "The importance of this constant in all attempts to deal with the problems of animal variation was first pointed out by Mr Galton...and I would suggest that the constant whose changes he has investigated and whose importance he has indicated, may fitly be known as 'Galton's function,'" p. 325.

‡ Edgeworth appears to realise this on p. 194, but he did not go back and correct his statement of p. 190.

Our object should be to find the 'best value' for r and not how it may be most easily determined at the obvious cost of accuracy.

Although I am unable to follow some of Edgeworth's notation, he undoubtedly reaches something like the correct value for the correlation surface of three variates. In his notation

$$R = \Delta \{(1-\rho_{23}^2)x_1^2 + (1-\rho_{31}^2)x_2^2 + (1-\rho_{12}^2)x_3^2 - 2x_1x_2(\rho_{12}-\rho_{13}\rho_{23}) - 2x_2x_3(\rho_{23}-\rho_{21}\rho_{31}) - 2x_3x_1(\rho_{31}-\rho_{32}\rho_{12})\},$$

where $$\Delta^{-1} = \{(1-\rho_{13}^2)(1-\rho_{12}^2) - (\rho_{12}\rho_{23}-\rho_{13})^2\}$$

according to him, but the factor $1-\rho_{13}^2$ should be replaced by $1-\rho_{23}^2$ I think.

Even with this change I am unable to reach the value he gives in Galton's case of

$$\rho_{12} = \cdot 8, \quad \rho_{13} = \cdot 9, \quad \rho_{23} = \cdot 8,$$

for these seem to give

$$\Delta = 9\cdot 9305,$$

whereas Edgeworth's value is 16·129.

I do not grasp his equation at the foot of p. 196, nor follow how the equation at the top of p. 197 follows from it.

Lastly we come to p. 201 where we should expect to find the general regression equation. Edgeworth tells us that the reasoning is quite general and accordingly we ought to anticipate that his results whatever they are would give our accepted values

$$p_{ss} = \frac{R_{ss}}{R} \text{ and } p_{ss'} = \frac{R_{ss'}}{R},$$

where R is the determinant of the correlations. Instead of this simple rule Edgeworth sums up in the middle of the page with equations

$$\Delta\rho_{13} = +\Delta^3(\rho_{24}\rho_{31}\rho_{42}),$$
$$\Delta\rho_{14} = -\Delta^3(\rho_{21}\rho_{32}\rho_{43}),$$
$$\vdots \qquad \vdots \qquad \vdots$$

There is no explanation of what the symbolism means, and I cannot interpret it, so as to provide the requisite generalisation for n variates.

On the other hand while unable to interpret Edgeworth's general analysis I agree in the case of four variates with the only two terms I have taken the trouble to test in his numerical illustration of this case,

$$\rho_{12} = \frac{1}{\sqrt{2}}, \quad \rho_{13} = \sqrt{\frac{1}{3}}, \quad \rho_{14} = \sqrt{\frac{1}{4}}, \quad \rho_{23} = \sqrt{\frac{2}{3}}, \quad \rho_{24} = \sqrt{\frac{2}{4}}, \quad \rho_{34} = \sqrt{\frac{3}{4}},$$

namely 2 as the coefficient of x_1^2 and $-2\sqrt{2}$ as the coefficient of x_1x_2, my R being $\frac{1}{24}$, R_{11} being $\frac{1}{12}$ and R_{12} being $\frac{-\sqrt{2}}{24}$. Edgeworth does not provide the needful external constant of the frequency surface, i.e.

$$\frac{N}{(2\pi)^n \sigma_1\sigma_2 \ldots \sigma_n} \frac{1}{\sqrt{R}}.$$

I should sum up Edgeworth's work of 1892 by saying that he left the problem of multiple correlation at least in a very incomplete state. He probably knew what he was seeking himself, but he did not give the requisite attention to the wording or printing of his memoir to make it clear to others, and accordingly in looking back at the matter now I am very doubtful whether in 1895 I ought to have called the problem of multiple correlation, "Edgeworth's Problem." He certainly did not put the answer to it in a form in which the statistician with a customary amount of mathematical training could determine the form of the surface for n variates, as soon as their S. D.'s and correlations had been calculated. I think I am justified in saying this for I have not to my recollection come across any treatment of multiple correlation which starts from Edgeworth's paper or uses his notation.

It will be seen from what has gone before that in 1892 the next steps to be taken were clearly indicated. They were, I think,

(*a*) The abolition of the median and quartile processes as too inexact for accurate statistics.

(*b*) The replacement of the laborious processes of dividing by the quartiles and averaging the deduced values of r, by a direct and if possible 'best' method of finding r.

(*c*) The determination of the probable errors of r as found by the 'best' and other methods.

(*d*) The expression of the multiple correlation surface in an adequate and simple form.

These problems were solved by Dr Sheppard or myself before the end of 1897.

Closely associated with these problems arose the question of generalising correlation. Why should the distribution be Gaussian, why should the regression curve be linear?

As early as 1893 I dealt with quite a number of correlation tables for long series and was able to demonstrate

(i) by applying Galton's process of drawing contours of equal frequency that most smooth and definite systems of contours can arise from long series, obviously mathematical families of curves, which are (*a*) ovaloid, not ellipsoid, and (*b*) which do not possess—like the normal surface contours—more than one axis of symmetry,

(ii) that regression curves can be quite smooth mathematical curves differing widely from straight lines,

(iii) that in cases wherein (i) and (ii) hold, homoscedasticity is not the rule.

I obtained differential equations to such systems, but for more than 25 years while often returning to them, have failed to obtain their integration.

This seems to me the desideratum of the theory of correlation at the present time: the discovery of an appropriate system of surfaces, which will give bi-variate skew frequency. We want to free ourselves from the limitations of the normal surface, as we have from the normal curve of errors.

As early as 1897 Mr G. U. Yule*, then my assistant, made an attempt in this direction. He fitted a line or plane by the method of least squares to a swarm of points, and this has been extended later to n-variates and is one of the best ways of reaching the multiple regression equations and the coefficient of multiple correlation†. Now while these methods are convenient or utile, we may gravely doubt whether they are more accurate *theoretically* than the assumption of a normal distribution. Are we not making a fetish of the method of least squares as others made a fetish of the normal distribution? For how shall we determine that we are getting a 'best fit' to our system by the method of least squares?

If we are fitting a curve $y = f(x, c_1, c_2, c_3)$

to a series of observations we can only assert that least square methods are theoretically accurate on the assumption that our observations of y for a given x obey the normal law of errors. That is the proof which Gauss gave of his method and I personally know no other. *Theoretically* therefore to have justification for using the method of least squares to fit a line or plane to a swarm of points we must assume the arrays to follow a normal distribution. If they do not, we may defend least squares as likely to give a fairly good result but we cannot demonstrate its accuracy. Hence in disregarding normal distributions and claiming great generality for our correlation by merely using the principle of least squares, we are really depriving that principle of the basis of its theoretical accuracy, and the apparent generalisation has been gained merely at the expense of theoretical validity. Take other distributions of deviations for the arrays and the method of least squares is not the one which will naturally arise from making the combined probability a maximum. It is by no means clear therefore that Mr Yule's generalisation indicates the real line of future advance.

I have endeavoured to indicate in this paper the broad outline of the early history of correlation which has now a most extensive literature. It is a long step from Francis Galton's 'reversion' in sweet pea seeds to the full theory of multiple correlation, which we now know to be identical with the spherical trigonometry of high-dimensioned space, the total correlation coefficients being the cosines of the edges of the polyhedra and the partial correlation coefficients the cosines of the polyhedral angles. But to find the correlation of the health of a child with the number of people per room while you render neutral its age, the health of its parents, the wages of its father, and the habits of its mother, is no less vital a problem than Galton's correlation of character in parent and offspring. It requires indeed more mathematics, but the mathematics are not there for the joy of the analyst but because they are essential to the solution. It is the transition from the mill as pestle and mortar to the mill with steam driven grain crushing steel rollers. But the inventor of milling was the person who bruised grain between two stones, and Galton was the man who discovered the highway across this new country with what he aptly terms "its easy descents to different goals."

* *Journal of Royal Statistical Society*, Vol. LX. Part iv, p. 3.

† *Biometrika*, Vol. VIII. p. 438. The method adopted in the paper is not that of fitting a generalised plane by least squares, but of making a generalised correlation coefficient take its maximum value. It appeals only to the rules of the differential calculus and not to the method of least squares, or indirectly to Gauss' law of errors.

THE HISTORICAL DEVELOPMENT OF THE GAUSS LINEAR MODEL

By HILARY L. SEAL

Summary

The linear regression model owes so much to Gauss that we believe it should bear his name. Other authors who made substantial contributions are: Cauchy who introduced the idea of orthogonality; Chebyshev who applied it to polynomial models; Pizzetti who found the distribution of the sum of squares of the residuals on the Normal assumption; Karl Pearson who linked the model with the multivariate Normal thereby broadening the field of applications; and R. A. Fisher whose extension of orthogonality to qualitative comparisons laid the foundations of the modern theory of experimental design.

1. Introduction

The rigid nineteenth century distinction between the Theory of Errors and Mathematical Statistics lingers on even today. Treatises on the 'combination of observations' by astronomers (e.g. Smart, 1958) and on 'compensation', 'adjustment', or 'graduation' by geodesists (e.g. Marchant, 1956; Tienstra, 1956; Rainsford, 1957; Grossmann, 1961) continue to expound the application of least squares to Gauss's linear model seemingly with little knowledge of the modifications introduced by statistical theory. On the other hand statistical texts on the 'analysis of variance', on 'regression analysis' or on the 'linear model' (e.g. Scheffé, 1959; Williams, 1959; Plackett, 1960; Graybill, 1961; Rao, 1965) seldom make more than a passing reference to Gauss and his nineteenth century followers.

Bortkiewicz's (1909) article on Statistics in the *Encyclopédie des Sciences Mathématiques* shows that on the continent of Europe the term mathematical statistics used to refer mainly to the treatment of observed relative frequencies. The estimation (as we would now say) of the interconnected probabilities of mortality and invalidity constituted the central problem of statistical theory and, in fact, the only two nineteenth century texts on 'mathematical statistics' (Wittstein, 1867; Zeuner, 1869) dealt almost exclusively with this topic. The Theory of Errors had by then been the subject of advanced mathematical treatment but essentially no overlap was envisaged with the theory of statistics (Bauschinger, 1908).

Yet the nineteenth century had seen the application of probability (frequency) distributions to measurements whose deviations from their mean could not be regarded as 'error'. Already Laplace (1812, Ch. VIII) had proved that the probability distribution of the expectation of life at any given age tends towards the Normal, though he did not illustrate this with any actual data. And before mid-century Quetelet (1846) had fitted Normal curves to 5738 chest measurements of Scots soldiers and to 10,000 measurements of the height of French conscripts. By the turn of the century Karl Pearson, inspired by the work of Francis Galton, had emphasized the diversity of observable frequency distributions (E. S. Pearson,

1956; papers 1, 2, 7 and 8). It no longer seemed proper to refer to such distributions as laws of 'error', though Bortkiewicz does so in the article referred to.

Why, then, could not the Theory of Errors be absorbed into the broader concept of statistical theory that was being evolved by Karl Pearson and his followers and, eventually, by R. A. Fisher? We suggest that the original reason was Pearson's preoccupation with the multivariate normal distribution and its parameters. The predictive regression equation of his pathbreaking 'regression' paper (1896; paper 3 in E. S. Pearson, 1956) was not seen to be identical in form and solution to Gauss's *Theoria Motus* (1809) model. Later, distaste for the 'innumerable treatises' on least squares caused him to misunderstand the model on which the subsequent 'lengthy analysis' is based (Pearson, 1920).* During the following 15 years R. A. Fisher and his associates at Rothamsted Experimental Station were rediscovering many of the mathematical results of least squares (or error) theory, apparently agreeing with Pearson that this theory held little to interest the statistician.

The purpose of this article is to trace the development of the present day theory of the linear model from its formulation by Gauss to its widespread application by the Fisherian school under the title 'analysis of variance'. We have found it convenient to terminate our account at the year 1935, the date of Fisher's text on the design of experiments and the commencement of a period of great activity in this subject. This year is also the date of Aitken's matrix formulation of the theory of least squares which, with several embellishments, is now the preferred approach by statisticians and 'errorists' alike.

We mention that in our search for priority of a method or a formula we have restricted ourselves to *published* work. For example, although Gauss claims in his *Theoria Motus* (1809) to have used the 'principle' of least squares before 1795 (when he was eighteen) we have not thereby removed Legendre from his position as the originator of this method. And although we have heard it stated that Fisher in private conversation showed himself familiar with the work of the earlier mathematicians on the continent of Europe, we have not used this information. Fisher published a number of historical vignettes but these did not include any references to the substantial literature of least squares that had appeared by the turn of the century. It is no discredit to the genius whose published output is large and continuous that he failed to take the time to review all previous literature.

In this connexion we mention that, apart from the cumulative review articles of the *Encylopédie*, there has been no difficulty in the last 100 years in checking whether an apparently new mathematical theorem has already seen the light of day. The *Jahrbuch über die Fortschritte der Mathematik* commenced publication in 1871 by reviewing all mathematical books and papers published in the year 1868. Even in this first volume there was a separate section for 'Wahrscheinlichkeitsrechnung'. Subsequent years' publications were reviewed in due course and the section on probability was changed to 'Kombinationslehre und Wahrscheinlichkeitsrechnung'. By the time the journal ceased publication with volume **66** for the year 1940 the relevant section had become 'Wahrscheinlichkeitsrechnung und Anwendungen' though, of course, non-mathematical journals only received accidental coverage.

2. Legendre's method of least squares

The linear model with a rather vague probabilistic framework (Eisenhart, 1964) was first subjected to general treatment by Legendre (1805). In the appendix to his work he writes that in problems where it is necessary to draw the most exact conclusions from

*Paper No. 14 in this volume

observational measurements one 'presque toujours' ends up with a system of equations of the form

$$e_i = \sum_{j=1}^{q} \beta_j z_{ji} - x_i \quad (i=1,2,\ldots,N;\ N>q). \tag{1}$$

Here the z_{ji} are known 'coefficients', the x_i are measurements, the β_j are the 'unknowns', and the e_i are the 'errors'. Note that in Legendre's illustration from astronomy there was a 'constant' term free of an 'unknown' in the subtractive term x_i and the 'measurement' was reduced by this before it appeared on the right of the equation. Gauss's numerical example in *Disquisitio Palladis* (1810) assumed a similar form.

Legendre's objective was to 'determine' (estimate) the q 'unknowns' (parameters) β_j if possible in such a way that each 'error' (residual) becomes 'très-petite' and the 'erreurs extrêmes', without regard to sign, are kept within narrow limits. The *principle* that he proposed for this purpose was the minimization, for variation in the β_j, of the sum of the squares of the 'errors' e_i.

Having stated this principle Legendre proceeded to obtain the (normal) equations, from which to estimate the β_j, by partial differentiation of Σe_i^2. As a numerical illustration he developed five observational equations of the form (1) in which three parameters were to be estimated, the first being in effect attached to a vector of unit 'constants', $z_{1i} = 1$. It should be added that Legendre did not distinguish between errors and residuals nor did he find the sum of the squares of his five calculated zero-sum residuals. In his numerical illustration this sum had only two degrees of freedom.

3. Gauss's linear model

The first discussion of the model (1) in which the probability distribution of the errors e_i was explicitly considered was in Gauss's astronomical text *Theoria Motus* (1809). We mention that Gauss's writings on least squares are conveniently available in the recently reissued compilation and translation by Börsch & Simon (1887). An excellent summary of them in English is given by Whittaker & Robinson (1926, Ch. IX). A beautiful exposition in modern terminology that goes only slightly further than Gauss himself is by van der Waerden (1957, Ch. 7).

In the *Theoria Motus* Gauss assumed that, with a uniform prior distribution of the parameter of location, the mode of the posterior distribution of N independent measurements (errors) was the arithmetic mean of those errors. He accordingly found that the distribution of errors, assumed continuous, was necessarily of Normal form (Whittaker & Robinson, 1926, §112). His linear model was thus appropriate for the case where the e_i are sampled independently from a Normal universe with zero mean and fixed variance σ^2 or, as Gauss preferred, a fixed precision h, where $2\sigma^2 = h^{-2}$. His slight extension to the situation where e_i has a variance $w_i^{-1}\sigma^2$ $(i=1,2,\ldots,N)$ with w_i known need not detain us.

It was only as the nineteenth century progressed that the Normal law, thus named by Galton, came to enjoy widespread acceptance. Its gradual recognition as the law of errors *par excellence* is detailed by Czuber (1891, 1899). Needless to say much of its popularity was based on the idea of a large number of 'elementary errors' combining to form the e_i of model (1).

But when Gauss wrote, the Normal law was still unfamiliar and in his *Theoria Combinationis* (1821–26) he only used it in the early paragraphs as one of three illustrative continuous probability distributions. Gauss's willingness to discard the Normal is explained by his discovery of a general inequality applicable to any continuous probability distribution symmetrical about its single mode μ. He stated this inequality as

$$P(|X-\mu| \geqslant \lambda\sigma) \leqslant \begin{cases} 1-\lambda/\sqrt{3} & (\lambda \leqslant 2/\sqrt{3}), \\ 4/(9\lambda^2) & (\lambda \geqslant 2/\sqrt{3}), \end{cases}$$

(Savage, 1961) and its use with $\lambda = 2$ shows that there is at least an 89 % chance that any such variate lies within two standard deviations of the mean and mode. Thus if any parameter estimate were accompanied by its standard deviation the use of an inverse argument would permit conclusions to be drawn about the probable range of values of the parameter itself.

If we suppose, with Gauss (1809), that h is known and that each β_j has a uniform prior probability distribution then the (posterior) probability of the system of errors $\{e_i\}$ is proportional to

$$\pi^{-\frac{1}{2}N} h^N \exp\left(-h^2 \sum_{i=1}^{N} e_i^2\right) \prod_{j=1}^{q} d\beta_j, \tag{2}$$

the constant of proportionality being determined by integrating this expression over all permissible values of $\beta_j \, (j=1,2,\ldots,q)$ and inverting the result. Clearly the foregoing expression will be a maximum when Σe_i^2 assumes its minimum value. This condition leads immediately to the simultaneous linear equations for the determination of the estimates of the β_j's.

In order to indicate succinctly the remainder of Gauss's theoretical procedure in his 1809 text and the subsequent computational algorithm of his *Disquisitio Palladis* (1810) lecture it is convenient to utilize matrix notation. Write the N-component vector of errors defined by (1) as

$$\mathbf{e} = \mathbf{Z}\boldsymbol{\beta} - \mathbf{x}, \tag{3}$$

where

$$\mathbf{Z} = ((z_{ji})) = \{\mathbf{z}_1 \mathbf{z}_2 \ldots \mathbf{z}_q\}$$

is an $(N \times q)$ matrix of 'coefficients', and $\boldsymbol{\beta}$ and $\mathbf{x}$ are the q-component and N-component vectors of parameters and observations, respectively. The normal equations for estimation of $\boldsymbol{\beta}$ are then

$$(\mathbf{Z}'\mathbf{Z})\,\hat{\boldsymbol{\beta}} = \mathbf{Z}'\mathbf{x}. \tag{4}$$

By combining the theoretical development of 1809 with the numerical procedure of the 1810 lecture we may restate Gauss's results as follows.

Let us premultiply the matrix $\mathbf{Z}'\mathbf{Z}$ by the $q \times q$ matrix $\mathbf{G} = \mathbf{G}_q \mathbf{G}_{q-1} \ldots \mathbf{G}_2 \mathbf{G}_1$, where each $\mathbf{G}_k$ is a lower triangular matrix with units in its principal diagonal (called hereafter lower unit-triangular),

$$\mathbf{G}_1 = \begin{bmatrix} 1 & & & & & \\ p_{21} & 1 & & & & \\ p_{31} & 0 & 1 & & & \\ \vdots & \vdots & \vdots & & & \\ p_{q1} & 0 & 0 & \ldots & 0 & 1 \end{bmatrix}.$$

$p_{i1} = -\mathbf{z}_1'\mathbf{z}_i/\mathbf{z}_1'\mathbf{z}_1$, namely minus the ith element of the first *row* of $\mathbf{Z}'\mathbf{Z}$ divided by the diagonal element in that row, and $\mathbf{G}_2$ is formed from $\mathbf{G}_1$ by moving the column of p-multipliers one to the right thus shortening it by one element. The new multipliers are to have the same form as those of $\mathbf{G}_1$ but are to be derived from the *second* row of the matrix $\mathbf{Z}'\mathbf{Z}$ *as it stands after premultiplication by* $\mathbf{G}_1$. The matrix $\mathbf{G}_3$ is formed in a similar manner after the matrix $\mathbf{G}_2\mathbf{G}_1(\mathbf{Z}'\mathbf{Z})$ becomes available. And so on, ending with $\mathbf{G}_q = \mathbf{I}_q$. The divisor of the multipliers in $\mathbf{G}_k$ is called the kth pivot. We write it as d_k so that, e.g. $d_1 = \mathbf{z}_1'\mathbf{z}_1$. Gauss proved that $d_k > 0$, all k, and we write them as the elements of the principal diagonal of the $(q \times q)$ diagonal matrix $\mathbf{D}$.

It can be shown (Fox, 1964, Ch. 3) that both $\mathbf{G}$ and $\mathbf{G}^{-1}$ are lower unit-triangular matrices. Furthermore, the method of constructing the multipliers of the successive $\mathbf{G}_k$ results in

$$\mathbf{G}(\mathbf{Z}'\mathbf{Z}) = \mathbf{D}(\mathbf{G}')^{-1}, \tag{5}$$

where $\mathbf{G}'$ is the transpose of $\mathbf{G}$ and is thus upper unit-triangular.

The application of $\mathbf{G}$ to the left-hand side of the normal equations (4) is no more than a formalization of the school-book method of successive elimination of unknowns in a set of simultaneous linear equations. But, using relation (5), Gauss introduced the linear transformation

$$\boldsymbol{\alpha} = \mathbf{D}(\mathbf{G}')^{-1}\boldsymbol{\beta} - \mathbf{G}\mathbf{Z}'\mathbf{x} \tag{6}$$

in which the upper triangular matrix $\mathbf{D}(\mathbf{G}')^{-1}$ has the pivots d_k in its principal diagonal. He then showed that

$$\begin{aligned} \mathbf{e}'\mathbf{e} &= (\mathbf{Z}\boldsymbol{\beta} - \mathbf{x})'(\mathbf{Z}\boldsymbol{\beta} - \mathbf{x}) \\ &= \boldsymbol{\alpha}'\mathbf{D}^{-1}\boldsymbol{\alpha} + \mathbf{x}'(\mathbf{I}_N - \mathbf{Z}\mathbf{G}'\mathbf{D}^{-1}\mathbf{G}\mathbf{Z}')\mathbf{x} \\ &= \frac{\alpha_1^2}{d_1} + \frac{\alpha_2^2}{d_2} + \ldots + \frac{\alpha_q^2}{d_q} + \mathbf{x}'(\mathbf{I}_N - \mathbf{Z}\mathbf{G}'\mathbf{D}^{-1}\mathbf{G}\mathbf{Z}')\mathbf{x}. \end{aligned} \tag{7}$$

The maximization of (2) for variations in the β_j's has thus been replaced by the minimization of (7) for variation in the α_j's which are linear functions of the β_j's. Hence the 'most probable' value of $\boldsymbol{\alpha}$ is $\hat{\boldsymbol{\alpha}} = \mathbf{0}$ and, from (6),

$$\hat{\boldsymbol{\beta}} = \mathbf{G}'\mathbf{D}^{-1}\mathbf{G}\mathbf{Z}'\mathbf{x}. \tag{8}$$

When relation (7) is inserted in expression (2) and $\Pi\, d\beta_j$ replaced by $\Pi\, d\alpha_j$ times a constant of proportionality (Gauss, 1809), it is seen that the 'variable' α_j is Normally distributed about a mean of zero with precision $h/d_j^{\frac{1}{2}}$. Hence, by relation (6), the 'variable' β_j is Normally distributed about a mean of $\hat{\boldsymbol{\beta}}_j$, given by (8), with precision $hd_j^{\frac{1}{2}}$. Now that uniform prior probability distributions are fashionable once again this theorem has become a textbook item (Lindley, 1965, Ch. 8).

We may note that by expressing $\boldsymbol{\beta}$ in terms of $\boldsymbol{\alpha}$ relation (6) permits the model (1) to be written in the form

$$\mathbf{e} = \mathbf{Z}\mathbf{G}'\mathbf{D}^{-1}\boldsymbol{\alpha} + (\mathbf{Z}\mathbf{G}'\mathbf{D}^{-1}\mathbf{G}\mathbf{Z}' - \mathbf{I})\mathbf{x}, \tag{9}$$

where the columns of the $(N \times q)$ matrix $\mathbf{Z}\mathbf{G}'\mathbf{D}^{-1}$ are now orthogonal to one another since $(\mathbf{Z}\mathbf{G}'\mathbf{D}^{-1})'\mathbf{Z}\mathbf{G}'\mathbf{D}^{-1} = \mathbf{D}^{-1}$, a diagonal matrix. This relation was not employed by Gauss but Thiele (1889, 1897, 1903—the latter a barbarous English translation of the second edition) made explicit use of it to illustrate his 'free functions', namely linear uncorrelated functions of the observations, and emphasized the computational advantages of selecting the vectors $\mathbf{z}$ in 'such a convenient way that all these sums of the products vanish'. Thiele's lengthy discussion of this problem is rather obscure and only hints at the advantage of balanced experimental designs. It was not until R. A. Fisher that these ideas were to be fruitfully applied.

As we have indicated Gauss abandoned the Normal law assumption in his *Theoria Combinationis*. In the first part (1821) of this work he proved that among all the unbiased estimates of β_j which are linear compounds of the observations $\mathbf{x}$, that produced by the least squares procedure has minimum variance. Gauss wrote

$$\boldsymbol{\beta} = \hat{\boldsymbol{\beta}} + \mathbf{A}'\mathbf{e},$$

where $\hat{\boldsymbol{\beta}}$, the most plausible vector of parameters ('valor maxime plausibilis'), is given by (8) and $\mathbf{A}$ is an $N \times q$ matrix of numbers such that $\mathbf{AZ}' = \mathbf{I}_N$. Since the expectation of $\mathbf{e}$ is $\mathbf{0}$ and each component has the variance σ^2, Gauss concluded that the variances of the components of $\hat{\boldsymbol{\beta}}$ are given by the diagonal elements of $\mathbf{A}'\mathbf{A}\sigma^2$, i.e. of $(\mathbf{Z}'\mathbf{Z})^{-1}\sigma^2$.

In the second part of this monograph (1823) Gauss extended the theorem to a linear combination of the β_j, a result thought to be new as recently as 1938 (David & Neyman, 1938), and derived the variance of the resulting estimate. He thus obtained what we would call the covariance of two $\boldsymbol{\beta}$ estimates. This second part of the *Theoria Combinationis* (1823) also contains: (*a*) the formula to determine S, the residual sum of squares, from the estimated β's; (*b*) a procedure to permit the addition of a further β to the model without having to recalculate the q β-estimates already determined, a problem later solved independently by Cochran (1938); (*c*) a proof that the expected value of S is $(N-q)\,\sigma^2$, so that the approximate value of σ may be taken ('accipere liceat') as $\sqrt{\{S/(N-q)\}}$, and (*d*) a formula for the variance of the corresponding estimate of σ^2. The Supplement (1826) is concerned with so-called 'adjustment by correlates' in which the N-component vector $\mathbf{x}+\mathbf{e}$ is supposed connected by $r < N$ known linear relations and it is desired to estimate $\mathbf{e}$.

Plackett (1949) follows Gauss's own ingenious proof of his (1821) theorem very closely and more pedestrian versions appear in several standard probability and least square texts (e.g. Helmert, 1872; Czuber, 1891; Czuber, 1903; Markov, 1912; Castelnuovo, 1919; Whittaker & Robinson, 1924). The Markov proof, essentially the same as Helmert's, was thought by Neymañ (1934) to be an original theorem and the Russian writer's name became attached to it. The extension of the theorem to a vector $\mathbf{x}$ of correlated variates, and incidentally the first formulation in terms of matrices, was published by Aitken (1935, 1945).

It is, perhaps, desirable at this point to remind readers how the modern treatment of the univariate linear model differs from that of Gauss. Reference to such texts as Kendall & Stuart (1961), Morgenstern (1964), Rao (1965), Schmetterer (1956), and Wilks (1962), all of which contain a chapter devoted to the linear model, shows that the real advance lies in the generalization of the concept of a submodel obtained by deleting one or more terms of the model and the distributional theory developed to test whether the discarded terms are significantly different from zero. It may be added that Gauss's own method of solving the normal equations is still widely, if not universally, used on desk calculators. The explicit use of ones and zeros in z-vectors to represent qualitative differences like 'low' *vs.* 'high' and the requirement that these components add to zero, to preserve $\hat{\beta}_1$ as the overall mean, in published work dates only from Irwin (1934).

4. Cauchy and Bienaymé

Although Gauss provided the means of calculating the standard error of any β estimate he does not seem to have been faced with the question of deciding whether one or more terms of the linear model could be legitimately discarded as of no significance. This problem was considered by Cauchy but without any reference to the probability distribution of the residual errors. In this respect Cauchy's work is a reversion to the less sophisticated ideas of Legendre.

Cauchy's lithographed memoir of 1835, later translated into English (1836), was expressed as a problem of interpolation. It is supposed that there are N observations x_i available $(i=1,2,\ldots,N)$ of a linear function of a number of known functions $z_1, z_2, \ldots$. These functions are assumed to be arranged in a converging series, for example, 'according to the ascending or descending powers of z, or according to the sines and cosines of the multiples of an arc z'; Cauchy writes x where we have written z.

If z_{ji} is the value of the function z_j when x assumes the value x_i, Cauchy's 'interpolation' formula is

$$x_i = \beta_1 z_{1i} + \beta_2 z_{2i} + \beta_3 z_{3i} + \ldots \quad (i=1,2,\ldots,N). \tag{10}$$

'Now, the question is, first, how many terms of the second member of the equation...are to be employed, in order to obtain a value of y [our x] so approximate that the difference between it and the exact value may be very small, and capable of being compared with the errors to which the observations are liable; secondly, to determine in numbers the coefficients of the terms retained...' Cauchy desired to solve this problem by using all the N observations x_i in the estimation of each of the parameters β_j. However, the sets of β's obtained in the 'different cases in which we should keep, first one term of the series, then two, then three, etc., are obtained by calculations almost independent of one another; so that each new approximation, far from being rendered easier, is more tedious and laborious than those which precede'.

Cauchy overcame this disadvantage by rewriting relation (10) in the form

$$x_i = \gamma_1 y_{1i} + \gamma_2 y_{2i} + \gamma_3 y_{3i} + \ldots, \tag{11}$$

where the y's are linear functions of the z's and where, for a given number of terms q, both forms must provide the same estimate of x_i. Since q is to assume the values $1, 2, \ldots$ in succession it is obvious that, for given q, the estimate $\hat{\gamma}_q$ must equal $\hat{\beta}_q$, except possibly for a constant factor. Furthermore, consideration of $q = 2, 3, \ldots$ shows that $\hat{\gamma}_{q-1}$ must be a linear function of $\hat{\beta}_q$ and $\hat{\beta}_{q-1}$; $\hat{\gamma}_{q-2}$ a function of $\hat{\beta}_q$, $\hat{\beta}_{q-1}$ and $\hat{\beta}_{q-2}$; and so on.

Using matrix notation for conciseness the equality of (10) and (11) for given q implies that $\mathbf{Z\beta} = \mathbf{Y\gamma} = \mathbf{x}+\mathbf{e}$. Multiplying by a $q \times N$ matrix $\mathbf{C}$ of coefficients chosen for their graduating power (when $q = N$ this multiplication has no effect on the solution) we obtain q equations to determine $\hat{\boldsymbol{\beta}}$ and $\hat{\boldsymbol{\gamma}}$, namely

$$\mathbf{CZ}\hat{\boldsymbol{\beta}} \equiv \mathbf{CY}\hat{\boldsymbol{\gamma}} = \mathbf{Cx}. \tag{12}$$

Now if $\hat{\gamma}_q$ is to be equal to $\hat{\beta}_q$; $\hat{\gamma}_{q-1}$ a linear function of $\hat{\beta}_q$ and $\hat{\beta}_{q-1}$ with the coefficient of $\hat{\beta}_{q-1}$ chosen as unity; $\hat{\gamma}_{q-2}$ a linear function of $\hat{\beta}_q$, $\hat{\beta}_{q-1}$ and $\hat{\beta}_{q-2}$ with the coefficient of $\hat{\beta}_{q-2}$ set equal to unity; and so on, $\hat{\boldsymbol{\gamma}}$ must be capable of being written as

$$\hat{\boldsymbol{\gamma}} = \mathbf{U}\hat{\boldsymbol{\beta}}, \tag{13}$$

where $\mathbf{U}$ is upper unit-triangular. Furthermore, if (12) is to produce a vector $\hat{\boldsymbol{\gamma}}$ the first q components of which remain unchanged when q is increased to $q+1$, $\mathbf{CY}$ must clearly be lower triangular. Relations (12) and (13) thus provide

$$\mathbf{CZ} \equiv (\mathbf{CY})\,\mathbf{U}, \tag{14}$$

where the matrix in parentheses on the right is lower triangular and $\mathbf{U}$ is upper unit-triangular.

In relation (14) Cauchy has provided what must be the first example of the decomposition of a square unsymmetric matrix into a product of two triangular matrices. As is well known (e.g. Fox, 1964) the relation is unique. Cauchy's recursive formulae for effecting the calculations involved in (14) became known as Cauchy's algorithm (Radau, 1891).

Cauchy's own choice for the elements of $\mathbf{C} \equiv ((c_{ji}))$ $(j = 1, 2, \ldots, q;\ i = 1, 2, \ldots, N)$ was

$$c_{ji} = \operatorname{sgn} y_{ij}, \quad \text{where} \quad ((y_{ij})) = \mathbf{Y}, \tag{15}$$

so that in passing from a given q to $q+1$ it was necessary to calculate all N values in the qth column of $\mathbf{Y}$. This was not considered a disadvantage since a feature of Cauchy's method was that the N residuals, namely

$$\mathbf{x} - \hat{\mathbf{x}} = \mathbf{x} - \mathbf{Z}\hat{\boldsymbol{\beta}} = \mathbf{x} - \mathbf{Z}\mathbf{U}^{-1}\hat{\boldsymbol{\gamma}} \tag{16}$$

had to be inspected after each component of $\hat{\boldsymbol{\gamma}}$ had been determined. In the application of the algorithm the lower triangular matrix $\mathbf{CY}$ was reduced to unit-triangular form by factoring out the elements in the principal diagonal, i.e. by writing $\mathbf{CY} \equiv \boldsymbol{\Delta}(\boldsymbol{\Delta}^{-1}\mathbf{CY})$ where $\boldsymbol{\Delta}$ is a diagonal matrix. Then (14) was, in effect, written as

$$\boldsymbol{\Delta}^{-1}\mathbf{CZ} = \boldsymbol{\Delta}^{-1}\mathbf{CYU}, \tag{17}$$

$\boldsymbol{\Delta}^{-1}\mathbf{CY}$ being lower unit-triangular. Once the elements of $\mathbf{U}$, $\mathbf{C}$ and $\mathbf{Y}$ had been calculated for a given q the next column of $\mathbf{U}$ was obtained from (17) and then, in order, the next column of $\mathbf{Y}$ from the relation $\mathbf{Z} = \mathbf{YU}$, the next row of $\mathbf{C}$ from (15), the next diagonal element of $\boldsymbol{\Delta}$, which is the vector product of this new row of $\mathbf{C}$ and the new column of $\mathbf{Y}$, the next component of $\hat{\boldsymbol{\gamma}}$ from (12), and the set of N residuals from (16). The procedure is terminated when these residuals 'seront devenues assez petites pour être comparables aux erreurs d'observation que comportent les diverses valeurs de k [our x]' (Cauchy, 1853).

The procedure just described was given an explicit form in a special difference notation by Cauchy himself and was arranged in a tableau convenient for numerical calculations by Villarceau (1849). However, Bienaymé (1853) pointed out that if $\mathbf{C} = \mathbf{Z}'$ the foregoing analysis would conform with least squares theory and the $\hat{\gamma}$'s would be least-squares estimates. Under these circumstances relation (14) becomes

$$\mathbf{Z}'\mathbf{Z} = \mathbf{Z}'\mathbf{YU} = \mathbf{G}^{-1}\mathbf{D}(\mathbf{G}')^{-1} \tag{18}$$

by (5) and the Cauchy and Gauss algorithms are thus connected by the relation $\mathbf{U} = (\mathbf{G}')^{-1}$. This simplifies the numerical calculations since it is no longer necessary to introduce $\mathbf{Y}$ explicitly and the elements of $\mathbf{D}$ are obtainable in succession from the equation of the outside members of (18). The modifications necessary to Cauchy's explicit formulae are indeed minor and were first published by Carvallo (1888) and simplified for the special case $z_{ij} = z_i^{j-1}$ by Radau (1891). The computational procedure of Wishart & Metakides (1953), stated to be 'the most convenient for use by the average computer' and supposedly novel, bears a strong resemblance to that of Cauchy. However, Yule's (1907) partial regression notation, used by Wishart & Metakides, is more explicit and thus more convenient than that of Cauchy as modified by Radau.

5. Chebyshev and Gram

Bienaymé (1853) did not follow up the consequences of putting $\mathbf{C} = \mathbf{Z}'$ in the Cauchy algorithm. One of them is that

$$\mathbf{Y}'\mathbf{Y} = (\mathbf{ZG}')'\,\mathbf{ZG} = \mathbf{G}(\mathbf{Z}'\mathbf{Z})\,\mathbf{G}' = \mathbf{D} \tag{19}$$

and on applying the principle of least squares to (11), or from (12),

$$\hat{\gamma} = (\mathbf{Y}'\mathbf{Y})^{-1}\mathbf{Y}'\mathbf{x} = \mathbf{D}^{-1}\mathbf{Y}'\mathbf{x}. \tag{20}$$

Writing $\mathbf{Y} = \{\mathbf{y}_1 \mathbf{y}_2 \ldots \mathbf{y}_q\}$, we may write the relation (20) alternatively as

$$\hat{\gamma}_k = d_k^{-1} y_k' \mathbf{x} \quad (k = 1, 2, \ldots, q),$$

while (19) is equivalent to

$$\mathbf{y}_k' \mathbf{y}_l = 0 \quad (k \neq l) \quad \text{and} \quad \mathbf{y}_k' \mathbf{y}_k = d_k \quad (k = 1, 2, 3, \ldots, q).$$

The N-component column vectors of $\mathbf{Y}$ thus form an orthogonal set. The equation $\mathbf{Z} = \mathbf{YU}$, or $\mathbf{Y} = \mathbf{ZU}^{-1}$, together with relation (19) provides a unique solution for the successive vectors $\mathbf{y}_k$, namely

$$\mathbf{y}_k = \mathbf{z}_k - \sum_{j=1}^{k-1} \frac{\mathbf{y}_j' \mathbf{z}_k}{\mathbf{y}_j' \mathbf{y}_j} \mathbf{y}_j \tag{21}$$

which was first written down by Schmidt (1908) and is called his orthogonalization process. Although (21) is sometimes ascribed to Gram (1883) that author did no more in this connexion than express Cauchy's formulae in an alternative, determinantal notation. The first use of (21) in the statistical literature was by Romanovsky (1925, 1927) and it was rediscovered as recently as 1949 by Rao.

Chebyshev (1858) was the first to make explicit use of (19) and (20) but in this and his four subsequent papers on the subject he limited himself to consideration of the case where $\mathbf{y}_k$ is a polynomial in z of degree $k-1$; here again we omit mention of the known 'weight' w_i attached to the observation x_i. In his introductory footnote to this French translation of an 1855 article in Russian on orthogonal polynomials Bienaymé shows that he is aware that the author's formulae are special cases of the Cauchy procedure.

It may be seen from Isserlis's (1927) summary of Chebyshev's rather long and repetitive papers that for us the novelty lies in the explicit requirement of orthogonality of the $\mathbf{y}_k$ and in the subsequent algebra to derive the recursive relation

$$y_{ki} = (z_i - b_k)\, y_{k-1,\,i} - a_k\, y_{k-2,\,i}, \tag{22}$$

where

$$a_k = \frac{\mathbf{y}_{k-1}' \mathbf{z}_{k-1}}{\mathbf{y}_{k-2}' \mathbf{z}_{k-2}} \quad \text{and} \quad b_k = \frac{\mathbf{y}_{k-1}' \mathbf{z}_k}{\mathbf{y}_{k-1}' \mathbf{z}_{k-1}} - \frac{\mathbf{y}_{k-2}' \mathbf{z}_{k-1}}{\mathbf{y}_{k-2}' \mathbf{z}_{k-2}}$$

and the ith component of $\mathbf{z}_k$ is z_i^{k-1} $(i = 1, 2, \ldots, N)$. In three of these five articles Chebyshev is concerned with the particular case where the weights are equal and the values of z are at equidistant intervals. Although the orthogonal polynomials for this case were obtained explicitly by Chebyshev up to the fifth degree and reproduced in the review articles of Radau (1891) and Selivanov & Bauschinger (1906) they do not seem to have been applied in practice, nor was an attempt made to tabulate them numerically, until Gram (1915). However, E. S. Pearson (1938) reports, in his Appendix VI, that Karl Pearson was lecturing on orthogonal polynomials in his first-year undergraduate courses on statistics in 1921–22. Their subsequent history is summarized by van der Reyden (1943).

6. Helmert and Pizzetti

The relative simplicity of Gauss's *Theoria Motus* derivation of the least squares procedure naturally caused it to be preferred in nineteenth century textbooks directed to its practical application (Czuber, 1899). This led to a number of theoretical papers on the distributions of statistics calculated from samples of N from an infinite Normal universe which we will take to be $N(0, \sigma^2)$.

For example, Czuber (1899) cites the derivation of the distributions of

$$x_i - \bar{x}, \quad \sum_{i=1}^{N} x_i^2 = \mathbf{x}'\mathbf{x}, \quad \sum_{i=1}^{N} (x_i - \bar{x})^2 = (\mathbf{x} - \bar{\mathbf{x}})'(\mathbf{x} - \bar{\mathbf{x}}),$$

and $S = (\mathbf{x} - \mathbf{Z}\hat{\boldsymbol{\beta}})'(\mathbf{x} - \mathbf{Z}\hat{\boldsymbol{\beta}})$, namely the sum of squares of the residuals after a least squares fitting, as well as the calculation of the mean and variance of the distributions of

$$\sum_{i=1}^{N} |x_i - \bar{x}|, \quad \left\{\sum_{i=1}^{N} (x_i - \bar{x})^2\right\}^{\frac{1}{2}} \quad \text{and} \quad \sum_{i \neq j} |x_i - x_j|.$$

Details of most of these proofs are provided in Czuber's (1891) text and it will be noticed that for samples that appear frequently to have run into the hundreds, the type III distributions that were obtained were close to the Normal and amply justified summarization through their means and variances.

Karl Pearson's (1931) reproduction of Helmert's derivation of the distribution of $(\mathbf{x} - \bar{\mathbf{x}})'(\mathbf{x} - \bar{\mathbf{x}})$ is well known but Pizzetti's (1891) extension of it to $(\mathbf{x} - \mathbf{Z}\hat{\boldsymbol{\beta}})'(\mathbf{x} - \mathbf{Z}\hat{\boldsymbol{\beta}}) \equiv \mathbf{r}'\mathbf{r}$, say, is less familiar.

In the notation of this review Pizzetti wrote the probability of a sample $\{x_i\}$ as

$$(h\pi^{-\frac{1}{2}})^N \exp(-h^2\mathbf{e}'\mathbf{e}) \prod_{i=1}^{N} de_i \tag{23}$$

and, using the discontinuity factor, the probability density of the random variable S representing the sum of the squares of the residuals as

$$\frac{1}{2\pi}\int_{-\infty}^{\infty} \exp\{-iu(\mathbf{r}'\mathbf{r} - s)\}\,du \int_{-\infty}^{\infty} de_1 \int_{-\infty}^{\infty} de_2 \ldots \int_{-\infty}^{\infty} h^{-N}\pi^{-\frac{1}{2}N} \exp(-h^2\mathbf{e}'\mathbf{e})\,de_N, \tag{24}$$

where $\mathbf{r}'\mathbf{r}$ must first be expressed in terms of $e_1, e_2, \ldots, e_N$.

In order to effect the integration Pizzetti noted that

$$\mathbf{e}'\mathbf{e} = (\mathbf{Z}\boldsymbol{\beta} - \mathbf{x}')(\mathbf{Z}\boldsymbol{\beta} - \mathbf{x}) = (\mathbf{Z}\hat{\boldsymbol{\beta}} - \mathbf{x})'(\mathbf{Z}\hat{\boldsymbol{\beta}} - \mathbf{x}) + \boldsymbol{\alpha}'\mathbf{D}^{-1}\boldsymbol{\alpha}, \tag{25}$$

by means of (6) and (8). Now

$$\mathbf{r} = \mathbf{Z}\hat{\boldsymbol{\beta}} - \mathbf{x} = \mathbf{e} - \mathbf{Z}(\boldsymbol{\beta} - \hat{\boldsymbol{\beta}}) = \{\mathbf{I}_N - \mathbf{Z}(\mathbf{Z}'\mathbf{Z})^{-1}\mathbf{Z}'\}\mathbf{e} \tag{26}$$

implying that any residual can be expressed in terms of the N errors. However, there are q linear relations, namely $\mathbf{Z}'(\mathbf{Z}\hat{\boldsymbol{\beta}} - \mathbf{x}) = \mathbf{0}$, connecting the N residuals, so that the last q components of $\mathbf{r}$ can be eliminated from $\mathbf{r}'\mathbf{r}$ which is thus a quadratic in the first $N-q$ components of $\mathbf{r}$ which we write as a vector $\mathbf{r}_0$. Actually, because of notational difficulties, Pizzetti developed his formulae for $q = 3$ and stated that they could be extended without difficulty to the general case.

Introduce the $N-q$ component vector $\mathbf{t}$ such that

$$\mathbf{t} = \mathbf{T}\mathbf{r}_0,$$

where $\mathbf{T}$ is an upper triangular matrix of order $N-q$, and choose the elements of $\mathbf{T}$ so that

$$\mathbf{t}'\mathbf{t} = \mathbf{r}'\mathbf{r}. \tag{27}$$

Considering the N-component vector

$$\begin{bmatrix} \mathbf{t} \\ \mathbf{D}^{-\frac{1}{2}}\boldsymbol{\alpha} \end{bmatrix} \equiv \mathbf{Pe},$$

where $\mathbf{P}$ is an $N \times N$ matrix, we see that

$$(\mathbf{t}', \boldsymbol{\alpha}'\mathbf{D}^{-\frac{1}{2}}) \begin{bmatrix} \mathbf{t} \\ \mathbf{D}^{-\frac{1}{2}}\boldsymbol{\alpha} \end{bmatrix} = \mathbf{t}'\mathbf{t} + \boldsymbol{\alpha}'\mathbf{D}^{-1}\boldsymbol{\alpha} = \mathbf{r}'\mathbf{r} + \boldsymbol{\alpha}'\mathbf{D}^{-1}\boldsymbol{\alpha} = \mathbf{e}'\mathbf{e} = \mathbf{e}'\mathbf{P}'\mathbf{P}\mathbf{e},$$

by (25) and (27). Hence $\mathbf{P}'\mathbf{P} = \mathbf{I}_N$.

In the integral (24) we now change the vector of variables $\mathbf{e}$ to

$$\begin{bmatrix} \mathbf{t} \\ \mathbf{D}^{-\frac{1}{2}}\boldsymbol{\alpha} \end{bmatrix}$$

and the Jacobian is $|D^{-\frac{1}{2}}|$. On integrating the result is

$$\frac{h^{N-q}}{\Gamma(\frac{1}{2}N - \frac{1}{2}q)} s^{\frac{1}{2}(N-q)-1} e^{-h^2 s}, \tag{28}$$

which, as the author remarks, means that the distribution of the sum of the squares of the residuals after estimating q parameters is the same as the distribution of the sum of the squares of $N-q$ 'true' errors.

We mention that in 1889 Pizzetti had solved the easier problem of finding the posterior probability density of h, from which he found its 'most probable', or modal, value. Replacing $\mathbf{e}'\mathbf{e}$ in (23) by the right hand member of (25) he assumed that the prior distributions of $h = (\sigma\sqrt{2})^{-1}, \beta_1, \beta_2, \ldots, \beta_q$ are all uniform and noting that

$$\int_{-\infty}^{\infty} \exp(-h^2\alpha_j^2/d_j)\, d\beta_j = h^{-1}\pi^{\frac{1}{2}} d_j^{-\frac{1}{2}}$$

reduced the q-fold integration of h^N times the exponential in (23) to

$$Bh^{N-q} \exp(-h^2 \mathbf{r}'\mathbf{r})\, dh, \tag{29}$$

where B is a constant found by integrating from $h = 0$ to ∞ and equating the result to unity.

7. The multivariate normal

The middle years of the nineteenth century saw the appearance of a new concept, the multivariate probability distribution and, in particular, what we now refer to as the multivariate-normal, or even the multinormal, law. The first explicit consideration of two- and three-variate probability distributions, as opposed to the incidental appearance of bivariate distributions (Walker, 1931), is given by Bravais (1846) and later, more generally, by Schols (1875). Both these articles are summarized in some detail by Czuber (1891).

Bravais considered two random variables X_1 and X_2 such that

$$X_1 - E(X_1) = x_1 = \sum_{i=1}^{n} a_i \epsilon_i, \quad X_2 - E(X_2) = x_2 = \sum_{i=1}^{n} b_i \epsilon_i,$$

where ϵ_i is distributed $N(0, \sigma_i^2)$ and is independent of ϵ_j $(i \neq j)$. We write

$$h_{x_1}^{-1} = 2\sum_{i=1}^{n} a_i^2 \sigma_i^2 \quad \text{and} \quad h_{x_2}^{-1} = 2\sum_{i=1}^{n} b_i^2 \sigma_i^2.$$

'La coexistence des mêmes variables...dans les équations simultanées en x et y [our x_1, x_2], amène une corrélation telle, que les modules h_x, h_y, cessent de représenter la possibilité des valeurs simultanées de (x, y) ....'

He found the joint probability density of X_1 and X_2 to be

$$K\pi^{-1}\exp\{-\tfrac{1}{2}K^2(\sigma_{11}x_1^2-2\sigma_{12}x_1x_2+\sigma_{22}x_2^2)\}, \tag{30}$$

where for $j = 1, 2$

$$\sigma_{jj} = E\{X_j-E(X_j)\}^2, \quad \sigma_{12} = E[\{X_1-E(X_1)\}\{X_2-E(X_2)\}],$$

and $K^2 = (\sigma_{11}\sigma_{22}-\sigma_{12}^2)^{-1}$. Hence points of equal probability in the X_1, X_2 plane lie on 'une série indéfinie d'ellipses semblables, dont l'aire augment proportionnellement à la constant' to which the exponent of e is equated. Of these ellipses Bravais named the '*ellipse fondamentale* celle dont la surface est égale à l'unité de surface'.

Bravais proceeded by transforming to new axes Y_1 and Y_2 such that

$$y_1 = x_1\cos\theta+x_2\sin\theta, \quad y_2 = -x_1\sin\theta+x_2\cos\theta, \tag{31}$$

then if θ is chosen so that $\tan 2\theta = 2\sigma_{12}/(\sigma_{11}-\sigma_{22})$, these axes coincide 'avec les axes principaux de l'ellipse fondamentale' and the exponent of e becomes $-\frac{1}{2}K^2(ay_1^2+by_2^2)$, where a and b are the roots of the quadratic equation

$$z^2-(\sigma_{11}+\sigma_{22})z+\sigma_{11}\sigma_{22}-\sigma_{12}^2 = 0; \tag{32}$$

'...dans ce cas, la probabilité des valeurs simultanées...est exactement la même que si les variables...étaient entièrement independantes l'une de l'autre'.

Bravais then introduced a third random variable X_3 such that

$$X_3-E(X_3) = x_3 = \sum_{i=1}^{n} c_i\epsilon_i$$

and found the joint probability density of X_1, X_2, X_3 to be similar to (30) with the quadratic in x_1, x_2 replaced by a quadratic in x_1, x_2, x_3. He was able to write the coefficients of this quadratic and its constant multiplier in terms of

$$\sigma_{ij} = E[\{X_i-E(X_i)\}\{X_j-E(X_j)\}] \quad (i, j = 1, 2, 3),$$

but confessed his inability to do this for the similar probability density in x_1, x_2, x_3, x_4. We may note, too, that in his discussion of the form assumed by the trivariate probability density when the co-ordinate axes were transformed to the principal axes of the ellipsoids of the exponent of e, Bravais did not write down the cubic equation corresponding to (32), nor did he provide an equivalent of the transformation (31).

On the other hand Schols (1875) wrote the principal axes of probability ('hoofdassen van waarschijnlijkheid') as three linear transformations of the original co-ordinates, the nine direction cosines of which were determined (a) three from the requirement that the squares of any set of three must add to unity, (b) three from the condition of orthogonality of each pair of axes, and (c) three from the requirement that the covariances of the three transformed random variables must be zero. Once these direction cosines had been found the three variances of the transformed variables could be written down. However, Schols made no attempt to generalize this principal components technique to four variates.

Bravais did not concern himself with the practical application of his results and assumed that the a_i, b_i and σ_i^2 of his model were known parameters and that n was very large, even infinite. This led Pearson (1920)*to conclude that no writers before Galton had considered directly measurable dependent variates X_1 and X_2. In this he was being less than just to Schols (1875) who discussed the application of the bivariate normal to artillery fire at a target and criticized military authors who had assumed that errors in the horizontal and vertical directions were independent. However Schols did not indicate how he would use a set of pairs of observations to estimate the angle of rotation to the principal axes and it is only later (Czuber, 1891) that we find the statement that if N, the number of observed points (X_{1i}, X_{2i}) is very great then σ_{12} is obtainable 'ohne einem beachtenswerten Fehler' from

$$N^{-1} \sum_{i=1}^{N} (X_{1i} - \bar{X}_1)(X_{2i} - \bar{X}_2),$$

where for $j = 1, 2$

$$\bar{X}_j = N^{-1} \sum_{i=1}^{N} X_{ji}.$$

It will not have escaped the reader that every reference we have made to important original work prior to 1892 has been to continental European authors. The first substantial contribution to our subject by an Englishman occurred in 1892 and, oddly enough, every significant advance thereafter until quite recent times must be ascribed to writers in English journals. What is even more surprising is that the English work was done largely without reference to the prior contributions of the continental writers.

Edgeworth (1892) was the first to provide a completely general statement of the multivariate normal distribution. Although marred by careless interchanges of suffices and expressed in a cumbrous determinantal notation no longer familiar, Edgeworth's statement that 'we may extend to four and higher numbers of variables the solution which has been given above for the case of three variables' is literally true. In fact Edgeworth himself correctly calculated *in extenso* the numerical values of the determinants appearing in a four-variate distribution. The explicit extension to the general case, in an improved notation, was effected by Pearson (1896) where the p-variate normal 'correlation surface', namely,

$$(2\pi)^{-\frac{1}{2}p} R^{-\frac{1}{2}} \exp\left(-\frac{1}{2R} \sum_{i,j}^{p} R_{ij} \frac{x_i x_j}{\sigma_i \sigma_j}\right) \tag{33}$$

with $(i, j = 1, 2, \ldots, p)$

$$R = |(\rho_{ij})|, \qquad \rho_{ij} = \rho_{ji}, \qquad \rho_{ii} = 1,$$
$$\rho_{ij} = E(x_i x_j)/(\sigma_i \sigma_j), \quad \sigma_i^2 = E(x_i^2), \quad E(x_i) = 0,$$

and R_{ij} the (i, j) cofactor of R, is referred to as 'Edgeworth's Theorem'. In our view Pearson's later (1920)*tergiversation is quite unjustified. Contrary to Pearson's statement Edgeworth certainly *did* 'put the answer...in a form in which the statistician with a customary amount of mathematical training [and knowledge of the 1866 notation for cofactors] could determine the form of the surface for n variates, as soon as their S.D.'s and correlations had been calculated'. Edgeworth's four-variate example extends without difficulty.

Before we revert to the main line of our inquiry we mention that Pearson (1901) was the first to generalize Schols's (1875) three-variate rigid rotation of the co-ordinate axes to their principal axes. Suppose $\mathbf{x}$ is a p-component vector representing the deviations from their

*Paper No. 14 in this volume

means of p normally correlated variate values and let $\mathbf{A}$ be the matrix of direction cosines of the p principal axes so that $\mathbf{y} = \mathbf{Ax}$, then Pearson determined $\mathbf{A}$ as the solution of

$$\mathbf{a}'\mathbf{a} = 1 \quad \text{and} \quad \mathbf{a}_i'(\mathbf{\Sigma} - \lambda_i \mathbf{I}) = \mathbf{0}, \tag{34}$$

where $$\mathbf{\Sigma} = (\sigma_{ij}), \quad \sigma_{ij} = E(x_i x_j) \quad (i, j = 1, 2, \ldots, p),$$

$\mathbf{a}_i'$ is the (row) vector of direction cosines of the ith principal axis and λ_i is the ith root of the determinantal equation

$$|\mathbf{\Sigma} - \lambda \mathbf{I}| = 0, \tag{35}$$

where $\lambda_1 < \lambda_2 < \ldots < \lambda_p$ (the case of equal roots not being considered). A similar transformation was made by McMahon (1923) using the correlation matrix instead of $\mathbf{\Sigma}$ and this is the case that later became known as a Principal Axes Factor Analysis.

8. Karl Pearson

During the 1880's Galton introduced the idea of measuring the correlation between the size of a parent's organ and the size of that organ in an offspring (Pearson, 1920).* With the help of J. D. H. Dickson he arrived at the bivariate normal surface as an expression of the simultaneous variation of such pairs of measurements. In the early 1890's Weldon introduced the correlation coefficient and suggested that this parameter might be constant for pairs of similar organs measured on different species (Pearson, 1920).* But it was Karl Pearson (1896) who gave a definitive mathematical formulation to these researches.

'The following assumption...lies at the basis of our present treatment of heredity. The variation of any organ in a sufficiently large population—which may be selected in any manner other than by this organ itself from a still larger population—is closely defined by a normal probability curve.' Extending this to p correlated organs Pearson derived the multivariate normal surface (33) and found that if the deviations from the means of $p-1$ correlated organs $X_1, X_2, \ldots, X_{p-1}$ assume the values $x_1, x_2, \ldots, x_{p-1}$, the distribution of X_p is (univariate) Normal about a mean (measured from the general, unconditional mean of X_p)

$$E(X_p | x_1, x_2, \ldots, x_{p-1}) \equiv x_{p.12\ldots(p-1)} = -\sigma_p \sum_{j=1}^{p-1} \frac{R_{pj}}{R_{pp}} \frac{x_j}{\sigma_j} \tag{36}$$

$$\equiv \sum_{j=1}^{p-1} \beta_{pj.23\ldots(j-1)(j+1)\ldots(p-1)} x_j$$

with variance

$$\sigma^2_{p.12\ldots(p-1)} = \sigma^2_p R/R_{pp}. \tag{37}$$

The expression (36) was called the regression of X_p on $X_1, X_2, \ldots, X_{p-1}$ and the β's are the regression coefficients in Yule's (1907) notation. In applying these formulae to numerical data the correlation coefficients in the determinants on the right-hand sides of (36) and (37) were computed by the method of maximum likelihood (not then given that name) applied to a sample of N observations from a bivariate normal distribution with known variances. The familiar result

$$r_{12} = \sum_{i=1}^{N} (x_{1i} - \bar{x}_1)(x_{2i} - \bar{x}_2)/(N\hat{\sigma}_1 \hat{\sigma}_2)$$

is actually correct only when ρ_{12}, σ_1 and σ_2, and $E(X_1)$ and $E(X_2)$ if desired, are estimated simultaneously by maximum likelihood (Kendall & Stuart, 1961; Example 18·14). These

*Paper No. 14 in this volume

variances were estimated as N^{-1} times the sum of the squares of the N deviations from the corresponding mean. Sample sizes were supposed fairly large: 'Of course 200 couples give graphically nothing like a surface of correlation....We assume *a priori* that 1000 couples would give a fair surface'.

Later, Yule (1897) showed that for $p = 2, 3$ and 4 equation (36) was the same as the result of estimating a value of X_p from a linear combination of (fixed) values of $X_1, X_2, \ldots, X_{p-1}$ by the method of least squares. Pearson (1898) proved that the right hand side of (36) was that linear function of $X_1, X_2, \ldots, X_{p-1}$ which had maximum correlation coefficient with X_p.

Now in deriving the above formulae Pearson (1896) was at some pains to emphasize that the Normality of X_p held good both for deliberately selected values of $X_1, X_2, \ldots, X_{p-1}$ and for any set of values randomly sampled from their $(p-1)$-variate normal distribution. It was the latter viewpoint that Pearson (1898) had in mind when he attempted to use the regression equation determined from a sample set of $X_1, X_2, \ldots, X_{p-1}$ and X_p to predict the X_p of a sample with different values of $X_1, X_2, \ldots, X_{p-1}$. As he writes (Stevenson, 1929): '...if these $[p-1]$ skeletal parts include *all* those which have been *directly* selected in the course of evolution, the regression equation for stature [our X_p] would remain the same in all races, although the means, standard deviations and correlations might change in a great variety of ways. The regression coefficients would be unaltered by the selection; in other words they have a stability far higher than that of means, standard deviations and correlations'. It is in this respect that (36) may be given a wider interpretation than the mathematically equivalent expression (1). In the latter the z-values could theoretically be a set of N highly improbable measurements; in the correlation surface interpretation the x's are thought of as a random sample even when the $(p-1)$-variate normality condition is relaxed.

The result of allowing the set of values of $X_1, X_2, \ldots, X_{p-1}$ to constitute a random sample from a $(p-1)$-variate normal distribution is that the distribution of $\hat{\beta}_{21}$ is non-Normal (Pearson, 1926) and the variance of $\hat{\beta}_{pj\,.\,23\ldots(j-1)(j+1)\ldots(p-1)}$ only agrees asymptotically with the Gaussian value $\sigma^2_{p\,.\,123\ldots(p-1)}/\{N\sigma^2_{j\,.\,23\ldots(j-1)(j+1)\ldots(p-1)}\}$ when $N \to \infty$. However, the predicted value of X_p in the estimated regression equation (36) is identical with that obtained by least squares (Yule, 1897; Pearson, 1905). Pearson never seems to have accepted this formal identity. His (1920)* description of Gauss's linear model refers to the $\hat{\beta}$'s as directly measured and the x_i as 'indirectly ascertained quantities'. And although Gauss himself derived an expression for the covariance between two $\hat{\beta}$'s Pearson concluded that there is a 'non-correlated surface for the directly observed variates'. This neglect of Gauss may be confirmed from the *Biometrika Index*: in the first 37 volumes Gauss as a subject appears once in connexion with a quadrature formula, while 'least squares' has three references in the first 32 volumes all of them in volumes **1** and **2**!

The random variability of the regressors implies the non-normality of the bivariate surface corresponding to a polynomial regression. Pearson (1905) thus refused to fit a higher degree polynomial than a quartic because of the large sampling variances involved in higher moments of the (marginal) distribution of the regressor z. Although he provided the orthogonal polynomials up to the fourth degree for fitting heteroscedastic arrays of x-values at non-equidistant z-values, assumed constant in any array, he did not appreciate that these were identical with Chebyshev's weighted, non-equidistant formulae (Pearson, 1921).

Pearson's non-normal bivariate sampling theory for polynomial regression prevented him in (1905) from dealing 'even superficially with the probable errors of the [regression]

*Paper No. 14 in this volume

constants involved'. Although the variances of the regression coefficients were known for p-variate normal sampling, even here no attempt seems to have been made to use them to judge whether one or more fitted β's could reasonably be deleted from a regression equation. Thus, for example, Macdonell (1901) used six other physical measurements made on 3000 criminals to 'reconstruct' the height of a given criminal. His method of deciding which of these six measurements would perform the most effective reconstruction was to compare the probable error of the six-variable estimate, based on equation (37), with the larger probable errors of the corresponding estimates utilizing only one, two or three measurements selected from the six. A similar comparison of probable errors was applied to the prediction by Stevenson (1929), four years before Pearson's retirement, of cadaver lengths from up to four measurements made on long bones. And, a little earlier, such a comparison was used to compare the effectiveness of straight line and cubic parabola fittings of one and the same data (Mumford & Young, 1923).

Karl Pearson must thus be given the credit for extending the Gauss linear model to a much broader class of problems than those of 'errors of measurement'. And he and his students applied it widely using a few 'selected' continuous regressors. Apart from Macdonell's (1901) paper we believe no article published in the first 22 volumes of *Biometrika* uses more than four regressors. If, at the same time, Pearson introduced a more complex theoretical basis it must be recognized that his sampling theory has a conceptual validity that may prove useful in practical applications.

9. R. A. Fisher

With Fisher (1922) the sampling theory of regression equations returned to the Gaussian model and two significant contributions were made: (i) the introduction of the idea of testing a group of β's for nullity and the technique for doing so, and (ii) the utilization of qualitative variables resulting in z-vectors with small integer or zero components and the choice of these to make the subsequent analysis numerically easy. The former procedure is known as the F-test, and the latter leads to the 'analysis of variance'. It is convenient to keep these innovations separate although, in fact, they developed together.

We have already mentioned the case of unequally weighted observations in the Gaussian model. At first sight one way such unequally weighted x-values could arise would be for a given x_k to have been observed w_k times for a given set of q z-values. Such an assumption would, of course, contradict the hypothesis of continuity of the distribution of X. Nevertheless, the least-squares literature provides many examples where, e.g. $N = 400$ observations have been grouped into $g = 8$ unequal classes within which the z-values are apparently constant, and a mean x-value has been calculated and used for each of these groups (Helmert, 1872). Having forced discreteness on X these authors then suppress the information about σ^2 contained in the variability within the groups and, instead, estimate σ^2 from the residuals of the g fitted means. There seems to be some notion here that the variability of X about its group means would be much smaller than the 'error' of the linear model.

This concept came to direct expression in Slutsky's (1914) attempt to test the 'goodness of fit' of a q-parameter regression equation in which g average x-values had been fitted by least squares with a residual r_k in group k ($k=1, 2, \ldots, g$). A perfect fit would be achieved if the error e_k, assumed replaceable by r_k, had the same variance as the variance within that group, namely σ_k^2. The latter being estimated from the w_k observations in group k, the prior probability of a well-fitted set of observations would be proportional to

$$\exp\left(-\frac{1}{2}\sum_{k=1}^{g}\frac{w_k r_k^2}{\sigma_k^2}\right) \equiv \exp\left(-\tfrac{1}{2}\chi_1^2\right) \tag{38}$$

and the probability of a set of residuals (errors) outside the ellipsoid defined by χ_1^2 is then given by the tail area of the chi-squared distribution with g degrees of freedom. Pearson (1917) criticized this derivation from the viewpoint of p-variate sampling ($p = q+1$). The sampling variance of X within group k is then only asymptotically σ_k^2/w_k and even in the multivariate (homoscedastic) normal case the distribution of X is 'markedly leptokurtic' for the small w_k found in the outlying groups. But this paper is more interesting as showing that Pearson was prepared to admit that when a 'physicist makes a few measurements of a variate A [our X] for each of a series of values of a variate B [our z_1]' he has fixed, not sampled, his z_1-values.

Fisher in his pathbreaking paper (1922) continued this line of thought by correcting (38) for the case where X is Normal with variance σ^2. In a few brief statements he argued that (i)

$$\sum_{k=1}^{g} w_k r_k^2$$

is a sum of squares of g Normally distributed variates subject to q linear relations and is accordingly distributed as $\sigma^2\chi_{g-q}^2$, where χ_{g-q}^2 represents a chi-squared variate with $g-q$ degrees of freedom, (ii)

$$\sum_{l=1}^{w_k} (x_{kl} - \bar{x}_{k\cdot})^2$$

is distributed independently of this and as $\sigma^2\chi_{N-g}^2$, where x_{kl} ($l = 1, 2, \ldots, w_k$) are the observed values of X in group k and $\bar{x}_{k.}$ is their mean, and (iii) the ratio of these two variates has a Pearson type VI distribution. This ratio multiplied by $(N-g)/(g-q)$ is now called $F_{g-q,\,N-q}$ and its (type VI) distribution is widely tabulated.

It will have been observed that Slutsky's criterion for judging the success of a linear model was the springboard for Fisher's F-distribution which soon came to play a central role in statisticians' applications of the model. In fact Fisher's (1922) paper continued by showing that the distribution of any single $\hat{\beta}$ being Normal about a mean of β and with variance $(\mathbf{Z'Z})^{-1}\sigma^2$, and the distribution of the residual sum of squares S being independent of $\hat{\beta}$ and distributed as $\sigma^2\chi_{N-q}^2$, the square root of the ratio $(\hat{\beta}-\beta)^2/\{(\mathbf{Z'Z})^{-1}S\}$ would be a Student-t variate with $N-q$ degrees of freedom, which two years later he identified as equivalent to $\sqrt{F_{1,\,N-q}}$.

Now the continental writers on least squares had already obtained the distributions of $\hat{\beta}$ and S, both of which involved the parameter σ^2. Pizzetti's (1891) proof had also established that these two distributions were independent. The real novelty of Fisher's (1922) paper, then, lay in the use of a ratio variate which eliminated the unwanted parameter σ^2 and provided a test of a hypothetical β-value (e.g. zero) by means of its value calculated from a sample.

This is not how statisticians of that time viewed Fisher's achievement. For although the distributional basis of his paper had been mentioned or summarized in the standard review articles of Czuber (1899) and Bauschinger (1908) and must have been known to university lecturers on the theory of errors, Fisher's writings imply that he, like Pearson before him, had made no detailed study of least-squares theory. This may explain why Fisher (1925) found it necessary to expand the arguments of his 1922 paper insofar as they concerned the simultaneous independent distributions of $\hat{\beta}_1 = \bar{x}$, $\hat{\beta}_2$ and S based on $q = 2$. His orthogonal analysis of a sum of squares of N variates only differs from that of Pizzetti (1891) by utilizing N-dimensional geometry instead of Fourier integration theory.

In his 1922 paper Fisher had stated that the difference between the means of two samples of n_1 and n_2 $(n_1+n_2=N)$ observations, respectively, from Normal universes having the same means and variances would, after division by the square root of $(n_1^{-1}+n_2^{-1})$ $(n_1+n_2-2)^{-1}$ times the sum of the squares of the deviations of each sample variate from its own observed mean, be distributed as Student's t with $N-2$ degrees of freedom. He had, in fact, already proved this in (1921) for $n_1 = n_2 = n$ by considering the so-called intraclass correlation coefficient calculated from the $\binom{n}{2}$ pairs of observations that can be formed by taking one member from each sample. By 1924 Fisher (1928) had perceived the connexion between these formulations and was using $F_{g-1, ng-1}$ as the variate to test whether the means of g groups each of n Normally distributed observations with one and the same variance σ^2 could be regarded as equal to one another. This problem had already been posed by von Bortkiewicz in 1895 as an extension of his earlier modification of the Lexis divergence coefficient for binomial observations (Seal, 1949). That author proposed to compare the mean square 'between' groups with the mean square of the ng observations about the common mean (Bortkiewicz, 1909). Fisher named the modification of this procedure which compares the 'between' and 'within' mean squares the 'analysis of variance'. It may be added that in (1928) Fisher pointed out that the Lexis coefficient would be approximately distributed as chi-squared, and in the numerous editions of his (1925) textbook he states that 'the discovery of the distribution of χ^2 in reality completed the method of Lexis'. However, Lexis actually proposed his coefficient after Helmert's two papers on the sum of squares of Normally distributed variates. And, prior to Fisher, Bortkiewicz (1922) had already given his qualified approval of the use of χ^2_{g-1} for the approximate distribution of the Lexis coefficient.

The 1924 (1928) paper was Fisher's only article on the mathematical basis of the analysis of variance. It was there presented as a method of testing the equality of group means, namely a one-way classification. At that time Fisher had just become actively involved in the design and analysis of agricultural experiments at Rothamsted Experimental Station. In a letter to W. S. Gosset ('Student'), the relevant portion of which is reproduced as a footnote in 'Student' (1923), he extended this technique to the yields obtained from a balanced two-way classification (variety and trial) and actually fitted the parameters ('quantities') of a linear model (without interaction terms) by least squares. The results were expressed as an analysis of variance. At about the same time Fisher published with Mackenzie (1923) the numerical analysis of a two-way classification in which the interaction terms were introduced as 'deviations from summation [additive] formula'.

Fisher's arguments in favour of the analysis of variance of balanced designs may be quoted from a paper with Eden (1927) in which balance was attained and which provides the first example of a 2^3 experiment, in randomized blocks, analyzed into individual degrees of freedom representing main effects and first and second order interactions: '...the conclusions from the results are drawn without any complex statistical analysis by simple arithmetical additions and subtractions. The only tests of significance required are those of which the theory is now fully understood, and for which adequate tables are available'. There is no hint here that in the corresponding Gauss linear model the interaction terms would appear as an arbitrary selection from the terms of a third degree polynomial if the factor levels were quantitative or if more than two levels were involved (Box & Wilson, 1951).

It may be added that in the discussion of Wishart (1934), Fisher said: 'the analysis of variance...is not a mathematical theorem, but rather a convenient method of arranging the arithmetic'. His insight into the proper constituents of each sum-of-squares item was indeed penetrating. Thus (Fisher & Wishart, 1930) he perceived that certain treatment vectors were linear compounds of (i.e. 'confounded' with) block vectors and extracted them correctly from the analysis. One of the great advantages of Fisher's balanced designs is that each group of β's implied by the analysis of variance has an interpretation that is independent of any other group. The interrelationships between the β estimates is one of the most vexing features of non-orthogonality.

However, this emphasis on simple arithmetic at the expense of a careful review of the underlying model and its possible lack of orthogonality led its author into the oversimplification of some complex situations. For example, the two-way classification of the Fisher–Mackenzie (1923) paper was a systematic design involving 6 (treatments) $\times$ 12 (potato varieties), each of the 72 combinations being replicated three times except that the third replicate was absent in three of the treatments of one variety. Fisher's procedure was to estimate each 'missing value' by the mean of the two replicates for the corresponding combination, and to proceed with an analysis of variance using the resulting 216 values. However, his residual ('variation between parallel plots') sum of squares was recognized to have only $69 \times 2 + 3 = 141$ degrees of freedom. This procedure produces the correct least squares residual but the main effects and interactions need recalculation and can no longer be shown as an analysis of variance. The adjustments necessary to the Fisher–Mackenzie figures are shown in Table 1. The total weight of the potatoes in each plot and the average weight per plant, the number of plants differing from plot to plot, were published in the Rothamsted *Report* for 1921–22. [Even after correcting one glaring misprint in the former we were unable to reproduce exactly the Fisher–Mackenzie sums of squares based on the plot weights.] While it may be argued, in this and subsequent illustrations, that Fisher was concerned more with developing a practical tool for agricultural science than with algebraic niceties many of the resulting discrepancies were not evident to the mathematical statisticians of the time.

Table 1. *Analysis of Fisher & Mackenzie's data*

		Sum of squares		
Variation due to	D.F.	Fisher	Recalculated	Least squares (non-additive)
Manures	5	6158	6176	5820
Varieties	11	2843	2848	2780
Interaction	55	981	981	944
Residual	141	1758	1729	1729

A similar criticism applies to the (1929) paper with Eden in which the eight z-vectors representing the interaction of nitrogen and quality of potash are not orthogonal to the eight block vectors. Rather than solve the 16 normal equations for these interaction and block β's, equations that would have simplified by judicious manipulation, Fisher regarded the interaction terms as 'error' and left them in the residual. Inadvertently, as Yates mentioned in his 1933 paper, he thus neglected an interaction of much more consequence than the two interaction parameters orthogonality allowed him to retain in his analysis. The relevant figures are as given in Table 2.

Table 2. *More detailed analysis of Fisher & Eden's data*

	Degrees of freedom	Eden–Fisher 'Sums of squares'	Quantity for significance test in 29 parameter model
Blocks	8	22,733	21,977·500
Nitrogen	2	49,905	43,361·625*
Potash *vs.* no potash	1		813·375*
Single *vs.* double potash	1		1,472·625*
Nitrogen–potash interaction	4		4,256·875*
Quality of potash	2	14,458	14,457·625*
Quantity and quality of potash interaction	2	1,005	1,005·125*
Nitrogen-quality of potash interaction	4	32,628	9,792·500
Nitrogen and quantity and quality of potash interaction	4		611·250
Residual	52		22,228·375
	80	120,729	

In the foregoing table we have corrected the original sum of squares for 'blocks', namely 21,442. Note that items marked with an asterisk in the last column are additive in a test of aggregate effects and that the interaction of nitrogen and quality of potash is highly significant.

Another result of Fisher's reliance on arithmetic and intuitive tests of hypotheses rather than on the development of an appropriate mathematical model was that some of the 'quantities' estimated in the least-squares solution were not conceived of as 'constants'. [F. J. Anscombe has suggested in conversation that Fisher may have abandoned a parametric model in favour of a co-ordinate-free formulation (Kruskal, 1961).] The earliest example of this occurs in Fisher's 1925 text-book where he reconsidered the balanced half of the Fisher–Mackenzie experiment on potatoes using, however, the plot-average weights per plant instead of the aggregate weights per plot. Formally this is now a three-way classification of 12 (varieties) $\times$ 3 (plot-groups) $\times$ 3 (treatments) which is to be analyzed by least squares or analysis of variance. However, since all three treatments were always applied to three contiguous plots of the field it was argued that comparisons of the treatments (and also of the interactions of treatments and varieties) would be subject to less 'error' than the direct comparisons between the varieties. The residual for the latter was the interaction between variety and plot-group (though in Fisher's example plot-group comparisons are included in this 'residual').

A further example occurs in a paper by Wishart & Clapham (1929), two of Fisher's colleagues at Rothamsted. Here the experiment was a 3×3 factorial design arranged in three randomized blocks, each of the 27 plots being 'split' into two subplots one of which received superphosphate. The calculation of separate 'whole plot' and 'sub-plot' errors is clearly described. Although astronomers had utilized such a 'random effects' model in the nineteenth century it was not explicitly recognized as such in experimental design until 1939 (Scheffé, 1956).

Wishart's extension of the 'split-plot' analysis to a more complex experiment arranged as a Latin square (1931) provides an excellent example of the difficulty of judging orthogonality without explicit reference to the z-vectors attached to the model's parameters. In this case

Wishart, and possibly Fisher too, overlooked the non-orthogonality of 33 interaction degrees of freedom with the 22 degrees ascribed to 'rows' and 'columns' of the square (Yates, 1933).

More serious, perhaps, was Fisher's own failure to realize that the introduction of a covariance term into an orthogonal linear model, namely a single vector $\mathbf{z}_q$ which is not orthogonal to any of the remaining z-vectors except $\mathbf{z}_1$, a vector of units representing the general mean, destroys the additive property of the individual (adjusted) sums of squares that are tested for significance against a residual sum of squares adjusted for the covariate. Fisher's first approach to this problem was to make a standard analysis of variance, not of $\mathbf{x}-\bar{\mathbf{x}}$, but of $(\mathbf{x}-\bar{\mathbf{x}})-\hat{\beta}_q(\mathbf{z}_q-\bar{\mathbf{z}}_q)$ (Sanders, 1930 and the 1932 edition of Fisher's text book). But this assumes that the estimate of β_q will not change when one or more β's are deleted from the linear model. It was only the divorce of the analysis of variance technique from the underlying linear least squares model that prevented this from being seen at once. The mistake was corrected by Bartlett (1934) and, apparently independently, by Fisher in the 1934 edition of his text but by then several papers using the incorrect technique, including one with two covariance terms, had already appeared in the literature. The corrected procedure still preserves the format of the analysis of variance but each constituent item is appropriately adjusted.

The first frank admission that a linear model analyzed by least squares was more fundamental than an intuitive analysis of sums of squares is due to Yates (1933), himself a geodesist before he went to Rothamsted, as he states in his contribution to the discussion of Wishart (1934). Nevertheless, even here the z-vectors of the linear model are not emphasized but, instead, 'constants' are introduced to represent the value of each product $\beta_j z_{ji}$ in the cases where z_{ji} is not zero. However, a year later Irwin (1934), also a worker at Rothamsted, provided an explicit vectorial exhibit of the standard randomized block and Latin square designs showing clearly how the orthogonality of these vectors analyzed the sum of squares about the mean into several constituent parts.

This brings us to the year 1935 which saw the publication of two important syntheses of least-squares theory by Aitken and Kolodziejczyk, respectively, and the first edition of Fisher's remarkable monograph on experimental design. To summarize the numerous mathematical accretions to the theory since that date and the proliferation of quasi-orthogonal experimental designs would require another article as long as this has been. Since 1935 is a landmark in both the theory and practice of the Gauss linear model, as we think it should be called, we here take leave of our readers.

References

AITKEN, A. C. (1935). On least squares and linear combination of observations. *Proc. R. Soc. Edinb.* **55**, 42–8.

AITKEN, A. C. (1945). On linear approximation by least squares. *Proc. R. Soc. Edinb.* **62**, 138–46.

BARTLETT, M. S. (1934). The vector representation of a sample. *Proc. Camb. Phil. Soc.* **30**, 327–40.

BAUSCHINGER, J. (trans. and revised by ANDOYER, H.) (1908). Théorie des erreurs. *Encycl. Sci. Math.* I **4** (2), 161–95.

BIENAYMÉ, J. (1853). Remarques sur les différences qui distinguent l'interpolation de M. Cauchy de la méthode des moindres carrés, et qui assurent la superiorité de cette méthode. *C.R. Acad. Sci. Paris* **37**, 5–13.

BÖRSCH, A. & SIMON, P. (1887, 1964). *Abhandlungen zur Methode der kleinsten Quadrate von Carl Friedrich Gauss.* Würzburg.

Bortkiewicz, L. von (trans. and revised by Oltramare, F.) (1909) Statistique. *Encycl. Sci. Math.* I **4**(3), 453–90.

Bortkiewicz, L. von (1922). Das Helmertsche Verteilungsgesetz für die Quadratssumme zufälliger Beobachtungsfehler. *Z. Angew. Math. Mech.* **2**, 358–75.

Box, G. E. P. & Wilson, K. B. (1951). On the experimental attainment of optimum conditions. *J. R. Statist. Soc.* B **13**, 1–45.

Bravais, A. (1846). Sur les probabilités des erreurs de situation d'un point. *Mém. Acad. Roy. Sci. Inst. France* **9**, 255–332.

Carvallo, E. (1888). Formules d'interpolation, and Sur l'application de la méthode des moindres carrés. *C.R. Acad. Sci. Paris* **116**, 346–9, 924–6.

Castelnuovo, G. (1919, 1925, 1945). *Calcolo delle Probabilità*. Bologna.

Cauchy, A. (1836). On a new formula for solving the problem of interpolation in a manner applicable to physical investigations. *Phil. Mag.* **8**, 459–68.

Cauchy, A. (1853). Sur la nouvelle méthode d'interpolation comparée à la méthode des moindres carrés. *C.R. Acad. Sci. Paris* **37**, 100–9.

Chebyshev, P. L. (1858). Sur les fractions continues. *J. Math. pures appl.* II **3**, 289–323.

Cochran, W. G. (1938). The omission or addition of an independent variate in multiple linear regression. *J.R. Statist. Soc. Suppl.* **5**, 171–6.

Czuber, E. (1891). *Theorie der Beobachtungsfehler*. Leipzig.

Czuber, E. (1899). Die Entwicklung der Wahrscheinlichkeitstheorie und ihrer Anwendungen. *Jber. dt. Math Verein* **7**, 1–279.

Czuber, E. (1903, 1908, 1911, 1923). *Wahrscheinlichkeitsrechnung*. Leipzig.

David, F. N. & Neyman, J. (1938). Extension of the Markoff theorem on least squares. *Statist. Res. Mem.* **2**, 105–16.

Eden, T. & Fisher, R. A. (1927). Studies in crop variation. IV. The experimental determination of the value of top dressings with cereals. *J. Agric. Sci.* **17**, 548–62.

Eden, T. & Fisher, R. A. (1929). Studies in crop variation. VI. Experiments on the response of the potato to potash and nitrogen. *J. Agric. Sci.* **19**, 201–13.

Edgeworth, F. Y. (1892). Correlated averages. *Phil. Mag.* Ser. 5, **34**, 190–204.

Eisenhart, C. (1964). The meaning of 'least' in least squares. *J. Wash. Acad. Sci.* **54**, 24–33.

Fisher, R. A. (1921). On the 'probable error' of a coefficient of correlation deduced from a small sample. *Metron* **1**, N. 4, 3–32.

Fisher, R. A. (1922). The goodness of fit of regression formulae and the distribution of regression coefficients. *J. R. Statist. Soc.* **85**, 597–612.

Fisher, R. A. (1925). Applications of 'Student's' distribution. *Metron* **5**, N. 3, 90–104.

Fisher, R. A. (1925, 58). *Statistical Methods for Research Workers*. Edinburgh: Oliver and Boyd.

Fisher, R. A. (1928). On a distribution yielding the error functions of several well known statistics. *Proc. Int. Congr. Math.*, Toronto (1924), 805–13.

Fisher, R. A. (1935, 60). *The Design of Experiments*. Edinburgh: Oliver and Boyd.

Fisher, R. A. & Mackenzie, W. A. (1923). Studies in crop variation. II. The manurial response of different potato varieties. *J. Agric. Sci.* **13**, 311–20.

Fisher, R. A. & Wishart, J. (1930). The arrangement of field experiments and the statistical reduction of the results. *Imp. Bur. Soil Sci., Tech. Comm.* no. 10, Harpenden, Herts.

Fox, L. (1964). *An Introduction to Numerical Linear Algebra*. Oxford University Press.

Gauss, C. F. (1809, 1810, 1821–26). See Börsch *et al.*

Gram, J. P. (1883). Ueber die Entwickelung reeller Funktionen in Reihen mittelst der Methode der kleinsten Quadrate. *J. reine angew. Math.* **94**, 41–73.

Gram, J. P. (1915). Über partielle Ausgleichung mittelst Orthogonalfunktionen. *Mitt. Verein. Schweiz. Ver.-Math.* **10**, 3–23.

Graybill, F. A. (1961). *An Introduction to Linear Statistical Models*, vol. 1. New York: McGraw Hill.

Grossmann, W. (1961). *Grundzüge der Ausgleichungsrechnung*. Berlin.

Helmert, F. R. (1872, 1907). *Die Ausgleichungsrechnung nach der Methode der Kleinsten Quadrate*. Leipzig.

Irwin, J. O. (1934). On the independence of the constituent items in the analysis of variance. *J. R. Statist. Soc. Suppl.* **1**, 236–51.

Isserlis, L. (1927). Note on Chebyshev's interpolation formula. *Biometrika* **19**, 87–93.

Kendall, M. G. & Stuart, A. (1961). *The Advanced Theory of Statistics*, vol. 2. London: Griffin.

Kolodziejczyk, S. (1935). On an important class of statistical hypotheses. *Biometrika* **27**, 161–90.

Kruskal, W. (1961). The coordinate-free approach to Gauss-Markov estimation, and its application to missing and extra observations. *Proc. 4th Berkeley Symp. Math. Statist. Prob.* **1**, 435–51.

LAPLACE, P. S. (1812). *Théorie Analytique des Probabilités*. Paris.
LEGENDRE, A. M. (1805). *Nouvelles Méthodes pour la Détermination des Orbites des Comètes*. Paris.
LINDLEY, D. V. (1965). *Introduction to Probability and Statistics from a Bayesian Viewpoint*. Cambridge University Press.
MACDONELL, W. R. (1901). On criminal anthropometry and the identification of criminals. *Biometrika* **1**, 177–227.
MARCHANT, R. (1956). *La Compensation des Mesures Surabondantes*. Brussels.
MARKOV, A. (trans. by Liebmann, H.) (1912). *Wahrscheinlichkeitsrechnung*. Leipzig.
McMAHON, J. (1923). Hyperspherical goniometry; and its application to correlation theory for n variables. *Biometrika* **15**, 173–208.
MORGENSTERN, D. (1964). *Einführung in die Wahrscheinlichkeitsrechnung und Mathematische Statistik*. Berlin.
MUMFORD, A. A. & YOUNG, M. (1923). The interrelationships of the physical measurements and the vital capacity. *Biometrika* **15**, 109–33.
NEYMAN, J. (1934). On two different aspects of the representative method. *J.R. Statist. Soc.* **97**, 558–625.
PEARSON, E. S. (1938). *Karl Pearson: An Appreciation of Some Aspects of His Life and Work*. Cambridge University Press.
PEARSON, E. S. (Ed.) (1956). *Karl Pearson's Early Statistical Papers*. Cambridge University Press.
PEARSON, K. (1896). Mathematical contributions to the theory of evolution. III. Regression, heredity and panmixia. *Phil. Trans.* A **187**, 253–318.
PEARSON, K. (1898). Mathematical contributions to the theory of evolution. V. On the reconstruction of the stature of prehistoric races. *Phil. Trans.* A **192**, 169–244.
PEARSON, K. (1901). On lines and planes of closest fit to systems of points in space. *Phil. Mag.* Ser. **6**, **2**, 559–72.
PEARSON, K. (1905). Mathematical contributions to the theory of evolution. XIV. On the general theory of skew correlation and non-linear regression. *Drap. Co. Res. Mem., Biom. Ser.* II.
PEARSON, K. (1917). On the application of 'goodness of fit' tables to test regression curves and theoretical curves used to describe observational or experimental data. *Biometrika* **11**, 239–61.
PEARSON, K. (1920). Notes on the history of correlation. *Biometrika* **13**, 25–45.*
PEARSON, K. (1921). On a general method of determining the successive terms in a skew regression line. *Biometrika* **13**, 296–300.
PEARSON, K. (1926). Researches on the mode of distribution of samples taken at random from a bivariate normal population. *Proc. Roy. Soc.* A **112**, 1–14.
PEARSON, K. (1931). Historical note on the distribution of the standard deviation of samples of any size drawn from an indefinitely large Normal parent population. *Biometrika* **23**, 416–8.
PIZZETTI, P. (1889). Sopra il calcolo dell'errore medio di un sistema di osservazioni. *Rend. Reale Accad. Lincei* **5**, 740–4.
PIZZETTI, P. (1891). *I Fondamenti Matematici per la Critica dei Risultati Sperimentali*. Genova. (Also printed in the 1892 *Atti. R. Univ. Genova*.)
PLACKETT, R. L. (1949). A historical note on the method of least squares. *Biometrika* **36**, 458–60.
PLACKETT, R. L. (1960). *Principles of Regression Analysis*. Oxford University Press.
QUETELET, A. (1846). *Lettres...sur la Théorie des Probabilités*. Brussels.
RADAU, R. (1891). Études sur les formules d'interpolation. *Bull. Astron.* **8**, 273–94, 325–51, 376–93, 425–55.
RAINSFORD, H. F. (1957). *Survey Adjustments and Least Squares*. London.
RAO, C. R. (1949). On a transformation useful in multivariate analysis computations. *Sankhyā* **9**, 251–3.
RAO, C. R. (1965). *Linear Statistical Inference and Its Applications*. New York: Wiley.
REYDEN, D. VAN DER (1943). Curve fitting by the orthogonal polynomials of least squares. *Onderstepoort J. Vet. Sci. Anim. Indust.* **18**, 355–404.
ROMANOVSKY, V. (1925). Sur une méthode d'interpolation de Tchebycheff. *C.R. Acad. Sci. Paris* **181**, 595–7.
ROMANOVSKY, V. (1927). Note on orthogonalising series of functions and interpolation. *Biometrika* **19**, 93–9.
SANDERS, H. G. (1930). A note on the value of uniformity trials for subsequent experiments. *J. Agric. Sci.* **20**, 63–73.
SAVAGE, R. (1961). Probability inequalities of the Tchebycheff type. *J. Res. Nat. Bur. Stand.* **65** B, 211–22.
SCHEFFÉ, H. (1956). Alternative models for the analysis of variance. *Ann. Math. Statist.* **27**, 251–71.

**Paper No. 14 in this volume*

SCHEFFÉ, H. (1959). *The Analysis of Variance*. New York: Wiley.

SCHMETTERER, L. (1956). *Einführung in die Mathematische Statistik*. Vienna: Springer.

SCHMIDT, E. (1908). Über die Auflösung linearer Gleichungen mit unendlich vielen Unbekannten. *Rend. Circ. Mat. Palermo* **25**, 53–77.

SCHOLS, CH. M. (1875). Over de theorie der fouten in de ruimte en in het platte vlak. *Verh. Nederl. Akad. Wetensch.* **15**, 1–75.

SEAL, H. L. (1949). Mortality data and the binomial probability law. *Skand. Aktuarietidskr.* **32**, 188–216.

SELIVANOV, D. & BAUSCHINGER, J. (trans. and revised by ANDOYER, H.) (1906). Calcul des différences et interpolation. *Encycl. Sci. Math.* I **4** (1), 47–160.

SLUTSKY, E. (1914). On the criterion of goodness of fit of the regression lines and on the best method of fitting them to the data. *J. R. Statist. Soc.* **77**, 78–84.

SMART, W. M. (1958). *Combination of Observations*. Cambridge University Press.

STEVENSON, P. H. (with Note by K. P.) (1929). On racial differences in stature long bone regression formulae, with special reference to stature reconstruction formulae for the Chinese. *Biometrika* **21**, 303–21.

'STUDENT' (1923). On testing varieties of cereals. *Biometrika* **15**, 271–93.

THIELE, T. N. (1889). *Forläsninger over almindelig Jagttagelsesläre*. Copenhagen.

THIELE, T. N. (1897). *Elementär Jagttagelseslære*. Copenhagen.

THIELE, T. N. (1903). *Theory of Observations*. London.

TIENSTRA, J. M. (1956). *Theory of the Adjustment of Normally Distributed Observations*. Amsterdam.

VILLARCEAU, Y. (1849). Premier mémoire sur les étoiles doubles. *Additions à la Connaissance des Temps* 1852, 1–197.

WAERDEN, B. L. VAN DER (1957). *Mathematische Statistik*. Berlin: Springer.

WALKER, H. M. (1931). *Studies in the History of Statistical Method*. Baltimore.

WHITTAKER, E. T. & ROBINSON, G. (1924, 1926). *The Calculus of Observations*. London: Blackie.

WILKS, S. S. (1962). *Mathematical Statistics*. New York: Wiley.

WILLIAMS, E. J. (1959). *Regression Analysis*. New York: Wiley.

WISHART, J. (1931). The analysis of variance illustrated in its application to a complex agricultural experiment on sugar beet. *Arch. PflBau* **5**, 561–84.

WISHART, J. (1934). Statistics in agricultural research. *J. R. Statist. Soc. Suppl.* **1**, 26–61.

WISHART, J. & CLAPHAM, A. R. (1929). A study in sampling technique: The effect of artificial fertilizers on the yield of potatoes. *J. Agric. Sci.* **19**, 600–18.

WISHART, J. & METAKIDES, T. (1953). Orthogonal polynomial fitting. *Biometrika* **40**, 361–9.

WITTSTEIN, T. (1867). *Mathematische Statistik und deren Anwendung auf National-Oekonomie und Versicherungs-Wissenschaft*. Hanover.

YATES, F. (1933). The principles of orthogonality and confounding in replicated experiments. *J. Agric. Sci.* **23**, 108–45.

YULE, G. U. (1897). On the theory of correlation. *J. R. Statist. Soc.* **60**, 812–54.

YULE, G. U. (1907). On the theory of correlation for any number of variables, treated by a new system of notation. *Proc. Roy. Soc.* A **79**, 182–93.

ZEUNER, G. (1869). *Abhandlungen aus der Mathematischen Statistik*. Leipzig.

ON THE EARLY HISTORY OF THE LAW OF LARGE NUMBERS

By O. B. SHEYNIN

Note—The author has amended the last two paragraphs on the following page (in contrasting type) for this reprint.

Summary

This paper is devoted to the early history of the law of large numbers. An outline of the prehistory of this law is given in § 1. The algebraic part of J. Bernoulli's theorem is presented in a logarithmic form and the lesser known role of N. Bernoulli is described in § 2. Comments on the derivation of the De Moivre–Laplace limit theorems by De Moivre, in particular, on the inductive character of his work, on the priority of De Moivre as to the continuous uniform distribution, on the unaccomplished possibility of Simpson having arrived at the normal distribution and on the role of Laplace are presented in § 3. The historical role of J. Bernoulli's form of the law of large numbers is discussed in § 4.

1. Prehistory of the law of large numbers

The most rudimentary form of the law of large numbers is to be credited to Cardano (Ore, 1963, p. 150), who held that 'in the long run' the number of occurrences of an event in n independent trials is approximately equal to

$$\mu = np, \tag{1}$$

where p is the constant probability of the occurrence of this event in one trial. This 'reasoning on the mean outcome', as Ore called it (p. 145), was systematically used by Cardano and, again as Ore pointed out, led him in some cases to erroneous results.

Halley (1693, p. 484) explicitly stated that the reason for the frequencies of mortality of different age groups being irregular 'seems rather to be owing to chance, as are also the other irregularities in the series of age, which would rectify themselves, were the number of years (of observation on the studied population of Breslaw) much more considerable....'

Such assertions were possibly also made by other scholars, but what distinguishes Halley is that he adjusted the frequencies concerned so that they would more nearly correspond to their general trend and therefore be applicable to populations other than the population of Breslaw; see also Graetzer (1883, pp. 77–8). More comment on Halley is made in § 3.

A 'reasoning on the mean', differing from Cardano's, occurs in a sixteenth-century commentary by Gańéśa on a still earlier Indian mathematical text. Commenting on the calculation of the volume of an irregular earth excavation considered to be equal to the product of the mean measures of the length, width and depth of the excavation, the measures being taken at different places, Gańéśa pointed out (Colebrooke, 1817, p. 97) that 'the greater the number of the places (of measurement), the nearer will the mean measure be to the truth and the more exact will be the consequent computation.'

The mean measures were introduced to compensate for inaccuracies in the mathematical model. The application of the arithmetic mean for the same purpose can be traced to ancient Babylonia where, for example (Veiman, 1961, p. 204), the area of a quadrangle was held to be equal to the product of the half-sums of its opposite sides. Though the commentary of Gańéśa should be specially noticed because of his reference to the increase of the number of measurements, it is to be inferred that, strictly speaking, his reasoning is of a deterministic (not probabilistic) kind and its formalization would have led to integral sums.

Qualitative assertions for the preference of the arithmetic mean of several observations as against a single observation in astronomy and geodesy are found in the works of the seventeenth to eighteenth centuries, notably of Cotes (1768, pp. 57–8). But these assertions, although undoubtedly of a probabilistic nature, are concerned only with a given set of observations and at least until the second half of the eighteenth century no mention was made of the effect of increasing the number of observations.

The general impression seems to be that the prehistory of the law of large numbers contains an understanding of the nature and the use of formula (1) and of the arithmetic mean of a given set of observations. The matching and further formalization of these separate ideas were however, due to Jacob Bernoulli.

2. Jacob and Nicholas Bernoulli

J. Bernoulli (1713, German version 1899) proved that as $n \to \infty$

$$\lim \operatorname{prob}\left(\left|\frac{\mu}{n}-p\right| < \epsilon\right) = 1.$$

His work is described notably by Todhunter (1863, §§ 123–4) and Pearson (1925). Bernoulli also gave an example from which it directly followed that the sum of $2n$ middle terms of the expansion of $(r+s)^{(r+s)n}$, $r = 30$, $s = 20$ (even excluding the middlemost term) will be more than $c > 0$ times the sum of the other terms of the expansion if the number of trials

$$nt \equiv n(r+s) \geqslant 25{,}500 + 5758 \log \frac{c}{1000} = 8226 + 5758 \log c. \tag{2}$$

After the death of J. Bernoulli, but prior to the publication of the *Ars Conjectandi*, N. Bernoulli (Montmort, 1713, p. 388) deduced an approximate estimate for the ratio of the middle part of a binomial series to its other parts and, taking Arbuthnot's (1712) data, used this estimate for probabilistic reasoning about the 'constant regularity observed in the births of both sexes'.

N. Bernoulli's estimate was rather crude, but it seems that he was the first to study how the probability that a random quantity falls in an interval depends on the length of that interval.

In more detail, let n be the annual number of births with the ratio of male and female births equal to m/f. Assuming a binomial distribution, N. Bernoulli gives the following estimate which, as is also the case with J. Bernoulli, is actually equivalent to a local limit theorem,

$$\operatorname{prob}(|\mu - rm| \leqslant l) \approx \frac{t-1}{t}, \qquad t \approx \left\{1 + \frac{l(m+f)^2}{mfn}\right\}^{\frac{1}{2}l}, \quad z = \frac{n}{m+f}.$$

It is to be emphasized that with $p = m/(m+f)$ and $q = n/(m+f)$ this may be written as

$$\operatorname{prob}(|\mu - rm| \leqslant l) \approx 1 - \exp\{-l^2/(2pqn)\}.$$

3. Derivation of the 'De Moivre–Laplace' limit theorems

3·1. *De Moivre*

A further development of the law of large numbers leading to the 'De Moivre–Laplace' limit theorems came with De Moivre. General information about De Moivre is to be found in many places, all of which, nevertheless, draw on the two main sources (Eloge..., 1759; Maty, 1760). His main work, *Doctrine of Chances*, appeared in 1718, 1738 and, posthumously, in 1756. The last two editions have recently been reproduced; this information was recently received from Dr C. Eisenhart. The relevant work of De Moivre (*Method of Approximating the Sum of the Terms of the Binomial...*) has been sufficiently described by Pearson (1924, 1925) and Archibald (1926); the present author has only a few comments to offer.

(i) De Moivre drew heavily on his book *Miscellanea Analytica de Seriebus et Quadraturis* (1730), an English translation of which is long overdue. It was there, partly in the *Supplement*, that all the main algebraic deductions were made.

(ii) It was also there, again in the *Supplement*, that De Moivre first published his 14-digit table of $\log n!$ for $n = 10(10)900$. The table is correct to 11–12 digits with a single misprint in the fifth digit of $\log 380!$. I have compared the table with several modern tables, notably that of Peters (1922).

Both Pearson and Markov (1924) consider the approximate formula for $n!$ to be derived by Stirling and De Moivre and therefore to be named the De Moivre–Stirling formula, while Archibald attributes it to De Moivre. If De Moivre's table is taken into account, as it should be, the opinion of Archibald is substantially strengthened.

(iii) The important appearance of the estimate of accuracy, $n^{-\frac{1}{2}}$, in De Moivre's work was originally occasioned by a purely algebraic fact: $x = n^{\frac{1}{2}}/2$ was a bordering value for two different ways of integrating

$$\int_0^l e^{-2x^2/n}\,dx.$$

De Moivre employed a power series when $l \leqslant \frac{1}{2}n^{\frac{1}{2}}$ and, because of the slow convergence of this series when $l > \frac{1}{2}n^{\frac{1}{2}}$, the approximate method of Newton–Cotes.

(iv) De Moivre experimentally checked the accuracy of his formula for $n = 900$ and 100 (*Method of Approximating...*, corol. 5). Although he gave no indication as to the nature of his experiments and although these experiments seem to be well within mathematics (not exact sciences at large), the fact that he made these checks is interesting. The opinion (Walker, 1934, p. 320) that 'even in his (De Moivre's) writings on the Doctrine of Chances his work is deductive and he does not set up experimental checks on the outcome' is hardly fair.

But a more important comment in this connexion is that the whole *Method of Approximating* was clearly intended for 'experimental checks', as is proved by the corollary to problem 87 in the 1738 edition of the *Doctrine of Chances* (this is the problem after which immediately follows the *Method*):

> ...if after taking a great number of experiments, it should be perceived that the happenings and failings have been nearly in a certain proportion...it may safely be concluded that the probabilities of happening or failing at any one time assigned will be very near in that proportion, and that the greater the number of experiments has been, so much nearer the truth will the conjectures be that are derived from them.
>
> But suppose it should be said, that notwithstanding the reasonableness of building conjectures upon observations, still considering the great power of chance, events might at long run fall out in a different

proportion from the real bent (according to) which they have to happen one way or the other; and that supposing for instance that an event might as easily happen as not happen, whether after three thousand experiments it may not be possible it should have happened two thousand times and failed a thousand, and that therefore the odds against so great a variation from equality should be assigned, whereby the mind would be the better disposed in the conclusions derived from the experiments:

In answer to this, I will take the liberty to say that this is the hardest problem that can be proposed on the subject of chance, for which reason I have reserved it for the last...I shall derive...some conclusions that may be of use to every body: in order thereto, I shall here translate (from the Latin) a paper of mine which was printed November 12, 1733, and communicated to some friends, but never yet made public, reserving to myself the right of enlarging my own thoughts, as occasion shall require.

This quotation means that De Moivre attempted a reconciliation of statistical and prior probabilities.

(v) A special comment on the theological reasonings of De Moivre's contemporaries is warranted. It is true that, as Pearson points out, De Moivre and other Fellows of the Royal Society were greatly influenced by Newton's theology. But at least one mathematician, Simpson, who, moreover, humbly refers to De Moivre in the introduction to *Nature and Laws of Chance* (1740), did not follow Newton's theology. On the contrary his error distribution (Simpson, 1757) was based on objective probabilistic properties of the errors of astronomical observations.

3·2. *Simpson*

It is hardly appropriate to describe the works of Simpson in detail in this paper. But it ought to be said that Simpson arrived at a continuous triangular distribution, proved that with this distribution the arithmetical mean is preferable to a single observation and went on to deduce the probability of a given error in the arithmetical mean in the limiting case. From this he could have arrived at a normal distribution (as pointed out in a private communication from L. N. Bolshev) although later than De Moivre and could have been the first to draw a graph of this distribution. His failure is evident in that his limit curve (Fig. 20) does not behave as a normal one.

General information about Simpson is given by Clarke (1929). Referring to his correspondence with the Royal Society, Penkov (1961, pp. 300, 302) states that there exists no portrait of Simpson.

3.3. *Continuous uniform distribution and De Moivre**

The tercentenary of the birth of De Moivre having occurred in 1967 we feel it opportune to add two short comments not connected with the law of large numbers.

(*a*) While considering one of his problems in games of chance De Moivre followed the same method which Poisson followed when he originally arrived at his distribution (Newbold, 1927, pp. 490–1)

(*b*) While calculating annuities on lives De Moivre arrived at the continuous uniform distribution and used the corresponding first moment. (The author has seen the 1756, posthumous edition of the *Treatise of Annuities on Lives*, incorporated in the *Doctrine of Chances*, 1725, and the 1743 edition, *Annuities on Lives*.) In problem 20, part 1 of the *Treatise*, he calculated the expectation of life being equal to

$$E(\zeta)-z_0=\int_{z_0}^{z_0+n}\frac{z\,dz}{n}-z_0=\tfrac{1}{2}n;$$

only the result is given. Here ζ is the random uniformly distributed duration of life (beginning from age z_0) and n is the complement of life. The uniform distribution appeared as an empirical law corresponding to Halley's data (1693).

*This distribution was originally employed by N. Bernoulli (Todhunter, 1863, §§338–40).

In Chapter 8 of Part 2, De Moivre calculates the probability of one person with a complement of life equal to n outliving another person whose complement of life is equal to $p < n$

$$\text{prob}\,(\zeta > \eta) = \int_0^p \frac{n-z}{n}\frac{dz}{p} = 1 - \frac{p}{2n}.$$

This result could have been arrived at geometrically: in fact geometrical probabilities, proportional to areas of corresponding figures, were considered by Halley (1693).

In problem 21, part 1, De Moivre calculated the expectation of two joint lives (again, only the result is given):

$$\text{prob}\,(\zeta \in [z, z+dz] \quad \text{or} \quad \eta \in [z, z+dz]) = \frac{n-z}{n}\frac{dz}{p} + \frac{p-z}{p}\frac{dz}{n}$$

and the expectation sought is

$$\int_0^p \left\{\frac{z}{n}(n-z)\frac{dz}{p} + \frac{(p-z)}{p}z\frac{dz}{n}\right\} = \frac{p}{2} - \frac{p^2}{6n}.$$

The reconstruction of the solution of this problem is due to Czuber who offers no further comment; see his Anmerkung (22) in De Moivre (1906), this being a translation of the 1756 edition.

3·4. *Laplace*

It has already been pointed out in § 3·1 that De Moivre had proved the theorems which are now known as the De Moivre–Laplace limit theorems; he did not introduce the concept of uniform convergence which is of a later origin. Laplace actually repeated these proofs, and, characteristically for his works in which his own contributions are not readily separated from those of his predecessors, did not refer to De Moivre, giving only a barest possible outline of the work of De Moivre in the historical part of his *Essai philosophique....*

It is worth noting that Laplace, who was really able, as no one else, to grasp and elaborate the ideas of his predecessors, in his first probabilistic memoir (1774) deduced an exponential distribution

$$\phi(x) = \tfrac{1}{2}m\,e^{-m|x|}.$$

A function of this kind might be deduced by reversing (2), the possibility of which was not mentioned by J. Bernoulli. But (2) is a deterministic not probabilistic formula and therefore this is an example, possibly superfluous, of the way Laplace extended the ideas of his predecessors.

Lastly, we note that the correspondence of Lagrange and Laplace (Lagrange, 1892; see letter to Laplace dated 30 December 1776 on p. 66) testifies that they both, independently, had contemplated translating De Moivre's *Doctrine of Chances* into French. Only Lagrange is mentioned by Todhunter in this connexion.

The appearance of the normal law in the work of De Moivre was not noticed until Eggenberger (1894, especially p. 165) referred to it possibly because this was not mentioned by Laplace. Although mentioned by Czuber (1899), Eggenberger's work seems to be little known, perhaps because Haussner called it unclear. Haussner's remark appears on pp. 158–9 of the *Ostwald Klassiker*, no. 108, see Bernoulli (1899). And Pearson (1924) independently of Eggenberger credited the normal law to De Moivre.

4. The role of the law of large numbers

4·1. *Statistics*

This particular law became the stepping stone between the theories of probability and statistics. The main problem of population theory after Quetelet, the problem of the stability of statistical frequencies, came down to the testing of the existence of preconditions for the law of large numbers in Bernoulli's form and was solved by W. Lexis and especially by Bortkiewicz (1898). The latter used the Poisson form of the law.

Bortkiewicz (1905, p. 140) praised Lexis as 'the first to establish the integral connexion' between theoretical statistics and the theory of probability. It seems, however, that initial connexions were made as early as in the 1850's. For example, Davidov (1855*b*) tested the statistical significance of various empirical inequalities. In another paper Davidov (1855*a*) stated that the 'excessive development' of statistics, and its deductions 'often unfounded', can smear statistics and that a discussion of the initial statistical data with probabilistic verification is the most reliable tool for eliminating 'immature deductions'. One of his probabilistic tools is the local De Moivre–Laplace limit theorem.

We note a special passage on the law of universal gravitation (Davidov, 1855*b*, p. 63) in a context devoted to philosophical problems of empirical proofs: 'Who is in a position... to state that this law is an exact expression of the law of nature, and that it is not a particular case of a more general principle, only approximately correct.'

For general information about Davidov (1823–85), see Zhoukovsky *et al.* (1890).

Similarly, Cournot (1843, or a German translation by Schnuse, 1849) whose book appears to be insufficiently known, reasons on the possibility of 'smearing' statistics (§ 103), tests the statistical significance of empirical inequalities (§§ 108–10) and considers the law of large numbers in J. Bernoulli's form a reliable base for connecting statistics with probability (§ 115).

4·2. *Classical theory of errors*

The classical theory of errors which originated in the middle of the eighteenth century, especially in the hands of J. H. Lambert (Sheynin, 1966) and T. Simpson, was based upon the law of large numbers only beginning with Laplace. But then, after Gauss, this law, as far as the theory of errors is concerned, had been almost forgotten and even the concept of random errors of observation happened to become divorced from the concept of random quantities in the theory of probability.

In our opinion the reason for this is that Gauss, while reasoning on the probabilistic properties of random errors (*Theoria Motus*, § 175; *Theoria Combinationis*, § 4), possibly having been satisfied with an empirical approach, did not refer to the law of large numbers.

A qualifying remark should be added: a definition of a random quantity did not seem to appear in the classical theory of probability. But we hold that in the second half of the nineteenth century a definition of a random quantity as 'dependent on chance' and possessing a certain law of distribution had become so natural, even if not definitely formulated, that this definition, just as the 'classical' definition of probability, ought to be called classical.

As to random errors, these were usually taken to be errors with certain probabilistic properties, their specific distribution especially following the *Theoria Combinationis* way of reasoning, being not so important.

It seems that Vassiliev (1885, p. 133) was the first who definitely held that random errors of observations are to be ranked among random quantities. As far as the theory of probability is concerned, A. V. Vassiliev (1853–1929) is known primarily as one of Markov's correspondents. It was in a letter to Vassiliev that Markov (1898) originally described his reasoning on the method of least squares. According to Markov, references to the law of large numbers in the theory of errors suddenly became even too numerous and the law was misused (p. 249). Although also connecting random errors with random quantities, in his other works, Markov stated (1898, p. 250) that 'nothing comes' from

$$\text{prob}\,(|\bar{x}-a| < \epsilon) \to 1 \quad \text{as} \quad n \to \infty,$$

because other linear estimates of the constant a possess the same limit property of consistency as the arithmetic mean $\bar{x}$ and because, furthermore, a method optimal in the case of a finite number of observations is needed. Such a method, according to Markov (p. 246) was the second (*Theoria Combinationis*) method of Gauss as opposed to his first (*Theoria Motus*) method.

On the work of Markov in the theory of errors, see also Plackett (1949). As to Markov's opinion described above, this is to be considered 'classical' as opposed to 'statistical'.

The above description of the law of large numbers in the theory of errors seems somewhat inconsistent, but it reflects different points of view held at different times. In particular, it is the present author's guess that the sudden interest in the law of large numbers which occurred at the turn of the nineteenth century, and to which Markov refers, was an inevitable result of the ideas of Lexis and other 'continental' statisticians and of the work of Bienaymé and Tchebychev becoming generally known.

4·3. *Probabilities proper*: *Markov versus K. Pearson*

Markov (1924) studied Bernoulli's work on the law of large numbers simultaneously with or somewhat prior to K. Pearson (1924, 1925). Markov held a high opinion on this law; he edited a Russian translation (1913) of Part 4 of the *Ars Conjectandi*, was the originator of a special sitting of the Russian Academy of Sciences in commemoration to the bicentenary of the law of large numbers (where speeches were given by Markov himself, Tchouprov (1914) and Vassiliev) and dedicated the third edition of his *Ischislenje veroyatnostey* (1913) to J. Bernoulli.

Pearson (1924) considered J. Bernoulli's estimate (the necessary number of trials) and, furthermore, the whole of part 4 of the *Ars Conjectandi* as unsatisfactory. This opinion is hardly fair: the law of large numbers in Bernoulli's form was of great significance for the whole development of the theory of probability at least until the time of Laplace and Poisson, the 'crude values' of his estimate 'with their 200 to 300 % excesses' (Pearson) being no serious obstacle. Similarly, this law proved itself of utmost importance in applications of the theory of probability (see §§ 4·1 and 4·2).

As opposed to Pearson, Markov, while modernizing Bernoulli's algebraic deductions and improving his estimate, did not use Stirling's theorem which remained unknown to Bernoulli. Consequently Markov's estimate turned out to be worse than K. Pearson's estimate, but because of this very reason his approach seems to be methodologically more correct. A special feature of the Markov's review is that he eliminated J. Bernoulli's tacit condition that the exponent $(r+s)\,n$ is divisible by the sum of the terms of the binomial $(r+s)$.

No attempt is made here to describe the original work of Markov in the field of the law of large numbers.

References

ARBUTHNOT, J. (1712). An argument for divine Providence, etc. *Phil. Trans. Roy. Soc.* **27**, no. 328 for 1710.

ARCHIBALD, R. C. (1926). A rare pamphlet of De Moivre and some of his discoveries. *Isis* **8**, no. 4 (28), 671–84. Also in 1926 a discussion with K. Pearson in *Nature, Lond.* **117**, 551–2, under the title 'A. De Moivre'.

BERNOULLI, J. (1899). *Wahrscheinlichkeitsrechnung* (*Ars Conjectandi*). Ostwald Klassiker no. 107–8. Herausgegeben R. Haussner. Leipzig: Wilh. Engelmann.

BORTKIEWICZ, L. (1898). *Das Gesetz der kleinen Zahlen.* Leipzig: Teubner.

BORTKIEWICZ, L. (1905). On statistical regularity. (Russian). *Vestnik prava* **35**, no. 8, 124–54.

CLARKE, F. M. (1929). *Thomas Simpson and his Times.* A thesis submitted to the Columbia University. New York. A microfilm of this was kindly sent to me by Dr A. E. Ritchie (Ames, Iowa, U.S.).

COLEBROOKE, H. T. (1817). *Algebra and Mensuration from the Sanscrit of Brahmegupta and Bhascara.* London: Murray.

COTES, R. (1768). Aestimatio errorum, etc. In: *Opera misc.* pp. 10–58. London: Typis Meyerianis. Originally published 1722.

COURNOT, A. A. (1843). *Exposition de la theorie des chances et des probabilités.* Paris: Hachette.

CZUBER, E. (1899). Die Enwicklung der Wahrscheinlichkeitstheorie, etc. *Jahresber. deustch. Math.-Vereinigung* **7**, no. 2 (separate pagination).

DAVIDOV, A. JU. (1855*a*). Application of the theory of probability to statistics. (Russian). In: *Scientific and Literary Papers of the Professors and Instructors of the Moscow Univ. Published on the Occasion of its Centenary Jubilee.* (Separate pagination for each paper.) Moscow.

DAVIDOV, A. JU. (1855*b*). Use of the deductions of the theory of probabilities in statistics. (Russian). *Journal Ministerstva Narodnogo prosveschenia* **88**, no. 11, pp. 45–109 of section 2.

DE MOIVRE, A. (1906). *Leibrente.* Herausgegeben E. Czuber. *Sonderheft der Versicherungswissenschaftlichen Mitt.*, Wien.

EGGENBERGER, J. (1894). Beiträge zur Darstellung des Bernoullischen Theorems, etc. *Mitt. Naturforsch. Ges. Bern* for 1893, no. 1305–1334, pp. 110–82. Also published as a separate edition (1906), Berlin: Fischer).

ELOGE DE DE MOIVRE (1759). *Hist. Acad. Roy. Sci. Paris* for 1754, pp. 175–84. (Incorporated in the same volume with the *Mem. Acad. Roy. Sci.*)

GRAETZER, J. (1883). *E. Halley und C. Neumann.* Breslau: Schottlaender.

HALLEY, E. (1693). An estimate of the degree of mortality of mankind, etc. *Phil. Trans. Roy. Soc. 1665–1800 abridged*, vol. 3. London, 1809.

LAGRANGE, J. L. (1892). *Oeuvres Complètes*, t. 14. Paris: Gauthier.

LAPLACE, P. S. (1774). Mémoire sur la probabilité des causes par les événements. *Oeuvres Completes*, t. 8. Paris: Gauthier (1891).

MARKOV, A. A. (1898). The law of large numbers and the method of least squares. (Russian). This is an extract of a letter to Vassiliev. Reprinted (1951) in Markov's *Selected Works*, 231–51. (Russian). Acad. Sci. U.S.S.R. Publ., S.l.

MARKOV, A. A. (1914). Bicentenary of the law of large numbers. (Russian). *Vestnik opitnoy physiki i elementarnoy mathematiki*, no. 603, 59–64.

MARKOV, A. A. (1924). *Ischislenie veroyatnostey*, 4th (posthumous) Russian edition. A German (1912) version of an earlier Russian edition is entitled *Wahrscheinlichkeitsrechnung*. An English translation was being arranged by F. M. Weida for the *Ann. Math. Stat. Suppl.* (stated by F. N. David & J. Neyman. (1938). Extension of the Markov theorem on least squares. *Stat. Res. Mem.* **2**, 105–117). The author is unaware if this was actually published.

MATY, M. (1760). *Mémoire sur la vie de De Moivre.* La Haye. This proved unavailable. An anonymous obituary of De Moivre is to be found in *J. Britannique* (vol. 18 for Sept.–Oct. 1755, pp. 1–51, La Haye, 1755) a journal edited by Maty. This source was originally found by A. De Morgan (1914). *Essays on the life and work of Newton*, Chicago and London: Open Court publ., p. 189, who states that (i) the author is Maty and that (ii) the 1760 memoir of Maty is a second edition of this.

MONTMORT, P. R. (1713). *Essay d'analyse sur les jeux de hazard.* Paris: Quilau. See letter of N. Bernoulli to Montmort dated 23 Jan. 1713.

NEWBOLD, E. N. (1927). Practical applications of the statistics of repeated events, etc. *J.R. Statist. Soc.* **90**, 487–547.

ORE, O. (1963). *Cardano, the Gambling Scholar.* Princeton University Press.

PEARSON, K. (1924). Historical note on the origin of the normal curve of errors. *Biometrika* **16**, 402–4.

PEARSON, K. (1925). James Bernoulli's theorem. *Biometrika* **17**, 201–10.
PENKOV, B. (1961) On the history of mathematics in 18th century England: Bayes and Simpson. (Bulgarian). *Phys.-Math. spisanie* **4** (37), no. 4, 292–303.
PETERS, J. (1922). *Zehnstellige Logarithmentafeln*, Bd. 1. Anhang, Taf. 6. 18-stellige log *n*! Berlin: Reichsamt für Landesaufnahme.
PLACKETT, R. L. (1949). A historical note on the method of least squares. *Biometrika* **36**, 458–60.
SHEYNIN, O. B. (1966). Origin of the theory of errors. *Nature, Lond.* **211**, 1003–4.
SIMPSON, T. (1757). An attempt to shew the advantage arising by taking the mean, etc. In: *Misc. Tracts on Some Curious Subjects*, etc., pp. 64–75. London: Nourse.
TODHUNTER, I. (1863). *History of the Mathematical Theory of Probability*. Cambridge and London: Macmillan.
TCHOUPROV, A. A. (1914). The law of large numbers in modern science. (Russian). *Statistichesky vestnik*, books 1–2, 1–21. Another version of this appearing in *Nord. stat. tidskr.* 1, no. 1 (1922) proved available only in a Russian translation (1960) in Tchouprov's *Selected Papers*. Moscow: 'Statistics' publ.
VASSILIEV. A. V. (1885). *Theory of Probabilities*. (Russian, a lithographic edition). Kazan.
VEIMAN, A. A. (1961). *Shumero–Babylonian mathematics of* 3*rd*–1*st millenia B.C.* (Russian). Moscow: Oriental liter. publ.
WALKER, H. M. (1934). A. De Moivre. *Scripta math.* **2**, no. 4, 316–33.
ZHOUKOVSKY, N. YE. *et al.* (1890). The life and works of Davidov. (An obituary.) (Russian). *Mathematichesky zbornik* **15**, no. 1, 1–57.

Later notes

In the present reproduction, the author has modified his original statement in the last nine lines of p. 232.

In regard to the third paragraph on p. 236 the author has recently found the same idea about the law of gravitation and, roughly, the same wording in a Russian translation of Quetelet's *Social Physics* (1869, book 1, §8). Possibly it is contained in an earlier work of Quetelet.

A NOTE ON AN EARLY STATISTICAL STUDY OF LITERARY STYLE

By C. B. WILLIAMS

In *Biometrika* for January 1939, G. Udny Yule discussed the frequency distribution of sentence length in samples of the writings of different authors. After showing that each author had a fairly characteristic distribution, he turned to the value of the method in cases of uncertain or disputed authorship. Thus, in the case of *De Imitationi Christi*, he showed that the frequency distribution of sentences with different numbers of words more closely resembled that of works by Thomas à Kempis than that of works by de Gerson.

In *Biometrika* for March 1940 I showed that the skew distribution found by Yule could be brought almost to a symmetrical form by using a geometric or logarithmic scale for the number of words per sentence, thereby simplifying the mathematical comparisons. In this note I mentioned that some years previously (about 1935) I had made a number of frequency distributions from different authors using the number of letters per word as the variable, but that I had not found any striking differences. I considered Yule's use of the number of words per sentence as a better technique, giving a greater range of possible variation and comparison.

In a letter written to me in June 1939 Yule said: 'I booked up some ten years ago a number of distributions of word-length by the number of syllables only. Monosyllables are always considerably in the majority (if I remember rightly I omitted "a" and "the"), and different authors diverged a good deal, but, so far as I can recall, the range from Bunyan to a *Times* Leader was not so very striking.'

Neither Yule nor myself was aware that quite extensive investigations in this line had been made and published in summary more than fifty years previously, giving frequency distributions of word lengths (by the number of letters per word) for several authors, and suggesting that similar distributions of the numbers of syllables per word, or the number of words per sentence, might well help to throw light on cases of doubtful or disputed authorship.

Through the kindness of Mr Rushworth Fogg of Glasgow I was put on the track of a paper published in 1901 by Thomas Corwin Mendenhall,* in which he gives a reference to a still earlier paper published in 1887, both of which I have been able to examine.

Mendenhall states in his first paper (1887) that five or six years previously he had seen a suggestion in a book by Augustus de Morgan, possibly his *Budget of Paradoxes*, that it might be possible to identify the author of a book, a poem or a play by the average length of the words used in the construction. Mendenhall, however, considered that the method which he had adopted in this publication, of using the frequency distribution of words of different lengths, was better, as while the average number of letters per word is easily obtainable from his data, the shape of the distribution provides considerably increased possibilities of comparison.

Augustus de Morgan was Professor of Mathematics at University College, London. His *Budget of Paradoxes* was first printed as weekly notes in the *Athenaeum* and republished in

* See biographical note on p. 248.

book form in 1872, after the author's death. I have examined the second edition (1915), but, although it contains a few references to cases of disputed authorship, I cannot find any suggestion about the use of the average number of letters per word. It may be in one of his earlier works, or perhaps in one of his *Athenaeum* notes that was not reprinted in book form.*

It is interesting to note than Mendenhall, who was primarily a physicist, was attracted to the frequency distribution technique by its resemblance to spectroscopic analysis, which in 1887 was much to the fore in scientific circles. He writes: 'It is proposed to analyse a composition by forming what may be called a "word spectrum" or "characteristic curve" which shall be a graphic representation of the arrangement of words according to their length and the relative frequency of their occurrence.' The mathematics of the comparison of frequency distributions was very little understood at the time when he was writing.

Mendenhall first discusses samples taken from different books by the same author to see if they resemble each other sufficiently closely to make comparisons between one author and another likely to be profitable. Most of the evidence is given in the form of thirteen diagrams and, unfortunately, in only very few cases are actual numbers presented.

Mendenhall's first seven graphs deal with various combinations of ten samples, each of one thousand words, from Dickens's *Oliver Twist* and Thackeray's *Vanity Fair*. In his second figure the distribution of five separate samples of 1000 words from *Oliver Twist* are shown superimposed, and there is no doubt as to their general resemblance. In another (Fig. 1 in this paper) he shows one graph for the whole 10,000 words from *Oliver Twist* and another for *Vanity Fair*. There is very little difference in the average length of the words (Dickens, 4·324; Thackeray, 4·481), but *Vanity Fair* has rather more words of 3 and of 7–10 letters, while *Oliver Twist* has more of 1, 2 and 4–6 letters. Mendenhall was somewhat disappointed by the lack of difference and commented 'it is certainly suprising that...so close an agreement should be found. This is particularly striking in the words of 11, 12 and 13 letters, the numerical composition of which is as follows:

Number of letters	11	12	13
Dickens	85	57	29
Thackeray	85	58	29'

Undaunted by this small difference he next tried two groups of words from John Stuart Mill's *Political Economy* and his *Essay on Liberty*, in which he 'expected to find more longer words than in the novelists'. 'But I confess to considerable surprise in finding from the very beginning that, although on the whole the anticipation was realized, the word which occurred most frequently was not the three-letter word, as with both Dickens and Thackeray, but the word of two letters.' The explanation he says 'is to be found in the liberal use of prepositions in sentence-building'. The results, given in two separate diagrams, are here combined into one (Fig. 2).

Mendenhall next studied two addresses given by a Mr Edward Atkinson on 'Labour Questions' to two different audiences, one consisting of working men and the other students of a Theological College. There was 'a marked difference in style', but the word-length distributions (from two samples of 5000 words) were very similar. He comments that 'Mr Atkinson's composition was remarkable in the shortness of the words used'. The average length was 4·298 letters; which is, however, only 0·044 shorter than the samples from Dickens.

*But see later note by R. D. Lord in following Paper.

For comparison with all the above studies of works in the English language, Mendenhall gave a distribution of the first 5500 words in Caesar's *Commentaries*, in Latin. He finds a mean word length of 6·065 letters and an entirely different form of curve with peaks at 2, 5 and 7 letters (see Fig. 3). This is of course connected with the Latin construction of adding to the main root for inflexions instead of using additional small words.

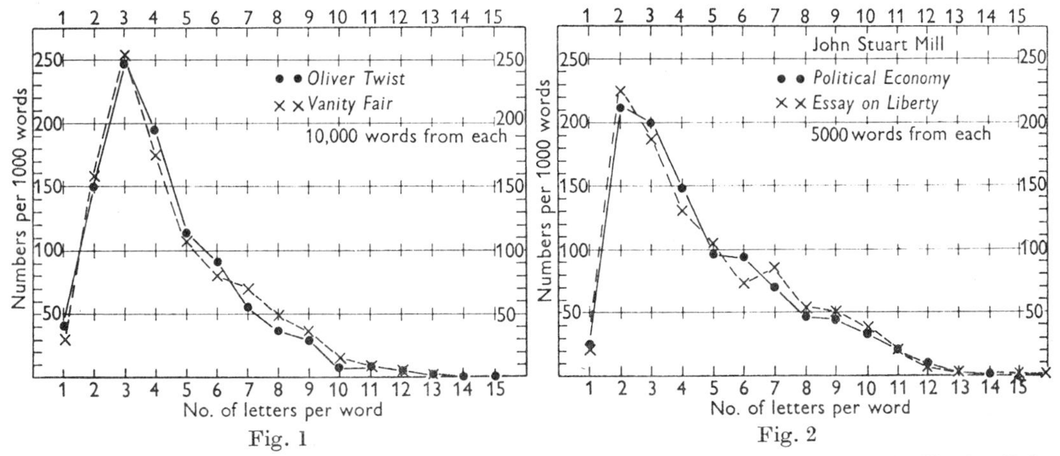

Fig. 1. Samples of 10,000 words each from Dickens's *Oliver Twist* and Thackeray's *Vanity Fair*. Redrawn from Mendenhall (1887, fig. 7).

Fig. 2. Samples of 5000 words each from two works of John Stuart Mill. Redrawn from Mendenhall (1887, figs. 8, 9).

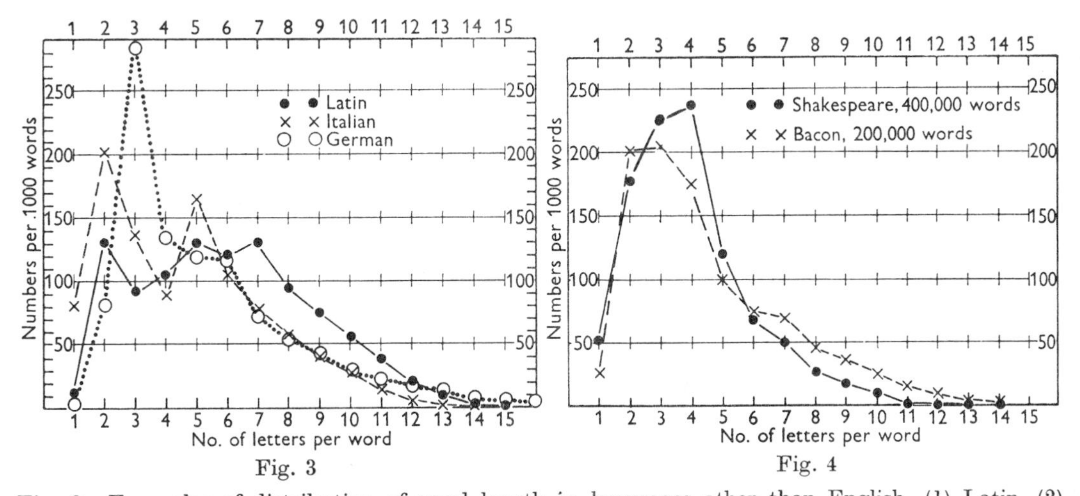

Fig. 3. Examples of distribution of word length in languages other than English. (1) Latin, (2) Italian, (3) German. Redrawn from Mendenhall (1887, fig. 13 and 1901, fig. 2).

Fig. 4. Comparison of frequency distribution of word length in two very large samples from the plays of Shakespeare and works of Bacon. Redrawn from Mendenhall (1901, fig. 7).

He considers that, for a really reliable estimate of the characteristic curve for an author, a sample curve of 100,000 words might be necessary, and concludes his first paper as follows:

'Many interesting applications of the process will suggest themselves to almost every reader; the most notable, of course, being the attempt to solve questions of disputed authorship, such as exist in reference to the letters of Junius, the plays of Shakespeare, and other less widely known examples. It might also be used in comparative language studies, in tracing the growth of a language, and in studying the growth of the vocabulary from childhood to manhood.'

'If striking differences are found between the curve of known and suspected compositions of any writer, the evidence against identity of authorship would be quite conclusive. If the two compositions should produce curves which are practically identical, the proof of a common origin would be less convincing for it is possible though not probable, that two writers might show identical curves.'

It was not until 14 years later than Mendenhall returned to the problem in a paper in *Popular Science Monthly*, published in December 1901. In this he repeats some of the discussion and diagrams from his earlier paper dealing with Dickens, Thackeray, John Stuart Mill and Mr Atkinson; but in addition to his earlier diagrams showing analysis of a Latin work he gives examples of single authors in Italian, Spanish, French and German (see Fig. 3). The French and Spanish curves, possibly by the idiosyncrasies of the authors chosen, have their peaks in the 2-letter words; the Italian has two peaks at 2 and 5 letters; while the German has a peak at 3 letters, but with more longer words, reaching one word of 27 letters.

After this introduction, he settles down to a discussion of the value of his technique in the study of the authorship of the plays of Shakespeare. For this the length of nearly two million words were counted from the works of Shakespeare and of some of his contemporaries. Most unfortunately the data are all condensed into half a dozen small diagrams, and not one table of the actual numbers is given. Does his evidence still exist, hidden away somewhere?

Mendenhall says: 'The result from the start, with the first group of 1000 words, was a decided surprise. Two things appeared from the beginning: Shakespeare's vocabulary consisted of words whose average length was a trifle below four letters, less than any writer of English previously studied; and his word of greatest frequency was the four-letter word, a thing never before met with' (Fig. 4).

A comparison of the diagrams with those of Thackeray and Dickens shows that Shakespeare had a higher proportion of words with 1, 2, 4 and 5 letters, and a lower proportion of words with 3 letters and of 6 letters upwards; which accounts for the fact that while his peak is higher, his average number of letters per word is lower. In modern terminology, he would have a smaller standard deviation.

Altogether about 400,000 words were counted including 'in whole or in part, nearly all his most famous plays', and it was found that this characteristic curve is most persistent—that based on first 50,000 words differing very little from the whole count. In a diagram giving two examples of 200,000 words each, it is practically impossible to separate the two lines, in spite of the fact that Mendenhall says the differences have been of necessity slightly exaggerated in order to make them show at all!

A comparison was next made of Shakespeare's prose and his poetry as exemplified by *The Rape of Lucrece* and *Venus and Adonis*. The prose gave more shorter words, particularly of 2 letters, and fewer words of 5, 6 and 7 letters; but both gave the characteristic peak at the 4-letter word. Mendenhall writes: 'At first this was thought to be a general characteristic of his time, but this was found not to be so.'

A study was then made of a number of works by Francis Bacon, including his *Henry VII* and his *Advancement of Learning*, with a total of nearly 200,000 words. The frequency distribution was quite different from that of Shakespeare (see Fig. 4) with the peak at the 3-letter word, with more 2-letter words, fewer with 4, 5 and 6 letters, and more longer words with 7–13 letters. Mendenhall here comments that 'the reader is at liberty to draw any conclusion he pleases from this diagram. Should he conclude that, in view of the extraordinary difference in these lines, it is clear that Bacon could not have written the things ordinarily attributed to Shakespeare...the question still remains, who did?'

An examination of the works of Ben Johnson, in two groups of 75,000 words, showed once more a peak at the 3-letter word; but an extensive study of the plays of Beaumont and Fletcher showed that on the final average the number of 4-letter words was slightly greater than those of 3 letters, although the excess was by no means persistent in smaller samples. The final curve was not unlike that of Shakespeare, and Mendenhall suggested that the 'lack of persistency of form among small groups' might be due to the dual authorship.

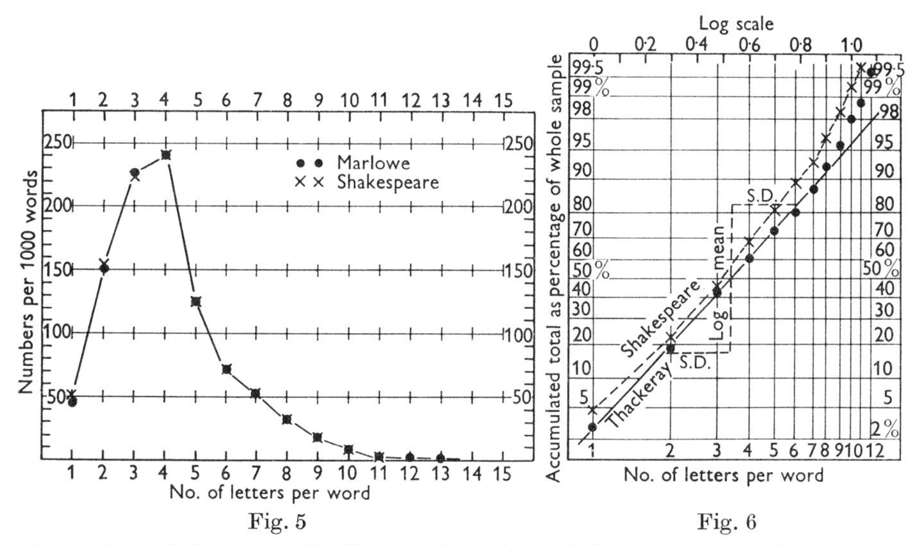

Fig. 5

Fig. 6

Fig. 5. Comparison of frequency distribution of word length in two large samples from the plays of Shakespeare and of Christopher Marlowe. Redrawn from Mendenhall (1901, fig. 9).

Fig. 6. The number of letters per word, on a logarithmic scale, from works of Thackeray and Shakespeare (as shown by Mendenhall) plotted against the accumulated total as a percentage of the whole sample on a probability scale. It indicates some resemblance to a log-normal distribution for words up to about 8 letters, but differing above this level.

When, however, he turned his attention to the plays of Christopher Marlowe 'something akin to a sensation was produced among those engaged in the work'. 'In the characteristic curve of his plays Marlowe agrees with Shakespeare about as well as Shakespeare agrees with himself' (Fig. 5).

Finally, Mendenhall pointed out that a dramatic composition *Armada Days* written by Prof. Shaler of Harvard, in which the author endeavoured to compose in the spirit and the style of the 'Elizabethan days', gave a curve (from only about 20,000 words) with 'excess of the 4-letter word and in other respects decidedly Shakespearian'. Mendenhall does not give this curve or any figures.

Discussion

We are not concerned here so much with the results that Mendenhall obtained, or with their repercussions, but rather with the general value of the technique. There appears to be little doubt that he was the first to act on the suggestion of de Morgan, and that his own method of using the frequency distribution, instead of merely the average length of word, was a distinct improvement, although the average length would not normally be given to-day without the standard deviation. The skew form of the curve makes this latter measure less reliable than it might otherwise be.

Mendenhall's sampling method was to take blocks of 1000 words each 'at the beginning of the volume and, after a few thousand words had been counted, the book was opened near the middle and the count continued'. This method is not above reproach, but, in view of the large number of samples and the general close resemblances, it is unlikely that a more randomized method would produce any measurably different result. In the case of the plays of Shakespeare the sampling was large enough to justify the statement that it included nearly all the most famous plays 'in whole or in part'.

That Mendenhall appreciated the difference between the statistical method and evidence based on selected phraseology believed to be characteristic is clear from the following quotations: 'The chief merit of the method consisted in the fact that its application required no exercise of judgement' and that 'characteristics might be revealed which the author could make no attempt to conceal, being himself unaware of their existence'; and again, 'the conclusions reached through its use would be independent of personal bias, the work of one person in the study of an author being at once comparable with the work of any other'.

That Mendenhall saw the wide range of possibilities is clear from his statement: 'it is hardly necessary to say that the method is not necessarily confined to the analysis of a composition by means of word-length: it may equally be applied to the study of syllables, of words in sentences, and in various ways.' And I have already quoted his suggestion as to its value in comparative studies.

Two additional comments may be of interest. The curve of the frequency distribution of words of different lengths is in every case skew, with the peak usually at 3 or 4 letters per word, and the tail running off generally to 15 or 16, but sometimes to higher than this. In my contribution to the study already mentioned, I showed that by the use of a logarithmic scale the skew distributions of sentence length became approximately symmetrical, and so the distribution resembled a log-normal. It is of interest to see if Mendenhall's figures for word length show a similar relation. We can, however, note beforehand that the length of a sentence is under the conscious control of a writer, who may stop when he pleases. The lengths of words are not so controlled and selection of words for reason of their length alone is not likely to occur.

Most unfortunately only three sets of numbers are given by Mendenhall, all in his first 1887 paper. They are for 1000 words in *Oliver Twist* and two sets of the same size for *Vanity Fair*. Taking the latter we find that the accumulated totals up to each successive number of letters per word, expressed as percentages of the whole, are as shown in Table 1.

When these results are plotted on to log-probability paper the result is as shown in Fig. 6. There is an approximately straight-line relation up to about 8 letters per word, but above that there is a definite departure. The straight-line portion suggests a log-normal distribution with a mean log at 0·53 and a standard deviation of approximately 0·26. On an arithmetic scale this is equivalent to a geometric mean of about 3·4 and a standard deviation of $\times$ or $\div$ 1·8.* The arithmetic mean is 4·5 letters per word.

* When a frequency distribution is skew on an arithmetic grouping of the data but approximately symmetrical when a geometric scale is used, the standard deviation cannot be expressed on an arithmetic scale as '+ or −'. The use of the expression '3·4, $\times$ or $\div$ by 1·8' implies that approximately 33% of the observations will be between 3·4 and $3{\cdot}4 \times 1{\cdot}8$; 33% will be between 3·4 and $3{\cdot}4 \div 1{\cdot}8$; and approximately 17% will be above, and below, these limits.

In his second paper Mendenhall gives five graphs showing Shakespeare's frequency distribution per 1000 words in comparison with other authors. With a lens and a fine scale it is possible to read the numbers to about three units, but unfortunately the results so obtained from the five diagrams do not agree. This is possibly because (as he admits in one case) Mendenhall exaggerated the differences in the diagrams in order to separate the two lines. I have made an estimate from each of the five diagrams, and the average values are given in Table 2.

Table 1

No. of letters per word	No. of words out of 2000	Accumulated no. of words total	Accumulated total % of 2000
1	58	58	2·9
2	315	373	18·7
3	480	853	42·7
4	351	1204	60·2
5	244	1448	72·4
6	154	1602	80·1
7	152	1754	87·7
8	100	1854	92·7
9	63	1911	98·9
10	43	1960	98·0
11	16	1976	98·8
12	15	1991	99·6
13	4	1995	99·8
14	5	2000	100·0

Table 2

Letters	Words	Letters	Words	Letters	Words
1	47·6	6	71·2	11	3·4
2	175·8	7	52·6	12	2·0
3	225·0	8	31·6	13	1·0
4	237·6	9	18·4	14	0·4
5	124·4	10	9·0	—	—

When the accumulated totals are plotted on log-probability paper as in the previous case, the result (Fig. 6) indicates a fairly regular departure from the straight line, although once again the break is more distinct above 7 letters per word.

I have also attempted to get, from Mendenhall's diagram of five samples of 1000 words each from *Oliver Twist*, some measure of the error of his results.

The frequency distribution of words of certain lengths in the five samples is approximately as given in Table 3.

If the size of the sample were increased sixteen times—to 80,000 words—without altering the pattern of the material sampled, the S.E. of the mean would be reduced to a quarter of the above—or approximately 2·1, 1·1, 2·1, 1·5 and 0·8 for words of 3–6 letters respectively. The error is smaller in the less frequent words, but greater in proportion to the mean.

Thus for the comparison of two samples of this size—assuming the same order of variation in each—the S.E. of the difference would be approximately 1·4 times the above or 2·9, 1·7, 2·1 and 1·1. Thus differences in number of words per 1000 would have to be of the order of 5·7, 3·3, 4·1 and 2·1 to be significant at the 1 in 20 level, and 7·5, 4·4, 5·4 and 2·8 to be significant at the 1 in 100 level. The five samples on which the above rough estimate is made were, however, consecutive samples of 1000 words from one work; when different works, written at different periods by the same author, are combined the error of the mean would almost certainly be greater.

A careful examination of Mendenhall's diagram giving the comparison of the distributions of Shakespeare and Marlowe suggests measurable differences only in words up to 5 letters, Marlowe differing from Shakespeare approximately as follows: 1 letter, 5 less; 2 letters, 3 less; 3 letters, 3 more; 4 letters, no difference; and 5 letters, 5 more. All the other

Table 3

No. of letters	Five samples	Mean	S.E. of mean for 5000 words
3	221, 232, 236, 254, 268	242·2	8·36
4	170, 175, 183, 186, 198	182·4	4·82
5	95, 102, 120, 122, 123	112·4	5·80
6	83, 92, 94, 97, 103	93·8	3·28

word lengths are indistinguishable in the diagram. These differences may have been exaggerated in the diagram. The numbers are in words per 1000 in large samples—in the case of Shakespeare over 200,000 words, but in the case of Marlowe the size of samples is not given. On the other hand, the comparison of Bacon with Shakespeare (see Fig. 5) shows a difference of nearly 60 words per thousand with 4-letter words.

It would seem likely that real differences between authors would show themselves as sequences of departures in the same direction for several consecutive word lengths. One can imagine one author differing from another in an unconscious preference for longer words or for shorter words, but it is unlikely that one author would prefer words of, say, 11 letters in preference to 12, while another author would prefer the 12 to the 11. Thus, rapid changes of departure directions in sequence would be less convincing than blocks of departures of similar sign as evidence of real differences, and would be more likely to be due to error.

Mendenhall, in his 1887 paper, calls attention to the fact that in an analysis of Dickens's *Christmas Carol*, words of 7 letters appeared to be unduly numerous, due to the fact that the character 'Scrooge', frequently referred to, is a word of this length. It would be desirable to leave names of persons and places out of any tabulation.

BIOGRAPHICAL NOTE

Thomas Corwin Mendenhall was born in Ohio on 4 October 1841 and died there on 22 March 1924. He was the descendant of a Benjamin Mendenhall who emigrated from England (probably from Wiltshire) in 1686 to join Penn's Colony and who settled at Concord,

Pennsylvania. T. C. Mendenhall spent some early years as a school teacher, but in 1873 became the first Professor of Physics and Mechanics at the newly founded Ohio Agricultural and Mechanical College. He was Professor of Physics in the University of Tokyo from 1878 to 1881 and in the Ohio State University from 1881 to 1886. He then became President of the Rosa Polytechnic Institute in Indiana and was elected the following year to the National Academy of Science. After a few years as Superintendent of the U.S.A. Coast and Geodetic Survey he became President of the Worcester Polytechnic Institute, where he remained until his retirement at the age of 60.

His biography by Henry Crow (*Biograph. Mem. Nat. Acad. Sci., Wash.*, **16**, 331–51), to which I am indebted for the above information, lists about sixty publications in physics, particularly geophysics, units of electrical measurement, state boundary lines in the U.S.A. and many other related subjects. The first of the two papers at present under discussion is listed, but not the second. There is no mention of his interest in the statistics of literary style, but it is said that he left in MSS. about 900 pages of an autobiography which has never been published. If still available, it might repay study.

REFERENCES

MENDENHALL, T. C. (1887). The characteristic curves of composition. *Science*, **9** (214, supplement), 237–49.

MENDENHALL, T. C. (1901). A mechanical solution of a literary problem. *Pop. Sci. Mon.* **9**, 97–105.

DE MORGAN, A. (1872). *A Budget of Paradoxes*. London (2nd edition 1915).

WILLIAMS, C. B. (1940). A note on the statistical analyses of sentence length as a criterion of literary style. *Biometrika*, **31**, 356–61.

WILLIAMS, C. B. (1952). Statistics as an aid to literary studies. *Penguin Science News*, no. 24, pp. 99–106.

YULE, G. U. (1939). On sentence-length as a statistical characteristic of style in prose. *Biometrika*, **30**, 363–84.

DE MORGAN AND THE STATISTICAL STUDY OF LITERARY STYLE

By R. D. LORD

C. B. Williams has recently given an account of two little-known papers published by T. C. Mendenhall in 1887 and 1901 on the statistical analysis of literary style, but was unable to trace Mendenhall's reference to de Morgan's suggestion that one might identify an author by the average length of his words. In his first paper Mendenhall said that he saw the suggestion 5 or 6 years earlier in a book by de Morgan, possibly his *Budget of Paradoxes*. I think it fairly certain that Mendenhall's memory was at fault, that the book was in fact *Memoir of Augustus de Morgan* by his wife Sophia, published in 1882, and that the suggestion occurs in a letter of 1851 to an old Cambridge friend, the Rev. W. Heald. The rest of the book has no hint that de Morgan ever followed up his idea. Had he done so he would probably have found, as did Mendenhall, Yule and Williams, that word-length is an unsatisfactory criterion compared with sentence-length. The letter can be allowed to speak for itself.

7 Camden Street,
Aug. 18, 1851

Dear Heald,

It has become quite the regular thing for the depth of vacation to remind me—not of you, for anything that carries my thoughts back to Cambridge does that,—but of inquiring how you are getting on, of which please write speedy word, according to custom, once a year....

* * * *

I wish you would do this: run your eye over any part of those of St. Paul's Epistles which begin with Παυλος —the Greek, I mean—and without paying any attention to the meaning. Then do the same with the Epistle to the Hebrews, and try to balance in your own mind the question whether the latter does not deal in longer words than the former. It has always run in my head that a little expenditure of money would settle questions of authorship in this way. The best mode of explaining what I would try will be to put down the results I should *expect* as if I had tried them.

Count a large number of words in Herodotus—say all the first book—and count all the letters; divide the second numbers by the first, giving the average number of letters to a word *in that book*.

Do the same with the second book. I should expect a very close approximation. If Book I. gave 5·624 letters per word, it would not surprise me if Book II. gave 5·619. I judge by other things.

But I should not wonder if the same result applied to two books of Thucydides gave, say 5·713 and 5·728. That is to say, I should expect the slight differences between one writer and another to be well maintained against each other, and very well agreeing with themselves. If this fact were established there, if St. Paul's Epistles which begin with Παυλος gave 5·428 and the Hebrews gave 5·516, for instance, I should feel quite sure that the *Greek* of the Hebrews (passing no verdict on whether Paul wrote in Hebrew and another translated) was not from the pen of Paul.

If scholars knew the law of averages as well as mathematicians, it would be easy to raise a few hundred pounds to try this experiment on a grand scale. I would have Greek, Latin, and English tried, and I should expect to find that one man writing on two different subjects agrees more nearly with himself than two different men writing on the same subject. Some of these days spurious writings will be detected by this test. Mind, I told you so. With kind regards to all your family, I remain, dear Heald,

Yours sincerely,

A. De Morgan.

REFERENCES

WILLIAMS, C. B. (1956). Studies in the history of probability and statistics. IV. A note on an early statistical study of literary style. *Biometrika*, **43**, 248–56.*

SOPHIA DE MORGAN (1882). *Memoir of Augustus de Morgan by his Wife Sophia Elizabeth de Morgan with Selections from his Letters.* London: Longman, Green and Co.

*Paper No. 17 in this volume

ISAAC TODHUNTER'S HISTORY OF THE MATHEMATICAL THEORY OF PROBABILITY

By M. G. KENDALL

Todhunter's *History* was published in 1865. I have had it in mind for some time to write a slight account of its author, and was intending to do so at the centenary of the publication date, expecting to be able to say that nothing comparable to his book had been produced in the intervening hundred years. Dr David has prevented this remark with her recent *Gods, Games and Gambling*. But it is almost true. There has been no other account in book form of the history of probability and even now the nineteenth century is unchronicled.

The *History of the Mathematical Theory of Probability* is distinguished by three things. It is a work of scrupulous scholarship; Todhunter himself contributed nothing to the theory of probability except this account of it; and it is just about as dull as any book on probability could be. What sort of man, then, was the author?

Isaac was the second of four sons of a congregationalist minister of Rye, and was born on 23 November 1820. His father died in 1825, leaving the family in poor circumstances. Isaac grew up the hard way. While acting as assistant master at a school in Peckham he attended evening classes at University College London, walking the five miles from Peckham to Gower Street every evening after a day's work at school.

Although it was recorded that he was an unusually backward child, he must have caught up with his studies at an early age. He was strongly influenced by de Morgan at University College, but shared his interests between mathematics and classics. He graduated B.A. at London in 1842 with a mathematical scholarship and gained a gold medal for his M.A. with prizes for Greek Testament and Hebrew. These were only stages on the way. For two years he taught at a school at Wimbledon and 'procured means' (which, I suppose, implies that he saved as hard as he could) for residence at Cambridge in 1844. It was said of him that he could have been senior wrangler as a freshman. Actually he became senior wrangler in 1848. In the same year he was awarded the Burney prize for an essay entitled 'The Doctrine of Divine Providence is inseparable from the belief in the existence of an Absolutely Perfect Creator'. Like many another mathematician, he seems to have been able to keep his theological logic and his mathematical logic in separate mental compartments.

Few anecdotes of his youth, or for that matter of his whole life, seem to have survived. Karl Pearson (writing in the *Mathematical Gazette* in 1936) has recorded that he was a strict disciplinarian with his students, but has left no opinion about his ability as a teacher. His life at St John's was as blameless as it was uneventful. As undergraduate, college lecturer and private tutor he led a secluded life amongst his books, studying among other things Hebrew, Arabic, Persian, Sanskrit and Russian. He was once heard to remark that the only title he coveted was that of D.D. And he himself said that he could do nothing with his fingers except write.

Write he did, to some effect. His text-books on Arithmetic, Euclid, Algebra, the Differential Calculus and the Integral Calculus were standard works for generations and ran, in some cases, to a dozen editions. In 1878, when examiner for the Indian Civil Service, he wrote: 'There is a library of mathematical books provided by the Civil Service Commission for the use of examiners. It consists of 14 volumes, ten of which are by myself.' And this fairly reflects his contribution to the mathematical studies of his day.

He was elected to Fellowship of the Royal Society in 1862, and was sufficiently well regarded by his colleagues to be a candidate for the Sadleirian chain in mathematics. When it was awarded to Cayley, he was the first to applaud the decision. He never, in fact, attained a chair, but continued to exert a strong influence at Cambridge. His efforts were not confined to mathematics; he was one of the founders of the Moral Sciences Tripos.

His lasting contributions to mathematics, however, were three histories: *The History of the Calculus of Variations during the Nineteenth Century* (1861); *The History of the Mathematical Theories of Attraction and the Figure of the Earth from the time of Newton to that of Laplace* (1873); *The History of the Mathematical Theory of Probability* (1865). We may add his unfinished *Treatise on Elasticity*, which was completed by Karl Pearson.

It comes as a slight surprise to find that in 1864 he married. 'You will not forget' he wrote to his future wife 'that I have always been a student and shall always be; but books shall not come into even distant rivalry with you.' All the same, he took Hamilton's *Quaternions* to read on his honeymoon. His family life was apparently very happy and there were several children. He kept canaries and cats, did not dance, had no artistic leanings and spend his holidays looking at cathedrals. Evidently a man of sober honest stuff, but a difficult subject for a biographer.

In 1880 he began to suffer with his eyes. The trouble spread, leading to paralysis and ultimately to his death in March 1884.

We should, perhaps, spare a few moments to consider this stiff drab Victorian figure, so unlike the colourful authors of whom he wrote, so meticulous in his attention to detail and so blind to the broad currents of his subject; for his *History of Probability* has stood for nearly a hundred years, without an imitator or a rival, and we are all indebted to it.

FRANCIS YSIDRO EDGEWORTH
(1845–1926)

FRANCIS YSIDRO EDGEWORTH, 1845–1926

By M. G. KENDALL

SUMMARY

An account is given of Edgeworth's life, details of which are rather scanty, and an appreciation of his work in statistics.

In the reign of Elizabeth I the Edgeworth family, originating at what is now Edgeware, transferred to County Longford in Eire and established themselves at Edgeworthstown. The head of the family at the end of the eighteenth century was Richard Lovell Edgeworth, a voluble Irishman who had four wives and twenty-one children, of whom one of the eldest, Maria Edgeworth, has an honoured name in the history of the English novel. There were strong literary and sociological interests in the whole family; Richard and Maria were both friendly with Ricardo and Jeremy Bentham.

Richard's sixth son, Francis Beaufort Edgeworth (1809–47) was on his way to Germany to study philosophy when, on the steps of the British Museum, he met Senorita Rosa Florentina Ercoles, aged 16. This was the end of serious philosophical studies for Frank. Within three weeks they were married and, after a few years in Florence, returned to settle in Edgeworthstown. It seems a great pity that more material wending its way towards Kant and Hegel was not redeemed in the same manner.

Francis Ysidro Edgeworth,* the fifth son of this marriage, was born in 1845 at Edgeworthstown and was brought up there. His father died shortly after his birth, and although the family doubtless could take care of his material needs in youth, his early years as a young man in London were spent in straitened circumstances. He entered Trinity College Dublin at the age of 17 and read languages—French, German, Spanish and Italian. A scholarship took him to Oxford. It appears that he entered Exeter College in January 1867, kept one term there and then transferred to Magdalen Hall for the rest of the year. In January 1868 he transferred to Balliol and was in residence for the next two terms. For some unknown reason he was not in residence for the next five terms (there were four terms to the year in those days), but returned for the Michaelmas term 1869. He sat his final schools in the summer term of 1869 and got a first in Literae Humaniores.

Edgeworth's command of ancient and modern languages is evident in all his writings. One finds a quotation from Horace or Plato on one page and a differential equation on the next. He was always ready to promote a new word, such as catallactics, or the science of exchange (*κατα* + *αλλαγή*), or to invent a new one, such as a brachistopone, a curve of minimal work (*βραχιστος* + *πόνος*). Few of them have survived, even in dictionaries. His predilection for polysyllables extended to his speech. In *Goodbye to All That* Robert Graves

* This is the way his name has always been written, and he signed himself as F. Y. Edgeworth; but the records at Somerset House give his Christian names as Ysidro Francis.

records a passage between Edgeworth and T. E. Lawrence, who had just returned to All Souls from a visit to London.

'I trust', said Edgeworth, 'that it was not too caliginous in the metropolis?'

'Caliginous,' replied Lawrence, 'but not inspissated.'

The roots of Edgeworth's thinking in relation to literature and the humanities are easy to trace. In contrast, we know very little about his grounding in mathematics. At some point before reaching the age of thirty he acquired a very respectable competence in algebra and calculus and was able to quote results in the calculus of variations. In this he seems to have been self-taught. Keynes (1933) remarks that whereas Alfred Marshall was second Wrangler at Cambridge and Edgeworth read Lit. Hum. at Oxford, Edgeworth, 'clumsy and awkward though he often was in his handling of the mathematical instrument, was in originality, in accomplishment and in the bias of his natural interest considerably the greater mathematician'. This is a very just comment, but does not really define Edgeworth's niche among mathematicians; for Marshall, so far as is known, merely translated some economics into mathematical language and, mathematically speaking, created nothing. Writing to Karl Pearson in 1915, Edgeworth said, *à propos* of a paper by Isserlis on the correlation coefficient: 'when I contemplate his stupendous mathematics I recall the apostle's remark on the relations between Abraham and Melchisidek "without controversy the less is blessed of the greater"—I do not feel competent to pronounce upon the stiffer parts of the mathematical work'. This is the kind of thing a man of 70 might well say; but I think it true that Edgeworth, though competent in mathematics, was not a creative mathematician. It would, nevertheless, be of considerable interest if we could discover how and when Edgeworth acquired his mathematics, and under what incentive.

Information about the period 1870–90 in his life is, in fact, very scanty. We know that he was called to the Bar in 1877, became Lecturer in Logic (evening classes) in 1880 at King's College, London, where he was made Professor of Political Economics in 1888 and held the Tooke chair of Economic Science from 1890 to 1891. More personal details are lacking. Edgeworth never married and I have not been able to trace any papers which he might have left. His will required the executors to destroy any manuscripts in his handwriting found in his rooms.

Apart from an article in *Mind* (1876) with the unexpected title of 'Matthew Arnold on the Butler doctrine of self-love', Edgeworth's first publication, in 1877 when he was already 32, was a little book of ninety octavo pages on *New and Old Methods of Ethics*. I doubt whether anyone but a historian would want to read it now, and I doubt whether many of Edgeworth's contemporaries read it then. None of his writings, at any time of his life, consisted of the kind of prose, or the orderly presentation of ideas, which give pleasure for their own sake; and in this particular work he actually writes down variational integrals, which must have put it beyond the understanding of most of those who were interested in ethical problems at the time.

However, we can discover in this booklet two elements which form the basis of much of his later work. First of all, he assumes that pleasure can be quantified, and that the amounts of pleasure experienced by different sentient beings can be added together. Consequently, at least in theory, a hedonic calculus can be set up in mathematical terms. Secondly, he asserts that most of the problems of physical ethics relate to optimization: for example, given a certain quantity of stimulus to be distributed among a given set of sentients, we may require to find the law of distribution productive of the greatest quantity of pleasure.

At a rather later point of time (1891) Edgeworth himself quoted Malthus to the effect that 'many of the questions both in morals and in politics seem to be of the nature of the problem *de maximis et minimis* in fluxions'. In one form or another the determination of stationary values was to occupy a good deal of his time henceforth. The field of application, however, changed. From the measurement of pleasure he was led to the measurement of utility and hence became an economist. From the measurement of utility he was led to the measurement of belief, and hence became a probabilist and a statistician.

Four years after the booklet on ethics there appeared *Mathematical Psychics* (1881), 'an essay on the application of mathematics to the moral sciences'. Here we have much more assurance, much greater clarity and material contributions to mathematical economics.

On several occasions Edgeworth was concerned with the difference between merely expressing economic ideas in mathematical language, which he deplored, and using mathematics to advance economic analysis. He began writing at a time when there was much scepticism about the use of mathematics in the social sciences. Attempts to turn economics into a science were particularly frowned upon. A writer in the *Saturday Review* of 1871 (11 November) puts it quite well:

'If we say that G represent the confidence of Liberals in Mr Gladstone and D the confidence of Conservatives in Mr Disraeli and x, y the number of those parties; and infer that Mr Gladstone's tenure of office depends on some equation involving dG/dx, dD/dy, we have merely wrapped up a plain statement in a mysterious collection of letters.'

Edgeworth was therefore concerned to establish the value of what he called 'unnumerical mathematics'. He has a good deal to say about maximum and minimum principles in physics and analogous principles in psychics. As late as 1915 he was still having to apologize for introducing mathematical nomenclature into the *Economic Journal*—'We would gladly imitate the parsimony of symbols practised by another of our authors. We should like to draw the line at partial differential coefficients. But unfortunately the reasoning turns mainly on the conceptions connoted by those coefficients... We can only practice temperance, not abstinence, in the matter of symbols.'

Utility, for Edgeworth, was the integral of pleasure over time and he divided the calculus of pleasure into two parts: economics, which investigated the equilibrium of a system of hedonic forces each tending to maximum individual utility; and utilitarianism, concerning the equilibrium of a system which maximized utility in a group as a whole. In *Mathematical Psychics* he proceeds to set up a series of definitions in economics and to prove a few theorems (e.g. that contract without competition is indeterminate). He invents indifference-curves and utility surfaces, which have become a standard part of economic thought. He then discusses the utilitarian calculus, proves a number of results, and ends with a number of appendices, ranging from 'Hedonimetry' to 'On the present crisis in Ireland'. It was typical of Edgeworth's thought that, even in a book, he had to have eight appendices. Nothing he wrote ever seemed to close a subject, even for himself.

It must have been about this time (1881) that Edgeworth became interested in probability. He was 38 when he published his first paper on the subject, one on the law of error. During the next five years he published twenty-five articles, mostly on probability and least squares, but with a sprinkling of work on index-numbers and variation in the value of money. Another booklet, *Metretike* (1887), or the method of measuring probability and utility, was concerned with probability as the basis of inductive reasoning, with weighted means and with the most advantageous mean.

In 1891 occurred the major event of an uneventful life. Edgeworth was appointed to the Drummond chair in Political Economy at Oxford, in succession to Thorold Rogers. He became a Fellow of All Souls in that year and retained both the chair and the fellowship until retirement in 1922. He was also the first editor of the *Economic Journal* and remained as such from 1891 to 1926, when he handed over to Keynes.

From 1891 until his death Edgeworth never again published a book, although some of his articles are as long as the three short books already noticed. Professor Harry Johnson expresses the opinion that a major influence on Edgeworth was Marshall's dismissal of him as not a serious economist, expressed in Marshall's review of *Mathematical Psychics*. This may have driven him from writing books to writing articles. His collected economic papers (1926), apart from a series of reviews, contain thirty-four articles written between 1891 and 1920. During the same period he published about the same number of articles on statistical subjects; and there were a few more added before his death in 1926.

Edgeworth's work as an economist was discussed by Keynes in one of his biographical essays (1933). Keynes knew him well and I must leave an appreciation of Edgeworth's economic work in his hands. Mathematical economists nowadays, I think, respect Edgeworth, much more than Marshall, for the precision and care of his analyses. He stands out among his contemporaries as a man who genuinely believed that mathematical analysis could *advance* economic theory, not merely express it concisely; and in this he was far ahead of his time. Edgeworth's work as statistician was the subject of an appreciation by Bowley (1928), who obviously went to a lot of trouble in preparing this work. It was often necessary to collate different papers to obtain complete proofs; some papers refer to others in a manner reminiscent of that style of legislation by reference which makes Acts of Parliament so hard to follow; there were frequent misprints which were corrected in later papers. Edgeworth also had the habit, infuriating to the bibliographer, of having some of his papers offprinted with supplementary material which was added at his own expense.

A good deal of Edgeworth's work in probability theory has been over-ridden by later writers and is of no special interest except as a record of the way in which the problems were considered at the time. Much the same is true of his work on the best form of average and on correlation. There are three topics, however, to which he made lasting and important contributions. One is his work on index-numbers. The other two concern what nowadays would be called the theory of frequency distributions, but were then thought of as generalizations of the law of error.

Edgeworth's main work on index-numbers was done while he was secretary of a committee of the British Association appointed 'for the purpose of investigating the best method of ascertaining and measuring variations in the value of the monetary standard'. The committee included Foxwell, Giffen, Marshall, Martin, Nicholson, Palgrave and Sidgwick. There are three documents recorded in the report of the Association, one for each of the years 1887–9. In the first the theoretical problems are discussed. The second deals with some practical issues and concludes with a long article by the secretary on the accuracy of the index numbers which the committee favoured. The third is again a long memorandum by Edgeworth on seven different methods of constructing index-numbers and a critical comparison. The author himself described this set of papers as a voluminous disquisition. It was indeed, but was the first study of the subject in depth—a tribute to the wide scholarship and penetrating mind of its author.

Edgeworth arrived at what would nowadays be called the Edgeworth expansion by an argument based on central limit considerations. The moments of a sum of independent variables can be expressed in terms of moments of the variables. Taking logarithms of the generating functions, he effectively defines the parental form of what are now known as cumulants. A further step, of which the genesis is obscure, leads him to the Edgeworth series. (Bowley, 1928, gives the details, but it is hard to see from the original papers how Edgeworth arrived at these methods.) He did some arithmetic checks to see how well the truncated series approximated to a given frequency function, but was concerned mainly with the determination of percentiles, not the evaluation of tail areas.

His second contribution to the theory of frequency distributions was the introduction of systematic variate-transformations or, as he described it, the method of translation. In actuality he considered in detail only quadratic transformations and accordingly did not produce a very flexible system. Much more recently N. L. Johnson has had greater success with a wider class of transformation. However, it is of some interest to note that Edgeworth considered the extension of the method of translation to bivariate surfaces, a topic which never seems to have been adequately followed up.

A good many of Edgeworth's numerical illustrations were, naturally enough, drawn from economic or social data. But every now and again some statistical regularity in another part of his world would attract his attention. Thus, we have papers on the statistics of bees, examination marks, bimetallism and Parliamentary elections. It was he who first remarked that the distribution of proportions of Liberal votes (to total Liberal plus Conservative votes) in British Constituencies was approximately normal.

By the kindness of Professor Egon Pearson I have seen forty-five letters or cards from Edgeworth to Karl Pearson, written between 1891 and 1915. (Pearson's side of the correspondence is unfortunately missing, but we can usually guess at what he wrote.) The two must have met while Edgeworth was in London, where he had rooms in Hampstead. The first two letters, interestingly enough, were written before Pearson had developed his interest in statistics. Edgeworth tried to engage Pearson in mathematical economics, but failed. Pearson enlisted Edgeworth's interest in biometrics, with rather more success.

Edgeworth to Pearson, 5th April 1891:

May I hope that you will contribute something to the pages of the *Economic Journal*?

A reasoned statement of considerations in favour of Socialism, such as few of that persuasion are so well qualified to furnish as you, is a *desideratum*. Or I might suggest some more definite point: such as the prospects of science under a socialist régime.

For my own part and speaking for myself I could wish that, as a prominent representative of an important interest, you belonged to the Association and took part in its direction.

I would have written sooner, but that I have been hoping to see you.

Let me add that I do not wish to restrict your literary activity to the class of subjects which I have indicated. Thus an estimate of the applicability of mathematics to Political Economy, made by you, would be very interesting.

Believe me,
Yours very truly,
F. Y. Edgeworth (signed)

Edgeworth to Pearson, 12 April 1891:

My dear Pearson,

We of the *British Economic* do not lay ourselves out for controversy. The method of rebuttal and rejoinder does not seem particularly suited to our subject. We have already indulged this tendency further than some approve in accepting from Mr Sidney Webb *Difficulties of Individualism*, in reply to Courtney. Hence I would rather that you omitted the second part of your title 'a rejoinder to Mr C.' and if possible consent to soften passages relating specially to him rather than generally to the subject.

At any rate, as you kindly offer, let me see the brochure when completed.

Pray remember that there is a subject in which you are a 'Fachmann', the application of mathematics. Why not glance at Marshall's *Principles of Economics* and see what you think of his use of mathematical conceptions? The matter is in short compass Marshall's *Appendix* suffices; perhaps you might add Cournot's *Recherches* (which I could lend you). No special knowledge of economics in detail is required.

Yours very truly,
F. Y. Edgeworth (signed)

Edgeworth always referred to Karl Pearson's work in terms of genuine admiration. Of χ^2 he wrote in 1900: 'I have to thank you for your splendid method of testing my mathematical curves of frequency. That χ^2 of yours is one of the most beautiful of the instruments which you have added to the Calculus.' Pearson, one feels, was less enthusiastic about Edgeworth's statistical work; but they continued to correspond and to exchange copies of papers on very friendly terms until Edgeworth's death in 1926. In a speech given at the Galton dinner in February 1926 Pearson said:

Another critic [he had previously spoken of William Bateson's death], dear old Edgeworth, has passed away also; him we can almost call a biometrician for he contributed to *Biometrika* and he expressed a wish to attend the meetings of our Society of Biometricians. Only last December he came and spoke as he had always spoken, only with a slightly greater sign of age, and his criticism failed as it had always failed, because he spoke not the language of the people. But he was, I think, happy to be there, and he only made us a trifle later than we might otherwise have been. And some of you may remember the gambolade he danced, *pas de seule*, down the cloisters last June, in front of the procession returning from the refectory to the laboratory after the dinner of the Society of Biometricians to Professor Westergaard. Aged 80 he was still a boy in his pleasures and spirits, if *Magister Obscurantissimus* in his dialectic. I should like to reckon him among the biometricians if he ploughed always right across the line of our furrows. Besides we owe him something; like a good German he knew that the Greek κ is not a modern c, and if any of you at any time wonder where the k in Biometrika comes from, I will frankly confess that I stole it from Edgeworth. Whenever you see that k call to mind dear old Edgeworth.

The bareness of the records of details of Edgeworth's life make it difficult to portray him as a person or a personality. If I were to take the way out of a good deal of current historical literature of the more popular kind I should write a novel and give my imagination more play. One could picture the boy, conversing with Aunt Maria on the basic principles of economics and reading widely in a number of languages; the Oxford scholar, becoming alive to mathematical discipline and the possibility of quantifying ethical principles; the penurious and rather lonely young man in London, exercising his logical mind in reading law; the ever-widening interests which rushed upon him and almost overwhelmed him; the growing recognition of his abilities resulting more from respect than sympathetic understanding; the long period of work as editor of the *Economic Journal*; his perpetual interest in so many aspects of life. I like to think of him in this way; but those who knew him well enough to tell us what he was really like have left no record.

Edgeworth was clearly an original thinker with unusual gifts and an unusual background. But it is hard to see his work in context. Perhaps he had no context; he was a disciple of no school; he founded no school; and Bowley was his only statistical heir. His statistical work is, as a general rule, very tedious to read and hard to follow. Its value was not clear to his contemporaries, including, possibly, Edgeworth himself. Nowadays, I should think, it is not read at all.

Nevertheless he was an important figure in our subject. Apart from the permanent contributions with which his name is associated, he was a great influence on his contemporaries and played a notable part in the development and acceptance of statistics as the subject was then understood. That he did not generate in the social sciences the same surge of interest that Karl Pearson generated in the biological sciences was as much the fault of the soil as of the seed or the sower.

To my delight I discovered, after this article was in draft, that Professor Harry Johnson and some colleagues had, a few years ago, compiled a complete bibliography of Edgeworth's writings. Unfortunately, the bibliography would take up more space than would be justified in this Journal, but copies have been deposited in the libraries of the British Museum, the Royal Statistical Society, the London School of Economics, All Souls, Oxford, the Bodleian and the University Library, Cambridge, and I have some spares should anyone want a copy.

I am indebted to Professor E. S. Pearson and Mrs H. S. Hacker for a sight of the Edgeworth–Pearson correspondence; to Lt.-Col. Kenneth Edgeworth for some information about the Edgeworth family; to the librarians of All Souls and Balliol College for some dates in Edgeworth's academic career; to the Royal Statistical Society for permission to reproduce the portrait at the beginning of this article; and to Professor Harry Johnson for some useful comments on a first draft of this paper.

References

Bowley, A. L. (1928). *F. Y. Edgeworth's Contributions to Mathematical Statistics.* London: Royal Statistical Society.
Edgeworth, F. Y. (1877). *New and Old Methods of Ethics.* London: James Parker.
Edgeworth, F. Y. (1881). *Mathematical Psychics.* London: Kegan Paul.
Edgeworth, F. Y. (1887). *Metretike.* London: Temple.
Edgeworth, F. Y. (1926). *Papers relating to Political Economy.* Three volumes. London: Macmillan.
Keynes, J. M. (1933). *Essays in Biography.* London: Macmillan.

Yrs affectionately

WALTER FRANK RAPHAEL WELDON. 1860—1906.*

A MEMOIR.

By KARL PEARSON

I. *Apologia.*

It is difficult to express adequately the great loss to science, the terrible blow to biometry, which results from the sudden death during the Easter vacation of the joint founder and co-editor of this journal. The difficulty of adequate expression is the greater, because so much of Weldon's influence and work were of a personal character, which only those who have enjoyed his close friendship can estimate, and which will only to some extent be understood should it ever be possible to publish his scientific correspondence. That correspondence is not only the most complete record of the development of the biometric conceptions, but the amplest witness to Weldon's width of knowledge, keenness of intellectual activity, and intense love of truth. It is marked by an extreme generosity to both friend and foe, which is not in the least incompatible with the use of frankly—perhaps it would be better to say playfully—strong language whenever the writer suspected unfair dealing, self-advertisement, or slipshod reasoning masquerading as science. Any form of publicity was very distasteful to Weldon; in particular he had a strong dislike for all forms of personal biography. The knowledge of this makes the writing of the present notice a peculiarly hard task. Yet Weldon's influence and activity must always be associated with the early history of biometry; if there be anything which can effectively aid younger workers in this field, it must be to realise that at least one man of marked ability and of the keenest scientific enthusiasm has devoted the most fertile years of his life to this new branch of science. Weldon's history is not written in a long series of published memoirs; much of his best work was unfinished at his death, and we can only trust that it will eventually be completed as the truest memorial to his life. But science, no less than theology or philosophy, is the field for personal influence, for the creation of enthusiasm, and for the establishment of ideals of self-discipline and self-development. No man becomes great in science from the mere force of intellect, unguided and unaccompanied by what really amounts to moral force. Behind the intellectual capacity

* I have gratefully to acknowledge much aid from Mr A. E. Shipley in the preparation of certain parts of this memoir. K. P.

there is the devotion to truth, the deep sympathy with nature, and the determination to sacrifice all minor matters to one great end. What after all helps us is not that "he settled *Hoti's* business"...

> "Properly based *Oun*—
> Gave us the doctrine of the enclitic *De*,"

but that the Grammarian had the strength of will which enabled him "not to Live but Know."

If there is to be a constant stream of men, who serve science from love as men in great religious epochs have served the Church, then we must have scientific ideals of character, and these do involve some knowledge of personal life and development. It is the abuse of the personal so prevalent in modern life, the mere satisfaction of a passing curiosity, which we have to condemn. But the personal which enables us to see the force of character behind the merely intellectual, is of value, because it moulds our working ideals. We see the environment—imposed and self-created—which favours scientific development, and we can with accumulating experience balance environment against heritage in the production of the highest type of scientific mind. From the standpoint that no man works effectively without a creed of life, that for width of character and healthy development there must ever be a proper balance of the emotional and the intellectual, it would be a distinct loss if the personal were removed from what we know of the lives of Charles Darwin and James Clerk-Maxwell. Science, like most forms of human activity, is occasionally liable to lose sight of its ultimate ends under a flood of controversy, the strugglings of personal ambition, or the fight for pecuniary rewards or less physical honours. The safety of science lies in the inculcation of high ideals among its younger votaries. A certain amount of purely human hero-worship is not to be condemned, and yet this is impossible without some knowledge of the personal. Weldon himself was no more free from hero-worship than the best of his contemporaries. Of the men whose influence tended most to mould his life and career—F. M. Balfour, T. H. Huxley, Francis Galton—the personal side was not the smaller element. There was enthusiasm, hero-worship in its best sense, unregarding self-sacrifice in the defence of the man who had become for Weldon not only an ideal thinker, but an ideal character. In the defence of hero or friend, Weldon belonged to a past age, he was out with his rapier, before considering the cause; it was enough for him to know that one he loved or admired was attacked. A criticism of Huxley was to the end inadmissible; if at any point apparently correct, this appearance of correctness was due solely to the inadequate manner in which the facts of his life had been reported by biographers,—the class who pandered to the public love of the petty. It was in this spirit that Weldon received with delight the request to write for the *Dictionary of National Biography*, a scientific appreciation of Huxley's work. From Weldon's standpoint that appreciation should have formed the "Life." It is a fine piece of work and it was a labour of love, but those who have ever watched the younger man with the old, will know that the Huxley of the appreciation was not all that

Huxley meant to Weldon; the feeling of affectionate reverence did not spring from intellectual appreciation. It had far more its source in the influence of a strong character on a sympathetic character. And when we turn to Weldon himself, his relation to his friends and pupils was not purely that of a keen strong intellect; his best and greatest influence arose from the strength of character, that subtle combination of force and tenderness, which led from respect for the master, to keenest affection for the man.

If then we are to realise his life, it cannot be by a strict adherence to an appreciation of his published work. Some account of his stock, his early environment, and his temperament becomes needful, and the value of such an account lies in the help with which any life spent in single-eyed devotion to the pursuit of truth provides us, when we have ourselves to form our creed of life, and to grasp that science is something more than one of the many avenues to a competency. It must be in this spirit, therefore, that Weldon's dislike to the biographical is in a certain sense, not forgotten, but frankly disregarded in these pages.

II. *Stock and Boyhood.*

It would be impossible in a journal like *Biometrika*, devoted to the consideration of the effects of inheritance and environment, to pass by the striking resemblance of Raphael Weldon to his father Walter Weldon. The facts of Walter Weldon's life are given in the *Dictionary of National Biography*. It appears to have been a resemblance not only in intellectual bent, but also in many respects in emotional character. Raphael Weldon's paternal grandparents Reuben Weldon and his wife Esther Fowke, belonged to the manufacturing middle class. Their son Walter Weldon was born at Loughborough, October 31, 1832. Of his childhood we know little, he was as reticent as his son about both his childhood and his home surroundings; there is reason to suppose they were not wholly happy, and that shadows from these early years may have cast themselves not only over the father, but in a lesser extent have moulded the thought and life of the son. Walter Weldon married Anne Cotton at Belper, March 14, 1854, and shortly afterwards, leaving his father's business, came to London, starting as a journalist, writing for the *Dial* and *Morning Star*. Here he first made the acquaintance of William and Mary Howitt, who proved long and intimate friends of the family. From 1860 to 1864 he edited Weldon's *Register of Facts and Occurrences relating to Literature, the Sciences and the Arts,* and had as contributors a number of men afterwards well known in the world of letters. Thus while Walter Weldon's real name was to be made in science, his first interests were in literature and art. The steps by which Weldon regenerated the manganese peroxide used in the manufacture of chlorine, and the extensions he made of his chlorine process up to his death have been well described by Dr Ludwig Mond in his address in 1896 to the Chemical Section of the British Association. They brought Weldon comparative wealth, though nothing compared with the three-quarters of a million pounds his process saved this country annually. They also brought him scientific reputation; a vice-presidency of the Chemical Society, and in 1882 the fellowship of the Royal Society. But for our present purposes the main point is this: that Walter Weldon made his discovery while totally unacquainted with the methods of quantitative chemical analysis and possibly because of this ignorance. He was accustomed to attribute the discovery to a peculiar source, but those who knew well the immense facility of his son for closely observing phenomena out of his own field of research, and rapidly studying their interaction, always probing things, whether in the physical universe, or in mechanism, to their basis in simple laws of nature, will at once realise the source of the father's inspiration, and the heritage to the son*.

If Walter Weldon's discovery brought him wealth, he was generous to a fault. Like his son he appears to have scarcely known the value of money, except as a means of giving pleasure to his friends. His early death in September, 1885, two years after his son's marriage, cut off a career far from completed. But his life had been lived to the full, each instant crowded with physical, intellectual, or emotional activity. It is impossible to regard Walter Weldon's character without seeing whence Raphael Weldon drew much of his nature. The intense activity, the keen sympathy and generosity, the reticence, the creative power in many channels, the artistic appreciation†, were common to father and son. Nay, perhaps to give

* Raphael Weldon delighted during his many voyages in spending days in the engine-room; he made a study of the various types of engines, and his knowledge in this respect was not without service to the Marine Biological Association. He even studied the use of indicator diagrams. His first plan with a new bicycle was to take it part from part, so that he could fully understand its working and the nature of possible repairs. The microscope was not merely an instrument to work with, but a familiar illustration of optical laws, so that he knew at once how to modify each detail to suit special needs. Over and over again, talking over physical problems he would say: "Well, I don't know what you people think, but it has always seemed to me that"—and then would come some luminous suggestion or apt criticism of a proposed investigation in a field wholly outside the biological. A striking instance of this occurred only in the autumn of last year. Many friends had already gone to see the eclipse, most people were talking about it, and Weldon was left in sultry Oxford, fighting out a theory of determinantal inheritance. It was settled that a holiday should be taken, the determinants put on one side and a continuous photographic record made of the eclipse. Neither Weldon nor his colleague knew anything about sun-photography, and miserable were their first attempts. But gradually the objective, the telephoto lens and the focal shutter were worked out; a camera which had done yeoman service in photographing snail habitats became a wonderful structure, and a whole series of colour screens prepared from biological sources were tested and criticised. It was Weldon who obtained the first clean cut photograph showing sun spots clearly and admitting of definite enlargement. But what is more, each developmental stage of his sun camera had been thought out physically, and he knew why he took it. The trained physical astronomer would have found the stages already made, and *a posteriori* each would have been obvious, but this was the case of a biologist with insight into other fields and a striking power of making things work.

† An interesting illustration of the relationship is given in *Mary Howitt, an Autobiography*, 1889 (p. 184). The child Raphael, then 10 years, had gone with his father and the Howitts to visit the Wiertz Gallery at Brussels. William Howitt writes: "On our first entrance I was quite startled, I did not think I should at all like the paintings, they appeared so huge, so wild and so fantastic. But by degrees I began to see a great mind and purpose in them....... Little Raphael came and took my hand as we left the gallery, and said: 'Mr Howitt, I think Wiertz could not be a good man.' I asked him why. He answered, 'I think he could not be a good man, or he would not have painted some things there.' I told him he might naturally think so, but that a vast deal was to be allowed for his education. No doubt Wiertz thought all was right, and that many of his pictures contained

expression to a paradox, their volume of life was too great to be compatible with its normal length. There are men—not the least favoured of the Gods—who live so widely and so deeply, that they cannot live long. Discussions on the inheritance of longevity now come back to the memory, wherein Weldon referred to stocks of short-lived but intense life, and the personal experience and its moulding effect on character are now clear, where at the time the mind of the listener ran solely on a correlation coefficient.

In one respect Raphael Weldon differed widely from his father. Walter Weldon turned naturally to the mystical to satisfy his spiritual cravings; he was a Swedenborgian, and *ipso facto* a believer in intercourse with another world. Whether owing to a difference of training or of temperament, these things were to Raphael Weldon uncongenial. He was through the many years the present writer knew him, like his hero Huxley, a confirmed Agnostic. Sympathetic as every cultured mind must be with the great creations of religious faith; knowing more than many men of religious art—painting, sculpture, and music—he yet fully realised that these things had for him only emotional, no longer intellectual value*. It may be that the difference of training made this distinction between father and son, for the latter's mind was keenly alive to spiritual influences. A solitary fortnight with the beloved Dante was not solely pleasure; the re-perusal of the *Inferno* left its sombre influence on Weldon's thoughts for long after, testifying not only to its author's supremacy, but to the spiritual impressibility of the reader's nature.

It may be that the difference was due to heritage from the mother's side. Of Anne Cotton we know little, she died in 1881, when Raphael Weldon had just taken his degree. She appears to have exercised a rather stern discipline, which had greater influence on Raphael, than on his brother Dante. She was a devoted companion to Walter Weldon, and a resourceful helpmate in his early struggling days†. A daughter Clara born in 1855 died in 1861. Of his childhood Weldon rarely spoke. He was born in the Highgate district, and shortly after his birth his parents removed to a three-gabled house on the West Hill still standing. Here we get occasional peeps of a solitary child who would retire for hours under the dining-room table with his Shakespere, learning whole acts by heart. At six years old he appears in Mary Howitt's letters as staying at Claygate near Esher.

great and useful lessons. His father came up and added that when Raphael was older he would see those lessons more clearly than he could now."

The prophecy was fulfilled, in perhaps rather a different way. The little Raphael became a big Raphael who did not look to art "for great and useful lessons," and who refused to study Ibsen because undiscerning critics made current the idea that his art was subservient to inculcating a lesson.

* The "fulness of life" admitted, nay demanded, many a visit to cathedral service, especially in Italy. Even a study of Gregorian music was entered upon, and the writer recollects many a summer's afternoon spent in visiting the churches of Oxfordshire and Berkshire,—the cycle ride, the keen eye on surrounding nature, not only from the standpoint of the biologist, but of the artist; then the break to the religious past, the "biometric tea" at the village inn; the return journey towards evening and the discussion which touched many things, from *Draba verna* to the Norsemen in Sicily. The "volume of life" was there, as it was in the midnight talks in Wimpole Street or in the discussions in the study at Merton Lea.

† See *R. S. Proc.* Vol. XLVI. "Obituary Notice of Walter Weldon," p. XIX. *et seq.*

"We find little Raphael Weldon one of the best of children. Secker is mowing the grass at this moment, and he harnessed like a pony is drawing the machine. The Pater calls him 'Young Meritorious.'" And again:

"[Agnes] and Raphael are the best of friends, and their ringing laughter comes to us in the garden through the open window, as they sit in the dining-room painting the Stars and Stripes and the Union Jack for each other's amusement.... Agnes is a little free-spoken American full of fun and *dash.* Raphael more silent and contemplative. They sit painting pictures together for hours at a time. I feel quite proud of them both."*

In 1870 comes the flying visit to Brussels; in 1872 a still more memorable first visit to Paris, where the destruction caused by the Commune to the Tuileries and other buildings much impressed the boy. The Weldons had moved meanwhile to The Cedars, Putney, and shortly afterwards went to the Abbey Lodge, Merton, near Wimbledon. The visits and the changes give one the impression of a rather broken education. We have no record of what school Raphael Weldon attended, if any, at Highgate. At Putney he had as tutor a neighbouring clergyman. In 1873 he was sent to a boarding-school at Caversham, and from this time onwards the educational career is more definite.

Even before 1870, however, we find in the boy the father of the man. His great pleasure was to organise lectures for his children friends, and the adult population, if it could be procured. The seats were formally arranged, tickets provided, and the boy would discourse on slug or beetle procured in the garden, observation and the scanty literature available providing the material. According to a surviving auditor the lectures were carefully prepared and good so far as they extended.

Of the school at Caversham we have some detailed information. Mr W. Watson, its headmaster, had been a private 'coach' in London to University College students. In 1865 he opened a school at Reading, which was transferred to the hill out of Caversham in 1873†. Mr Watson's daughter Ellen Watson had a brief but brilliant career as a mathematician and pupil of W. K. Clifford's. Her life has been written by Miss Buckland. It is possible that she first stirred Weldon's mathematical tastes, as he spoke with admiration of her powers; she does not, however, appear to have taught in the school. The pupils were chiefly sons of Nonconformists of some eminence. Among the earlier scholars were Viriamu Jones, Alfred Martin, and E. B. Poulton, and among the later pupils Owen Seaman, F. W. Andrewes, P. Jacomb-Hood, and W. F. R. Weldon; names afterwards distinguished in literature, science, or art. The headmaster appears to have been a clever man of wide knowledge and sympathy, but there was little to specially encourage biological tastes in the school. It is reported of one under-

* *Loc. cit.* p. 162 *et seq.*

† As an illustration of Weldon's reticence I may state that we had passed this house several times together, before he mentioned it as his old school.

Raphael Weldon aged 10.

master that he protested against the study of insects, asking: "How do you think that such pursuits will put a leg of mutton on your table?" and the ability that proceeded from the school has been attributed by one of its former pupils to the special class from which it drew its chief material.

III. *Lehrjahre.*

Weldon did not remain fully three years at this school. It was followed by some months of private study and he matriculated at 16 (1876) in the University of London. In October of '76 we find him at University College taking classes in Greek, English, Latin, and French, with two courses of pure mathematics. In the summer term of 1877 physics and applied mechanics were studied. During this whole session he also attended Daniel Oliver's general lectures on botany and Ray Lankester's on zoology. He used to come up to town for Oliver's 8 o'clock lectures, getting his breakfast at a bun-house on the way*. Of his education at University College he especially praised in after years Olaus Henrici's lectures on mathematics. They were he held most excellent, and he considered Henrici the first born teacher under whom he came. Later in the Christmas vacation of 1879, after he had gone up to Cambridge, he researched for some weeks under Ray Lankester, who set him to work out the structure of the gills of the mollusc *Trigonia*. This completes Weldon's relations as a student to University College.

The difficulty of access, or possibly Walter Weldon's strong views, led Raphael Weldon in the autumn of 1877 to transfer himself to King's College. Here he stayed for two terms attending classes in chemistry, mathematics, physics, and mechanics, beside the zoology course of A. H. Garrod and the biology of G. F. Yeo. Divinity under Barry, at that time I believe compulsory, was also taken. At this time Weldon had the medical profession in view. He was only entered on the Register of Medical Students on July 6, 1878, but there can be no question that his course on the whole was directed towards the Preliminary Scientific Examination of the London M.B. This examination he took in December, 1878, after he had gone up to Cambridge; he was coached for it by T. W. Bridge, now Professor of Zoology in Birmingham, but he had already completed the bulk of the work in his London courses. With the Preliminary Scientific, Weldon's relation to London ceased. His student career there was not of quite two years' duration and it dealt with a variety of subjects, dictated as much by Weldon's catholic tastes, as by the discursiveness of the London examination schedule. But in his case, as in that of others, the grounding he received in physics and mathematics became a valuable asset, and the taste for languages, afterwards so emphasised, was to some extent trained and coordinated with literary knowledge. Yet Weldon's earlier instinct to study biology was not substantially modified either by the choice of medicine as a profession or by the diversity of his London studies. In 1877 he attended the Plymouth Meeting of the British Association, and there he was generally to be found in Section D.

* Weldon states in his applications for the Jodrell and the Linacre Chairs that he commenced the study of zoology under Lankester in 1877.

The presence of a life-long friend, who had already gone to Cambridge, was at least one of the causes which led to Weldon's entering himself as a bye-term student at Cambridge, and probably his choice of St John's College was due to Garrod's influence. He was admitted on April 6, 1878, as a pupil of S. Parkinson's. In the record his father is given as a "Journalist," although the chlorine process had now become a success, and his reference is to the Professor of Mathematics at King's College, then W. H. Drew*.

At Cambridge Weldon soon found his work more specialised and he rapidly came under new and marked influences. His first May term and Christmas term were devoted to his preparation for Little-Go and the London Preliminary Scientific. For the classical part of the former he seems to have worked by himself. After these examinations were over reading for the Tripos was begun and, under the influence of Balfour, Weldon's thoughts turned more and more to zoology, and the medical profession became less and less attractive. During the years 1879 and 1880 Weldon worked steadily for his Tripos; in the first year he was given an exhibition at St John's, and almost the only break in his work was the York Meeting of the British Association. In the second year a little original investigation on beetles was started; in May he took, for a month, Adam Sedgwick's place and demonstrated for Balfour. Overwork led to a serious breakdown, and resulted in insomnia and other ills, which occasionally troubled him again in later life. At the annual British Association holiday, this year in Swansea, Weldon saw for the first time Francis Galton, but an actual friendship was not begun till some years later.

The Tripos work was continued in spite of ill-health, till the Easter of 1881, when Weldon was unable to enter for the college scholarship examinations. By the influence of Francis Balfour, however, Weldon's real ability was recognised and a scholarship was awarded to him. A three months' holiday had become necessary, and Weldon went to the south of France, returning only shortly before his Tripos examination. At the very start of this, in itself all-sufficient, mental strain, Dante Weldon, who had joined Peterhouse, died suddenly of apoplexy. It says much for Weldon's self-control that the terrible shock of his brother's death, though it greatly affected him, did not interfere with his place in the first class of the Natural Sciences Tripos. The distress he had felt at his brother's death was redoubled a few weeks later when his mother passed away. She had never recovered from the blow resulting from the tragic death of her younger son. Of these things Weldon did not speak, but they undoubtedly influenced immensely his deeply emotional nature. Balfour's untimely death in the following year, and the early death of Weldon's father a few years later, left also their indelible impresses, a certain tinge of melancholy, a doubt whether he too would live to finish his work, and a tendency to take the joy and fulness of life while it was there. Few who saw the almost boyish delight in work and in play, the energy which spent itself for hours at a problem, or cycled eighty or a hundred

* There are errors in the entries in the Register, Weldon's mother's maiden name is erroneously given as Chester, not Cotton. Weldon was actually born at Suffolk Villa, Highgate.

miles in the day, the activity in debate, the vigour in lecture, the flow of thought and talk to the midnight hours, realised that the man was not of iron physique, and had indeed but small reserves of strength. To see Weldon keen over a piece of work was to believe him robust and ready for any fray; but looking back on the past one can see what each piece of work cost him, and the strain on a highly nervous temperament began in even those early Cambridge days.

IV. *Wanderjahre.*

With the Tripos Weldon's *Lehrjahre* closed and, as his nature directed, the *Wanderjahre* began without any interval of rest. Immediately after his Tripos, Weldon started for Naples to work at the Zoological Station. We have seen that at Cambridge he had been a pupil of F. M. Balfour's, whose death from an Alpine accident in the July of 1882 was the greatest loss British zoology had sustained for years. The charm of Balfour's personality had aroused the affection of all who attended his classes, and had awakened a keen desire to follow, even if but a long way behind, in his footsteps. In those days the stimulus given by Darwin's writings to morphological and embryological researches was still the dominating factor amongst zoologists, and Weldon threw himself at first with ardour into the effort to advance our knowledge by morphological methods. In Naples he began his first published work, a "Note on the early Development of *Lacerta muralis*" (1), but at the same time did much miscellaneous work on marine organisms. The lizard paper was finished in the winter at Cambridge, Weldon gratefully acknowledging the help of Adam Sedgwick, in whose laboratory he was then working. Anticipation in the publication of some of the results by C. K. Hoffmann, who had been working at the same points, caused a not unnatural disappointment.

In September Weldon was back in England at the Southampton meeting of the British Association. Here Adam Sedgwick, who had succeeded to the teaching work of Francis Balfour, invited Weldon to demonstrate for him. Thus the winter found Weldon in Cambridge again, and from Sedgwick's laboratory was issued the next piece of work: "On the Head-Kidney of *Bdellostoma*, with a suggestion as to the Origin of the Suprarenal Bodies" (2). Weldon hoped to show that "at all events in Reptiles and Mammals, the connection between the Wolffian body and the suprarenal is much more intimate than has generally been supposed," and he followed the matter up in the next year by publishing his paper "On the Suprarenal Bodies of Vertebrates" (3).

Meanwhile a great change had come over Weldon's personal life. On March 14, 1883, the anniversary of his parents' wedding-day, he was married to Miss Florence Tebb, the eldest daughter of William Tebb, now of Rede Hall, Burstow, Surrey, which formerly, after he left Merton, had been the house of Walter Weldon. The Weldons and Tebbs had been intimate friends for many years, and Miss Tebb had been at Girton while Raphael Weldon was at St John's. At Cambridge the new Statutes had just come into force, marriage was the order of the day, and houses were even difficult to procure. The Weldons on their return from a tour in France took Henry Fawcett's furnished house and settled down in Cambridge for the May term. Raphael Weldon still had his scholarship, and he was demonstrating for Sedgwick. He was now compelled to undertake "coaching,"—work which he gave up as soon as his means would allow of it, for his whole heart was then as afterwards in research. Still this coaching work brought him in touch with many men who afterwards distinguished themselves in biological or other fields.

After the death, on the 14th January, 1883, of Forbes, a fellow Johnian, Weldon for four months—June 15 to October 15—acted as *locum tenens* for the Prosector at the Zoological Gardens, London, and during that time he read the following papers before the Zoological Society: "On some points in the Anatomy of *Phoenicopterus* and its Allies" (4); a "Note on the Placentation of *Tetraceros quadricornis*" (5), and "Notes on *Callithrix gigot*" (6). Weldon did not succeed Forbes—it was rumoured that some of the electors doubted the fitness of his physique for the work and considered that the post was not without danger. But the temporary work into which he threw his usual energy gave him increased insight into vertebrate anatomy and had the further advantage of making him personally known to the active workers in zoology of the metropolis.

In the following year (1884) the paper above referred to on the development of the suprarenal bodies was published in the *R. S. Proceedings*. Weldon was now demonstrating in comparative anatomy at Cambridge, and the holidays were devoted to collection. At Easter Banyuls was visited, and the summer vacation found Weldon in Naples again for three months preparing his fellowship dissertation. In Naples the cholera had broken out, and the Weldons experienced not only difficulty in getting the precious dissertation back to England, but in returning themselves. This was done by an Orient liner, the last allowed to call. Thus began the long series of holidays in Italy with the sea passage to or fro. The summer heat of Naples seemed to suit Weldon, and he could work and think under circumstances which only allow mere existence to an ordinary Englishman. On returning to Cambridge, Weldon was elected to a fellowship at St John's College on November 3rd, and was shortly afterwards appointed University Lecturer in Invertebrate Morphology. About this time the Weldons took a permanent home at No. 14, Brookside, which soon became a centre for Cambridge workers in biology*. The Weldons' home, whether in Cambridge, London, or Oxford, was always a centre, where not only the right people met, but whence actual profit came by the right people interchanging ideas and planning work.

On his return to Cambridge in November 1884 Weldon had taken up again his invertebrate work. His next memoir "On *Dinophilus gigas*" (7) dealt with the anatomy and affinities of *Dinophilus*, at that time a very little known Annelid. A. E. Shipley had been fortunate enough to collect a number of these minute worms at Mounts Bay, Penzance, and had handed them over to Weldon for

* "The house in Brookside in which he lived after his marriage until he left Cambridge was a delightful and hospitable centre, where all sorts of subjects were discussed, attacked and defended until all sorts of hours in the morning." A. E. S.

description. The latter gave a full account of their anatomy and added a careful discussion on the affinities of the genus, expressing his belief that while it is "related on the one hand to *Archiannelida*, it retains on the other many features characteristic of the ancestor common to those groups (especially Chaetopods, Gephyreans, Mollusca, Rotifers and Crustacea) which possess a more or less modified trochosphere larva."

The next few years of Weldon's life were—if it be possible to make any comparative where all were intense—more active than ever. He had now given up coaching, and as he only needed to be in Cambridge two terms of the year, travel and research could occupy the time from the beginning of June to January. On May 8, 1885, Weldon gave his first Friday evening lecture at the Royal Institution on "Adaptation to surroundings as a factor in Animal Development." No report is published in the *Proceedings* of this lecture, but there are those who still remember the impression caused by the youthful lecturer of 25 years of age. And here may be a fitting place to say something of Weldon's lecturing power. There are two distinct sides to lecture work; the instruction of small or large classes of students and the public oration. Success in the one field does not necessarily connote success in the other. In the former case the eye must be kept on the average student, the lecturer must realise what the individual auditor is feeling, he must expand his exposition or must contract it to meet the carefully observed needs of his audience, for he knows that he can take up the subject again on the next occasion exactly where he has left off. In this form of lecturing Weldon was an adept, it brought out all his force and enthusiasm as a teacher. As a writer in the *Times* (April 18, 1906), says:

"Seldom is it given to a man to teach as Weldon taught. He lectured almost as one inspired. His extreme earnestness was only equalled by his lucidity. He awoke enthusiasm even in the dullest, and he had the divine gift of compelling interest."

In public lecturing on the other hand, with a time limit and an unknown audience, the personal touch with individuals is impossible. There is no time to elaborate points, the whole matter must be *a priori* fitted to the time, and if the audience is not grasping an idea, then the lecturer must put both explanation and disappointment on one side; he must make his audience jump gladly, and trust to better luck in his exposition of the next stage of his thesis. Shortly, he must feel his audience with him *as a whole* and pay no regard to the individual.

Weldon's own intense thoroughness made him only too conscious when a portion of his audience were not following him; his highly nervous temperament made it a necessity that he should have a sympathetic grip on the individual. This made for success in his lectures to students; but it brought also a factor of uncertainty into his public lectures. The most carefully prepared discourse, and no man gave more time and energy than Weldon to preparation*, might be

* Drafts and re-drafts were written, elaborate diagrams painted, or lantern slides made and coloured by Weldon himself.

spoiled by Weldon's consciousness that certain members of the audience were not following him. He would then turn his exposition into explanation of minor points, so that the lecture would not be completed, or he would settle down to speak to the few he realised to be following him, and neglect the audience as a whole. If a portion of his audience were hostile or actively unsympathetic, this always prevented Weldon from reaching his best; it formed a strain on the lecturer's nervous temperament, which could only be realised by those of like fibre, and in some cases left its permanent mark. Thus it came about that the success of a public lecture by Weldon could not *a priori* be measured; it depended far too much on the audience. Individual lectures at the British Association, the London Institution, at University College or elsewhere were brilliant achievements, but at the same places on other occasions, Weldon was not so successful, for no man was ever more responsive to immediate environment than he was. To do his best and to be at his best he needed essentially a sympathetic environment. Weldon has been spoken of as an eager, ready and dramatic debater, keen to see a weak joint in his opponent's armour and quick in putting his own case with telling effectiveness. This is undoubtedly true, but it needs the qualification that this intellectual readiness when in full action meant a high pressure; it was a strain the less oft repeated the better. A torpedo-boat destroyer is associated with a 26 knot speed, and such speed differentiates it from other vessels of war; but the less it is run at this rate, the longer undoubtedly it will last. Controversy was not an atmosphere in which Weldon rejoiced*; it came to him because he felt bound to criticise what he held to be error, because he must defend a friend, but it was—running the destroyer at 26 knots!

This digression may be justified on the ground that we have reached the period when Weldon began to exercise a personal influence over his students at Cambridge, and the sources of that influence are to be found first in the lecture-room and then in strong personal sympathy. In the lecture-room he always impressed his hearers with the importance of his topic. You could not listen to him lecturing on a flame-cell or on the variations in the carapace of *Pandalus annulicornis* without sharing his intense conviction of the importance of the matter in hand. He aroused a consciousness in his students that things were worth studying for their own sake, apart from their examination value.

The summer months of 1885 were spent in Guernsey, and the death of Walter Weldon in the September of this year kept Raphael Weldon at other things than research. Christmas, however, found the Weldons at Rome. The Lent and May terms (1886) were spent in Cambridge as usual. In June came a visit to the south of France on Chlorine business, but in July came freedom, the crossing to America and the visit to the Bahamas in August to collect. From his

* Actual experimental work which upset another man's views, Weldon declined to publish. "Yes, I know he is wrong, but I don't want merely to controvert him, I want to get at the truth of these things for myself." And when he had satisfied himself he would pass on to a new point of investigation and never publish at all.

headquarters in the Bahamas Weldon went with two friends to North Bimini in the Gulf stream and enjoyed immensely his first experience of tropical or at least semi-tropical seas. He made considerable collections, but his published results were confined to "*Haplodiscus piger*; a new Pelagic organism from the Bahamas" (8), and a "Preliminary Note on a *Balanoglossus* Larva from the Bahamas" (9). Haplodiscus was netted near the Island of New Providence. It is a member of the Acoela, the most simplified of the class Turbellaria, and for some time Weldon's account was one of the most complete we had of any member of this group. Working at the *Balanoglossus* material in 1887, Weldon found that his results differed from those reached by Professor Sprengel. He accordingly went to Giessen at Easter,—his second visit to Germany, the first having been at Christmas, 1886—and finally handed over to Professor Sprengel the whole of the *Balanoglossus* material he had collected in the Bahamas. During the Lent and May terms Weldon came up from Cambridge and gave a course on Economic Entomology to the forestry students at the Royal Engineering College, Cooper's Hill. The summer and autumn of this year involved a meeting of the British Association at Manchester, a visit on Chlorine business to France, and later, collecting and working in Guernsey. The Christmas was spent at Plymouth.

In 1888 the buildings of the Marine Biological Laboratory in Plymouth were nearly completed, and the visits to Plymouth now replaced those to Guernsey. To the Marine Biological Association Weldon gave both time and sympathy during the rest of his life. His annual visits of inspection to Lowestoft during the last few years were always a great pleasure to him, and he was preparing for and talking of this year's visit only a few days before his death. Lent and May terms, 1888, were spent as usual in Cambridge, but June to December were given up to Plymouth, with a brief Christmas holiday in Munich. And here we must note the beginning of a new phase in Weldon's ideas. His thoughts were distinctly turning from morphology to problems in variation and correlation. He has left on record the nature of the problems he was proposing to himself at this time and they are summed up as follows:

(1) The establishment of a new set of adult characters leading to the evolution of a new family has always been accompanied by the evolution of a new set of larval characters leading to the formation of a larval type peculiar to the newly established family; the two sets of characters having as yet no demonstrable connection one with the other.

(2) The evolution of the adult and that of the larval characters peculiar to a group advance *pari passu* one with the other, so that a given degree of specialisation of adult characters on the part of a given species implies the possession of a larva having a corresponding degree of specialisation and *vice versa.*

The next year was to place in Weldon's hands a book—Francis Galton's *Natural Inheritance,* by which one avenue to the solution of such problems, one quantitative method of attacking organic correlation, was opened out to Weldon;

and from this book as source spring two of the friendships and the whole of the biometric movement, which so changed the course of his life and work. In 1889, the year of the issue of this book, another change also came. Weldon found that his dredging and collecting work separated him from his books for half his time. Accordingly, he applied for a year's leave from Cambridge, and the Weldons settled down in a house of their own at Plymouth. This period of hard work lasted through 1890, and was broken only by flying visits to Dresden in September and at Christmas, 1889, and an autumn visit in 1890 to Chartres and Bourges. The intellectual development and the experience and knowledge gained in this period were far more important than the mere published work would indicate. In 1889, Weldon investigated the nature of the curious enlargement of the bladder associated with the green, or excretory, glands in certain Decapod Crustacea, and published in October of the same year his paper on "The Coelom and Nephridia of *Palaemon servatus*" (10). The result of his investigation was to confirm "the comparison so often made (by Claus, Grobben, and others) between the glomerulus of the vertebrate kidney and the end-sac of the Crustacean green gland." A little later, June 1891, Weldon published the results of more extended researches in this field in what proved to be his last strictly morphological paper. It was entitled: "The Renal Organs of certain Decapod Crustacea" (11). In this he showed that in many Decapods spacious nephro-peritoneal sacs "should be regarded rather as enlarged portions of the tubular system...than as persistent remnants of a 'coelomic' body cavity into which the tubular nephridia open."

One further paper of a year later may be best referred to here, Weldon's only piece of work on invertebrate embryology, "The Formation of the Germ Layers in *Crangon vulgaris*" (12). This contains a clear account of the early stages of segmentation and the building up of the layers of the shrimp, illustrated by excellent figures. And here it may be mentioned that Weldon's power with the pencil was not that of the mere draughtsman, accurate in detail, but too often lifeless. Weldon was an artist by instinct, and he had the keenest pleasure in drawing for its own sake. His brilliant blackboard drawings will be remembered by all his students; some correspondents will remember elaborately beautiful sketches sent merely to illustrate a passing question, where a rough diagram would have sufficed; a delicately pencilled shell to please a child friend; carefully copied architectural details to gratify himself and made to be destroyed;—all signs of a real artistic power of creation. And the sense he enjoyed in himself, he appreciated in others. Nothing refreshed him so much as a visit to the National Gallery, or to a lesser extent the sight of more modern art. Weldon, smiling before one of his own pictures, unconscious of his environment, was good to behold, and made one realise how for him pictures were still differentiated from furniture. In the last two years of his life, when he had become an ardent photographer, the artistic feeling came to play a prominent part as the difficulties of the craft were one by one mastered.

(*a*) "L'Apparition: Le Café Orleans."

(*b*) H. Hortensis, from a letter.

Plate III

December, 1890, closed the Cambridge work* and concluded the *Wanderjahre*. Weldon now succeeded Ray Lankester in the Jodrell Professorship at University College. In June he had been elected a Fellow of the Royal Society largely on the basis of his first two biometric papers, which will be considered more in detail in the next section.

It will be seen that the years between Weldon's degree and his first professoriate were years of intense activity. He was teaching many things, studying many things, planning many things. His travels perfected his linguistic powers, and his fluency in French, Italian and German was soon remarkable. But while this added immensely to his delight in travel, it opened to him also those stores of literature, which appealed so strongly to his artistic temperament. From the mediaeval epics to Balzac he was equally at home in French literature; and the Italian historians were read and carefully abstracted, that he might understand Dante without the aid of a commentator, and appreciate Italian towns without the help of a guide-book. In German he had a less wide knowledge of the earlier literature and history, but he spoke the language with an accent and correctness remarkable in an Englishman. In later years he had commenced the study of Spanish, the Romance tongues and literatures being always more sympathetic to him than the Scandinavian or Teutonic. His remarkable thoroughness in science reappeared as a form of scholarly instinct when he approached history and literature, and the present writer remembers Weldon's keen pleasure and exactitude in following up more than one historical enquiry. His delight in knowing spread far beyond the limits of natural science.

V. *London and the First Professoriate*, 1891–1899.

A word must here be said as to the transition which took place during the *Wanderjahre* in Weldon's ideas. He had started, as most of the younger men of that day, with an intense enthusiasm for the Darwinian theory of evolution; it threw open to him, as to them, a wholly new view of life with its possibility of seeing things as a connected whole. Weldon realised to the full that the great scheme of Darwin was only a working hypothesis, and that it was left to his disciples to complete the proofs, of which the master had only sketched the

* A note may be added as to the general influence of Weldon at Cambridge. At the time Weldon began lecturing there were a considerable number of students largely attracted to Cambridge by Balfour's fame and remaining there to mourn his loss. Mr W. Bateson of St John's, Dr Harmer of King's, Professor Sherrington of Caius, Professors D'Arcy Thompson and J. Reynolds Green of Trinity, Professor Adami and Mr A. E. Shipley of Christ's, graduated in 1883 and 1884, and all, to some extent, came under his influence. For six years (1884–1890) he gave advanced lectures to the candidates for Part II of the Natural Sciences Tripos. During these few years the number of men in his class who have since done much to advance science was considerable. The following is by no means a complete list. Among botanists, F. W. Oliver, C. A. Barber, W. B. Bottomley; among geologists, T. T. Groom, P. Lake, S. H. Reynolds, H. Kynaston and H. Woods; among physiologists, pathologists and medical men, A. E. Durham, H. E. Durham, J. S. Edkins, W. B. Hardy, A. P. Beddard, E. H. Hankin, H. Head; and among zoologists, H. Bury, G. P. Bidder, W. F. H. Blandford, R. Assheton, F. V. Theobald, T. H. Riches, E. W. MacBride, H. H. Brindley, A. T. Masterman, C. Warburton, and Malcolm Laurie.

outline. Naturally he turned first to those methods of proof, morphological and embryological, which were being pursued by the biological leaders of the period, and it was only with time that he came to the conclusion that no great progress could be attained by the old methods. We have already seen that even before the appearance of *Natural Inheritance,* Weldon's thoughts were turning on the distribution of variations and the correlation of organic characters. He was being led in the direction of statistical inquiry. The full expression of his ideas is well given in the first part of the "Editorial" with which *Biometrika** started:

"The starting point of Darwin's theory of evolution is precisely the existence of those differences between individual members of a race or species which morphologists for the most part rightly neglect. The first condition necessary, in order that any process of Natural Selection may begin among a race, or species, is the existence of differences among its members; and the first step in an enquiry into the possible effect of a selective process upon any character of a race must be an estimate of the frequency with which individuals, exhibiting any degree of abnormality with respect to that character, occur. The unit, with which such an enquiry must deal, is not an individual but a race, or a statistically representative sample of a race; and the result must take the form of a numerical statement, showing the relative frequency with which various kinds of individuals composing the race occur."

It was Francis Galton's *Natural Inheritance* that first indicated to Weldon the manner in which the frequency of deviations from the type could be measured. A mere catalogue of exceptional deviations seemed to him of little value for the study of Natural Selection. But this description of frequency was only the first stage. How did selection leave the distribution? and How was the intensity of selection to be measured? naturally arose as the next problems. These problems led at once to the even greater question of the influence of selection on correlation. What is the relation between organs in the same individual, and how is this changed, if at all, by the differentiation of species, or at least by the establishment of local races? Nor could the problem of evolution be complete without ascertaining the manner in which deviations were inherited. The modern biometric methods of discussing these problems, if very far from fully developed, were at least suggested in Galton's great work, and that book came as a revelation not only to Weldon, but to others who were preparing to work on similar lines†.

In Plymouth, 1890, Weldon started his elaborate measurements on the Decapod Crustacea and soon succeeded in showing that the distribution of variations was closely like that which Quetelet and Galton had found in the case of man. So far as the present author is aware, the paper "The Variations occurring in certain Decapod Crustacea I. *Crangon vulgaris*" (13) was the first to apply the methods of Galton to other zoological types than man‡. In this paper Weldon shows that different measurements made on several local races of shrimps give frequency distributions closely following the normal or Gaussian law. In his next paper,

* Vol. I. p. 1.

† The present writer's first lecture on inheritance was given on March 11, 1889, and consisted of an exposition and amplification of Galton's theory.

‡ Galton had dealt with the weights of sweet pea seeds, Merrifield with the sizes of moths, but they had not published fitted frequency distributions.

"On certain correlated Variations in *Crangon vulgaris*" (14), Weldon calculated the first coefficients of organic correlation, i.e. the numerical measures of the degree of interrelation between two organs or characters in the same individual. It is quite true that the complete modern methods were not adopted in either of these papers, but we have for the first time organic correlation coefficients—although not yet called by that name—tabled for four local races. These two papers are epoch-making in the history of the science, afterwards called biometry.

It is right to state that Weldon's mathematical knowledge at this period was far more limited than it afterwards became. The first paper was sent to Francis Galton as referee, and was the commencement of a life-long friendship between the two men. With Galton's aid the statistical treatment was remodelled, and considerable modifications made in the conclusions. But the credit of making the vast system of measurements, of carrying out the necessary calculations (now with the aid of his wife, who was for years to assist in this part of the work), of seeing *a priori* the bearing of his results on the great problems of evolution, must be given to Weldon. Nor must we forget the rich suggestiveness of these papers. Weldon was on the look-out for a numerical measure of species. He was seeking for something constant for all local races, and although his suggestion that the correlation coefficient was a constant for local races has not been substantiated—the "selection constant," the quantity uninfluenced by racial differentiation, being of a much more complex nature—yet his suggestion directly led up to the investigation of correlation in man, animals and plants, and has given us immensely clearer ideas on the inter-relationship of organic characters. And Weldon realised this also:

"A large series of such specific constants would give an altogether new kind of knowledge of the physiological connexion between the various organs of animals; while a study of those relations which remain constant through large groups of species would give an idea, attainable in no other way, of the functional correlations between various organs which have led to the establishment of the great sub-divisions of the animal kingdom*."

The defect in mathematical grasp, which Weldon had realised in his first paper, led him at once to seek to eliminate it. He sought first to 'enthuse' a mathematician with his project of demonstrating Darwinian evolution by statistical enquiries. A visit was paid to Cambridge with this end in view, but it did not lead to the required result. Weldon then set about increasing his mathematical knowledge by a thorough study of the great French writers on the calculus of probability. He did not turn to elementary text-books but with his characteristic thoroughness went to the fountain head. Turning over his papers now, it is astonishing to notice the completeness of his studies as evidenced by his notes and abstracts. He thus attained to a very great power of following mathematical reasoning and this power developed with the years. He never reached a high wrangler's readiness in applying analysis to the solution of new questions, possibly this requires years of training in problem papers; but he was able to follow and

* *R. S. Proc.* Vol. LI. p. 11.

criticise extremely complicated algebraical investigations, and to reproduce and often simplify them for the use of his own students. He had, however, a touch with observation and experiment rare in mathematicians. In problems of probability he would start experimentally and often reach results of great complexity by induction. Thus he was able to find out a number of problems relating to the correlation between a throw of n dice, and the result obtained when a re-throw of m out of the number n was made, and others relating to the mixture of n packs of cards and the throwing out of random portions*. In all these cases Weldon was illustrating by a game of chance a definite biological process.

From 1890 onwards, Weldon's knowledge, theoretical and experimental, of the theory of chance increased by bounds. Weldon and the present writer both lectured from 1 to 2, and the lunch table, between 12 and 1, was the scene of many a friendly battle, the time when problems were suggested, solutions brought, and even worked out on the back of the *menu* or by aid of pellets of bread. Weldon, always luminous, full of suggestions, teeming with vigour and apparent health, gave such an impression to the onlookers of the urgency and importance of his topic that he was rarely, if ever, reprimanded for talking 'shop.'

It is difficult now, after fifteen years of common work and continuous interchange of ideas, to distinguish where one or other idea had its source, but of this the writer feels sure, that his earliest contributions to biometry were the direct results of Weldon's suggestions and would never have been carried out without his inspiration and enthusiasm. Both were drawn independently by Galton's *Natural Inheritance* to these problems, but the papers on variation and correlation in shrimps—which in rough outline are types of all later biometry—were published before their friendship had begun.

Weldon's work at University College commenced in 1891. The house in Wimpole Street was taken and, if possible, life became more intense. Easter was spent at Chartres. In the summer came the annual visit to Plymouth, where work on crabs was now to replace that on shrimps. September gave some rest with a sea trip to Malta. In October came the college inaugural lecture for the session, Weldon taking as his subject the statistical treatment of variation. At Christmas there was a break for opera in Munich and Dresden. This year and the next were strenuous years in calculating. The Brunsviga was yet unknown to the youthful biometric school; the card system of correlating variables was still undeveloped, we trusted for multiplication to logarithms and Crelle, and computors trained to biometric work had to be created. The Weldons toiled away at masses of figures, doing all in duplicate. At Easter, 1892, they went to Malta and Naples, and the summer was spent over crab-measurements at the zoological station in the

* In the summer of 1905 a great deal of work was done by the present writer in conjunction with Weldon on mixtures of card packs, the main features of the work having been already outlined by Weldon. The results are summed up in a theory of determinantal inheritance which, it is hoped, will be eventually published.

latter city, and the first biometric crab paper "On certain correlated Variations in *Carcinus moenas*" (15) was issued in this year. In this paper Weldon confirms on the shore crab his results for the common shrimp. The distributions of characters are closely Gaussian with the exception of the relative frontal breadth, which Weldon considered dimorphic in Naples, a problem which led to the present writer's first paper in the *Contributions to the Mathematical Theory of Evolution*. It is right to say that Weldon had reached a moderately accurate solution by trial and error before he proposed the problem to his colleague. He does not refer to this fact in his memoir. As for shrimps the correlations again came out closely alike for the Plymouth and Naples races. Weldon was not dogmatic on the point; he considered the constancy as at least an "empirical working rule" and this it has certainly proved.

"The question whether this empirical rule is rigidly true will have to be determined by fuller investigation on larger samples; but the value of a merely empirical expression for the relation between abnormality of one organ and that of another is very great. It cannot be too strongly urged that the problem of animal evolution is essentially a statistical problem: that before we can properly estimate the changes at present going on in a race or species we must know accurately (*a*) the percentage of animals which exhibit a given amount of abnormality with regard to a particular character; (*b*) the degree of abnormality of other organs which accompanies a given abnormality of one; (*c*) the difference between the death rate per cent. in animals of different degrees of abnormality with respect to any organ; (*d*) the abnormality of offspring in terms of the abnormality of parents and *vice versâ*. These are all questions of arithmetic; and when we know the numerical answers to these questions for a number of species we shall know the deviation and the rate of change in these species at the present day—a knowledge which is the only legitimate basis for speculations as to their past history, and future fate."

These concluding words were surely epoch-making; they formulated the fundamental principles of biometry. We may criticise the memoir in that the index measurements selected by Weldon overlooked the question of spurious correlation, or because the growth law of the indices had not been previously determined. But these are minor matters compared with the general ideas involved in the memoir. It is a paper which biometricians will always regard as a classic of their subject. It first formulated the view that the method of the Registrar-General is the method by which the fundamental problems of natural selection must be attacked, and that is the essential feature of biometry.

Besides biometry a new bond drew Weldon and the present writer together. Since 1884, a strong movement for the reform of the University of London had been in progress, association followed on association, royal commission on royal commission. Few people had distinct ideas of what they themselves wanted, scarcely any one had a notion of what a real university must connote. At University College, after severe crises, the teaching staff had won direct representation on the governing body, and was beginning to insist upon being heard in the question of university reform. One of the most vigorous protagonists in this matter was Lankester, and his removal to Oxford threatened the little group who had definite notions of academic reform with complete defeat. Luckily Weldon joined us and his energy and enthusiasm were of immense service. We had to fight our own College authorities as well as outside influences. It is not now the fitting occasion to tell the complete history. A joint letter to the *Times* roused the authorities, there were rumours of dismissal from chairs, and of wiser counsels prevailing solely at the instance of a distinguished Liberal statesman then on the College Council. The authorities were supporting a scheme which would have united King's and University Colleges in a second-rate and duplicate London University to be termed the Albert University, and rebellion had to go to extremes, if this project was to be defeated. Weldon with the help of one or two colleagues circularised every member of the College, and the night before the discussion of the charter a widely signed petition against it was in the hands of every member of the House of Commons; the Albert charter was dead, and the College Council hopelessly defeated.

The destructive attitude was now dropt; at a meeting in Wimpole Street, Weldon, G. Carey-Foster and the present writer drafted the scheme, afterwards accepted with small modifications, of an "Association for Promoting a Professorial University for London." The idea was to bring all the London teachers into one camp, to get them to accept a common ideal, and to enlist support for it among thinking men outside. The ideal was the foundation of a university in which teaching should be done by the university professors only, who should largely control the university; the separate colleges were to be absorbed. The aim was thus expressed:

"The creation of a homogeneous academic body with power to *absorb*, not to federate existing institutions of academic rank, seems the real solution of the problem. An academic body of this character might well be organised so far as teaching is concerned on the broad lines of a Scottish University. Such a corporation may be conveniently spoken of as a professorial university to distinguish it from a collegiate or federal university."

The Association met a real need, the London teachers to our surprise and joy joined readily. We got the support of great names in literature and science. We produced a distinct effect on public opinion and by our witnesses even on the Royal Commission. But we considered that we ought to have a leader of great name, and we asked Huxley to be president. Huxley accepted, and came to us with views diverging to some extent from those of the initiators of the Association. Instead of holding up an ideal of academic reform, his plan was to find the minimum which would be accepted by various opposed interests and compromise on the basis of this. The alternatives were a long campaign to impress the powers that be with true notions of academic life, or the immediate acceptance of a teaching university, which should be an *omnium gatherum* of all the teaching institutions in London. The present writer resigned the secretaryship of the Association, and was succeeded by Weldon. It was only after very anxious consideration that the open letter of the former to Professor Huxley of December 3, 1892, was sent to the *Times**. It was a course which Weldon strongly condemned.

* *Personal* requests to join the Association had been made to many on the basis of a circular containing the words cited above, the spirit of which was directly repudiated by Huxley.

At a general meeting of the Association held on December 21, 1892, the report of the Executive Committee was received, and after a strong speech from Huxley, adopted. It was then moved by Pearson, and seconded by Unwin: "That the Association trusts that its Executive Committee will persevere in its efforts to establish as far as possibly may be a professorial as distinguished from a federal university." This was carried. At the meeting of the Executive Committee on January 24, 1893, the President presented his own scheme for a teaching University for London; a vague motion to prepare a scheme to be submitted to the Association was, at the instance of Pearson, seconded by Lankester, amended as follows: "That the Committee prepare a scheme to be submitted to the Association in general accordance with the proposals adopted by the Association." This was carried. On January 25, 1893, Huxley wrote withdrawing his scheme on the ground that the amendment moved by Pearson and seconded by Lankester was "incompatible with any progress towards attainable ends." At the following meeting of the Committee in February, Professors Carey-Foster, Rücker, and Pearson were asked to prepare a scheme embodying the principles of the "proposals" of the Association as a basis for the charter of the proposed university. Mr Dickens and Professor Weldon were added to this committee. The scheme was actually prepared and Weldon aided with yeoman service in the drafting of it*. But the influence of the Association was dead; it never recovered from the divisions thus manifested in its executive. The spirit of compromise and the fatal easiness of federation dominated the situation and the present University emerged out of the chaos. No one felt more bitterly than Weldon the contrast between the original ideal and the result achieved. In fact, it is not too much to say that the greatest hopes for the University, and its most progressive steps since its incorporation, lie in the endeavours made to carry out in part the ideal of a homogeneous professorial university, as it was originally developed one Sunday evening in the house in Wimpole Street, and later substantially reproduced in the proposals of the Association.

This account of one movement, however, with which Weldon was closely concerned would not be justified here, did it not illustrate strongly a marked characteristic of the man. He found his great leader attacked, as he and some others believed, unwarrantably. He wrote one very strong private letter on the point, and never referred to the matter again; not the slightest breach was caused in his friendship, and the biometric talks, the common work and plans for work were resumed a day or two afterwards as if no source of friction had for a moment arisen. Yet Weldon always felt deeply, and felt this attack on Huxley more than many men would feel a direct personal attack on themselves.

With the death of Huxley in 1895, the Association practically came to an end. Weldon succeeded his hero in 1896 as Crown nominee on the Senate of the University; here, as on the Board of Zoological Studies of the later reconstituted

* The scheme was printed and adopted by the Association, March 23, 1893.

University, he continued to work and fight for truer ideals of academic administration.

As an administrator and committeeman Weldon combined geniality with strong convictions; he saw at once through flimsy pretexts, and expressed clearly and concisely his own point of view,—"An impulsive loveable man going to the heart of any subject immediately, and always speaking up with great feeling for what he thought right," is how one of his former colleagues aptly describes him. But he lacked several of the essentials which go to make the completely effective committeeman. He was always full of the current piece of research and he grudged all time taken from it; to carry through his own projects he did not adopt the manner of the bull and crush down all opposition; some few men can do this, but it needs not only physique, but its combination with very dominant intellectual power; nor had he the persistency of the corncrake, to wear down his colleagues by continual nagging; nor silent in committee would he molelike be active underneath, "lobbying" his men, and thus more effectually work his will. These types I have known and each was less loved, but more successful than Weldon. He "played the game," threw firmly and well the lance for the cause he thought right, and went his way. He remained to the end the public school or 'varsity lad, whom the idea of "good form" controlled; but unfortunately the type is not so persistent in practical life that it dominates scientific or academic politics. From this standpoint Weldon's death removed from the field a healthy administrator, who acted as a tonic upon weaker colleagues. It was in this sense that he did excellent work, not only on various bodies connected with University College and the University of London, but on the Council of the Royal Society (1896–8), and on its Government Grant and Sectional Committees.

To the biometrician, perhaps, the most interesting committee with which Weldon was associated in these years was that which came later to be called the Royal Society Evolution (Animals and Plants) Committee. It is somewhat difficult to give the full history at present, but some attempt at a sketch of Weldon's connection with it must be made here. Weldon's papers on variation and correlation in shrimps and crabs had brought him closely in touch with Francis Galton, and both were keenly interested in the discovery of further dimorphic forms such as had been suggested by the frontal breadths of the Naples crabs. Weldon was full also of other ideas ripe for investigation. He had started his great attempt at the measurement of a selective death-rate in the crabs of Plymouth Sound; experiments on repeated selection of infusoria were going on in his laboratory; he was gathering an ardent band of workers about him, and much seemed possible with proper assistance and that friendly sympathy which was ever essential to him.

The idea that a group of men can achieve more than a single investigator, if true in some forms of social work, is rarely applicable to scientific committees; but such committees have often been tried in the past, and will no doubt be again attempted in the future. If used as instruments of research, the work done is too often a compromise between different methods and divergent personalities; if merely administrative they are successful or not, according to the width of view of some dominating temperament. If run in the interests of one school, still more of one individual, a committee may no doubt do good work, but it is likely, at the same time, seriously to damage the reputation of any larger body in whose name it works, by too markedly connecting it with one aspect of a problem or one side of an unsettled controversy. These difficulties of the situation seem only by degrees to have come home to the founders of the Evolution Committee.

The project was first discussed informally by R. Meldola, Francis Galton, and Weldon, at a meeting held on the 9th of December, 1893, at the Savile Club. Francis Darwin, A. Macalister, and E. B. Poulton had expressed themselves willing to assist such a project. It was settled that a proposal should be made to the Royal Society for the formation of a committee "For the purpose of conducting Statistical Enquiry into the Variability of Organisms," the members suggested being F. Darwin, F. Galton, A. Macalister, R. Meldola, E. B. Poulton, and W. F. R. Weldon, to whom "it may afterwards be desirable to add a statistician." It was resolved further to ask for a grant of money to obtain material and assistance in measurement and computation.

A Committee* consisting of these members was finally constituted by the Council of the Royal Society, with Francis Galton as chairman and Weldon as secretary, the Committee being entitled: "Committee for conducting Statistical Inquiries into the Measurable Characteristics of Plants and Animals." The use of the words *statistical* and *measurable*, somewhat narrowly, but accurately, defined the proposed researches of the Committee. It went on until 1897, with these members, the same title and scope. Then in the early part of that year its scope was much extended by adding to its objects the "accurate investigation of Variation, Heredity, Selection, and other phenomena relating to Evolution," and W. Bateson, S. H. Burbury, F. D. Godman, W. Heape, E. R. Lankester, M. Masters, Karl Pearson, O. Salvin and Thiselton-Dyer were added to its number. But at present our account must deal with the earlier biometric period of the Committee. Looking back on the matter now, one realises how much Weldon's work was hampered by this Committee. It is generally best that a man's work should be published on his own responsibility, and when he is a man of well-known ability and established reputation, grants in aid can always be procured. In this case Weldon had a sympathetic committee, but the members were naturally anxious on the one hand for the prestige of the Society with whose name they were associated, and secondly, they were desirous of showing that they were achieving something†. Both conditions were incompatible with tentative researches such as biometry then

* First meeting, January 25, 1894.

† "Of course these considerations only make the problem more interesting than it was before: and I very much want to solve it. But the committee may say that it requires a problem which is reasonably certain to yield an adequate solution in a fairly short time, and that so risky an attempt as this is not suitable for its present work." Letter of Nov. 13, 1894, relating to the secretion of a specific poison by *Daphnia*.

demanded. Trial and experiment were peculiarly needful in 1893; the statistical calculus itself was not then even partially completed; biometric computations were not reduced to routine methods, and the mere work of collecting, observing, experimenting, and measuring was more than enough for one man. Weldon with his "volume of life" was eager to do all these things, and run a laboratory with perhaps sixty students as well. He was impatient because the probable errors of biometric constants, on which tests of significant differences depend, were not at once forthcoming; he wanted the whole mathematical theory of selection, the due allowances for time and growth, the treatment of selective death-rates and the tests of heterogeneity and dimorphism settled in an afternoon's sitting. The Committee did not possess a mathematician to put on the brake, and Weldon attempted too much in too short a time. Each week Weldon had new and exciting problems, he thrust them upon his friends, demanded solutions, propounded solutions, and was never discouraged when difficulties were pointed out and time asked for.

One of the first subjects to be taken up by the new Committee was to test whether the method of resolution into two Gaussian curves, which suggested dimorphism in the Naples crabs, would be helpful in confirming a similar dimorphism said to exist in the herring. Several thousand herrings arrived at University College, a measurer was trained to deal with them, and the variability of a wide series of characters determined. The distributions came out skew, and Weldon was intensely hopeful that statistical evidence of dimorphism would be forthcoming. Instead of this, the analysis showed dimorphic Gaussian components to be impossible. This result was a great disappointment to him, and, I believe, to the Committee. I could never understand why. A most extensive and valuable series of measurements had been made, which in themselves were well worth publishing. It had been shown that simple dimorphism of a Gaussian kind certainly did not hold for these herrings; in all probability it was a typical case of skew frequency, which would have been most valuable as adding to the known instances, and aiding statisticians eventually to classify such occurrences. But Weldon, and, I presume, the Committee were disheartened, they had been searching for dimorphism and had not found it. The herring data were put on one side by Weldon, and as far as I know have never been published. It is much to be hoped that they may some day be resuscitated from the archives of the Committee (16).

The next point that I personally became aware of in relation to the Evolution Committee was Weldon's attempt to solve the problem of subraces in the case of the ray florets of ox-eyed daisies. I am unaware who brought the material before the Committee, but it was obviously heterogeneous in the highest degree. There was no evidence at all that any attempt had been made to allow for seasonal and environmental effects, and whatever truth there may be in a tendency of the modes to fall into Fibonacci groups, we now know that varying season and period will produce within a certain range almost any mode in this flower*. To break

* *Biometrika*, Vol. I. pp. 305, 309 *et seq.*

up such a heterogeneity even into Gaussian components was a problem not then solved, and one which has not since been solved. It was cruel fate that thrust such a problem on Weldon, and kept him over it for weeks. He was struggling with most highly complex mathematical difficulties, and actually beginning with a problem which a more highly trained mathematician would certainly have put on one side in the then state of statistical analysis*.

The next portion of the Committee's work was far more successful—the "Attempt to Measure the Death-rate due to the Selective Destruction of *Carcinus moenas*, with respect to a Particular Dimension" (17). This formed the first report of the Committee, and was presented to the Royal Society in November, 1894. Weldon's general project in this case was, I believe, absolutely novel at the time, and embraces, I consider, the best manner still of testing the truth of the Darwinian theory. It consists in determining whether the death-rate is correlated with measurable characters of the organism, or, as he himself put it, "in comparing the frequency of abnormalities in young individuals at various stages of growth with the frequency of the same abnormalities in adult life, so as to determine whether any evidence of selective destruction during growth could be discovered or not."

Thus stated the problem might appear an easy one, but it is the very reverse. How is the 'abnormality,' i.e. what we should now term the deviation from type, to be measured at each stage of growth? What is to determine 'adult' life? What measure is there of the time during which the individual adult life has been exposed to the selective destruction? Weldon undoubtedly chose the crab because of the facilities it offers for measurement. But its age then becomes an appreciation based merely on the obviously close, but probably imperfect correlation between age and size. Further, the law of growth, complicated rather than simplified by the moults, and the question as to how far the variability of the characters dealt with is affected by growth combine, in the case of crabs, to form an exceedingly difficult problem. It is practically impossible to keep a sufficiently large series of crabs through the whole period of adolescence, and if it were possible, it is far from certain that the claustral environment necessary would not sensibly affect their law of growth.

Looking back now on Weldon's paper of 1894, one realises its great merits; it formulates the whole range of problems which must be dealt with biometrically before the principle of selection can be raised from hypothesis to law. Almost each step of it suggests a mathematical problem of vital importance in evolution, which has since been developed at length, or still awaits the labour of the ardent biometrician. On the whole, I think, Weldon came very near to demonstrating his point, but whether he did or did not scarcely affects the suggestiveness of the paper†.

* We now know that some of the most skew distributions are given by the parts of flowers, and the problem propounded to Weldon was to resolve into a number, probably five or six, of such skew components a strikingly irregular frequency distribution for ray florets!

† Reading through the criticisms I communicated to him at the time, criticisms written purely from

Unfortunately the paper, as well as the suggestive "Remarks on Variation in Animals and Plants" (18), with its memorable words:—"The questions raised by the Darwinian hypothesis are purely statistical, and the statistical method is the only one at present obvious by which that hypothesis can be experimentally checked"—fell on very barren soil. The paper produced a mass of criticism—folios were written to the Chairman of the Committee, showing how this, that or the other vitiated entirely the results. The very notion that the Darwinian theory might after all be capable of statistical demonstration seemed to excite all sorts and conditions of men to hostility. Weldon, instead of being allowed to do his own work in his own way, had to be constantly replying to letters, some even eighteen sheets long, addressed to the Chairman of the Committee. These letters were not sympathetic and suggestive, but mostly purely controversial. The need for further investigation of the law of growth had been frankly admitted by Weldon in the "Remarks" issued at the discussion on the "Report," but the critics declined to wait for answers till further results were published. This attack lasted for the next three years, during which further researches on the selective death-rate and growth of crabs were carried out, and it formed a serious impediment to calm progressive investigation. A further instructive report (19) on the growth at two moults of a considerable number of crabs was made to the Committee in 1897, but I believe has never been published. Later, an account of work on Natural Selection in crabs was given by Weldon in his "Presidential Address to the Zoological Section of the British Association," Bristol, 1898 (20).

In this paper Weldon returns to the problem of whether frontal breadth in crabs is correlated with a selective death-rate, but he now deals with type and not variability. He first approaches the problem from the consideration of whether for this character the crabs in Plymouth Sound are remaining stable, and he shows from measurements made by Sir Herbert Thompson and himself during the years 1892 to 1896, that the population is unstable. He next seeks a cause for this secular change, and he finds it in the turbid state of the water in Plymouth Sound, due to the continual carriage into it of large amounts of china clay and sewage. Direct experiments were then made on the selective death-rate of crabs kept in water with suspended china clay and on another occasion in foul water. In all cases the survivors were found to have a smaller frontal breadth relatively to their carapace length. Confirmatory experiments showed that after the first shock of confinement was passed this selection did not occur among crabs kept in pure sea water. A reasonable explanation of this selective action was provided in the character of a crab's breathing apparatus. Thus, after several

the mathematical standpoint, I still think them valid, but I realise also how much of my own work flowed directly from the suggestiveness of this paper. In fact it was the starting-point of the whole of the work on the influence of selection on the correlation and variability of organs. The sequel to that work, the influence of selection during growth, flows equally from Weldon's paper, but although we know much more than we did ten years ago as to the laws of growth, no sufficiently general formula of growth can yet be applied to allow of the completion of Weldon's work in this direction.

A "Crabbery" at Plymouth.

Plate IV

years of discouragement and much hard labour, Weldon succeeded in demonstrating that natural selection was really at work, and further that it was at work at a very sensible rate*. The labour involved was excessive. One "crabbery" consisted of 500 wide-mouthed bottles, each with two syphons for a constant flow of sea water, and each crab had to be fed daily and its bottle cleaned. During the summer of 1897 Weldon spent the whole of his days at the aquarium, and his wife hardly left him except to fetch the needful chop. The sewage experiment was "horrible from the great quantity of decaying matter necessary to kill a healthy crab." In 1898 the china clay experiments were continued at Plymouth. But in the autumn a rest came. The Address was written and Weldon thoroughly enjoyed his presidency of Section D of the British Association at Bristol.

It may not be out of place here to note the great aid Weldon's artistic instinct and literary training gave to his scientific expression. His papers are models of clear exposition, his facts are well marshalled, his phraseology is apt, his arguments are concise, and his conclusions tersely and definitely expressed. The result, however, was not reached without much labour. I do not mean that it was an effort to him to write well and clearly, but that his standard was so high, that having written a memoir, he would to please his own sense of the fitting rewrite the whole of it and possibly redraw all the diagrams. Nor was the remodelled memoir necessarily in its final form. A third or fourth reconstruction might follow to satisfy his own standard of right expression. To him a paper was a literary whole, which had not only to convey new facts, but to play its part on the scientific stage,—and he was not satisfied until it was in his judgment artistically complete. There was never any artificial brilliancy introduced in the process; rhetoric in the service of science was intolerable to Weldon. It was simply an attempt to choose the suitable form and the right words for a given purpose. It was comparable with Weldon's sense of sound, with his extraordinary gift of appreciating and reproducing the exact intonation of a foreign tongue. Both were the result of observation and experiment—not manifest in the final product—guided by a trained artistic sense.

Considerable changes were soon to take place in Weldon's environment and scheme of work. Lankester had been appointed director to the British Museum (Natural History), and in February, 1899, Weldon succeeded him in the Linacre Professorship at Oxford. In the February of 1897 the Royal Society Evolution Committee received a large increase of membership; it ceased henceforth to "conduct statistical inquiries into the measurable characteristics of plants and animals." It became transformed into an Evolution (Plants and Animals) Committee. At first there were great hopes of achievement, there was a possibility of securing Charles Darwin's house as a centre for breeding experiments, and a considerable sum of money was promised in aid. Francis Galton struggled bravely for a great idea. He wanted to see the numerous bodies engaged in horticulture

* The 60,000,000 years or so, which the physicist then allowed the evolutionist, were at that time a little more of an incubus than they are now!

and zoology coordinated in at least one aspect of their work, and that research of a scientific kind should be introduced into the proceedings of each of them. He strove to make two schools, widely diverse in method and aim, understand each other. He wanted to keep individuals and societies up to their work, and prevent overlappings. But it was not to be. The members were pulling in opposite directions, there was too much friction, and too little compromise. A false antithesis was raised between what was termed "natural history" and any sort of statistical inquiry leading to numerical results. The biometric members ceased to attend regularly and finally resigned towards the end of 1899. Thiselton-Dyer and Meldola also left the Committee, which became from that date confined to one special school and one limited form of investigation. From beginning to end the Committee has, in the opinion of the present writer, been a mistake; not only because at first it distinctly forced the pace and hampered Weldon's work, but because experience shows that such a committee can only work effectually in the interests of one school of ideas, and this, whatever safeguards may be taken, has at least the appearance of destroying the impartiality of the parent body, a matter of very grave importance.

During the eight years of Weldon's London professoriate his development was great; he became step by step a sound mathematician, and gained largely in his power of clear and luminous exposition. His laboratory was always full of enthusiastic workers, and over forty memoirs were published by his students, who included E. J. Allen, E. T. Browne, F. Buchanan, G. H. Fowler, E. S. Goodrich, H. Thompson, E. Warren, and others of known name. The following lines, provided by a friend, graphically recall Weldon in his early London days:—

"In so vivid a personality it is hard to point to the period of greatest mental activity, but of the nineteen years in which I knew him I should select the first few years of his Professorship at University College, London. Fresh from contemplative research at the Plymouth Laboratory of the Marine Biological Association, and with his mind full of the new problems to which the study of marine life had introduced him, he threw himself into teaching with renewed zest. The effect on his students was amazing; most of them began a zoological course as a compulsory but annoying preliminary to a degree; Weldon soon changed that. Without ever forgetting the requirements of examinations, he made the subject alive and absorbing; his advanced classes soon filled up; and while on the one hand, the scholarships at London University were always claimed by his students, on the other the output of original investigation published by his department was one of which no university need have felt ashamed. Besides all this his students loved him; he was so intensely human....Into the question of remodelling the University and the defence of his College, Weldon threw himself as if unencumbered with arduous teaching and research; his notably lofty ideals and vigorous championship were far from being wasted; but his removal to Oxford at the time of the birth of the new University was a severe loss to the cause of real education in London. Gentle with ignorance, he was fiercely intolerant of educational shams and cants."

As the present writer has indicated, the stress during these London years was very great—the struggle with new mathematical processes, the wear of incessant calculation, the worry of unending controversy to a man fully occupied with research and teaching, all told on Weldon. The holidays were more limited in extent, but were very varied in character. In 1893, Easter was spent in the Sieben Gebirge; the Weldons were up at six, calculating till one, and starting a great tramp at two, from which they returned at eight. The autumn they spent in Venice, going by sea, and the Christmas at Brussels, with opera each night and walks to Waterloo most days. In 1894 it was Siena for the Palio, with a knapsack tour from Stresa to Alagna by Orta and Varallo. In 1895 Easter found them fossil-collecting in the Eiffel, and, after the hard summer at Plymouth, in the Apennines, winding up in Florence. Bicycling was the rule in 1896, even cycling from Wimpole Street to Plymouth, and the only holiday a cycling tour in Normandy. In 1897 there was an Easter visit with architectural sketching to the cathedrals of North-East France, and after the specially hard summer at Plymouth a trip to Perugia and a return from Genoa by sea. The last year in London included a butterfly and moth collecting expedition to Ravenna at Easter, but no summer holiday abroad; the British Association, followed by a study of Wells Cathedral, occupied its place. The restlessness of work seemed to have overflowed into the holidays, and Weldon's friends knew that it was telling upon him, and trusted that Oxford life might be quieter than the London life had been.

VI. *Oxford and the Second Professoriate,* 1900—1906.

The removal of Weldon from the London field of work, while an incalculable loss to his colleagues, was not without compensation to his nearest friends. They knew that the life of the last few years had been one of great tension, that Weldon's time had been too much encroached upon by committee work, that the separation between the locus of his teaching and of his research work was very undesirable, that even the social life of London involved too much expense of energy. Oxford, in some respects, would present a narrower field of administrative duties; it would provide a roomy and amply equipped laboratory, where experiments hitherto shared between Plymouth and Gower Street could be carried out, and remain under control while ordinary teaching work was going forward. Even the social life in Oxford had more regular hours and was less over-stimulating. It is true that Weldon occasionally regretted the contact with many minds working on kindred topics, and even the stimulus of keen men working on quite different subjects, which is characteristic of the metropolis. He would speak with great affection of "dear old Gower Street, where everybody was working and everybody wanted to work"—and he would be vexed that so many of "these nice Oxford boys" had no *res angusta domi* to force them from the river and the playing field into the laboratory and the lecture room. "They are so nice, they come to my lectures because they think it would be rude to leave me alone." The lad who would not make a sacrifice to his love of science—accept an Asiatic appointment of the merest bread and butter value, or take passage in a tramp steamer to collect in South America—was anathema to him. He wanted everywhere an infant Huxley, realising the value of tropical or semi-tropical observation and experience and anxious to seize the opportunity of it at any slight personal inconvenience. Weldon did not grasp that it was largely his own personality which had created the band of earnest workers round him in London, and that with time it would be effective in more conservative Oxford. He did not realise that the over-stimulus of the London period, with its midnight hours and incessant interchange of ideas, would be better replaced by the more leisurely intellectual and more regulated social life of Oxford.

There is another point which emphasises the value of this change. Weldon's taste, his whole emotional nature, made him essentially a field naturalist. It was no innate taste for figures or symbols, no pleasure in arm-chair work, which drew him to statistical research. Nor was it the influence of any personality. On the contrary, he was impelled to it by the feeling that no further progress with Darwinism could be made until demonstration from the statistical side was forthcoming. His biometric friendships arose from the direction he felt his work must take. He distrusted mathematicians as much as any good Mendelian might do; they were persons who neither observed nor experimented, who had "a true horror of a real measurement." Acceptance of each stage of biometric theory could only be won from Weldon by a tough battle; it had first to justify its necessity, and next to justify its mathematical correctness. He was not drawn into actuarial work by his sympathies or his friendships, he was *driven* into it by the looseness he discerned in much biological reasoning; he felt an *impasse*, which could only be surmounted by the stringency of mathematical logic. Those who have known Weldon collecting on the shore, dredging at sea, or in later days sampling ponds and wells for his *Crustacea* book, photographing snail environments in Sicily, or hunting for *Clausilia* in the woods at Risborough or Plön, realise that he was in the first place the open-air naturalist. If further evidence be indeed wanting, let the following words provide it:

[*April*, 1903.]

"Just back, and have just read your letter. I will play with the spanner and talk of it to-morrow.

I did not telegraph because our office was shut. It was a great disappointment to miss you; but the ride was the one thing I enjoyed out of the last three weeks. I have felt nothing like it since the old days when I used to lie in a fishing boat dodging the squalls off Rame Head or the Deadman, when we were all young and arithmetic was not yet. That is all gone. The good old man I used to sail with went to haul lobster pots in one of the March gales, and his boat was found bottom upwards.

He was a good soul. 'Yes, my dear,' he used to say in a breeze; 'we'll shake out all them reefs if you like. You'll get wet, but I'm only a fisherman and wet don't hurt me.' Then he would sing Devonshire songs while the water came over the gunwale, till you went on your knees to him to ask for at least one reef back again.

Really, even Basingstoke railway station looked good with the squalls climbing round it. *Ride* home. It will do you no end of good. Go by Farnham, Basingstoke, not by Guildford. Sandro and I rode home to-day. We had no snow, and no rain, and not half the fun of Monday. A sober, middle-aged ride on a good road in good weather.

Nevertheless, my head is so full of chalk-downs and clouds, and things, I can't write biometry to-night. Always, when I have been with the country, the feeling breaks out that the other folk have the best of it. The other way you live with the country and become part of it; and you dredge, or fish, or shoot something wonderful, and you describe it, and everyone sees that it is wonderful, and you all enjoy the wonder. And there is no solution, and if there were, it would not be worth the shadow of a shower flying across the country.

And this is all wicked nonsense, and I am going to bed. Yours affectionately,

W. F. R. WELDON."

Weldon was a child of the open air and the breezes, and we hoped that he might have more of them, if not in lowland Oxford, at least on the hills around. There was space and air too for the experimental work that had been so cramped in Gower Street. The *Daphnia* studies, which had occupied so much energy under unfavourable conditions in London, were at once resumed on broader lines in the ponds and ditches round Oxford. Weldon, with a basket of bottles attached to his cycle handle, and a fishing creel, filled with more bottles, on his back, might be met even as far as the Chilterns, collecting not only *Daphnia*, but samples of the water in which they lived. His University College work had shown him how widely *Daphnia* are modified by their chemical and physical environment, and how this modification is largely due to selection. There exist elaborate drawings of the *Daphnia* from the Oxfordshire ponds, indicating their differentiation into local races, and notes on the peculiarity of their habitat and the chemical constitution of the water:

"In the meantime I have been led into a non-statistical work for the moment. Get out of the library and read Klebs: *Bedingungen der Fortpflanzung bei einigen Algen und Pilzen.*

By tricks of nutrition, light, etc., Klebs can make simple algae reproduce either a-sexually or sexually, or parthenogenetically, as he pleases. In cases where every textbook tells you that a regular alternation of sexual and a-sexual generation is the rule, he can make *either* form recur as often as he likes.

If one can by similar tricks throw *Daphnia* into this condition, then the measuring machine can again come into play, and one can compare parthenogenetic inheritance with sexual inheritance as often as one pleases.

That is a *Nebensache.*—The *Hauptsache* here is the great variation in the chemical composition (pardon the phrase) of the water in the little rivers. Their percentage of dissolved salts varies enormously, and I hope to go about as I have begun, with a large fisherman's creel tied to the handle-bar of my bicycle, learning the correlation between the salts in the waters and the fauna.—Then again comes the measurement, and the attempt to derive one local form from the other under controlled conditions by direct selective destruction due to the conditions."

This was precisely the same problem which a study of Kobelt's *Studien zur Zoogeographie*, 1897–9, led him later to take up with regard to land snails. What is the meaning of the slight but perfectly sensible differences in type to be found in shells from adjacent valleys, or even from different heights of the same mountain? Weldon attacked the problem in his usual manner; he spent two Christmas vacations collecting Sicilian snails of the same species from habitats extending over a wide area, the local environments were described, and the snails often photographed with their immediate surroundings. Innumerable shells were brought back to Oxford, and Weldon delighted to discourse on the significant differences in local type, and yet the gradual change of type to type from one spot to another. No rapidly made measurement on the outside of the shell would satisfy Weldon; the shell must be carefully ground down through the axis, and measurements made on the section thus exposed. Perhaps four or five snail shells could be ground and measured in a day, and at the time of his death, not more than a few hundred of the Sicilian thousands had even been ground. Like the *Daphnia,* the Sicilian snails remain as an indication of the way—the path of absolute thoroughness—the master would have us follow. "Life is not long enough for biometry," murmurs the superficial critic. But the man of deeper insight replies:

> That low man goes on adding one to one,
> His hundred's soon hit:
> This high man, aiming at a million,
> Misses an unit.

But these attempts to get to the kernel of selection in its action on local races were far from occupying the whole of Weldon's thoughts in these early days. In conjunction with his assistant, Dr E. Warren, he had commenced at University College his first big experimental investigation into heredity.

"The Oxford rivers have had to rest during the last few weeks, because of the pedigree moths. These are apparently going on very well indeed. There are at present about 3500 caterpillars, belonging to thirty-eight families forming the third domesticated generation."

The characters to be dealt with consisted of the number of scales in particular colour patches, and the work of counting these was very laborious. A little later (16 July, 1899) Weldon writes:

"The caterpillars are hatching by hundreds and I hope the clean air will help them to do better here than in London. From egg to moth, poor Warren, in spite of magnificent efforts, had a death-rate of over eighty per cent.; and that seems to me a rather serious thing... because one cannot be sure that the death-rate was not partly selective with respect to things in the caterpillar which are correlated with colour in the moth. The influence of climate is shown by the fertility of the eggs—Warren got forty per cent. of fertilised eggs from his pairing and an average of over one hundred eggs per batch. Nearly all my pairs lay fertile eggs and those I have counted give an average of one hundred and sixty-five eggs per batch."

And again, on the 14th August of the same year:

"I want to come and talk to you, especially about death; but I cannot come till my caterpillars are safely turned into pupae. For the sake of these caterpillars I have, at the risk of personal liberty and reputation, stolen from the roadside one hundred square feet of clover turf, the property of the Lords of various Manors in this neighbourhood. The little ruffians have now eaten all this clover and for the last day or two of their existence have to be fed by hand.—Therefore I have to pluck fresh clover (which is not stealing if you do not do it in an enclosed pasture) every day.—My bicycle is nearly worn out from carrying extra weight. Riding down a steep hill, with your brake smashed, and with five or six feet of heavy turf on your back is like playing at Attwood's machine. You get very near to the theoretical acceleration too!"

In the course of three years many hundreds of pedigree moths were dealt with and the observations were reduced. But *no definite inheritance at all of the character selected for consideration was discovered.* Weldon, I believe, thought that there had been some fatal mistake in the selection of pairings, and undoubtedly in some cases parents of opposite deviation had been mated, so that a rather influential negative assortative mating resulted. But from other series of pedigree moth data that I have since seen, it seems to me probable that there is some special feature in heredity in moths, or possibly in those that breed *twice* in the year, and that the vast piece of work which Weldon and Warren undertook in 1898—1901, may still have its lesson to teach us. At the time it formed another link in that chain of apparent failures which for a time, but only for a time, disheartened Weldon.

In these three first years at Oxford, Weldon's intellectual activity was intense. The letters to the present writer, which in 1899 averaged one a week, in 1900 and 1901 reached an average of two, and in some weeks there were almost daily letters. These letters not only teem with fruitful criticism and suggestion with regard to the recipient's own work, but contain veritable treatises—drawings, tables, calculations—on the writer's own experiments and observations. To the pedigree moth experiments was added in the summer of 1900 an elaborate series of Shirley Poppy growings, 1250 pedigree individuals being grown and tended in separate pots; Weldon's records were the most perfect of those of any of the cooperators, and his energy and suggestions gave a new impetus to the whole investigation. They were ultimately published in *Biometrika* under the title, *Cooperative Investigations on Plants*, I. *On Inheritance in the Shirley Poppy* (22). As Weldon himself expressed it, the moths and poppies meant "a solid eight hours daily of stable-boy work through the whole summer, and through the Easter vacation, with decent statistical work between." The autumn of 1899 provided no proper holiday, but Christmas found the Weldons in Rome. After the Shirley Poppies were out of hand in the summer of 1900, the Weldons went to Hamburg and thence to Plön. The object of this visit was to collect *Clausilia* at Plön and Gremsmühlen for comparison with the race at Risborough. The same aim—the comparison of local races—led Weldon at Christmas to collect land snails in Madeira. Thus he slowly built up a magnificent biometric collection of snail shells—i.e. one sufficiently large to show in the case of many local races of a number of species the type and variability by statistically ample samples. Of this part of Weldon's work only two fragments have been published, "A First Study of Natural Selection in *Clausilia laminata* (Montagu)" (23), and "Note on a Race of *Clausilia itala* (von Martens)" (24). In the first of these memoirs Weldon shows that two races of *C. laminata* exist, in localities so widely separated as Gremsmühlen and Risborough, with sensibly identical spirals, although no crossing between their ancestors can have existed for an immense period of time, and although there are comparatively few common environmental conditions. At the same time, while no differential secular selection of the spiral appears to have taken place during this period, there yet seems to be a periodic selection of the younger individuals in each generation, the variability of the spirals of the young shells being sensibly greater than that of the corresponding whorls of adults. In other words stability to the type is preserved by selection in each new generation.

In the second memoir Weldon sought for demonstration of a like periodic selection in the *C. itala* he had collected from the public walks round the Citadel of Brescia. He failed, however, to trace it, and was forced to conclude that *C. itala* is either not now subject to selective elimination for this character, or is multiplying at present under specially favourable conditions at Brescia, or again, as both young and old were gathered in early spring, after their winter sleep, that elimination takes place largely during the winter, and "that individuals of the same length, collected in the autumn, at the close of their period of growth, might be more variable than those which survive the winter."

Quite apart from the results reached, Weldon's papers are of the highest suggestiveness. Does selection take place between birth and the adult or reproductive stage? This is the problem which everyone interested in Darwinism desires to see answered. But to answer it we need to compare the characters of the organisms at the same stage of growth, for these characters are modified by growth. How is it possible to compare a sample of the race at an early stage, with its adult sample? The problem of growth, to be studied only under conditions of captivity, possibly modifying the natural growth immensely, had made the crab investigation an extremely complex one. Weldon solved the difficulty by the brilliant idea that the snail carries with it practically a record of its youth. If the wear and tear of the outside of the shell to some extent confuses the record there, a carefully ground axial section will reveal by the lower whorls the infancy of the organism. Hence the days given to experimental grinding, the training in manipulation and the final success, and then the steady work, grinding and measuring a few specimens a day, till the necessary hundreds were put together; the laborious calculations not in the least indicated in the papers—the arithmetical slips with bad days of depression, and the completed result: the illustration of how shells may be used—by those who will give the needful toil—to test the truth of the Darwinian theory.

The summer of 1899 found the present writer at Great Hampden on the Chilterns, working at poppies and developing a theory of homotyposis, namely, the quantitative degree of resemblance to be found on the average between the like parts of organisms. Weldon, who came over from Oxford to dredge the ponds and to discover *Clausilia* by the White Cross at Monks Risborough, provided the criticism, suggestion, and encouragement, in which he never failed*:

"You have got hold," he wrote, after returning to Oxford, "of the big problem which all poor biologers have been trying for ever so long. I wish you good luck with it."

The collection and reduction of material were on a larger scale than had been previously attempted, and the memoir was not presented to the Royal Society until the October of the following year (1900). It was soon evident that the attitude of the Society with regard to biometry was undergoing considerable change. The meeting of November 15 and the discussion that took place on the

* His aid in the second part—Homotyposis in the Animal Kingdom, shortly to be published,—was even more substantial.

homotyposis paper was the immediate cause of the proposal to found this Journal. A little later a detailed criticism of the paper by one of the referees was actually printed by the Secretaries and issued to Fellows at a meeting, before the fate of the criticised paper itself had been notified, and before the paper itself was in the hands of those present. This confirmed the biometric school in their determination to start and run a journal of their own.

On November 16 Weldon wrote:

"The contention 'that numbers mean nothing and do not exist in Nature' is a very serious thing, which will have to be fought. Most other people have got beyond it, but most biologists have not.

Do you think it would be too hopelessly expensive to start a journal of some kind?......

If one printed five hundred copies of a royal 8vo. once a quarter, sternly repressing anything by way of illustration except process drawings and curves, what would the annual loss be, taking any practical price per number?... If no English publisher would undertake it at a cheap rate, the cost of going to Fischer of Jena, or even Engelmann, would not be very great."

This was the first definite suggestion of the establishment of *Biometrika*. On November 29, the draft circular, corresponding fairly closely to the first editorial of the first number (25), reached me from Oxford with the words: "Get a better title for this would-be journal than I can think of!" The circular went back to Oxford with the suggestion that the science in future should be called Biometry and its official organ be *Biometrika*. And on December 2nd, 1900, Weldon wrote:

"I did not see your letter yesterday until it was too late for you to have an answer last night. I like *Biometrika* and the subtitle. Certainly we ought to state that articles will be printed in German, French, or Italian. One may hope for stuff from anthropologers, and —— for instance, ought to be allowed to use his own tongue."

Thus was this Journal born and christened. The reply to circulars issued during December was sufficiently favourable to warrant our proceeding further. A guarantee fund sufficing to carry on the Journal for a number of years was raised at once; good friends of Biometry coming forward to aid the editors. By June of 1901 its publication through the Cambridge University Press had been arranged for, and the sympathetic help of the Syndics and the care given by the University Printers enabled us to start well and surmount many difficulties peculiar to a new branch of science*. Those of us who believe that *Biometrika* came to stay and to fulfil a definite function in the world of science hope that the name of the man who first formulated a definite proposal for a biometric organ may always continue to be associated with our title-page. During the years in which Weldon was editor he contributed much, directly and indirectly, to its pages. He was referee for all essentially biological papers; and his judgment in this matter was of the utmost value. He revised and almost rewrote special articles. He was ever ready with encouragement and aid when real difficulties arose. For the mechanical labour

* A special feature has been of course the masses of tabulated numbers. It deserves to be put on record that on more than one occasion 15 or 20 pages of figures have been set up without a single printer's error.

of editing, for proof-reading, for preparation of manuscripts and drawings for the press, or for interviews with engravers, he had little taste or time. He was too full of his current problem to undertake work of this kind regularly; proofs might remain for weeks unopened, until the number was printed off, and manuscript might disappear into a drawer, when the co-editor imagined it was safely on its way back to the author! But Weldon was always delightful when such "laches" were discovered; to meet Weldon when he was in an apologetic frame of mind meant that you must apologise to him yourself for the very thought of scolding him! It was all over before he had shaken hands, sat down, and lighted the inevitable cigarette; you were not talking of proofs, but of Kobelt, Mendel, Maeterlinck, the *Kritik der reinen Vernunft*,—anything and everything but dull editorial matters. And you felt a freshness and a tonic, and a sense of the healthy joy and pleasure of life, and you wondered how it was possible to do anything but love this man and rejoice in the clearness of his vision and the suggestiveness of his thought.

Starting on October 16, 1900, and extending throughout the early *Biometrika* letters, runs a flood of information with regard to Mendel and his hypothesis.

"About pleasanter things, I have heard of and read a paper by one, Mendel, on the results of crossing peas, which I think you would like to read. It is in the *Abhandlungen des naturforschenden Vereines in Brünn* for 1865. I have the R.S. copy here, but I will send it to you if you want it." [*October* 16, 1900.]

Then follows a resumé of the first of Mendel's memoirs, and for months the letters—always treatises—are equally devoted to snails, Mendelism and the basal things of life. It is almost impossible to give an idea by sampling of the crush of keen and vital interest these letters represent. Some attempt must, however, be made:

"Have you ever been up here? It is not at all a bad little country when you are tired.—We started simply to see the architecture at Lübeck, because neither of us knew the North German brick and wood church work. That was very interesting, then we came here for fresh air and quiet—and we found SNAILS.

I have rather more than 5000 snails all properly pickled, with localities recorded.......There are so many points about snails, if one could only measure and breed them!...Also I have been greatly impressed with the way in which they are dependent upon conditions of environment, so that one quickly learned to know almost exactly what species one would find in any place one passed through. I think that by going from here, which is almost the eastern limit of several species, to a very different country, such as Oxford, one might almost hope to make a good shot at some of the essential conditions which determine the distribution of some of the species.... It is ridiculous that such abundant material as snails afford everywhere (except at Danby End?) should be left useless because one cannot see how to take advantage of it. Send me some "tips" for trials. (To Oxford,—we go home to-day)." [Plön, 5/9/90.]

"You ought to see Lübeck some day. You know so much about German art that I suppose the pathetic ugliness of it does not hurt you any more?......

You can't get a beautiful art in a climate where people must wear clothes. Just as the northern idea of a portrait is a round face stuck on top of a heap of fine clothes, so the northern idea of a building is a thing with all its good simple lines disguised by silly excrescences. If you want to see really majestic brickwork go to Siena or Pistoja, where you can see naked men and women in the streets on a summer's day. Lübeck is very earnest, and very interesting, and so on and so on, but it is *not* beautiful!... I am sending you a parcel of snails that you may see the sort of thing.

In one beech wood, on the trunks of the trees, we collected rather more than 3000 snails, most of them *Clausilia biplicata* and *C. laminata* (see the parcel), but some *Helix lapidicida.* There are certainly one thousand each of the two *Clausilia* species from this locality, and four or five hundred of each from another wood."

Then follow pages of minute description of each type of shell in the parcel and discussion as to the possibility, by grinding or by taking a melted paraffin cast of the inside, of measuring biometrically this or that character.

"I fancy want of moisture must have more to do with the absence of snails about you than want of chalk. What are you on? Surely you have nearly the same Oolites and Lias that we have here?...... Have you committed the sin of digging up a bit of your moor, and looking among roots?" [Oxford, 13/9/1900.]

"A happy New Year to you! I send in another envelope specimens of the problematic snail, which has been found in sufficient numbers already. The pattern cannot, I think, be treated as due to lines of growth, and I hope it will be possible to find some way of estimating its variability.—It occurs more or less in a whole series of species here, and here only; and the hills here are so separated by deep valleys, with great climatic differences at different elevations, that there are well isolated local races.

It is rather hard for me to collect many races, because I have to look after my sick laboratory boy, and to teach him sea work, which takes time and produces only isolated examples of pretty museum things, which are a joy to see, although they teach one very little*. To do this and also to find time for a 2000 or 3000 feet climb after snails makes a very good day, and one goes to bed very fit, and full of beautiful remembrances. As one walks up hill, the impression is very absurd. Here the garden is full of bananas and sugar canes; in an hour's climb one gets into a wood of pine trees and heather, and looks down on to all this tropical valley. The contrast is very curious, and I have not got accustomed to it.

It seems rather a bad year for land beasts. The normal rainfall in December and January is said to be about thirty inches; and this year practically no rain has fallen since the spring. I suppose as a result of this every live thing gets under the biggest block of basalt it can find. This makes snail hunting rather exciting, because when you get to the top of a kopje where the beasts are you find the sides so steep that any stone you disturb rolls down, unless you take great care. My first day's hunt resulted in such a roll. A stone which I could only just lift rolled down into a sugar cane bed some three or four hundred feet below. I have never felt so ill as I felt until I found that stone and made sure that it had not smashed up an innocent Portuguese peasant!

* One of the blows to Weldon, which resulted from his biometric view of life, was that his biological friends could not appreciate his new enthusiasms. They could not understand how the Museum "specimen" was in the future to be replaced by the "sample" of 500 to 1000 individuals, not to be looked at through a glass, but to be handled, used, and if necessary *used up*. They warned his pupils solemnly to give up this sort of fooling and take to the real business of the "biologer," if they wished for success. "I told —— about these snails," Weldon wrote on Oct. 11, 1900, "and he wrote me an earnest letter, urging me to return to the pleasant way of describing beasts for the delight of the faithful. That is the real thing if you want to be popular. Go to sea, and have a good time, and bring back a jelly fish which is bright blue."

There is much missionary work yet to be done by biometricians, and Weldon's loss will make it still harder!

If one knew anything about natural history one might do a great service to these people. The whole place is covered with the tracks of a little black ant, introduced with some South American sugar canes five or six years ago. The ant cultivates a number of aphides, which produce serious diseases on all the fruit trees; also it attacks the newly hatched birds and all beasts which shelter under stones. Under a big stone, where some dozen snails have sheltered, about half the shells (which look quite fresh) are eaten empty by ants; so it is with the beetles, grasshoppers, and other things. The only good thing they have done is to eat the cockroaches. Every kitchen is now full of ants, and contains no cockroaches at all.

How snails make their shells here is hard to understand; there is not a scrap of limestone in the place: all basalt and beds of gorgeously coloured volcanic sands. Yet when one finds the right place, one finds that snails swarm and their shells are rather harder than usual!

I should very much like to know whether the habit of hiding under stones is as general in all seasons as it is now. You know Wallace points out that most of the beetles here have lost their wings; and he regards it as probable that flying beetles would be blown out to sea in storms. Now first of all there are practically no storms, and secondly, if there were, the valleys are so deep and their sides so precipitous, that there is abundant shelter against winds. But the loss of wings might well be correlated with the habit of walking under stones to get out of the sun. You find a patch of bare hot sand, so steep that you can hardly stand on it, with a stone here and there, and no sign of any living thing. If you turn over a stone you find a number of snails, a lizard, twenty or thirty beetles, a grasshopper or two, and armies of millipedes....... The man we see most of in this inn is a splendid creature. A captain in the Canadian frontier police, who volunteered for service in South Africa, and is recovering from a bullet through his right lung. Because he has a colonial accent, —— cannot see any merit in him.......It is only another sample of the difficulty I feel every day at Oxford. The boys there are so occupied with silly superficial things that one can never bring them to think of fundamental matters."

[Funchal, 29/12/00.]

"I am glad you are disgusted with the Life. I was afraid you were not.—You cannot judge the man from the bits of his letters. I do not think one ought to try to have an opinion about a man's conduct towards his wife, or indeed about his ethical value at all. One cannot possibly get hold of evidence enough, and the little bits of bad journalism which people give one are only sufficient to disturb one's mind. Take the old man as one knows him by his work, without troubling to guess at his motives, and there is not much the matter with him. I quite agree with you in loving Darwin and —— more; but a man may be a great deal lower than these two, and yet be high enough for reverence." [Oxford, 2/12/00.]

The earlier part of 1901 was chiefly occupied by snails, but a new factor had come into Weldon's many-sided occupations. It was settled that *Biometrika* should have in an early number a critical bibliography of papers dealing with statistical biology. Weldon undertook this task as his study of Mendel had led him to a very great number of such papers dealing with inheritance, and the section on Heredity was to be published first. Like all Weldon's projects, it was to be done in so thorough and comprehensive a manner that years were required for its completion. A very full list of titles was formed, especially in the Inheritance section, and many of the papers therein were thoroughly studied and abstracted (26). But such study meant with Weldon not only grasping the writer's conclusions, but testing his arithmetic and weighing his logic. Thus Weldon's Note on "Change in Organic Correlation of *Ficaria ranunculoides* during the Flowering Season" (27) arose from this bibliographical work and the erroneous manner in which he found Verschaffelt and MacLeod dealing with correlation. A further result of this work was that his confidence in the generality of the Mendelian hypotheses was much shaken. He found that Mendel's views were not consonant with the results formulated in a number of papers he had been led to abstract, and that the definite categories used by some Mendelian writers did not correspond to really well-defined classes in the characters themselves. It was a certain looseness of logic, a want of clear definition and scale, an absence of any insight into how far the numbers reached really prove what they are stated to prove, that moved Weldon when he came to deal with Mendelian work. And his attitude has been largely justified. The simplicity of Mendel's Mendelism has been gradually replaced by a complexity as great as that of any description hitherto suggested of hereditary relationships. This complexity allows of far greater elasticity in the deduction of statistical ratios, but the man in the street can no longer express a judgment upon whether the theory really accounts for the facts, and the actual statistical testing of the numbers obtained, as well as the logical development of the theory, will soon be feasible only to mathematical power of a high order. The old categories are, as Weldon indicated, being found insufficient, narrower classifications are being taken, and irregular dominance, imperfect recessiveness, the correlation of attributes, the latency of ancestral characters, and more complex determinantal theories are becoming the order of the day. If Weldon's papers "Mendel's Laws of Alternative Inheritance in Peas" (28), "On the Ambiguity of Mendel's Categories" (29), and "Mr Bateson's Revisions of Mendel's Theory of Heredity" (30), be read with a due regard to the dates of their appearance, it will be seen that they served, and that they continue to serve, a very useful purpose: they enforce the need for more cautious statement, for more careful classification, and for greater acquaintance with the nature of the inferences which are logically, i.e. mathematically, justifiable on the basis of given statistical data. The need will become the more urgent if the complexity of Mendelian formulae increases at the present rate.

To those who accept the biometric standpoint, that in the main evolution has not taken place by leaps, but by continuous selection of the favourable variation from the distribution of the offspring round the ancestrally fixed type, each selection modifying *pro rata* that type, there must be a manifest want in Mendelian theories of inheritance. Reproduction from this standpoint can only shake the kaleidoscope of existing alternatives; it can bring nothing new into the field. To complete a Mendelian theory we must apparently associate it for the purposes of evolution with some hypothesis of "mutations." The chief upholder of such an hypothesis has been de Vries, and Weldon's article on "Professor de Vries on the Origin of Species" (32) was the outcome of his consideration of this matter. During the years 1902 to 1903 an elaborate attempt was made to grow the numerous sub-races of *Draba verna*, with the idea that they might throw light on mutations. The project failed, largely owing to difficulties in the artificial cultivation of some of the species. But for a time all other interests paled before *Draba verna.*

"Where are you going at Easter? Stone wall country is very good, and if you find a place with delightful old stone villages and pretty churches, *Draba verna* will be there! Come into this region, with the bike, and learn to know and love the dear Dog !* Also explain to me how without thrashing to teach that same animal that lambs are not made to be eaten by puppies. There must be a way. I have taught him to walk at heel past the most tempting of other dogs, and even past chickens, and I have not yet beaten him at all. Cows and sheep will I suppose make one or two beatings necessary ?" [Oxford, 7/2/02.]

And a little later comes a letter which shall be cited because it may induce another to take up a form of biometric work, which must some day be pushed to a successful issue; in the fifteen years of letters there are many problems like this of *Draba verna*, which are discussed month after month with specimens, drawings, and tables, some merely schemed, but in surprising detail, others reaching the experimental stage, some in part solved, others but records of failure, one and all suggestive.

"*Draba verna*, or its earliest race is in full flower. I have four model types from a certain wall.

Now can you and Mrs Pearson give us the week end, so that your eyes may see the glory of this plant?

If you can turn up on Friday (I finish lecturing at 6) we can go for a tramp on Saturday, and see *Draba* at home on its walls. A gentle 7 or 8 miles all told, in a decently pretty country, with a variable plant in the middle and a really Perfect Dog all the way makes a very good combination;—only bring some knickerbockers, because Oxford Clay goes over one's ankles in places just now.

We can bring home our spoil, and discuss the very difficult question of descriptive units.

I think it a good and important thing to try. All the problems of treating mixed races come in; and above all I am curious to see what comes of statistical treatment applied to characters which have been chosen by "naturalists." They all say we choose anything which is easy to measure, and neglect the real points of "biological" importance; and there is a little truth in this reproach.

For *Draba* we want units of "habit," of shape and colour of leaves, of hairiness, of shape and colour of petals, sepals and fruit. We want to treat leaves which are very distinctly differentiated according to their relative time of appearance, and I think having tabulated all these characters, we want to break up the plants on a wall which you shall see on Saturday into about four races.

Do come if you possibly can. I saw one plant yesterday with all its seed capsules ripe and open; so that the first lot of little races will very soon be out of flower.

* The great Borzoi Sandro, henceforward to be a marked feature of the Weldon household, at home and away from home. Sandro pursuing sheep over the Yorkshire moors, Sandro pursuing game in the Buckinghamshire beechwoods, Weldon pursuing Sandro with every tone of affectionate persuasion, on the track the stacked cycles and the co-editor pursuing the deserted biometric problems in solitude, Weldon returning with the unchastised dog, after any interval of from 10 to 40 minutes, the chase being fully completed, the apologies for the Borzoi, his sustentation on chocolate and the human need for cigarettes, the return to the cycles, to the experiment that was to be crucial, to the colour and the sunset, these are all memories, the like of which others will have shared, which helped to form the atmosphere about the man. Sandro achieved his purpose, he kept his master out in the air—such wolf hounds can follow a cycle for miles—and to exercise him was held up as a moral duty. But his limited intelligence led to his own disablement and he had to become a partaker only in biometric "at homes." For two years, however, he was a great feature of our joint expeditions.

As *Nebensachen*, there are the mice and I very much want you to group the snails* in your own way—to see how far your grouping will bring them into a better form for the curve!

But *Draba verna* ought to be an example of the whole bearing of statistical methods upon systematic problems. I think it is rather important to begin this spring by collecting material for an analysis of the races round here, and a comparison of these with the types recognised by systematists. Learning these and their variability in this neighbourhood this spring, we can learn cotyledon characters and the characters of the first formed leaves in the Autumn—basing upon these a first set of hereditary correlations.

Again, the statement that each of Jordan's species can be recognised *at any stage in the life-history of the plant* makes it necessary to work correlations between cotyledon characters, autumn leaf characters, and spring characters. All this is work for lots of folk, and it is most important to get it properly planned now. Therefore, and for lots of other reasons try to come on Friday!

The Dog shall be washed for you!" [Oxford, 22/2/02.]

The reference in this letter to the mice indicates that that great piece of experimental work in heredity was now started. A study of the work of von Guaita had convinced Weldon early in 1901 that the cross between the European albino mouse and the Japanese waltzing mouse was not one which admitted of simple Mendelian description. In May, 1901, his letters contain inquiries as to Japanese mice dealers. During the summer and autumn the collection of Japanese mice was in progress. These mice were to be bred to test the purity of the stock; during December about forty does had litters, and pure breeding went on until the autumn of 1902, when hybridisation commenced. The work on these mice was for two years entrusted to Mr A. D. Darbishire, but the whole plan of the experiments, the preparation of the correlation tables, and the elaborate calculations were in the main due to Weldon. On Mr Darbishire's leaving Oxford, Weldon again resumed personal control of the actual breeding arrangements, and from second hybrid matings carried on the work to the sixth hybrids' offspring. The work was nearing completion at his death, and through the energy of Mr Frank Sherlock the skins of the 600 pedigree mice forming the stud at that time have been dressed and added to those of the earlier generations. The reduction and publication of this material will, it is hoped, be not long delayed (33). Weldon had this work much at heart, and his letters during 1904 and 1905 give many indications of the points he considered demonstrated. The experimental part of the work would have been nearly completed had not his whole thought and energy been directed from November, 1905, into another channel.

From 1901 it is harder for the present writer to give a detailed account of Weldon's life, the co-editorship of *Biometrika* and common work brought them so continually into contact. In the early part of the summer there had been a hurried visit to Gremsmühlen for young *Clausilia*; Weldon on his return visited his co-editor at Througham in Gloucestershire bringing his Brunsviga†, and there

* *H. nemoralis* and *H. hortensis* of which many hundreds had now been collected from various parts of England by Weldon and his helpers.

† The familiar mechanical calculator of the biometrician, the grinding sound of which (emphasised by the want of oiling in Weldon's case!) is the music which tells him how much his labours can be lightened.

was calculation and reduction of *Clausilia* data. Later there was a hurried visit to the Tegernsee and to Munich for opera. At Easter, 1902, there was a noble missionary effort (with the Brunsviga) to Parma; the missionary carried a memoir, which he had spent some weeks in rewriting in biometric form, but his efforts to show that a science of statistics exists were unavailing. In the summer *Biometrika* was edited from Bainbridge in Wensleydale, and accompanied by Sandro, the co-editors cycled to the churchyards of the Yorkshire dales, collecting material for their joint paper "On Assortative Mating in Man" (34). From Bainbridge the Weldons went to the British Association meeting at Belfast, where an evening lecture on Inheritance was given. At Christmas came one of the above-mentioned visits to Palermo to collect Sicilian snails. An event of this year (1902) was the publication of Mr Bateson's *Mendel's Principles of Heredity*. The origin of Weldon's first paper on Mendel has been described in this memoir; it was an expansion of a part of the promised bibliography for this Journal, and was written without any *arrière pensée* or knowledge of Mr Bateson's not then published experimental work. It is impossible for one who has been and again may be a combatant in this field to say more than that the tone of Mr Bateson's defence deeply pained Weldon, and rendered it difficult for a finely strung temperament to maintain—as it did maintain to the end—the impersonal tone of scientific controversy.

In the spring of 1903 Weldon was busy, as were the whole available members of the biometric school, in studying the influence of environment and of period of season on the variation and correlation of the floral parts of Lesser Celandine.

"Give my love to the Brethren who are cooperating in the matter of Celandines, and beseech them to make a better map of their country than the enclosed." [Oxford, 23/2/03.]

Weldon threw his whole energy and love of minute exactitude into the task, and his letters are filled with an account of the almost daily changes in the type and variability of the Celandine flowers from his selected stations. The result of this enquiry was the collection of an immense amount of data showing that environment and period in the flowering season affected the flower characters to an extent comparable with the differences attributed to local races. The reduction of the material has gone on progressively, if intermittently since, and it is hoped that a memoir, which will be a sequel to that issued in *Biometrika**, may be published shortly (35). The wider standpoint of this second memoir will be chiefly due to Weldon's initiative and critical mind. At Easter of 1903 a series of mishaps prevented the common holiday, but this was more than compensated for by the summer vacation. The Weldons started with a sea trip to Marseilles and back. They then returned to Oxford, that work on the article *Crustacea* for the Cambridge Natural History might be carried on, and an eye kept on the mice. But a biometric camp was formed at Peppard on the Chilterns; here the "Consulting Editor" and one of the co-editors had established themselves, and the Weldons took a week-end cottage. The three Oxford members of the party arrived partly on cycles and partly on four

* Vol. II. pp. 145—164, *Cooperative Investigations on Plants.* II. Variation and Correlation in Lesser Celandine from Divers Localities.

feet, and were often met *en route* by the residents in the uplands, the numbers being swelled by the addition of biometricians from the London or Oxford schools. Hence arose a series of Friday "biometric teas," for the discussion of the week's work and plans for the next two days. Saturday and Sunday morning were given to steady calculating and reducing work, and much was got through. The data on assortative mating in man collected in the previous year were reduced and a joint paper sent to press; the immense amount of calculation and reduction involved in the mouse-paper was got through; a joint criticism of Johannsen's *Ueber Erblichkeit in Populationen und in reinen Linien* was written by the co-editors under the title "Inheritance in *Phaseolus vulgaris*" (36); the Huxley Lecture was written with yeoman help from the Oxford contingent, and lastly, a joint study was made, at Weldon's suggestion, of the relationship between Mendelian formulae and the theory of ancestral heredity. It was shown that there was no essential antagonism between the two methods of approaching the subject, and the results were published ultimately as Part XII of the *Mathematical Contributions to the Theory of Evolution,* Weldon persistently declining to allow it to appear as a joint memoir, because he had taken no part in certain portions of the more complicated algebraic analysis. Christmas found two-thirds of the party reunited in Palermo, and Weldon on the snail quest. His letters thence to his co-editor teem with the freshness of the sky and the joy of open-air work:

"Out between five and six, in the dark, without any breakfast, sunrise up in the hills, a day's tramp on a piece of bread and a handful of olives, and home at seven, laden with snails. Then after dinner to clean the beasts. That is not work, and it makes one very fit, but one gets tired enough to sleep when the snails are cleaned!

The camera works all right, and I think there is a very marked correlation between the general character of the limestones and the character of the shells; but developing in one's bedroom does not make for negatives which will "process"! Also it involves heavy subsidies to such chambermaids as do not understand what new form of madness this particular foreigner has developed!

I have repeated all last year's collections, and have tried hard to get a series of forms, such as Kobelt describes, intermediate between the rounded and the flat keeled forms, but I cannot at present find these. They ought (according to him) to live in a certain wilderness of beautiful mountains twenty miles away. I have several times tramped without any result. I hope to try again. I feel sure something worth having will come out of these shells; they illustrate local races and the general problem of what is a species splendidly. But the question of their markings comes in also; and you, or Galton, or someone, will have to make a scale of patterns for me, I expect. They will be the most perfectly hopeless things to draw!

It is, of course, just conceivable that the intermediate, slightly keeled forms described in 1879 by Kobelt have been exterminated since his time? He is very precise in his localities, and everywhere except in his transitional region I find *exactly* what he describes, but in this region I find so far only rounded forms.

The only difficulties about tramping in this country are the *carabinieri.* Every high road is patrolled by groups of two or three, so that even in a desolate place, so long as you keep to the road, you are rarely out of rifle shot; but these men come and solemnly warn you that the people round are ruffians, who would cheerfully cut your throat for a soldo; and if you simply grin, they make a great pretence of falling in behind you and guarding you....Now collecting snails with an armed guard becomes ridiculous after a time and there is no danger at all; the men only want tips. When one gets off the road into the hills the goatherds and other ruffians are most friendly. They want to see one's camera, and one's knife, and of course they want half one's bread, but they never ask for tips, and my throat is still uncut.

We have so far had two wet days; to-day, and one other.—We have had several inches to-day, and shall have some more; but between we have had the most glorious sun. I look as if I washed in strong coffee every morning." [Palermo, 31/12/1903.]

At the beginning of 1904 the work on the Brescia *Clausilia* was in progress, the mice were multiplying after their kind and Weldon's thoughts were turning more and more to a determinantal theory of inheritance, which should give simple Mendelism at one end of the range and blended inheritance at the other. Easter was spent in common, one editor at Rotherfield Greys and the other at Peppard, with the usual flow of suggestion on Weldon's part and the bi-weekly cyclings to Oxford to look after the mice. Now and then the fear would strike Weldon's friends that life was being lived at too fast a pace, but the constant intellectual and physical activity was so characteristic of the man that there was no means of calling halt, and to many when Weldon was most active he seemed most fit and well. The summer found the Pearsons twelve miles from Oxford, at Cogges, near Witney, and the Galtons twenty miles further, at Bibury; there was much cycling too and fro, and the plan of a new book by Weldon on Inheritance was drafted, and some of the early chapters written. The vacation was broken by the visit to Cambridge—Weldon cycling in one day from Oxford—for the British Association. The Presidential Address in the Zoological Section was chiefly an attack upon biometric work and methods, and the discussion which followed culminated in the President dramatically holding aloft the volumes of this Journal as patent evidence of the folly of the school, and refusing the offer of a truce to this time-wasting controversy. The excitement of the meeting, as earlier contests at the Zoological and Linnean Societies, seemed to brace Weldon to greater intellectual activity and wider plans, but the torpedo boat was being run at full speed.

The book on Inheritance occupied most of the remainder of the year, and to aid it forward and help those of us who were not biologists to clearer notions, I suggested to Weldon a course of lectures in London to my own little group of biometric workers. The project grew, other departments of the College desired to attend, and ultimately the lectures were thrown open to all members of the University and even to the outside public. Weldon had a good audience of more than a hundred, and enjoyed the return to his old environment. But it may be doubted whether his vitality responded as quickly as of old to the additional stress; there were special elements of difficulty, and I believe now that it would have been kinder and more helpful had we limited the audience to my own small body of sympathetic students.

"It will be a great pleasure to me to come and talk, and to feel that you cared to ask me; the lectures will do far more good to me than to anyone else ... and I owe U. C. L. a bigger effort than this anyhow." [Oxford, 16/10/04.]

And again:

"It makes me more than ever glad I am coming back to Gower Street where there are live people to talk to! Surely thirty people* is a great many. Try talking for five years to an audience of from three to nine, and see how the thought of thirty will cheer you! And none of these excellent folk are sent by their tutors!" [Oxford, 7/11/04.]

The letters of Weldon to both Francis Galton and myself during the years 1904 and 1905 are full of inheritance work, the details of the great mice-breeding experiment, the statement and the solution, or it might be the suggested solution, of nuclear problems leading to determinantal theories of inheritance. Occasionally there would be a touch of conscience, and the drawings for the *Crustacea* would be pressed forward:

"I ought to give my whole time to the Cambridge Natural History for a while. They have been very good to me, and I have treated them more than a little badly. I am rather anxious to get them off my conscience." [Oxford, 15/2/05.]

But only the chapter on *Phyllopods* got completed, figures and all, and set up. Many figures were prepared for other parts; beautiful things, which gave Weldon not only scientific but artistic pleasure, he had made, but the text remains the veriest fragment. In the same way but little was absolutely completed of the article on *Heliozoa* for Lankester's *Natural History*. It was not Weldon's biometric friends that kept him from these tasks, it was solely his own intense keenness in the pursuit of new knowledge. It was occasionally with a feeling of great responsibility that the present writer would propound to him an unsolved problem with which he might himself be struggling. There was absolute certainty that if the problem was at all an exciting one, Weldon would leave his scent and follow the new trail with his whole keenness and at full speed. All else would be put on one side, and he could only be recalled to natural history or biometrics by an appeal to his conscience. Like Sandro, the chase must be completed before he returned to the humdrum trot behind a cycle on the highway.

The fascination of inheritance problems kept Weldon, however, for months at a time at the Heredity Book. At Easter, 1905, he went to Ferrara†, because that place had a university, and as such must have a library, where work could be done. The contents of the library were perfectly mediaeval, a characteristic appropriate in the castle, but hardly helpful in heredity‡. Still, portions of the manuscript came to England for comment and criticism, and we were hopeful that the end of the year would see the book completed.

* The number I already knew would certainly attend.

† "The town is worth a lot, and the fields are full of a little speedwell, which varies most delightfully. I have so far resisted the temptation to chuck the wretched book and tabulate the variations of its flowers, and I hope I shall do to the end. But it is a temptation......I feel out of the world, an absolute blank, with only a slight interest in newts' tails and an even slighter in a statue of Savonarola which looks at me all day through the window." [Ferrara, 3/4/05.]

‡ From Ferrara came back if not the speedwells, masses of silkworms' eggs of different local races, but providentially they failed to hatch out in sufficient numbers owing to the May frosts and no new scent took Weldon off the book and the mice during the summer.

It must not be thought for a moment that Weldon was desultory in his work. As E. R. Lankester says in a letter to the writer: "His *absolute thoroughness* and unstinting devotion to any work he took up were leading features in his character." He pursued science, however, for sheer love of it, and he would have continued to do so had he been Alexander Selkirk on the island with no opportunity for publication and nobody to communicate his results to. He never slackened in the total energy he gave to scientific work, but having satisfied *himself* in one quest, he did not stay to fill in the page for others to read; his keen eye found a new problem where the ordinary man saw a cow-pasture, or a dusty hedgerow, and he started again with unremitted ardour to what had for himself the greater interest. The publication of his researches will show that it is not we who are the losers, because he went forward, regardless of publication and finality of form. The true function of such a man is not to write text-books or publish treatises, it lies in directing and inspiring a school, which will be trained by completing the work and carrying out the suggestions of its master. The curse of the English educational system is that it leaves such men to solitude, and throngs the chambers of those who cram all nature into the limits of the examination room.

In the summer the present writer was at East Ilsley, some seventeen miles from Oxford, and there was cycling out several times a week; the writer's chief work was on other than biometric lines and broken by other claims on his time, but there was steady joint work on the determinantal theory of inheritance as outlined by Weldon, and it is hoped that it is sufficiently advanced to be completed and published (37). Weldon had in August, 1905, given to the Summer Meeting of the University Extension in Oxford a lecture on *Inheritance in Animals and Plants* (38), and this had taken up some of his energy during the summer vacation. On the whole, however, he worked persistently at the Inheritance Book. It is too early yet to say definitely how far it can be considered ready for press, but a considerable number of chapters are completely ready, and there are drafts for several others. We can only hope that this, the work he was in many respects best fitted for both by direct experiment and by study of the labours of others, will be issued in his name and show the full measure of his activities during these last few years.

It cannot be denied that those who were often with Weldon during the last two years were occasionally anxious—the pace had been too great—but at no time had one definitely realised that there was an immediate anxiety. His intellectual activity was never apparently diminished, and his long cycling rides were maintained to the end. It was an occasional, but never long persistent, lack of the old joyousness in life which was noticeable. At East Ilsley he was full of keenness over his photographic work; he enjoyed an antiquarian investigation into the probable final *locus* of the bones of St Birinus with a view to testing a local legend; we examined carefully a human skeleton dug up from under a sheepfold, the authorities having determined that no inquest was needful, the bones being those of an old man who died "hundreds of years ago." "And you think?" said

To the Editor of Biochronicle

Sir, — In the last number of your most instructive Journal you publish an account of the way in which Snails are collected by the energetic naturalists of Brush. It may ~~be~~ interest to some of your numerous readers to follow up the thoughts suggested by this account, and to know something of the way in which snails grow up before they are collected.

As you, Sir, know very well, a snail is unlike most of the animals found in the neighbourhood of Brush, in its shape. ~~And~~ Most of the animals in the Brush collection, whether Bats, or Butterflies, or Crayfishes, or others, can be divided into two halves, which are curiously alike. If you take a thing like the drawing in the margin, which is meant to be like half a cray-fish, and put the straight edge of the drawing against a looking-glass, the reflection of this drawing in the glass will be very like the other half of the crayfish. All the legs of a crayfish are in pairs, so that there are right and left legs, as curiously like each other as ~~the hands~~ your Editorial Hands are like each other.

But except the two halves of its brain, and its eyes, and one or two other organs on its head, the parts of a grown up snail are not arranged in pairs in this way. When a snail is quite young, inside its egg-shell, you can recognise a right and a left half of it, just as you can in a cray-fish: but as it grows up, its left hand side grows very much faster than its right hand side, so that the right-hand side nearly disappears. I would ask you to consider, Sir, what the effect of such a mode of growth would be upon your Editorial Self; and I trust you will not think me wanting in respect if I submit for your consideration a few diagrams indicating the effect of this upon your person.

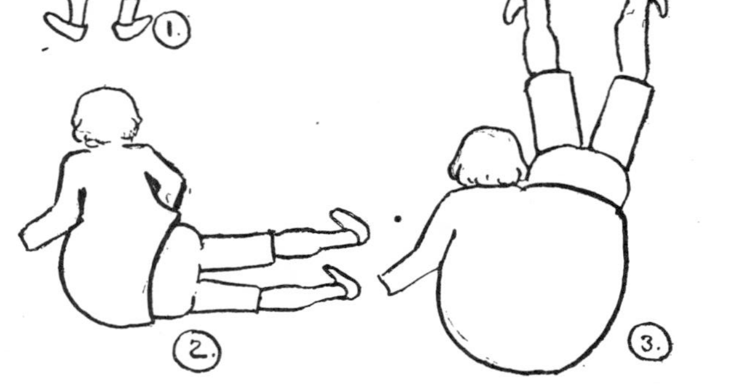

You see that the effect would be to make your legs point forwards, towards your head, and to make the leg which is now on your right side look as if it belonged to your left side, or at least to make it lie to the left of the other leg.

In the region of your waistcoat you would have a most uncomfortable swelling, which I apologise for suggesting.

Now a snail never has any legs; but it has a number of paired organs, right and left, in the hinder part of its body, when it is young. As it grows, it twists the hind end of its body forwards as I have described; and while it is doing this, the organs which lay on its left side, its left kidney, the left half of its heart, and other things, disappear. So that it only has one kidney, and the piece of its heart which lay on its right side when it was young and had not yet got a large hump on its left side.

Now this left-sided hump, which is produced by the growth of the left side, gets twisted like a screw, and the shell is formed on it. So that when you look at a full-grown snail you see this kind of thing:

This is the hump formed out of the left side of the snail, and then screwed up.

over here lie the right hand organs of the young snail.

This is the hind end of the young snail, turned forwards.

These are the horns of the snail, in front of the place where its left side begins to swell up, so (that the right & left sides are alike.)

Nearly all snails screw their hump, after they have formed it, the same way round. They screw it so that if you stand on the top of the hump, and imagine yourself walking down the twisted hump towards the place where it joins the body of the snail, you will always have to turn to the right. But some snails screw their humps the other way, so that in going from the top to the bottom of their hump you would always have to turn to the left. Among the snails found near Brush, every one has a right handed hump, except Clausilia, which has a left handed hump. Only every now and then a snail of a sort which has usually a right-handed hump makes a left-handed hump for itself. Nobody knows what the children of such wrongly twisted snails are like; whether they are like their parents, or whether they return to more usual ways of twisting their humps. And nobody knows how a snail is better off for going through all this process of enlarging one side and twisting its hump at all; so that certainly no one knows whether, if a snail must have a twisted hump, it is better to have a right-handed twist or a left-handed twist.

Because no one knows these things, some very proud people, who dislike Biochronicles, say that it does not matter to a snail which way its hump is twisted. They say that it is absurd to suggest that such a trifle, the meaning of which they cannot understand, can have any effect upon a snail. I venture to hope, Sir, that as Biochronicle progresses you will take an ever increasing interest in snails, but I hope you will never be led to express in your influential columns the opinion that nothing matters to a snail unless you can yourself understand it.

I am, Sir, Your obedient servant, WFRWeldon

Plate V

Weldon on our homeward way. "Having no anatomical training I think they are those of ——." "A young woman, who has not been buried so very long," he interrupted, with a responsive twinkle in his eye. "Let us have a smoke and consider the scientific education of the English medical profession." His sense of humour was always keen, whether with word or pencil, and it remained with him to the end. The joy of life which in the early days led him to dance and sing on the completion of a heavy bit of work, made him in later manhood ripple over with quiet humour in talk and letter when problems were going well.

Thus to Francis Galton:

"I enclose the best I can do with one of the negatives you were kind enough to let me make. Please forgive me for caricaturing you in this way.—You know enough about the lower forms of man to know that respect and affection show themselves in strange ways:—look upon this as one of them and pardon it." [Oxford, 27/7/05.]

Nor did he spare a quiet joke at a friend:

"Your work on dams has filled the Italian papers with horror. They say you threaten the safety of all existing dams, *however long they have stood.*" [Ferrara, 7/4/05.]

In November, 1905, Weldon was unfortunately taken off from the work on his inheritance book by the presentation to the Royal Society of a paper by Captain C. C. Hurst: *On the Inheritance of Coat-Colour in Horses.* He had had no proper summer holiday, but he threw himself nine hours a day into the study of *The General Studbook**.

"I can do nothing else until I have found out what it means....The question between Mendel and Galton's theory of Reversion ought to be answered out of these. Thank God, I have not finished that book. There must be a chapter on Race Horses!"

Weldon felt himself in a difficult position; as Chairman of the Zoological Committee, he had at once directed the printing of Hurst's paper. But the subject being one in which he personally was keenly interested, he had immediately attacked the original material and to his surprise came to views definitely opposite to those of Hurst. He felt bound to report this result at once to the Society, and he did so on December 7, when the original paper was read. His results were provisional, as could only be the case considering the short period of preparation that had been possible. He promised to communicate a note to the Society involving more details of his inquiry. This was done on January 18, 1906 in a "Note on the Offspring of Thoroughbred Chestnut Mares" (39).

* I cannot resist citing a last illustration of Weldon's humour: "What volumes of Weatherby have you? I have found in Bodley 17—20. To show you what Bodley is, I looked in the catalogue vainly under: *Weatherby* (found here and not under Racing, *Racing Calendar*), *Jockey Club* (found here pamphlets about the J. C. but *not* its own publications), *Horses, Race Horses, Racing, Studbooks* (found here only Clydesdale Studbook, Pigeon Studbooks, and Dog Studbooks), *Turf, Sport, Race*, all suggested by assistants in the Library. For a whole day I raged, and came back despairing. Next day I raged worse, and captured a man who knew something. He smiled and said: 'Oh, Yes, The General Studbook is entered under *General* of course.' I said, 'Why not under *The*?' and he thought that unseemly!"

"The object of the present note is partly to fulfil my promise and partly to call attention to certain facts which must be considered in the attempt to apply any Mendelian formula whatever to the inheritance of coat-colour in race-horses."

It is impossible at present to say more on this point, for the whole subject is likely to be matter for further controversy. Even one authenticated case of a non-chestnut offspring to chestnut parents is sufficient to upset the theory of the 'pure gamete,' but if studbooks are to be taken as providing the data, the whole question must turn on whether one in sixty of the entries of the offspring of chestnut parents can be reasonably considered as a misprint or an error in record.

Here it can only be said that Weldon took up the subject with his usual vigour and thoroughness. But he was overworked and overwrought and a holiday was absolutely needful. He went to Rome, but the volumes of the Studbook went with him:

"Will you think me a brute, if I take the Studbook to Rome? I really want a holiday, but I cannot leave this thing unsettled."

And then from Rome:

"I think it will be worth while to deal for once with a whole population, not with a small random sample. Only I could find it in my heart to wish one need not do it in Rome! To sit here eight hours a day or so, doing mere clerk's work, seems rather waste of life?"

And again:

"I have really been working too hard to write, or to do anything else. I have seen *nothing* of Rome....I want to know what these horses will lead to, but it would not interest me at all to know that my paper on them would or would not be printed. More important is the enormous time these horses will take. It seems clear that one ought to carry these arrays back to another generation of ancestors—and that means a very long job. I wish I had a pupil! A mere clerk would be no good, but a pupil, such as one had in good old Gower Street, would help with the drudgery, and then he might stick his name all over the paper, if he liked."

[February, 1906.]

The letters are filled with Studbook detail till Easter, there is hardly a reference to anything else. Re-reading them now one sees how this drudgery with no proper holiday told on Weldon. Hundreds of pedigrees were formed and a vast amount of material reduced. At Easter the Weldons went to the little inn at Woolstone, at the foot of the White Horse Hill, and his co-editor came down later to Longcot, a mile away, for the joint vacation. Weldon was still hard at work on the Studbooks, but he was intellectually as keenly active as of old; he was planning the lines of his big memoir on coat-colour in horses (40) and showing how they illustrated the points he had already found in the mice. He was photographing the White Horse, and rubbing mediaeval idlers' scrawlings on the church pillars. He projected the despoiling of a barrow, and planned future work and rides.

On Sunday, April 8, he rode into Oxford to develop photographs, and the present writer rode some miles of the way with him; the joint ride terminated with the smoke by the roadside and Weldon's propounding the problem which was to be brought solved for him on Tuesday. On Tuesday I found him in bed, with what appeared to be an attack of influenza. He had expressed himself tired after his ride on Sunday, an almost unique admission. But on Monday he went a long walk over the Downs, getting home late. He came down to breakfast on Tuesday but had to return to bed. In the afternoon when I came he insisted on smoking and wanted the solution of the problem, saying he was better. I begged him, as one still closer did, to stay in bed on the morrow and give up a projected journey to Town. But there was a dentist to be seen, preparations for a visit to the M.B.A. Laboratory at Lowestoft to be made, and a wonderful picture-gallery to be visited to free him from the atmosphere of the Studbooks. His will was indomitable; he went up to Town and went to the pictures on Wednesday, he went to the dentist on Thursday, but from the dentist's chair he had to be taken to a doctor's, and thence to a nursing home. The summoning telegram reached his wife on the same afternoon, and he died of pneumonia on Good Friday, April 13. So passed away, shall I say not unfitly—for it was without any long disabling illness and in full intellectual vigour—a man of unusual personality, one of the most inspiring and loveable of teachers, the least self-regarding and the most helpful of friends, and the most generous of opponents.

As for his life, I think it was to him what he would have wished it. There were moments of discouragement and depression, he felt occasionally a want of sympathy for his life-work in some of his former colleagues, and while he was born to be the centre of an enthusiastic school, he found at times somewhat scanty material for its maintenance in pleasure-loving Oxford. But every stone he lifted from the way became gold in his hands; each problem he touched became a joy which absorbed his whole being. The artist in his nature was so intense that he found keen pleasure in most men and in all things. Only meanness or superficiality fired him, and then, considering how the world is built, sometimes to almost an excess of wrath. But he had no personal hate; he could make the graceful amend, and had he ever a foe, that foe, I veritably believe, could have won Weldon's heart in the smoking of a cigarette.

If we pass from himself to those whose fortune brought them in close contact with him—to his friends and pupils—their loss can only be outlined, it is too intimate and personal for full expression. There was a transition from respect to reverence, a growth from affection to love; to such a tenderness as some bear for a more delicate spiritual nature, to even such feeling as the Sikh is reputed to hold for the white man's child in his charge.

And lastly as to science, what will his place be? The time to judge is not yet. Much of his work has still to be published, and this is not the occasion to indicate what biometry has already achieved. The movement he aided in starting is but in its infancy. It has to fight not for this theory or that, but for a new method and a greater standard of logical exactness in the science of life. To those who condemn it out of hand, or emphasise its slightest slip, we can boldly reply, You simply cannot judge, for you have not the requisite knowledge. To the biometrician, Weldon will remain as the first biologist who, able to make his name by following the old tracks, chose to strike out a new path—and one which carried him far away from his earlier colleagues. It is scarcely to be wondered at if those he joined should wish to see some monument to his memory; for he fell, the volume of life exhausted, fighting for the new learning.

Is what he gave science small? That depends on how it is measured. He was by nature a poet, and these give the best to science, for they give ideas. They follow no men, but give that which another generation may study from and be inspired by. He was the enthusiast, but the enthusiasm was that of the study, trained to its task; and when the time comes that we shall know, or that those who come after us shall know, whether Darwinism is the basal rule of life or merely a golden dream which has led us onwards to greater intellectual insight, then the knowledge, so biometricians have held and still hold, will be won by those actuarial methods which he first applied to the selection of living forms. If there be aught else to be said, let another say it.

Step to a tune, square chests, erect each head,
'Ware the beholders!
This is our master, famous, calm and dead,
Borne on our shoulders.

Description of Plates.

Plate I. W. F. R. Weldon.

Plate II. Raphael Weldon, aged 10.

Plate III. (*a*) Rapid pencil caricature by W. F. R. W. "Apparition: Le Café Orleans."

(*b*) Sample of Illustration to letters. Description of bands of *H. hortensis* in letter to a lady collector. "Has it occurred to you that a lady of artistic ability, and so enlightened that she likes snails, would have great joy and do great service by drawing them? There is a good inexorable severity about their lines which one would enjoy, I should think, if it were not so unattainable (to me!) on paper. And it would be nearly as good fun as real engraving to get all their lights and shadows put in with curved lines which also indicate the growth lines on the shell? Think how Bewick liked it."

Plate IV. A "crabbery" at Plymouth.

Plate V. Contribution to a manuscript magazine run by a youthful friend.

LIST OF MEMOIRS, ETC., BY W. F. R. WELDON.

(1) Note on the early Development of *Lacerta muralis*. *Q. Jour. Mic. Sci.* Vol. XXIII, pp. 134—144, 1883.

(2) On the Head-Kidney of *Bdellostoma*, with a suggestion as to the Origin of the Suprarenal Bodies. *Q. Jour. Mic. Sci.* Vol. XXIV, pp. 171—182, 1884.

(3) On the Suprarenal Bodies of Vertebrates. *Q. Jour. Mic. Sci.* Vol. XXV, pp. 137—150, 1885.

(4) On some points in the Anatomy of *Phoenicopterus* and its Allies. *Proc. Zool. Soc. Lond.* 1883, pp. 638—652, 1883.

(5) Note on the Placentation of *Tetraceros quadricornis*. *Proc. Zool. Soc. Lond.* 1884, pp. 2—6, 1884.

(6) Notes on *Callithrix gigot*. *Proc. Zool. Soc. Lond.* 1884, pp. 6—9, 1884.

(7) On *Dinophilus gigas*. *Q. Jour. Mic. Sci.* Vol. XXVII, pp. 109—121, 1886.

(8) *Haplodiscus piger*; a new Pelagic organism from the Bahamas. *Q. Jour. Mic. Sci.* Vol. XXIX, pp. 1—8, 1888.

(9) Preliminary Note on a *Balanoglossus* Larva from the Bahamas. *R. S. Proc.* Vol. XLII, pp. 146—150, 1887.

(10) The Coelom and Nephridia of *Palaemon serratus*. *Journal Marine Biol. Assoc.* Vol. I, pp. 162—168, 1889.

(10) *bis* Note on the Function of the Spines of the Crustacean *Zooea*. *Journal Marine Biol. Assoc.* Vol. I, pp. 169—170, 1889.

(11) The Renal Organs of certain Decapod Crustacea. *Q. Jour. Mic. Sci.* Vol. XXXII, pp. 279—291, 1891.

(12) The Formation of the Germ Layers in *Crangon vulgaris*. *Q. Jour. Mic. Sci.* Vol. XXXIII, pp. 343—363, 1892.

(13) The Variations occurring in certain Decapod Crustacea. I. *Crangon vulgaris*. *R. S. Proc.* Vol. XLVII, pp. 445—453, 1890.

(14) Certain correlated Variations in *Crangon vulgaris*. *R. S. Proc.* Vol. LI, pp. 2—21, 1892.

(15) On certain correlated Variations in *Carcinus moenas*. *R. S. Proc.* Vol. LIV, pp. 318—329, 1893.

(16) [On Variation in the Herring. Unpublished measurements and reductions presented to the Evolution Committee.]

(17) Attempt to measure the Death-rate due to the Selective Destruction of *Carcinus moenas* with respect to a Particular Dimension. Report of the Committee...for conducting Statistical Inquiries into the Measurable Characteristics of Plants and Animals. *R. S. Proc.* Vol. LVII, pp. 360—379, 1895.

(18) Remarks on Variation in Animals and Plants. *R. S. Proc.* Vol. LVII, pp. 379—382, 1895.

(19) [Report to the Evolution Committee on the Growth of *Carcinus moenas* at successive moults. 1897. Unpublished.]

(20) Presidential Address to the Zoological Section of the British Association. *B. A. Transactions*, Bristol, 1898, pp. 887—902.

(21) [Researches on Pedigree Moths, 1899-1901. Unpublished.]

(22) Cooperative Investigations on Plants. I. On Inheritance in the Shirley Poppy. *Biometrika*, Vol. II, pp. 56—100, 1902. [A joint paper with others.]

(23) A First Study of Natural Selection in *Clausilia laminata* (Montagu). *Biometrika*, Vol. I, pp. 109—124, 1901.

(24) Note on a Race of *Clausilia itala* (von Martens). *Biometrika*, Vol. III, pp. 299—307, 1903.

(25) The Scope of *Biometrika*. Editorial. *Biometrika*, Vol. I, pp. 1, 2, 1901.

(26) [Critical Bibliography of Memoirs on Inheritance. Unpublished.]
(27) Change in Organic Correlation of *Ficaria ranunculoides* during the Flowering Season. *Biometrika*, Vol. I, pp. 125—8, 1901.
(28) Mendel's Laws of Alternative Inheritance in Peas. *Biometrika*, Vol. I, pp. 228—254, 1902.
(29) On the Ambiguity of Mendel's Categories. *Biometrika*, Vol. II, pp. 44—55, 1902.
(30) Mr Bateson's Revisions of Mendel's Theory of Heredity. *Biometrika*, Vol. II, pp. 286—298, 1903.
(30) *bis* Mendelism and Mice. *Nature*, Vol. LXVII, pp. 512, 610, Vol. LXVIII, p. 34, 1903.
(31) Albinism in Sicily and Mendel's Laws. *Biometrika*, Vol. III, pp. 107—109, 1904.
(32) Professor de Vries on the Origin of Species. *Biometrika*, Vol. I, pp. 365—374, 1902.
(33) [On the Results of Crossing Japanese Waltzing with Albino Mice. Unpublished.]
(34) On Assortative Mating in Man. *Biometrika*, Vol. II, pp. 481—498. A joint memoir, 1903.
(35) [Measurements and observations on Lesser Celandine. Unpublished.]
(36) Inheritance in *Phaseolus vulgaris*. *Biometrika*, Vol. II, pp. 499—503. Joint review, 1903.
(37) [A Determinantal Theory of Inheritance. Unpublished.]
(38) Inheritance in Animals and Plants. *Lectures on the Method of Science.* Edited by T. B. Strong, Oxford, 1906.
(39) Note on the Offspring of Thoroughbred Chestnut Mares. *R. S. Proc.* Vol. 77B, pp. 394—398, 1906.
(40) [Material for an extensive memoir on the Inheritance of Coat-colour in Thoroughbred Horses. Unpublished.]
(41) Article on *Crustacea* for the *Cambridge Natural History*—fragmentary, except for a chapter on the *Phyllopods* already set up.
(42) [A Treatise on Inheritance, largely completed.]
(43) A portion of an account of the *Heliozoa* for the *Oxford Natural History.*
(44) Account of Kölliker's scientific work. *Nature*, Vol. LVIII, pp. 1—4, 1898.
(44) *bis* Dreyer's Peneroplis, eine Studie zur biologischen Morphologie und zur Speciesfrage. (A review.) *Nature*, Vol. LIX, pp. 364—5, 1899.
(45) Account of Huxley's scientific work for the Supplement to the *Dictionary of National Biography*, 1900.
(46) Article on *Variation* in the "Times" Supplement to the *Encyclopaedia Britannica.*

SOME INCIDENTS IN THE EARLY HISTORY OF BIOMETRY AND STATISTICS, 1890–94*

By E. S. PEARSON

1. Introduction

Perhaps the two great formative periods in the history of mathematical statistics were the years 1890–1905 and 1915–30. In both, the remarkable leap forward was made in answer to a need for new theory and techniques to help in solving very real problems in the biological field. In the earlier years the original questions posed concerned the interpretation of data bearing on theories of heredity and evolution; in the later period the first call was to sharpen and develop the tools used in agricultural experimentation.

There is considerable fascination in trying to find out how things looked at the time to the men concerned in such pioneer movements, from what background they started and what was the combination of circumstances which lead to the particular lines of advance which they followed. Below I shall try to describe some of the history of the first few years of the 1890–1905 period. A good deal of this has already been put on record, for example, by K. Pearson (1906,‡ 1930) and I shall draw freely on this material, but the availability of certain letters between Francis Galton (1822–1911), F. Y. Edgeworth (1845–1926), Karl Pearson (1857–1936) and W. F. R. Weldon (1860–1906)† has made it possible to add some illuminating personal touches to what is already on record.

The final event which brought about the association of Pearson and Weldon, leading 10 years later to the founding of *Biometrika*, was Weldon's election in 1890 to the Jodrell Chair of Zoology at University College, London where Pearson had been Professor of Applied Mathematics since 1884. But to understand the basis of the co-operation between these two men we must look still further back. The threads were gathered from many sources.

In the second half of the 1880's Pearson was by profession an applied mathematician, a good deal of whose teaching was to students of engineering. Between 1884 and 1893 he had first prepared for the press W. K. Clifford's unfinished manuscript of *The Common Sense of the Exact Sciences* and had then undertaken the far more arduous task of completing Isaac Todhunter's *A History of the Theory of Elasticity;* the second volume of this, containing some 1300 pages were almost entirely Pearson's contribution.

But throughout the '80's his research energies were also occupied in a quite different direction, in the study of mediaeval and Renaissance German literature and folk lore. This work is recorded in the series of essays, most of them first given as lectures, which were later published in *The Ethic of Freethought* (1888) and *The Chances of Death* (1897). The

* This article is enlarged from a talk originally given to students at University College in 1960 and later, in 1961, at Princeton University from where it was issued as No. 45 in the *Statistical Techniques Research Group Reports* (1962).

† Many of the letters are in possession of University College London with whose permission they are quoted.

‡Paper No. 21 in this volume

link between these investigations and *The Grammar of Science* of 1892 lay as he wrote himself in a 'fundamental note of the author's thought, namely: the endeavour to see all phenomena, physical and social, as a connected growth, and describe them as such in the briefest formula possible'. The manner in which his applied mathematics could best be harnessed in this endeavour was at first less obvious; but several events helped to shape the course of his activities.

Early in 1889, shortly after its appearance, he had read Galton's *Natural Inheritance*. From the paper which he presented soon afterwards to a small discussion club* it would seem that he was then not altogether ready to accept Galton's approach and was perhaps rather critical of the popular way in which Galton expounded his subject. But it is clear from what he wrote at a much later date (see E. S. Pearson, 1938, pp. 18–19) that from this time onwards he began to be aware of a door which might be opened into new and exciting fields.

Another lead opened up at this juncture. Pearson had attended Todhunter's classes when at Cambridge between 1876 and 1879, and his later connexion with *A History of the Theory of Elasticity* may have increased his interest in Todhunter's earlier publication, *A History of the Mathematical Theory of Probability.* At any rate when in 1890 he applied for the vacant Lectureship in Geometry† at Gresham College in the City of London, he included among the subjects which he offered to present, 'graphical statistics' and 'the theory of probability', as subjects likely 'to supply a want felt by clerks and others engaged during the day in the City'.

The first eight of the Gresham Lectures, given in March and April 1891, fell under the heading: 'The Scope and Concepts of Modern Science'; they contained the material later developed and enlarged in *The Grammar of Science* (1892). The first edition of this book had chapters on 'Cause and Effect—Probability' and on 'Life', but the treatment follows Pearson's philosophic approach to scientific concepts built up during the 1880's, i.e. belongs to what may be termed the pre-Galton–Weldon period of his development.

In the same way the second series of Gresham Lectures under the general heading 'The Geometry of Statistics and the Laws of Chance' began with twelve lectures (November 1891, January and May 1892) which appear from the syllabuses (E. S. Pearson, 1938, pp. 142–53) to have been concerned with a somewhat formal account of methods of presentation of descriptive statistics.

It is only later, from November 1892 through 1893 and 1894, that we begin to see the subject taking a new life: the introduction of experiment; the comparison of theory with observation, whether the latter resulted from coin tossing or measurements taken on organs of certain groups of animals; use of the mean, the standard deviation and the coefficient of correlation; frequency curves, symmetrical, skew and double humped; problems of evolution; of differential death rates and selection; illustrations made on data of Galton and Weldon; the study of racial differences through measurement of human skulls.

* This small forward-looking club had been formed in 1885 by a group of men and women who were convinced that some of the social problems of their day could only be furthered by a deeper understanding, based partly on historical research, of the relationship between the sexes. It was for the light which it might throw on the laws of heredity that Pearson had picked on Galton's *Natural Inheritance* for review. Gaussian distributions and the calculus of probabilities were not of primary interest to the Club members.

† The duty of the Lecturer seems to have been to give a dozen end-of-the-day lectures a year to an extra mural audience.

It seems to me that reading through these old syllabuses of 1891–94 we can get in summary form the picture of how under the stimulus of contact with Weldon at University College—and here I would place Weldon's influence before that of Galton—the applied mathematician in Pearson had at last discovered what he was needing, a field in which his special powers could be brought into action in solving some of the problems of life—*metron* applied to *bios*.

Weldon approached the unborn subject of biometry from a quite different angle. He had taken the Cambridge Natural Science Tripos in 1881, with zoology as his principle interest. After a period occupied in research and demonstrating he had been appointed in 1884 to a University Lectureship in Invertebrate Morphology. Having much of the outlook of a field naturalist, he spent many of his vacations when at Cambridge and later in collecting and 'dredging', sometimes abroad and often, after its completion, at the Marine Biological Laboratory in Plymouth.

He had started as most of the younger men of his day with an immense enthusiasm for the Darwinian theory of natural selection and sought the opportunity to contribute to the proofs of what could be described as only a working hypothesis. He had turned first to the current morphological and embryological methods of attack, but by the late 1880's he was beginning to doubt whether much progress could be made this way. So it was that his thoughts began to turn to the study of variation and correlation in organic characters. It was at this juncture that Galton's *Natural Inheritance* brought him sudden illumination. Here, he felt, were the statistical methods of measurement, description and analysis which might help to establish evidence supporting the Darwinian theory.

His paper of 1890, showing that the distributions of four different measurements (expressed as ratios to total length) made on several different local races of the shrimp (*Crangon vulgaris*) closely followed the normal or Gaussian law, was almost certainly the first paper in which statistical methods were applied to biological types other than man. In the statistical treatment of the data he had received help from Galton as a referee, but the credit for making the vast number of measurements and for seeing the bearing of such results on the problems of evolution was his own. A second paper in the series (1892) gave the coefficients or correlation between organs in the same individual and compared these for the four local races.

These two early papers were but first steps, showing that the simple models of statistical theory then current, the univariate and bivariate normal distributions, were relevant to zoological data. The further programme of research which was taking shape in his mind was to be set out in the third paper (1893, p. 329), where he wrote:

> It cannot be too strongly urged that the problem of animal evolution is essentially a statistical problem: that before we can properly estimate the changes going on in a race or species we must know accurately (*a*) the percentage of animals which exhibit a given amount of abnormality with regard to a particular character; (*b*) the degree of abnormality of other organs which accompanies a given abnormality in one; (*c*) the difference between the death rate per cent. in animals of different degrees of abnormality with respect to any organ; (*d*) the abnormality of offspring in terms of the abnormality of parents and vice versa. These are all questions of arithmetic; and when we know the numerical answers to these questions for a number of species we shall know the deviation and the rate of change in these species at the present day—a knowledge which is the only legitimate basis for speculations as to their past history, and future fate.

To handle these questions with any degree of assurance required in fact something more than arithmetic—the development of a more advanced theory of mathematical statistics.

Weldon's own mathematical knowledge at this period was limited and he realized that much more would be needed if the new tools were to deal adequately with the problems he began to see ahead. He set about removing this disadvantage in two ways; by himself beginning a study of the great French writers on the calculus of probability; and by seeking the co-operation of a mathematician in this project of demonstrating Darwinian evolution by statistical inquiry. After failing to get help from Cambridge he turned very naturally to his colleague Karl Pearson at University College.

In this sketch of antecedents I have not attempted to give any account of Francis Galton. His position as friend and counseller of the two younger men cannot be questioned; it was he who had taken the initial step in developing the ideas of correlation and regression which were to hasten the introduction of quantitative analysis into fields of biological, medical and sociological research. But in 1890 Galton was 68 years old, a man of established reputation whose long history of achievement has been told elsewhere. It was Weldon and Pearson who were to bring a fresh impetus into the field and it is their approach to our period of history which is therefore of special interest.

In the pages which follow I shall try to bring out some of the human side of the venture: the conflict of view-points of the biologist and the mathematician; the time so often taken, when on the fringe of the unknown, to see that next step forward which now seems so obvious to us, years afterwards, the enthusiasm which could lead the protagonists, though attached to the same College, to speed a second letter by the midnight post with some modified calculations after a first, sent on the same evening. I shall deal almost entirely here with problems coming under the head of variation, hoping at a later date to introduce some of the discussions on correlation.

2. The place of the normal curve

Whether we turn to *Natural Inheritance*, to the correspondence between Galton, Weldon and Pearson or to the Notes of Pearson's statistics lectures taken down in 1894–96 by Udny Yule (1871–1951), we see the central, predominating place which the Error Curve and the Binomial held in statistical thinking in 1890. As a result some effort was needed to break free from certain traditions. Starting from the work of De Moivre in 1733,* what we now call the Normal Curve was first derived as a mathematical approximation to the point binomial. Later on, when the Normal Curve was found to give an admirable fit to numerous recorded distributions of errors of observation, it had become customary to explain this good graduation in terms of a theory of the super-position of a series of elementary errors; text books on the Theory of Errors contained, for example, so called 'proofs' of the normal law.†

Galton had given to this idea a visual significance, by making a small mechanical model, often termed his Quincunx, and fully described with a diagram on pp. 63–5 of *Natural Inheritance*. In this model, lead shot from a funnel are dropped onto a succession of rows of equally spaced pins and collected below in a number of vertical compartments. In the

* Published in the *Supplementum* to his 1730 *Miscellanea Analytica*.

† As a reaction to this view among astronomers I remember how Sir Arthur Eddington in his Cambridge lectures about 1920 on the Combination of Observations used to quote the remark that 'to say that errors must obey the normal law means taking away the right of the free-born Englishman to make any error he darn well pleases!'

happy phrasing which he used when appealing to his reader's imagination, Galton wrote

> 'The shot passes through the funnel and issuing from its narrow end scampers deviously down through the pins in a curious and interesting way; each of them darting to the right or left, as the case may be, every time it strikes a pin. The pins are disposed in a quincunx fashion, so that every descending shot strikes against a pin in each successive row....The outline of the columns of shot that accumulate in successive compartments approximates to the Curve of Frequency, and is closely of the same shape however often the experiment is repeated....
>
> The principle on which the action of the apparatus depends is, that a number of small and independent accidents befall each shot in its career. In rare cases, a long run of luck continues to favour the course of a particular shot towards either outside place, but in the large majority of instances the number of accidents that cause Deviation to the right, balance in a greater or less degree those that cause Deviation to the left....'

Possibly for use in his Gresham College Lectures on the Theory of Probability given in 1893, Pearson constructed a modification of Galton's model, in which the shot fell onto a succession of rows of small triangles projecting from the back-board, which could be progressively stepped sideways from row to row. As a result the chance of a shot bouncing to the right (p) was not equal to that of it bouncing to the left (q). The distribution collected in the compartments should then not be symmetrical, but correspond roughly to the terms of the binomial $(q+p)^n$ with $p \neq q$.

The emphasis which Galton placed on the Normal curve can be best illustrated by quoting some further passages from his *Natural Inheritance*.

> I need hardly remind the reader that the Law of Error upon which these Normal Values are based, was excogitated for the use of astronomers and others who are concerned with extreme accuracy of measurement, and without the slightest idea until the time of Quetelet that they might be applicable to human measures. But Errors, Differences, Deviations, Divergencies, Dispersions, and individual Variations, all spring from the same kind of causes. Objects that bear the same name, or can be described by the same phrase, are thereby acknowledged to have common points of resemblance, and to rank as members of the same species, class, or whatever else we may please to call the group. On the other hand, every object has Differences peculiar to itself, by which it is distinguished from others.
>
> This general statement is applicable to thousands of instances. The Law of Error finds a footing wherever the individual peculiarities are wholly due to the combined influence of a multitude of 'accidents', in the sense in which that word has already been defined. All persons conversant with statistics are aware that this supposition brings Variability within the grasp of the laws of Chance, with the result that the relative frequency of Deviations of different amounts admits of being calculated, when those amounts are measured in terms of any self-contained unit of variability, such as our Q (pp. 54–5).

Again, in a section headed *The Charms of Statistics*, we find

> It is difficult to understand why statisticians commonly limit their inquiries to Averages, and do not revel in more comprehensive views. Their souls seem as dull to the charm of variety as that of the native of one of our flat English counties, whose retrospect of Switzerland was that, if its mountains could be thrown into its lakes, two nuisances would be got rid of at once. An Average is but a solitary fact, whereas if a single other fact be added to it, an entire Normal Scheme, which nearly corresponds to the observed one, starts potentially into existence.
>
> Some people hate the very name of statistics, but I find them full of beauty and interest. Whenever they are not brutalized, but delicately handled by the higher methods, and are warily interpreted, their power of dealing with complicated phenomena is extraordinary. They are the only tools by which an opening can be cut through the formidable thicket of difficulties that bars the path of those who pursue the Science of man (pp. 62–3).

Finally, under *Order in Apparent Chaos*, he writes

> I know of scarcely anything so apt to impress the imagination as the wonderful form of cosmic order expressed by the 'Law of Frequency of Error'. The law would have been personified by the

Greeks and deified, if they had known of it. It reigns with serenity and in complete self-effacement amidst the wildest confusion. The huger the mob, and the greater the apparent anarchy the more perfect is its sway. It is the supreme law of Unreason. Whenever a large sample of chaotic elements are taken in hand and marshalled in the order of their magnitude, an unsuspected and most beautiful form of regularity proves to have been latent all along. The tops of the marshalled row form a flowing curve of invariable proportions; and each element, as it is sorted into place, finds, as it were, a pre-ordained niche, accurately adapted to fit it. If the measurement at any two specified Grades in the row are known, those that will be found at every other Grade, except towards the extreme ends, can be predicted in the way already explained, and with much precision (p. 66).

This enthusiastic placing of the Normal distribution in the forefront of the study of what he terms the Science of man was not made without considerable background investigation. At the International Exhibition held in London in 1884, Galton had had an Anthropometric Laboratory and he includes in *Natural Inheritance* a summary of some of the data which he had collected, by taking measurements on visitors to the Exhibition. Thus he measured nine physical characters in men and women, the frequencies for the 18 distributions varying between 212 and 1013. For each distribution he had found the standardized deviates (measured from the median in terms of the probable error, not the standard deviation) to the eleven percentage points shown below. He found the mean of these standardized deviates for the 18 distributions and compared them with the Normal curve values as follows:

Cumulative % from lower tail		...	5	10	20	30	40	50	60	70	80	90	95
Mean of 18 deviates	Observed		−2·44	−1·87	−1·24	−0·77	−0·40	0	0·38	0·75	1·21	1·92	2·4
	Normal		−2·44	−1·90	−1·25	−0·78	−0·38	0	0·38	0·78	1·25	1·90	2·4

Over 70 years ago these results must have appeared as remarkable pointers in a hitherto unexplored field. It is hardly surprising that Weldon, when collecting the measurements of physical characters in animal populations should start with the assumption that these would be normally distributed within a homogeneous race.

3. The double humped curve

In this early work there appeared to be one noticeable exception to the fit of Normal curves to Weldon's distribution, that of the relative frontal breadth of the Naples crabs. He thought that this might arise from the mixture of two local races, providing perhaps evidence of the beginning of some process of selection at work. With much arithmetical labour, using trial and error, he graduated the distribution with the sum of two Normal curves. On the 27 November 1892, he wrote to both Galton* and Pearson telling of his achievement; to the latter he says:

Out of the mouths of babes and sucklings hath He perfected praise! In the last few evenings I have wrestled with a double humped curve, and have overthrown it. Enclosed is the diagram... [He adds numerical results]. If you scoff at this, I shall never forgive you.

It was this problem of dissection of a frequency distribution into two Normal components which led to Pearson's first statistical memoir, presented to the Royal Society in the autumn of 1893 and published in the *Philosophical Transactions* in the following year. In the most general case, the method involved the determination of the roots of a nonic whose

* At the end of his letter to Galton he remarked: 'Therefore, either Naples is the meeting point of two distinct races of crabs, or a 'sport' is in process of establishment. You have so often spoken of this kind of curve as certain to occur, that I am glad to send you the first case which I have found.'

parameters were derived from the first five moments of the observed composite distribution. Note that the introduction of the method of moments into the fitting of frequency curves is here described as giving a utilitarian answer to a practical problem; the possibility of other better solutions is admitted. Pearson recognized that some objective method of measuring goodness of fit had yet to be found; a possible method suggested was the comparison of the next, here the sixth, moments of the observed and fitted distributions. In fact, of course, the standard error of a sixth moment was so large that little could be achieved from this method of attack, but the derivation of the sampling errors of high moments lay in the future.

The question 'does a Normal curve fit this distribution and what does this mean if it does not?' was clearly prominent in their discussions. There were three obvious alternatives:

(*a*) The discrepancy between theory and observation is no more than might be expected to arise in random sampling.

(*b*) The data are heterogeneous, composed of two or more Normal distributions.

(*c*) The data are homogeneous, but there is real asymmetry in the distribution of the variable measured.

The conclusion (*c*) may have been hard to accept, such was the prestige surrounding the Normal law. But if it was accepted, it was still possible to retain the concept of a finite number of under-lying contributory causes, by deriving a continuous asymmetrical frequency curve from an asymmetrical binomial, on the lines of Pearson's Gresham College shot-model referred to above.

Weldon explored the position in an empirical way, both by tossing dice and by calculating the terms of a number of binomial distributions, $N(q+p)^n$, with $p \neq q$. The following extract from a letter to Pearson refers to this latter type of investigation:

23 April 1893

I have had a shock!

It seemed to me that the apparent symmetry of variation in animals showed that every 'accident' occurred about as often as any other, and that there was not, in any animal I had seen a 'tendency', as biologists have it 'to vary in one direction rather than in the opposite'.

But certain words of yours remained in my mind; and on Friday night I expanded $(0{\cdot}6+0{\cdot}4)^{20}$ and $(0{\cdot}7+0{\cdot}3)^{20}$ with most alarming results. I enclose a diagram of the appreciable terms in $(0{\cdot}7+0{\cdot}3)^{20}$. I should certainly not appreciate so slight a degree of asymmetry in an experimental curve zig-zag-ing about the diagram. But I should say that the observations varied symmetrically about the 7th term, $(p^{14}q^6)$.

So that no result of a kind which I had fondly hoped for can be drawn from these curves! I hoped that if an organ varied in a particular direction—that is if p became greater than q—the asymmetry of the curve would give some sort of measure of the difference between the two; and so a sort of kinetic of variation might be built up.

But if p may be more than twice as great as q with the abominable result which I enclose, that little hope goes to pieces.

It may seem curious to us today that so much weight was given to the idea of an underlying model in which contributory causes or factors led to discrete distributions which could be approximated by a continuous curve. As far as I know neither Weldon or Pearson made any serious attempt to identify the parameters of the discrete distributions with any biological phenomena. That all the component 'accidents' of Weldon's model would have a common p seems most unlikely.

As is well known Pearson (1895) obtained the fundamental differential equation of his generalized frequency curves as the limit of the slope/ordinate ratio of a hypergeometric series, and this was the approach which he still followed in his lecture presentation of the 1920's. In theoretical development, the 200-year-old tradition of deriving the Normal curve as an approximation to the binomial had a lasting hold on the imagination, though it is doubtful whether in practice, even in early days, he was much concerned about the physical meaning of his hypergeometric parameters.

The immediate stimulus for the development of theory leading to a system of skew frequency curves seems, however, to have come from Edgeworth, not from Galton or Weldon. Writing to his friend W. H. Macaulay of Kings College, Cambridge on 18 August 1895, Pearson remarks:

> There is a long tale as to the skew curves. Edgeworth came to me with some skew price curves nearly 18 months or two years ago [letters from Edgeworth suggest it was in the autumn of 1893] and asked me if I could discover any means of dealing with skewness. I had come to skewness also in my Gresham lectures. I went to him in about a fortnight and said I think I have got a solution out, here is the equation, and told him my chief (assumed) discoveries. I further said I don't intend to publish till I have illustrated every point from practical statistics....

In this connexion we find him writing to Galton on 19 November 1893:

> If you will suggest any type of statistics which you think ought better than another to give Macalister's curve,*...my Demonstrator Mr Yule and I will endeavour to fit them as an example of a type or class of asymmetrics.

The first Brunsviga calculator was not purchased until 1894, so that moment-calculation and curve fitting required in illustrating theory from 'practical statistics' proved rather laborious procedures involving a number of numerical slips which had later to be corrected. In writing to Yule in November 1894, Pearson remarked: 'I want to purchase a Brunsviga calculating machine before anything else, and am making inquiries about it. I think it would make moment-calculating fairly easy.' And later: 'Henrici† speaks well of the Brunsviga, but doubts whether it will not wear out with a few years' use'. But Henrici, of course, was wrong! To the end of his life, Pearson used as a spare machine at home a Brunsviga which must have been of the beginning-of-the-century vintage, while Maurice Kendall has told me that he has and still uses the Brunsviga which Yule purchased for work on his first (1910) edition of *An Introduction to the Theory of Statistics.*

4. Wanted: A test for goodness-of-fit

It seems appropriate to quote at the head of this section some remarks which Pearson wrote long afterwards about Galton in the third volume of his *Life* (1930, 3A, p. 6).

> Again, if the reader anticipates that Galton was a faultless genius, who solved his problems straight away without slip or doubtful procedure, he is bound to be disappointed. Some creative minds may have done that, or appear to have done it, because, the building erected, they left no signs of the scaffolding; but the majority of able men stumble and grope in the twilight like their lesser brethren, only they have the persistency and insight which carries them on to the dawn.

Weldon's extensive dice tossing had another object in view. No valid criterion existed at this date by which to judge whether the differences between a series of group frequencies

* This was the log-normal curve which did not of course belong to his system, although agreeing very closely in shape with a corresponding four-moment Type VI curve.

† O. Henrici had been Pearson's predecessor in the Applied Mathematics chair at University College; he had moved from there to the City and Guilds of London Institute.

and a theoretical law, taken as a whole, were or were not more than might be attributed to the chance fluctuations of random sampling. Weldon, with his flair for empiricism, therefore decided to explore the kind of random fluctuations which one might expect to get in sampling, by comparing his tossing results with theoretical binomial expectations.

It was at this juncture that an incident occurred leading to the series of letters quoted below; unfortunately scarcely any of the letters written *to* Weldon have been preserved.

1. *Weldon to Galton*, from 30a Wimpole Street, W., 4. ii. 94.

Dear Mr Galton,

Will you be kind enough to give me your opinion on the following point?

I have collected 26,306 tosses of groups of 12 dice, for use at the Royal Institution. In each group the event recorded is the number of dice with 5 or 6 points, so that the chance of success with each die is 1/3. I enclose the result, which seems to me good.

A certain set of 7000 tosses, forming part of this result, was made for me by a clerk in the office of University College, whose accuracy in work of another kind I have had occasion to test by asking him to copy 24,000 numbers of 3 figures each, with excellent results.

A day or two ago, Pearson wanted some records of the kind in a hurry, in order to illustrate a lecture,* and I gave him the record of the clerk's 7000 tosses, together with some others.

I gave him the 7000 separately from the rest, and on examination he rejects them, because he thinks the deviation from the theoretically most probable result is so great as to make the record intrinsically incredible.

You will see how serious a matter this is. On the one hand I feel that I have no right to reject an experimental result for this kind of reason; and on the other, the result itself does not seem to me incredible.

I am anxious, however, not to rely upon my own judgement in so difficult a matter—I have therefore resolved to consult as many people as possible. Last night I saw Greenhill, whose experience in target practice at Woolwich makes him know this kind of thing statistically as well as mathematically—he is of opinion that the record is perfectly credible, and that I have no shadow of reason to disregard it.

Today I am sending it to you and to Edgeworth.†

Forgive me for troubling you, when I know how busy you are; but my need is very great.

Yours very truly,

W. F. R. WELDON

The 'most probable' result in each of the enclosed tables is obtained from the expansion of

$$(\tfrac{1}{3}+\tfrac{2}{3})^{12}$$

and not from any form of approximation.

We can get a clue as to why Pearson considered 'the record as intrinsically incredible' from Yule's notes, taken down at Pearson's lectures in the autumn of 1894. In column 2 of the table below are presumably the result of the clerk's dice throwing, 7006 (not 7000) tosses of 12 dice. Column 3 shows the binomial expected frequencies as given by Yule, from the expansion of $(\frac{2}{3}+\frac{1}{3})^{12}$. Note that these figures add to 7010·2 not 7006; Yule does not however quote totals. He remarks:

> The fit is good except at the ordinates four and five; 4 is 98 too low, 5 is 69 too high. This is odd: can the experimenter have inadvertently booked his results in the wrong column? Let us find out first what the chance against the above combined occurrence is.

* This lecture is almost certainly one of the four lectures on 'The Geometry of Chance' given that winter by Pearson at Gresham College.

† For the correspondence with Edgeworth which resulted, see pp. 333 – 335 below.

No. of 5's or 6's	Observed frequency	Expected frequency	$O-E$	$\frac{(O-E)^2}{E}$
0	45	54·0	−9·0	1·50
1	327	323·7	3·3	0·03
2	886	890·2	−4·2	0·02
3	1475	1483·7	−8·7	0·05
4	1571	1669·2	−98·2	5·78
5	1404	1335·3	68·7	3·53
6	787	788·9	−1·9	0·00
7	367	333·8	33·2	3·30
8	112	104·3	7·7	0·57
9	29	23·2	5·8	1·45
10	2	3·5		
11	1	0·4	−0·9	0·21
12	0	0·0		
Totals	7006	7010·2	−4·2	16·44 = χ^2

The argument used in the Notes, which was presumably the lecturer's, was based on the following analysis in which I have made the totals 7000 to fit in with Yule's arithmetic.

Group	Observed		Expected	
$x = 4$	1571		1669	
$x = 5$	1404	5429	1335	5331
Remainder	4025		3996	
Total	7000		7000	

Directing attention to the most exceptional group, that containing four '5's or 6's', Yule gives the standard error of the observed frequency as

$$\left(7000.\frac{1669}{7000}.\frac{5331}{7000}\right) = 35{\cdot}65$$

which gives $(1571-1669)/35{\cdot}65 = -2{\cdot}75$ as the ratio of the deviation from expectation to its standard error. Using the normal approximation to the sum of binomial terms, the probability of an absolute deviation (positive or negative) from expectation as large or larger than that observed is 0·0060 or $\frac{1}{167}$.*

The argument of the Notes now runs as follows:

> But now what is the chance of this being combined with the deviation of 69 in the next ordinate? We must be careful how we proceed here. We cannot simply work out, *as above*, the chance of the deviation 69 and multiply by the chance above, for the two are *not* independent. If the 98 has already been lost (or gained) it must be made up *somehow* by the other ordinates. We must remove the 1571 whose position has already been allotted and deal only with the rest.

He now finds a standard error for the 'five' group of

$$\left(5429.\frac{1335}{5331}.\frac{3996}{5331}\right) = 31{\cdot}92,$$

an expectation in the group of $5429 \times 1335/5331 = 1359$ and a ratio of deviate to standard error of $(1404-1359)/31{\cdot}92 = 1{\cdot}39$. The probability of an absolute deviation exceeding

* Quoting what he says is a table given in Pearson's 'Gresham College Lecture Notes', Yule gives the probability as $\frac{1}{136}$.

1·39 is, from the normal approximation, $0·16\overline{5}$. It is now argued that the probability of a pair of deviations, having *opposite signs*, as large or larger than those observed is

$$\tfrac{1}{2} \times 0{\cdot}0060 \times 0{\cdot}165 = 0{\cdot}00049$$

or approximately $\frac{1}{2000}$.* Yule adds that 'if we went through the whole lot of all other deviations, the total chance for them might be somewhat smaller'. 'Consequently it appears not unreasonable' his Notes add, 'to conclude that the experimenter has made some slip or other in entering results in the wrong column'.

To these quotations from Yule's Notes must be added some correspondence with F. Y. Edgeworth which occurred in February 1894 when the argument was still at its height.

2. *Edgeworth to Weldon*, from All Souls College, Oxford, 7. ii. 94.

Dear Weldon,

The tests which I have applied to the cases with *four* and *five* dice do not yield a result which excites much suspicion. I shall be curious to know your final decision.

Yours very truly,

F. Y. Edgeworth

From some loose notes with the letters it appears that Edgeworth's calculations had been rather perfunctory. He had: (*a*) correctly found the separate moduli (standard error $\times\sqrt{2}$) for the 'four' group and the 'five' group as 50·4 and 46·4 (S.E.'s of 35·6 and 32·8); (*b*) divided these into the observed deviations from expectation of 98·2 and 68·7; (*c*) commented that in the first case: 'the ratio C is not quite 2, corresponding to odds against of rather more than 200 to 1, which can't be thought prohibitive I think'; (*d*) in the second case, remarked that the ratio $C = 1{\cdot}4$ corresponded to a very 'moderate improbability' and added: 'I write without a Table by me. But I know that the odds are nothing out of the way, say 50 or 100 to 1'.

We do not know what Pearson had first written to Edgeworth but he may well have proposed a procedure on the lines set out in Yule's Notes. There are, however, three further letters which have survived and are of interest in showing how at the beginnings men 'stumble in the twilight'.

3. *Edgeworth to Pearson*, from All Souls College, Oxford, 9. ii. 94.

My dear Pearson,

Your method would be all right as long as you are given only *two* results of the kind operated on. But it is not open to you I think to apply twelve (independent) tests to a composite event such as that considered; to select *two* which accuse, as the French say, a cause other than accident; and multiply the (im)probability of each to find the (im)probability of the system. To take a simple case, suppose each result presented one of two alternatives, either (*a*) ordinary, (*b*) improbable in the degree 1:100. Suppose that having *fifty* returns such as those before us (or any similar data) you look through them and find two events the probability of each of which is only 1/100, I don't think it is open to you to say that the probability of the system is 1/10,000. You should consider the growing likelihood, as you increase the number of your trials, that such extraordinary results will be presented. You surely would not make the same assertion if there had been *fifty thousand* data.

So it seems to me at present; but I know how kaleidoscopic these problems are.

In haste

Yours very truly,

F. Y. Edgeworth

* Through what was possibly a numerical slip afterwards corrected in red ink, the original calculations in the notes gave 1/14,416. Whether it was this probability, rather than the 1/2000, which induced Pearson to tell Weldon that the record was 'intrinsically incredible', one cannot now say!

4. *Pearson to Edgeworth*, from 7 Well Road, Hampstead, N.W., 10. ii. 94

My dear Edgeworth,

Probabilities are very slippery things and I may very well be wrong, but I do not clearly follow A| your reasoning or illustrations. You say take 50 returns of 1:100 degree of probability. If two occurred should I calculate the chance of the system as 1/10,000? Certainly *not*. Following the B| method I applied to the dice, I should ask what is the chance of one 1:100 event occurring *in 50 throws*. This is $\frac{1}{2}$, and after its having occurred what is the chance of another like event in the remaining 49 occurrences. It is 49/100. The combined chance is $49/100 \times \frac{1}{2} = 0{\cdot}245$, strikingly close to the $0{\cdot}25 = \frac{1}{4}$, which I assume you to mean to be the probability of the event you suggest.

Now look at the dice problem in the same way. Disregard all but 5 and 6 occurring 4 times in the 12 dice. I calculated the chances that in 7000 throws there should be a defect of 98 or more. Chance = 1/270. Now make another experiment with the same 12 dice, take 7000 throws and inquire how often there will be 5 and 6 occurring 5 times a certain number of times in excess. Suppose C| this came out 1/40. Surely the combined chance against the two experiments would be $1/270 \times 1/40$? They are quite independent.

Now what I contended was this, that admitting a defect of 98, this defect ought to be distributed theoretically along the whole line of groups and that having done this, the distribution of the C| remaining number of throws among the 5, 6, 7, 0, 1, 2, etc., was a practically new and independent experiment.

Your method of looking at the matter leads me into difficulties in the following way. Suppose we are quite certain that a population follows a normal frequency curve. We have, we will say, discovered this by measurements on several 100,000's. Now we take a sample 10,000, and draw its frequency curve, with a result when compared with the normal curve like this:

[Here Pearson sketched roughly a normal curve and a frequency polygon with a single 'hump' rising well above the curve at X.]

Here the hump at X is counterbalanced by proportionate diminution of all the other ordinates. The chance of this hump we will take to be 1/270. Now suppose instead of this, which fits the curve very well except at A, we had a result like that over page, with a marked deficit at Y.

[Here is another diagram in which the frequency polygon has a marked dip at Y as well as the hump at X, the defect at Y being rather less than excess at X.*]

D| Are you prepared to say that both these systems are equally probable and both improbabilities are to be measured, hump X being more improbable than dip Y by the improbability of X? It seems to me that the appearance of another anomaly like Y which almost counterbalances X must much increase the improbability of this second system as compared with the first. You say, No! E| Chance of $X = 1/270$, chance of $Y = 1/40$, and chance of whole system is the greater of these $= 1/270$, F| and is not touched by whether Y exists or not. This does not seem to me at all satisfactory. I quite agree X and Y are not independent. Well, make them so by cutting X off and distributing it in C| proper proportions round the curve. Y will take some of it, but *not* all. Having done this the reduced Y is an independent event, is an independent discrepancy in the normal frequency curve.

You say but there are other defects besides Y. Certainly, but when we have filled up Y again they are of such minor importance that they are hardly worth considering—they fall, continuing the process, so low down in the scale of fractions of the corresponding S.D.S. Here seems to me our G| difference. You appear to calculate the improbability of a given distribution by its chief irregularity. I assert that weight must be given to other irregularities. Now if dice gave a symmetrical curve we should compare the theoretical and experimental Standard deviations and thus get a test of the H| system as a *whole*. Surely the experimental S. deviation pays attention to Y as well as to X. It diverges more from the theoretical, because X is not proportionately distributed but is collected largely at Y. Now a similar result, it seems to me, must hold for these skew systems. Accordingly I contend that your 1/270—mind you in itself a somewhat improbable result—does not represent I| the badness of the experiment, that irregularity in other columns is also to be taken into account. J| I admit that the proper way of doing this is quite open to question, but I think whatever way it is done the great dip at Y will immensely increase the odds. Please note that a *second* 7000 experiments by the same clerk, calculated from the two worst columns only give a chance of 1/250 instead of 1/3240 showing a marked improvement. While 19,000 experiments give a chance of 1/160 (1st 7000 K| *excluded*) by same method.

Yours very sincerely,

KARL PEARSON

* I have changed Pearson's original letters, A and B, to X and Y in order to avoid confusion with the letters A–K which Edgeworth added to Pearson's letter for reference purposes when he returned it with his own answer.

5. *Edgeworth to Pearson,* from 5 Mt. Vernon, Hampstead, 12. ii. 1894.

Dear Pearson,

Excuse my returning your letter for convenience of reference.*

A. I mean *fifty* data such as your numbers of *fours, fives,* etc.; or rather the fifty *independent* observations such as those which you derive these from.

For C, (p. 4 and after), I quite agree as to the 'independence' of which you speak.

B. The chance of an event 'of 1:100 degree of probability' (p. 1) occurring (at least) once in fifty (independent) trials is $1-(99/100)^{50}$. The chance of its occurring as much as twice is

$$1-\left\{\left(\frac{99}{100}\right)^{50}+50\times\left(\frac{99}{100}\right)^{49}\times\frac{1}{100}\right\}.$$

D. Certainly not. I am not prepared to take *no* account of B (Y in your figure). I only say it is very *difficult* to take account of it where the case is not so simple as my 1/100.

The significance of X and Y is modified according to the number of 'trials' or independent events there may be (I have tried to indicate these by so many intervals).

F. I here meant that—exactly. 'Not much suspicion' are I think my words to Weldon.

G. I never emphasized the *chief.*

H. Surely. But what I complain [of] is that you don't take the system as a whole but content yourself with a method proper to a *single datum or pair.* Having n observations you look out for one or two improbable results; and of course you will find them if n is large. There is our difference. You seem to think the size of n makes no difference. I say then as before let n be 50,000. It is chock sure that you will have two of the events (considered as independent) occurring, although the probability of each is 1/100. See your A. The improbability is *not* 1/10,000 but

$$1-\left\{\left(\frac{99}{100}\right)^{50,000}+50,000\left(\frac{99}{100}\right)^{49,999}\times\frac{1}{100}\right\}.$$

There is the issue; I subscribe to J.

K. Well then it is no wonder that we should have some deviations just to keep up the average. If you are at University College tomorrow afternoon I may see you. I shall be there about 6.15.

Yours very truly,

F. Y. EDGEWORTH

I write in great haste and may well have made slips.

There are two further letters from Edgeworth to Pearson written later that February which show that the methods of dealing with multiple discrepancies were still being discussed. But the topic has shifted from the immediate problem of Weldon's 7000 dice throws, and I have not space to quote them here.

The interested reader must choose his own method of tackling this problem. Edgeworth was clearly criticizing Pearson for having picked out the largest and second largest of a set of differences from expectation without making proper allowance for the fact that they were the largest in a group of 13. But even if this were recognized, no theory existed for dealing with the largest of a set of mutually but unequally correlated differences.

Pearson's (1900) χ^2-test for goodness-of-fit, derived some five years later after the theory of multiple correlation had been developed, leads to the calculations shown in columns 4 and 5 of my table. The probability of obtaining a χ^2 based on 10 degrees of freedom, exceeding 16·44, if the discrepancies in the table taken as a whole were due to chance, is 0·088. Certainly this does not seem exceptional, but of course the χ^2-test does not take account of certain kinds of coincidence. Perhaps the position of the two large discrepancies next to each other in the table might justify some suspicion.

Had there been prior grounds for thinking that Mr Hull, the clerk, might have mixed up

* The letters A–K refer to the corresponding passages which Edgeworth had marked on Pearsons' letter of 10 February.

entries in these two particular columns, of 'fours' and 'fives', a sensitive test would consist in applying χ^2 to the following table.*

	Observed	Expected	Difference
$x = 4$	1571	1669	-98
$x = 5$	1404	1335	$+69$
Remainder	4031	4002	$+29$
Total	7006	7006	

Here, $\chi^2 = 9{\cdot}53$ and $P\ (\chi^2 \geqslant 9{\cdot}53 | \nu = 2) = 0{\cdot}009$, so that the result appears more exceptional, though hardly so even here on the basis of critical odds used by Edgeworth. However, we have been given no prior reasons for supposing that Mr Hull would confuse these two columns, and the selection of this most unfavourable three-category table would seem to be without justification. Theory, in fact, would be hard put to it to disprove Weldon, Greenhill and George Darwin's instinctive reaction to the figures; but the discussion must have emphasized the need for more thought and more mathematical research.

Yule's lecture notes mention an empirical measure of goodness-of-fit which was in use at the time. This was the ratio of (a) the area between the theoretical curve and the observed frequency polygon (not the histogram), taken everywhere as positive,† to (b) the total area under the curve. This may be set down roughly as

$$R = \Sigma|O-T|/\Sigma T$$

where O is the observed and T the theoretical frequency in a group. Yule quotes numerical values of this ratio, expressed as a percentage, for the cases of normal curves fitted to eleven different frequency distributions. The coefficients range between 5·85 and 13·5%, with a mean value of 8·0%. He remarks that when skew or compound curves were fitted to the same data, much better fits resulted, figures of '4% or so' being obtained.

During 1894 Weldon was much occupied with proposals for work which he planned to submit to a small, newly formed Royal Society 'Committee for conducting statistical inquiries into the measureable characteristics of plants and animals', of which he was secretary and Galton chairman. But among his letters to Galton discussing experiments to be sanctioned by the committee, there are many references to Pearson's asymmetrical frequency curves. It seems that neither Galton nor Weldon felt at home in the mathematics of Pearson's second Royal Society memoir (published in 1895); they were perhaps looking for a physical explanation of the fundamental differential equation, which they could not discover. Pearson also, with the enthusiasm of the creator of so elastic a system, was perhaps excusably trying out his curves on any set of non-normal data which came his way, without considering very deeply the biological meaning of the asymmetry. Thus, as Weldon pointed out, a skew distribution of measurements on the breadth of foreheads of Crabs or the stature of St Louis schoolgirls, might only reflect the fact that growth was going on within the age limits covered by the data.

The final letter which I shall quote, of a year later, bears on this theme and again well illustrates the slow process of bringing together the approaches of the biologist and the mathematician.

* I have amended Yule's reduced table given on p.332 above so that the totals now agree with the original data quoted.

† The use of the Drawing Office planimeters made this an easy quantity to measure.

6. *Weldon to Galton*, from 30*a* Wimpole Street, W., 6. iii. 1895.

Dear Mr Galton,

Let me congratulate you heartily upon your recovery; I shall look forward with great pleasure to seeing you next week.

Pearson *does* admit that he omits to consider the moving mean in his theory of skew curves—or he did so nearly a fortnight ago, when I charged him with it—we had a delightful afternoon, abusing each other in a friendly way about this point for some hours; he promised more consideration of it; but since then he has been in bed with influenza—I hope he will be well enough to take a short holiday in a day or two, but he cannot work for some time.

Ten of our men at University College, and in many classes half the students, are in bed: so that I, who never get anything worse than a bad cold, feel like the Wandering Jew in time of plague.

About the mathematicians. I feel the force of what you say, naturally. But I am horribly afraid of pure mathematicians with no experimental training.

Consider Pearson. He speaks of the curve of frontal breadths, tabulated in the report, as being a disgraceful approximation to a normal curve. I point out to him that I know of a few great causes (breakage and regeneration) which will account for these few abnormal observations: I suggest that these observations, because of the existence of exceptional causes, are not of the same value as the great mass of the others, and may therefore justly be disregarded. He takes the view that the curve of frequency representing the observations, must be treated as a purely geometrical figure, all the properties of which are of equal importance; so that if the two 'tails' of the curve, involving only a dozen observations, give a peculiarity to its properties, this peculiarity is of as much importance as any other property of the figure.

For this reason he has fitted a 'skew' curve to my 'frontal breadths'. This skew curve fits the dozen observations at the two ends better than a normal curve; it fits the rest of the curve, including more than 90 % of the observations, *worse*. This sort of thing is always being done by Pearson, and by any 'pure' mathematician.

Greenhill, to whom I took my troubles, laughs at the whole thing. You know that his chief business is to teach the properties of probability surfaces to artillery officers in connection with target practice; and he has a good deal of experience of curves made with your quincuncial screen of pins.

Greenhill is quite ready to admit the necessity of ignoring the few aberrant observations. George Darwin says that Pearson pays too much attention to the higher moments (which of course depend chiefly on the character of extreme observations.)

Now these are the two men working at the applications of Probability who know, not only mathematics, but the degree of approximation to be expected from an experiment. This sort of instinct as to what may be expected of an experiment and what may not is a quality very rare among young mathematicians, so far as I know them.

But enough of them—I shall look forward to your proposals next week.

Yours very truly,

W. F. R. Weldon

The Herring, which makes a skew curve are very heterogeneous. The mean value of the length from snout to anus, on 717 males, was very widely different from that given by 990 males—the extra 270 being obtained by opening another of the cases of herrings. I have not the figures at hand, because I sent them to Pearson, as basis for his curve; but he says that '*the material is homogeneous, with skew variation about one mean*'. I don't believe it!

5. Concluding remarks

These letters are in several ways revealing; it would have been easy for the 'young mathematician' and the younger zoologist (Weldon was three years Pearson's junior) to drift apart, but the compelling urgency of the field for exploration which lay ahead, a field in which they were in so many ways fitted to co-operate, held them together. No doubt Galton played an important part in bridging the gap between them, so that six years later with friendship firmly established they were planning the first issue of *Biometrika*.

In the context of the 1894 discussions, they were still 'stumbling in the twilight'. There were many defects in Weldon's crustacean data: breakage and regeneration in individuals (as he himself pointed out); lack of homogeneity; the unknown effect of age on relative growth of parts and other disturbing factors. These particular series of observations were indeed unlikely to lead to any clear evidence, based on the fit of frequency curves, bearing on the process of natural selection. But the arguments which arose undoubtedly helped to bring out the need for more thorough investigation, both experimental and theoretical.

Pearson, too, showed an unsureness in the handling of the theory of probability. I suppose that he was never really interested in this calculus for its own sake, as a pure mathematician might have been; he needed its help as a tool in the solution of problems which *did* hold his interest. It was perhaps typical of the early British approach to mathematical statistics that he could write 'probabilities are very slippery things' in answer to Edgeworth's 'I know how kaleidoscopic these problems are'!

But reflexions of this kind cannot conceal the fact that out of these arguments developed the ever expanding structure of the theory of mathematical statistics which we know today. That seems ample justification for putting these incidents on record.

References

GALTON, FRANCIS (1889). *Natural Inheritance.* London: Macmillan and Co.

PEARSON, E. S. (1938). *Karl Pearson. An Appreciation of Some Aspects of His Life and Work.* Cambridge University Press.*

PEARSON, KARL (1888). *The Ethic of Freethought, a Selection of Essays and Lectures.* London: T. Fisher Unwin. Republished (1901) by A. and C. Black.

PEARSON, KARL (1892). *The Grammar of Science.* London: Walter Scott. (1900, 1911), with additions, republished by A. and C. Black. (1937), first edition, republished in the Everyman Library (no. 939) by J. M. Dent and Sons Ltd.

PEARSON, KARL (1894). Contributions to the mathematical theory of evolution. *Phil. Trans.* A, **185**, 71–110.

PEARSON, KARL (1895). Contributions to the mathematical theory of evolution. II. Skew variation in homogeneous material. *Phil. Trans.* A, **186**, 343–414.

PEARSON, KARL (1897). *The Chances of Death and other Studies in Evolution.* London: Edward Arnold.

PEARSON, KARL (1906). Walter Frank Raphael Weldon. 1860–1906. A Memoir. *Biometrika*, **5**, 1–52.†

PEARSON, KARL (1930). *The Life, Letters and Labours of Francis Galton.* **3***A*. Cambridge University Press.

WELDON, W. F. R. (1890). The variations occurring in certain Decapod Crustacea. I. *Crangon vulgaris. Proc. Roy. Soc.* **47**, 445–53.

WELDON, W. F. R. (1892). Certain correlated variations in *Crangon vulgaris. Proc. Roy. Soc.* **51**, 2–21.

WELDON, W. F. R. (1893). On certain correlated variations in *Carcinus moenas. Proc. Roy. Soc.* **54**, 318–29.

* This book, now out of print, put together two articles previously published in *Biometrika* (1936), **28**, 193–257 and (1937), **29**, 161–248. Unfortunately two Appendices, one giving the Gresham College Syllabuses and the other G. U. Yule's summary of his 1894–96 lectures notes, are not included in the *Biometrika* articles.

†Paper No. 21 in this volume

SOME REFLEXIONS ON CONTINUITY IN THE DEVELOPMENT OF MATHEMATICAL STATISTICS, 1885–1920

By E. S. PEARSON

SUMMARY

Some aspects of the work of Galton, Edgeworth, Weldon, K. Pearson, Gosset and Fisher are reviewed, with stress on continuity of development.

1. INTRODUCTION

The origin of this paper lies in a wish to put on record some of the history of the development of mathematical statistics and biometry, roughly contained in that formative period, 1890–1905. In a previous article (1965)* I described some incidents which occurred when W. F. R. Weldon and Karl Pearson first made contact in the early 1890's. On going further into the mass of correspondence and many published papers of the ensuing years, it became evident that the subject was a more formidable one to tackle than I had originally imagined. It was clearly necessary to go back a little in time, if only the better to understand the background from which the chief contributors in my period started. Again, I had to explore forward towards the present, in an endeavour to trace what happened to the ideas and methods which then saw birth and what part they played in the continuing process of historical development.

What struck me in making this survey was the essential continuity of development over a wide span of years in what might be termed the English school of mathematical statistics. It is not surprising to find this continuity within a single country, but I think that a tendency to concentrate on the discoveries and advances made by particular individuals has helped to play down the importance of the links connecting them. This aspect of continuity I tried to present in a talk, which forms the basis of this paper, given to the Birmingham section of the Royal Statistical Society in January 1967. The justification for publishing what is hardly an exhaustive presentation lies in the hope that my sketch of relationships may encourage others to fill in gaps which I have left and to confirm or reject the impressions I have formed.

While wide and diligent reading may enable the historian to assign origins and priorities, the influence of one man's thought or writing upon another is far harder to establish and must often be a matter of personal opinion. Sometimes a writer has recorded in a letter or publication the debt which he has owed to a predecessor or contemporary; sometimes he may be almost unconscious of this influence; sometimes the only clue may lie in the preservation of a book from his library, with a date of purchase and perhaps pencilled annotations in the margin.

An interesting case is that of the relation between F. Y. Edgeworth and Karl Pearson. They were both influenced by Francis Galton, they were certainly in personal contact and

*Paper No. 22 in this volume

were working at the same time on asymmetrical frequency curves and multiple correlation. They were, however, both of them clearly originators with very distinct lines of approach and any attempt to assess priorities would be meaningless. In the case of Pearson and R. A. Fisher, whose periods of greatest output were separated by many years, the existence of continuity is clear. There can be no doubt that Fisher's study in his early twenties of Pearson's *Philosophical Transactions* papers of the 1890's influenced the direction of a good deal of his early work. Coming with a fresh mind and great mathematical powers, Fisher started in several directions from the points where Pearson and his school had left off. What is less clear is the extent to which Fisher was familiar with some of Edgeworth's work, where somewhat obscure presentation has put off many from its detailed study.

Two examples of my own illustrate the chancy nature of the record of what one man has owed to another. The idea of defining the class of 'alternative hypotheses' formed an essential part of Jerzy Neyman's and my approach to the testing of statistical hypotheses. The germ of the idea which we formalized was almost certainly given to me by W. S. Gosset, though I doubt whether he was aware of this. However, I kept a letter of his written to me in May 1926 and was able to publish an extract from it in my obituary article of 1938. Without the letter, this particular link with Gosset would have been unrecorded.

Again, in a less important matter I happened to re-read recently a *Bell Telephone Bulletin* of 1929 in which W. A. Shewhart presented four criteria for detecting the presence of assignable causes underlying the variation in manufactured products. 'We shall see later' he wrote '... that the two criteria to be described are less powerful than the two already given'. Did this sentence, I wonder, suggest to me the use of the word 'power' in Neyman's and my work of that period?

But however difficult it may be to be sure of the connecting links, I think one may sum up by saying that the history of mathematical statistics in the nineteenth and early twentieth centuries was characterized by a series of widening horizons to which many contributed, whether in parallel or sequence, as more and more fields of application were brought within the scope of statistical treatment. Imagination was often called for to see that a new field could be opened up; mathematical skill as well as imagination was necessary to develop the new tools which might be required. All the time, because data often tended to present themselves in certain standard forms whatever the field, there was a steady development of accepted techniques. Then too, the logically trained mind faced with broadly similar situations evolved, often more than once and independently, certain methods of data handling and certain principles of inference which were accepted as relevant in drawing conclusions from observations. It is some of these ideas which I shall now try to illustrate with more concrete facts.

2. The theory of errors. Francis Galton and correlation

One may ask, with a referee of this paper, how far the British statisticians of 70–80 years ago were familiar with the work of the continental mathematicians in the fields of probability and error theory. Certainly, if they did not go to original sources they could and did read Todhunter's *A History of the Mathematical Theory of Probability* (1865) which covered the period from Pascal to Laplace. But Todhunter did not describe the work of Gauss and his successors, and this may have been one of the reasons why English-speaking statisticians have seemed unfamiliar with it. It seems likely, for example, that with the pressure of work at University College in the 1890's and in the excitement of fashioning statistical techniques to tackle the new demands of biometry, Karl Pearson's study of continental literature was not at this time very thorough. This, indeed, he admits in a 1921 paper quoted below. A recent account by Seal (1967)* in the present *Biometrika* series of 'history' papers describes some of the results which Pearson and his collaborators seem to have overlooked.

The development of the theory of errors was particularly associated with the needs of the astronomers. To these men there would have seemed little occasion to go beyond the use of the Gaussian distribution in developing their theory of combination of observations. But they early became aware of the need to choose between alternative estimators, e.g. of a measure of dispersion, and we owe to them the first derivation of the distribution of a sum of squares $\Sigma(x_i - \bar{x})^2$. For many years this has been attributed to Helmert (1876) but very recently a Russian, O. B. Sheynin (1966), has pointed out an earlier derivation of this χ^2 distribution in a published Thesis of Ernst Abbe, presented to Jena University in 1863. Neither of the derivations was known in England at the time. The astronomers also had gone some way in formulating simple variance-components models; see Scheffé (1956).

The existence of correlation between functions of independent observable quantities had, of course, been recognized in the theory of combination of observations. Thus Bravais had in 1846 derived bivariate and trivariate Gaussian error distributions for the position of a point (x, y, z) in space, where x, y and z were estimated separately from linear functions of a certain number of independent observable variables, e.g. lengths, angles, etc. Later, as Seal reports, Schols had pointed out (1875) the existence of correlation between the horizontal and vertical errors in firing at an artillery target. But in this connexion a measure of the correlation was not of primary importance. It was Francis Galton who made the real step forward in the field of correlation by realizing its importance in the study of heredity. Here, it seems, was one of the big acts of imagination with far-reaching consequences, although, as sometimes happens in pioneer work, the hatching out of the idea was spread over several years.

Galton did not approach the subject from the aspect of error theory, but being a man of inquiring mind he collected data on inheritance and plotted in a two-way table the corresponding character values of parent and offspring. He also collected measurements of various physical characters, e.g. stature and span, from the same adult individual and made similar plots. He found that in his tables the marginal distributions were Gaussian or 'normal' (as Quetelet had found before him taking one variable at a time) but further, he noticed that the array means fell nearly on a straight line and that the variation within the arrays was approximately constant. From this he developed the idea of regression as applied to human inheritance, not realizing at first that this phenomenon of regression towards mediocrity in the offspring of a selected parent represented a characteristic to be found in any correlation table.

Some time later he noticed that the contours of equal density in his tables were elliptical and so he came in 1886 with the help of a Cambridge mathematician, Mr Hamilton Dickson, to the equation of the bivariate normal surface. [In fact the elliptic contours of his x, y distribution had been discussed by Bravais (1846) who showed that by a rotation of axes two new independent linear functions of the observable variables could be found.] It was a little after this, in 1888, that Galton reached the conception of the correlation coefficient as a measure of the intensity of relationship between two characters, suitable for application to all living forms. At the same time, after obtaining from his ancestral data other correla-

*Paper No. 15 in this volume

tions beyond that between parent and child, he began to feel his way towards his Law of Ancestral Heredity, later to be generalized and given more precision by Pearson in the form of a multiple regression formula.

3. F. Y. Edgeworth

The problems which Galton opened up in his *Natural Inheritance* of 1889 had a stimulating and exciting effect on several younger men, but before following down the main line of descent I should like to say a few words about F. Y. Edgeworth.

The writings of Edgeworth are interesting because apart from his own original contributions they provide a sidelight on some of the statistical ideas and practices recognized by the comparatively few persons who could be termed mathematical statisticians towards the close of the nineteenth century. We find that concepts which have been developed in a more thorough way at a later date, when they were sometimes thought to be original, were certainly current at this time. It was from the foundations laid by Laplace and Gauss that Edgeworth was himself, in his own way, tackling a number of the same problems which Karl Pearson was shortly to be concerned with. These sidelights exist partly because from time to time Edgeworth wrote articles in the *Journal of the Royal Statistical Society*, surveying current statistical activity. Probably in those days the Society provided rather stony ground on which to sow his seed! But he, and later Yule, made brave attempts which ultimately bore fruit.

The first of these survey articles was written as early as 1885 under the title: 'Methods of Statistics'. Here Edgeworth is largely concerned with what he terms the Science of Means, where the latter word includes measures of both location and dispersion. He says that this science comprises two main problems:

(1) To find how far the difference between any proposed Means is accidental or indicative of a law.

(2) To find what is the best kind of Mean to use.

As an illustration he compares the root-mean-square, the mean deviation and the inter-quartile distance estimators of the modulus, $c = \sigma\sqrt{2}$. He remarks that the last two may be favoured for convenience, but if the distribution is not Gaussian the relations between these estimators, which are appropriate for that distribution, no longer hold so that the root-mean-square estimator should be used to assess the reliability of a sample mean.

He points out that though the original distribution may be far from Gaussian, the distribution of means in quite small samples tends rapidly to this form. As an illustration he takes 280 samples of the sums of 10 random digits and shows by a diagram how nearly these approach the Gaussian form, although the individual digits follow a discrete uniform distribution. This may have been one of the first random sampling experiments in statistical literature. He also compares the use of the sign test to detect a shift in population means with the standard test based on measurement and mean values. Finally, he remarks: 'no originality is claimed for these principles'.

A few years before this Edgeworth had written a number of papers in which he derived new results, starting from a basis of prior probabilities. For example, in 1883 he foreshadowed the 'student' distribution. Starting with a sample, $x_i\,(i = 1, 2, \ldots, n)$ from a normal population, $N(\mu, \sigma)$, he took as the prior distribution for μ and σ

$$g(\mu, \sigma)\, d\mu\, d\sigma = C\sigma^{-2} d\mu\, d\sigma.$$

This implies that the precision constant, with which he then measured dispersion, i.e. $h = 1/(\sigma\sqrt{2})$, has a uniform prior distribution within the limits 0 and ∞. He derived the posterior distribution of μ in the form

$$p(\mu) = K\{1+t^2/(n-1)\}^{-\frac{1}{2}(n+1)},$$

where in present day notation

$$t = (\bar{x}-\mu)\sqrt{n}/\{\Sigma(x_i-\bar{x})^2/(n-1)\}^{\frac{1}{2}}.$$

This is essentially Student's distribution though reached by a quite different approach. Edgeworth seems, however, to have attached little importance to this result (of which Student was certainly unaware) because he realized that in small samples alteration in the form assumed for $g(\mu, \sigma)$ would have a decisive influence. For some recent comments on the relation of this early work of Edgeworth's to Student's later derivation, see Welch (1958).

Proceeding in his own way to tackle problems a little in advance of the main forward sweep, we find him in 1886 attempting to derive a correction for grouping to the mean square measure of dispersion, and in 1887 discussing the treatment of discordant observations. His paper of 1892 in the *Philosophical Magazine* entitled 'Correlated averages' starts with a reference to Galton's work on correlation and Weldon's application of this to measurements on shrimps and shows how the coefficients in a multivariate normal correlation distribution can be derived from the determinant of the basic correlations between the variables. He deals with the cases of 2, 3 and 4 variables.

We know that in 1893 both Edgeworth and Pearson were working on the derivation of asymmetrical frequency curves. Pearson's first paper appeared in 1895. Two of Edgeworth's contributions were published in 1896 and 1898. [The problem was tackled much more fully by Edgeworth later in a number of papers published between 1914 and 1924 in the *Journal of the Royal Statistical Society*.] The 1896 paper rather tentatively discussed what has since been termed the Edgeworth expansion, going only as far as the cubic term. In the second paper he introduced what he termed the 'method of translation' under which an observed variable, x, was regarded as a function of a unit normal variable, X. As an example he takes a polynomial function, e.g.

$$x = a(b+X)^2 \quad \text{or} \quad x = a(b+X)+c(d+X)^3.$$

If we can determine the parameters in the relationship, the advantage of this method is that it brings us back to the well-tabled normal function, X. This form of transformation is seen in the logarithmic distribution; it was used much later very successfully by Fisher in his transformation of the distribution of the correlation coefficient and more recently still by N. L. Johnson in deriving his S_B and S_U frequency distributions.

His three papers of 1908 under the title 'On the probable errors of frequency constants' are perhaps less informative than their title suggests. At one point he describes his discussion as 'reflections' largely suggested by Pearson & Filon's 'general theorem' on the same subject published in 1898 and referred to below. His approach, as always, was from the angle of inverse probability. He showed that he had himself no doubt that Pearson & Filon's results were applicable only to maximum-likelihood estimators and he broke new ground in indicating why their standard errors would be expected to be smaller asymptotically than those of any alternative estimators.

It is curious that a man of Edgeworth's ingenuity seems to have had so little effect on the main line of development of mathematical statistics. Perhaps the reason for this may be found in the following circumstances:

(1) His method of approach involved the use of inverse probability which others were not prepared to accept.

(2) His presentation was often obscure and expressed in a language which did not appeal to the mathematician.

(3) He was a man of charming but retiring nature who seems not to have created any school of students working along his own lines.

(4) He did not concentrate on any special fields of application nor show beyond doubt that the methods he put forward obtained valuable results.

A summary of his statistical writing, which was spread over more than 40 years, will be found in A. L. Bowley's *F. Y. Edgeworth's Contributions to Mathematical Statistics*, published by the Royal Statistical Society in 1928.

4. W. F. R. Weldon and Karl Pearson

Certainly Galton's development of correlation had some influence on Edgeworth, but its impact was more fruitful in another direction. W. F. R. Weldon, a Cambridge zoologist, had the imagination to realize that Galton's methods of exact measurement and statistical analysis might be used in searching among animal and plant populations for evidence bearing on Darwin's theory of evolution by natural selection. He first turned his attention to a study of the differences and similarities between local races of shrimps, and, later, of shore crabs. As he wrote in 1893: 'It cannot be too strongly urged that the problem of animal evolution is essentially a statistical problem: that before we can properly estimate the changes going on in a race or species we must know accurately' four things. These may be summarized as:

(*a*) the frequency distribution of a character;
(*b*) the correlation between characters in an individual;
(*c*) the link between the death rate and the value of a character;
(*d*) the inheritance of characters.

He emphasized the need to find answers to these questions for a number of species in numerical terms, before speculating further on their 'past history or future fate'.

In following out his project, Weldon's work immediately threw out new problems requiring a mathematical attack: the description of frequency distributions which were not normal; the need for standard errors of new forms of estimate; problems of growth, of multiple correlation, and of the effect of selection on one or more of a large number of correlated characters. These problems were of a kind which did not have to be faced in the more classical theory of errors.

It was just at this juncture, in 1891, that Weldon became Professor of Zoology at University College London, where Pearson had already for some years been Professor of Applied Mathematics, and a partnership began which grew in closeness until Weldon's early death at the age of forty-six in 1906.

I shall only refer to two main aspects of Pearson's work during the next 10 years:

(1) his development of a system of frequency curves;
(2) the exploitation of the theory of multiple correlation and regression in many fields.

His system of curves, derived as solutions of a single differential equation, was possibly neater than any of Edgeworth's suggestions and did not suffer from certain disadvantages, e.g. that of giving in some circumstances negative ordinates and subsidiary maxima. Also, by chance, for Pearson was quite unaware of this at the time, all but one of his main types of curve were later found to be extremely relevant as distributions occurring in what may be termed 'normal sampling theory'. We may prefer now to speak of a t-distribution, a gamma or inverted gamma distribution, a beta or inverted beta distribution, rather than use Pearson's classification of Types VII, III, V, I or VI respectively, but all these distributions are in fact derivable from his basic differential equation. Again the *Tables of the Incomplete Gamma* and *Incomplete Beta Functions* which he planned as providing the probability integrals of his frequency curves, have played an even more important role in statistical methodology than he could originally have imagined.

Equally important, though perhaps less spectacular because of the heavy algebra in which the work often became involved in his hands, were the uses which Pearson found for the methods of multiple correlation and regression. The development of the basic theory cannot be assigned to any single person. In his published paper of 1896 Pearson attributed it to Bravais and Edgeworth, but he seems to have reworked the results independently, and, as he wrote in 1920 more than 25 years later, had been 'far too excited to stop to investigate properly what other people had done'. In 1896 G. U. Yule, who was working as Pearson's assistant, derived the coefficients of partial correlation, first mentioned in a paper on pauperism. Also in 1897, Yule showed that a least squares approach gave the same multiple and partial regression equations, even if the true regression was not linear nor the distributions normal as Bravais and Edgeworth had assumed.

As I see it, Pearson's big contribution was to grasp within the space of a few years the number of uses to which this basic theory could be put.

(1) He saw that only through the application of multiple regression could an answer be found to Weldon's questions about the effect of selection on correlated organs (1896).

(2) He realized that to determine the numerical values of the multiple regression coefficients it would be necessary to collect extensive data on the basic means, standard deviations and correlations for all relevant variables. It became clear that Weldon's crab investigations and similar investigations on animal populations laboured under the handicap that age was not determinable. This meant that the interpretation of results was often obscured by the unknown effect of growth on the relation between parts of the body. Pearson, therefore, with Galton's encouragement, concentrated first on the collection of data for physical measurements on adult man.

(3) Thus the importance of collecting data before theorizing, which became a feature of the work of the later Biometric and Galton Laboratories, was early established and led to a long series of publications such as the *Treasury of Human Inheritance*.

(4) Among other investigations initiated during this period were:

(*a*) comparative racial studies of measurements on the human skull;

(*b*) the use of correlation in meteorology.

(5) Multiple correlation theory led, in a paper of 1898 with L. N. G. Filon, to the use of the likelihood function in deriving the joint asymptotic distribution of the sample estimators of a group of population parameters. The basis of the approach used here is a little obscure and there seems to be implicit in it the classical concept of inverse probability. However, the authors obtained what would now be termed the information matrix from

which the variances and covariances of the estimators were to be derived by inversion. They did not, however, as Edgeworth did later, seem to realize fully that the theory only applied to maximum-likelihood estimators. In the case of the bivariate normal distribution they correctly derived the large sample standard error of the maximum likelihood estimator, $r = \Sigma xy/\{\Sigma x^2 \Sigma y^2\}^{\frac{1}{2}}$, of the population correlation coefficient, ρ. But in dealing with the simultaneous estimation of the three parameters of the Pearson Type III (or gamma) distribution, they failed to note that the standard errors obtained from their theory for two of these parameters were not appropriate to the moment estimators used in their example. The oversight was probably due to a lack of appreciation of the order of magnitude of the first derivative term in a Taylor expansion of the likelihood function.*.

(6) Again, it was a multiple correlation approach which led to the χ^2 test of goodness of fit, published in 1900.

5. Biometry and Mendelism

There is one aspect of this work which seems to justify a little more detailed discussion, because of the part it played in the later controversy between Weldon and Pearson on the one hand and William Bateson on the other, over the application of Mendel's principles of heredity. Using his collected material on human inheritance, Galton in 1889 had made an attempt to give a linear prediction formula showing the average contribution to a character such as stature received from each generation of a man's ancestors. He was undoubtedly looking for some genetic explanation of the remarkable constancy of the correlations which he found. His presentation, however, suffered inevitably because at that date his understanding of the relations involved in what was essentially a multiple regression problem was only intuitive.

In 1898 Pearson had expressed Galton's formula in the form of a multiple regression equation whose parameters were functions of the individual ancestral correlations and of what he termed the coefficient of assortative mating, i.e. the correlation between the character values in husband and wife. This equation he termed Galton's *Law of Ancestral Heredity*, and he emphasized that he regarded it as a descriptive statistical law only, which applied to populations mating at random. As he put it in a paper of 1896: 'In the first place, we must definitely free our minds, in the present state of our knowledge of the mechanism of inheritance and reproduction, of any hope of reaching a mathematical relation expressing the degree of correlation between individual parent and individual offspring....'

When in the autumn of 1900 the rediscovered papers of Gregor Mendel were seen to offer a possible model of a mechanism of inheritance, the biometricians were at once interested. 'About pleasanter things' wrote Weldon to Pearson on 16 October 1900 after commenting on some University of London affairs, 'I have heard of and read a paper by one Mendel on the results of crossing peas, which I think you would like to read...'. But they considered it essential to test how far the large-scale working of this mechanism would lead to results consistent with the statistical evidence. Quite clearly the original simple Mendelian model was not capable of explaining a large number of the observed facts of inheritance. However, the idea that techniques of a mathematical character could disclose inadequacy in a genetic theory of inheritance was inevitably at that date repugnant to many biologists. Hence the prolonged controversy.

* I am particularly grateful to Professor B. L. Welch for some helpful discussion on the interpretation of parts of this paper.

The very open approach of the biometricians to these problems is illustrated by the following extract from the first chapter of an uncompleted book on Inheritance which Weldon had made several attempts at drafting in the few years before his death:

The student of heredity has two main objects; the first is to discover what degree of stability is actually exhibited by the various races of animals or of plants, and to determine the extent to which deviation from the average character of parents or other ancestors is associated with deviation in their descendants; the second object is to acquire such knowledge of the changes which occur during the growth and maturation of the germ-cells, their fusion and subsequent development, as may serve to indicate the process by which the observed relation between parental and filial characters is brought about. The first object is to make a purely descriptive statement of the actual relation between the visible bodily characters of living things and those of their ancestors or their descendants; the second is to learn the process to which this relation is due. These two objects are pursued by different methods, and as it happens they are generally pursued by different men, so that few attempts have been made to consider the bearing of what we actually know concerning the relation between the visible characters of parents and those of their offspring upon the possible interpretation of structural changes revealed by minute study of the germ-cells and of the embryonic processes in general.

A descriptive statement of the relation between the visible somatic characters of parents and those of children involves no biological methods of compilation. When we compare the distribution of statures in two races of men, or determine the relation between the stature of fathers and of sons, we have to compare two sets of objects which differ in length; and the only thing we ought to consider while making the comparison is the number of inches, or millimeters or other units of length, which each of these objects contains...[Afterwards] we ought to remember that we have been measuring men, and to look for something in the phenomena of human development which may help us to see the process by which the observed relation between parental and filial stature has been brought about; but the two operations are perfectly distinct. It is perfectly possible to make an accurate and useful statement of the laws which express the inheritance of any visible character without going on to consider the process by which the hereditary transmission of that character may have been effected, and the collection of such statements is the necessary first step to a fuller knowledge of heredity. Descriptive statements of this kind may suggest a conception of the vital process by which inheritance is brought about;...

It is the purpose of this essay first of all to describe the two principal theories of inheritance* which have been based upon direct comparison between the characters of living things and those of their ancestors or their descendants, and afterwards to see how far the facts of development and regeneration as well as the facts of inheritance support one theory or the other.

In the years 1901–6 a great amount of attention was given to this problem of propounding a Mendelian-type mechanism of inheritance which would be consistent with observed biometric correlations. We see this in Yule's review of Bateson's *Mendel's Principles of Heredity*, published in *The New Phytologist* in the autumn of 1902; in Pearson's 'The law of ancestral heredity' (*Biometrika*, 1903) and in his Royal Society paper 'On a generalized theory of alternative inheritance, with special reference to Mendel's Laws' of 1904. In this last paper, using a particular generalized model in which it was supposed that a number of factors combined their effects by simple addition, he obtained agreement up to a point between the model's prediction and the observed correlations in man and horse. He concluded, however, that the model chosen was 'not elastic enough to account for the numerical values of the constants of heredity hitherto observed'.

Again, in 1905, Weldon was working on another scheme with a cytological basis under which he supposed that the chromosomes carried one or more 'determinants' controlling the somatic character in the individual. He died in 1906 before his presentation was complete, and we have only a summary put together by Pearson in 1908, giving what he believed was the basis of Weldon's theory. There seems little doubt that had Weldon lived the Mendelian and biometric approaches would have been welded together at a much earlier date.

* Those of Galton and Mendel.

A myth has grown up that these two approaches were essentially contradictory and that the latter was in error. Careful reading of the literature cannot possibly substantiate this claim. The early biometricians seem to me to have had open minds, but they insisted that any model of individual inheritance must lead to conclusions statistically consistent with large-scale observation. In the few years after the rediscovery of Mendel's work, changes through the inevitable need to expand his theory made it difficult to pin down a model which could be tested against biometric data.

As far as I know the next really promising attempt at this welding process was Fisher's. His paper of 1918 on 'The correlation between relatives on the supposition of Mendelian inheritance' is carrying on this earlier exploration. He acknowledges the earlier pioneer attempts though he does not refer to Weldon's 'Mathematical theory of determinantal inheritance'. However, he uses very fully the accumulated correlation data of the biometric school. Indeed the verification of his own theory depended on the use of results which his predecessors had had the foresight and perseverance to collect. The continuity in thought is obvious. He writes

> Several attempts have already been made to interpret the well-established results of biometry in accordance with the Mendelian scheme of inheritance. It is here attempted to ascertain the biometrical properties of a population of a more general type than has hitherto been examined, inheritance in which follows this scheme.

6. W. S. GOSSET ('STUDENT')

I think the first great formative biometric period ended with Weldon's death. The next step forward was taken by W. S. Gosset, writing under the nom de plume of 'Student', though at the time this step of his was scarcely recognized as such. Gosset had gone in 1899 to Arthur Guinness Son and Co. Ltd. in Dublin joining the small group of young university scientists who the firm had begun to appoint to their staff as 'brewers' a few years before. He had taken only Mathematical Moderations at Oxford before proceeding to a degree in Chemistry. However, after leaving the routine work of the Brewery it was natural that, as the most mathematically minded among his colleagues, he should have been turned on to the analysis of the Brewery's mass of data and asked whether this could be made to throw more light on the relations between the quality of the raw materials for beer, such as barley and hops, the conditions of production and the quality of the finished product.

His first tools were those of the standard error theory, found in Airy's *Theory of Errors of Observations* and Merriman's *The Method of Least Squares.* By 1904 he had become sufficiently familiar with the problems involved to write his first *Report* on 'The application of the "Law of Error" to the work of the Brewery'. By 1905 he was already conscious of certain difficulties which the classical theory did not suffice to answer. In the first place he was not happy about using a sample standard deviation, based perhaps on 10 or fewer observations, to assess the reliability of a mean. Secondly, he realized that some of his observations were not independent and that he needed some way to measure this association and to judge its significance. He had a flair for getting to the fundamentals of a problem, and if in July 1905 he had not arranged to consult Pearson it is likely that he would have rediscovered the correlation coefficient, by a quite original and independent route!

His line of attack was as follows: suppose $A_i\,(i = 1, \ldots, n)$ are a set of daily observations on a variable, A, and B_i the corresponding daily observations on another variable, B.

Then after reducing (or increasing) the B observations so that the standard deviation of B was roughly equal to that of A, he formed the two series

$$A_i + B_i, \quad A_i - B_i \quad (i = 1, \ldots, n).$$

He then found the variance of each series (actually he worked in terms of probable errors) and if these two quantities were markedly different he concluded that this was a sign of relationship between A and B. He was not, however, sure how to measure this nor how to judge its significance.

The way he would have proceeded on his own we do not know, but it is interesting to speculate that if he had taken the ratio

$$\frac{\text{Diff. between s.s.}}{\text{Mean s.s.}} = \frac{\Sigma(A_i+B_i)^2 - \Sigma(A_i-B_i)^2}{\frac{1}{2}\{\Sigma(A_i+B_i)^2 + \Sigma(A_i-B_i)^2\}},$$

where the observations are referred to their means and S_B has been adjusted to equal S_A, then this ratio equals $2r_{AB}$.

After the meeting with Pearson he was of course made familiar with the coefficient of correlation, its large sample standard error and much else current in the literature of the day. It was arranged that he come to spend a part of the year 1906–7 at the Biometric Laboratory in London. Out of this period came three remarkable papers:

(1) a paper in which he rediscovered Poisson's limit to the binomial (1907);

(2) the famous paper on 'The Probable Error of a Mean', in which he surmized the χ^2 distribution for the sample estimate of variance and, hence, derived the t-distribution (1908*a*);

(3) a paper (1908*b*) in which he guessed correctly the distribution of the correlation coefficient, r, in normal samples from a population in which the correlation coefficient $\rho = 0$.

All these results were of fundamental importance and the incentives to discovery were the requirements of experimental work in the brewery. Referring to the inadequacy for certain purposes of the statistical techniques available at the time, Gosset wrote in the introduction to his paper on the mean:

> There are other experiments, however, which cannot easily be repeated very often; in such cases it is sometimes necessary to judge of the certainty of the results from a very small sample, which itself affords the only indication of the variability. Some chemical, many biological, and most agricultural and large scale experiments belong to this class, which has hitherto been almost outside the range of statistical inquiry.

Earlier, in his Brewery Report of 1905 written after meeting Pearson, he had written:

> Correlation coefficients are usually calculated from large numbers of cases, in fact I have only found one paper in *Biometrika* of which the cases are as few in number as those at which I have been working lately.

Can we find the reason for this lack of interest among the early biometric school in small sample theory? Yes, I think we can. No one who wished to get to the bottom of some knotty statistical problem would choose to deal in small numbers of observations if more could be collected and the early biometricians, who were looking for small differences which might provide evidence that the forces of natural selection were at work, or seeking to determine accurate estimates of ancestral correlations, dealt wisely in samples of 100's rather than 10's. Quite simply, they were not themselves faced with the type of problem to which Gosset refers in the quotation I have given. They were also, no doubt, anxious to discourage the biologist or the medical man from believing that he had been supplied with an easy method of drawing conclusions from scanty data.

What may seem more surprising *after* Gosset had worked for nearly a year in contact with him at University College is the little interest which Pearson, who had earlier pioneered the way into new fields, appeared to show in this uncharted field of experimentation. Perhaps the explanation was as follows. The Biometric and Galton laboratories were at the time overflowing with problems of quite a different kind. In 1906 Pearson was in very low spirits; his great friend Weldon had died a few months before Gosset's arrival and the years 1893–1906 of shared excitement in discovery seemed to lie behind him. He perhaps had not the energy to enter into a new and so far unfamiliar field. When in 1915 Fisher confirmed Student's inspired guesses regarding the distributions of the variance and of r, and obtained the general distribution of r for $\rho \neq 0$, Pearson was extremely interested in the results as marking the elegant solution of so far unsolved distributional problems. But as to the possible uses of t and r in small samples he always remained sceptical.

There is, however, no question at all about the continuity in development. Gosset obtained from the biometric school:

(1) the theory of correlation and regression;

(2) correlation data which he used in the random sampling experiments he carried out to check his theory;

(3) the method of moments and the Pearson frequency curves, which were essential to his method of derivation of the distribution of the variance;

(4) a free year for research and the opportunity then and in later years of what he once described as the interchange of views between 'professor' and 'student'. There was for years a firm friendship and flow of correspondence between the two men.

Now the Dublin Brewery, as a very large consumer of barley, was naturally interested in agricultural problems including certain large-scale experiments carried out under the Irish Department of Agriculture. Gosset was not concerned in giving advice in these experiments until a later date and his first real interest in agricultural work probably arose from his contact with the maltster, E. S. Beaven. Beaven had begun experimental work on strains of barley in the 1890's, testing his varieties in a cage at Warminster, and from 1905 until Gosset's death in 1937 there was a continual exchange of ideas between them.

Because of Gosset's known interest in the subject, the two classical papers on the analysis of uniformity trial data—that by Wood & Stratton (1910) from Cambridge and that by Mercer & Hall (1911) from Rothamsted—passed through his hands. To the latter paper he wrote an Appendix showing how it was possible to bring the changing fertility level of the field into service,

(*a*) by scattering the varieties to be compared in small plots over the fields, and

(*b*) then taking as the statistical variable for analysis the difference between the characteristics of two varieties on neighbouring plots.

He emphasized that it was far better to determine the residual sampling error between plots from the experiment itself, rather than to use some established figure derived from a series of previous experiments, as Wood & Stratton seemed to suggest. His t-distribution could then be used to test significance. These suggestions may seem commonplace now; they were not in 1911.

We can see that the progress towards a science of agricultural experimentation was already taking place, if slowly and often by trial and error. In the years between 1911 and 1923, with a long break during the war period, Gosset's work with Beaven had gone a good deal further. In the spring of 1923, working on a balanced design in which eight varieties of barley were compared in a replicated block experiment in one of Beaven's cages, Gosset obtained a pooled estimate of the error variance by subtracting the variance due to varietal differences and that due to block differences from the overall variance; see, for an account of Gosset's work of this period, Pearson*(1938, pp. 234–7). He showed his tenative analysis to Yule and to Fisher (now established at Rothamsted) with both of whom he had been in correspondence for some time. Exactly at this moment Fisher had completed his paper with W. A. Mackenzie on 'Studies in crop variation', in which another balanced chess-board design with potatoes, led to a similar analysis. To this Fisher added a method of testing the significance of the treatment sum of squares, taken as a whole. It is doubtful if Gosset's mathematics could have tackled this bit of distribution theory. In any case the first period of exploration was over and a new chapter was opening.

7. R. A. Fisher

I do not propose to go further into the work of R. A. Fisher than to point out that it illustrates again the two aspects of advance which I have emphasized:

(1) a broad continuity in the problems tackled and in many of the methods used;

(2) an exercise of the imagination which results in the sharpening of existing techniques, the invention of new ones and in the forcing of an entry into new fields.

Both during his life time and since there has been a tendency when writing of Fisher to overlook the importance to him of the foundations upon which he built and, as a result, certain myths have grown up. All this it seems to me does not really do best service to the memory of a great statistician whose achievements were in any case quite outstanding.

That there has been continuity, that Fisher did build upon what was there before, can hardly be questioned by any one who has time to read the literature with an open mind. Consider a few of the subjects which he tackled. He gave increased precision to the properties of the likelihood function and turned it to uses unconsidered previously, but this function had been used by others before him and its properties have had to be given greater precision by later hands. He accepted the fact that in many cases long established moment estimates of parameters would be needed as a first approximation to his maximum-likelihood solutions. Pearson's frequency curves with their known properties appeared again and again, as it were uninvited, as the sampling distribution functions which Fisher rapidly derived in the early 1920's; he did not *need* the earlier work but his discoveries gave it a new relevance. Again he recognized the well established need for determining the sampling moments of moments, but while in many ways simplifying the procedure by the introduction of cumulants and k-statistics.

In his paper of 1918 on the correlation between relatives on the supposition of Mendelian inheritance he took up the problem where Weldon and Pearson had left it 13 years before. His work on discriminant functions seems to have been partly devised to improve the methods of classifying racial types which he found in existence in the field of craniometry. His great work on the use of statistical methods in agricultural experimentation started from the limit to which Gosset's mathematical powers had been able to take it.

*Paper No. 24 in this volume

Undoubtedly the science of mathematical statistics was in need of new vision and new stimulus in the early 1920's, just as it had been in the early 1890's. On both occasions it found the man to put life into the subject.

References

Abbe, Ernst (1863). Dissertation zur Erlangung der Venia Docendi bei der Philos. Fabultät Jena. (Library of Congress, Washington, D.C. No. QA 275 A 12.)

Bowley, A. L. (1928). *F. Y. Edgeworth's Contributions to Mathematical Statistics*. London: Royal Statistical Society.

Bravais, A. (1846). Sur les probabilités des erreurs de situation d'un point. *Mém. Acad. Roy. Sci. Inst. France* **9**, 255–332.

Edgeworth, F. Y. (1883). The methods of least squares. *Phil. Mag.* **16**, 360–75.

Edgeworth, F. Y. (1885). Methods of statistics. *J. R. Statist. Soc.* Jubilee Volume, pp. 181–217.

Edgeworth, F. Y. (1886). On the determination of the modulus of errors. *Phil. Mag.* **21**, 500–7.

Edgeworth, F. Y. (1887). On discordant observations. *Phil. Mag.* **23**, 364–75.

Edgeworth, F. Y. (1892). Correlated averages. *Phil. Mag.* **34**, 190–204.

Edgeworth, F. Y. (1896). The asymmetrical probability curve. *Phil. Mag.* **41**, 90–9.

Edgeworth, F. Y. (1898). On the representation of statistics by mathematical formulae. Part I. *J. R. Statist. Soc.* **61**, 670–700.

Edgeworth, F. Y. (1908). On the probable errors of frequency constants. *J. R. Statist. Soc.* **71**, 381–97, 499–512, 652–78.

Fisher, R. A. (1918). The correlation between relatives on the supposition of Mendelian inheritance. *Trans. Roy. Soc. Edin.* **52**, 399–433.

Fisher, R. A. & MacKenzie, W. A. (1923). Studies in crop variation. II. *J. Agric. Sci.* **13**, 311–20.

Galton, Francis (1886). Family likeness in stature. *Proc. Roy. Soc.* **40**, 42–73.

Galton, Francis (1888). Co-relations and their measurement, chiefly from anthropometric data. *Proc. Roy. Soc.* **45**, 135–45.

Galton, Francis (1889). *Natural Inheritance*. London: Macmillan and Co.

Gosset, W. S. (1907). On the error of counting with a haemacytometer. *Biometrika* **5**, 351–64.

Gosset, W. S. (1908*a*). The probable error of a mean. *Biometrika* **6**, 1–25.

Gosset, W. S. (1908*b*). Probable error of a correlation coefficient. *Biometrika* **6**, 302–10.

Gosset, W. S. (1923). On testing varieties of cereals. *Biometrika* **15**, 217–93.

Helmert, F. R. (1876). Die Genauigkeit der Formel von Peters zur Berechnung des wahrscheinlichen Beobachtungsfehlers direkter Beobachtungen gleicher Genauigkeit. *Astr. Nachr.* **88**, no. 1096.

Mercer, W. B. & Hall, A. D. (1911). The experimental error of field trials. *J. Agric. Sci.* **4**, 107–32.

Pearson, E. S. (1938). 'Student' as statistician. *Biometrika* **30**, 210–50.*

Pearson, E. S. (1965). Studies in the history of probability and statistics. XIV. Some incidents in the early history of biometry and statistics, 1890–94. *Biometrika*, **52**, 3–18.†

Pearson, Karl (1895). Contributions to the mathematical theory of evolution. II. Skew variation in homogeneous material. *Phil. Trans.* A **186**, 343–414.

Pearson, Karl (1896). Mathematical contributions to the theory of evolution. III. Regression, Heredity and Panmixia. *Phil. Trans.* A **187**, 253–318.

Pearson, Karl (1898). Mathematical contributions to the theory of evolution. On the law of ancestral heredity. *Proc. Roy. Soc.* **62**, 386–412.

Pearson, Karl (1900). On the criterion that a given system of deviations from the probable in the case of a correlated system of variables is such that it can reasonably be supposed to have arisen from random sampling. *Phil. Mag.* **50**, 157–75.

Pearson, Karl (1903). The law of ancestral heredity. *Biometrika* **2**, 211–29.

Pearson, Karl (1904). Mathematical contributions to the theory of evolution. XII. On a generalised theory of alternative inheritance, with special reference to Mendel's laws. *Phil. Trans.* A **203**, 53–86.

Pearson, Karl (1908). On a determinantal theory of inheritance, from notes and suggestions by the late W. F. R. Weldon. *Biometrika* **6**, 80–93.

Pearson, Karl (1921). Notes on the history of correlation. *Biometrika* **13**, 25–45.‡

Pearson, Karl & Filon, L. N. G. (1898). Mathematical contributions to the theory of evolution. IV On the probable errors of frequency constants and on the influence of random selection on variation and correlation. *Phil. Trans.* A **191**, 229–311.

**Paper No. 24 in this volume*

†Paper No. 22 in this volume

‡Paper No. 14 in this volume

SCHOLS, C. M. (1875). Over de theorie der fouten in de ruimte en in het platte vlak. *Verh. Nederl. Akad. Wetensch.* **15**, 1–75.
SEAL, H. L. (1967). Studies in the history of probability and statistics. XV. The historical development of the Gauss linear model. *Biometrika* **54**, 1–24.*
SCHEFFÉ, HENRY (1956). Alternative models for the analysis of variance. *Ann. Math. Statist.* **27**, 251–71.
SHEYNIN, O. B. (1966). Origin of the theory of errors. *Nature, Lond.* **211**, 1003–4.
TODHUNTER, ISAAC (1865). *A History of the Mathematical Theory of Probability*. London: Macmillan and Co.
WELCH, B. L. (1958). 'Student' and small sample theory. *J. Am. Statist. Ass.* **53**, 777–88.
WELDON, W. F. R. (1893). On certain correlated variation in *Carcinus moenas*. *Proc. Roy. Soc.* **54**, 318–29.
WOOD, T. B. & STRATTON, F. J. M. (1910). The interpretation of experimental results. *J. Agric. Sci.* **3**, 417–40.
YULE, G. U. (1896). On the correlation of total Pauperism with proportion of out-relief. *Econ. J.* **6**, 613–23.
YULE, G. U. (1897). On the theory of correlation. *J. R. Statist. Soc.* **60**, 812–54.
YULE, G. U. (1902). Mendel's laws and their probable relations to intra-racial heredity. *New Phytologist* **1**, 193–207, 222–40.

**Paper No. 15 in this volume*

On Dartmoor during a fishing holiday, April 1936.

On the bowling green, May 1937.

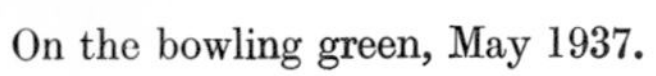

William Sealy Gosset

Plate I

WILLIAM SEALY GOSSET, 1876–1937

The two appreciations which follow have been written from somewhat different angles. The first is by a younger colleague and friend at the St James' Gate Brewery, who for a number of years was in close contact with Gosset in Dublin, both in and out of the brewery. The friendship of the second writer is one which grew through a correspondence that roved at length over statistical methods and theories. If in some places the articles overlap, this will only help to emphasize certain events or characteristics which independently we have felt impelled to record.

Both of us would like to express our warmest thanks to the many friends who have helped us, and in particular to Mrs W. S. Gosset and Mr E. Somerfield.

L. McM., E. S. P.

(1) "STUDENT" AS A MAN

By L. McMULLEN

William Sealy Gosset was the eldest son of Colonel Frederic Gosset, R.E., and was born at Canterbury in 1876. In 1906 he married Marjory Surtees Phillpotts, daughter of the late headmaster of Bedford School, and they had one son and two daughters. He died on 16 October 1937, and was survived by both his parents, his wife and children, and one grandson.

He was educated at Winchester, where he was a scholar, and New College, Oxford, where he studied chemistry and mathematics.

He entered the service of Messrs Guinness as a brewer in 1899.

It is not known exactly how or when "Student's" interest in statistics was first aroused, but at this period scientific methods and laboratory determinations were beginning to be seriously applied to brewing, and it is obvious that some knowledge of error functions would be necessary. A number of university men with science degrees had been taken on, and it is probable that "Student", who was the most mathematical of them, was appealed to by the others with various questions and so began to study the subject. It is known that he could calculate a probable error in 1903. The circumstances of brewing work, with its variable materials and susceptibility to temperature change and necessarily short series of experiments, are all such as to show up most rapidly the limitations of large sample theory and emphasize the necessity for a correct method of treating small samples. It was thus no accident, but the circumstances of his work, that directed "Student's" attention to this problem, and so led to his discovery of the distribution of the sample standard deviation, which gave rise to what in its modern form is known as the t-test. For a long time after its discovery and publication the use of this test hardly spread outside Guinness's brewery, where it has been very extensively used ever since. In the Biometric school at University College the problems investigated were almost all concerned with much larger samples than those in which "studentizing", as it was sometimes called, made any difference. Nevertheless, although their lines of research diverged somewhat rapidly, the close statistical contact and personal friendship between Karl Pearson and "Student", which began during his year at University College, were only terminated by death.

The purpose of this note is not however to give an account of "Student's" statistical work, but to try to give a more general impression of the man himself. Although his public reputation was entirely as a statistician, and he was acknowledged to be one of the leading investigators in that subject, his time was never wholly and rarely even mainly occupied with statistical matters. For one who saw enough of him to know roughly how his time was spent both at work and at home, it was very difficult to understand how he managed to get so much activity into the day. At work he got through an enormous amount of the ordinary routine of the brewery, as well as his statistics. Until 1922 he had no regular statistical assistant, and did all the statistics and most of the arithmetic himself; later there was a definite department, of which he was in charge till 1934, but throughout he did a great deal of arithmetic and spade-work himself. It might be supposed from the amount he did in the time that he was unusually good at arithmetic and the arrangement of work; such, however, was not the case, for his arithmetic frequently contained minor errors. In one of his obituary notices a tendency to do work on the backs of envelopes in trains was mentioned, but this tendency was not confined to trains; even in his office much work was done on random scraps of paper. He also had a great dislike of the tabulation of results, and preferred to do everything from first principles whenever possible. This preference led in certain instances to waste of time in routine work, but was of assistance in maintaining that flexibility and speed of attack on new problems which was so characteristic of him. An actual example would need too much explanation of relevant circumstances, but I can vouch for the analogical truth of the following. If a body performs simple harmonic motion with acceleration μ per unit displacement, it may readily be shown that the period of a complete oscillation is $2\pi/\sqrt{\mu}$. Hence, in the case of a simple pendulum $t=2\pi\sqrt{(l/g)}$ and $l=gt^2/4\pi^2$, where l is the length of the pendulum and g the acceleration due to gravity. If it were necessary to calculate the lengths of pendulum corresponding to different periods as a routine matter, most people would evaluate $g/4\pi^2$ for their locality and always multiply t^2 by this numerical constant, which would be about 24·85. "Student" would probably have started from $2\pi/\sqrt{\mu}$ every time. If therefore he had suddenly wanted to calculate the period of oscillation of a weight on a stretched spring he could have done it, whereas the man who only remembered that $l = 24{\cdot}85t^2$ for a pendulum would be unable to tackle the problem without much more preliminary work.

His method was, of course, not necessarily the most suitable for others not aspiring to the same degree of versatility. Perhaps it is not altogether fanciful to compare the two methods with the organic evolution of, say, he human hand, the most versatile object known, and the construction of some highly efficient but absolutely specialized piece of machinery. I do not mean to imply that he gave this explanation, or was even altogether conscious of it. When he handed over to me a routine calculation which he had done for many years, I was astonished to find that he had written out every week an almost unvarying form of words with different figures. To my question, "Why ever don't you get a printed form?" he did not reply, "Doing it from first principles every time preserves mental flexibility". He would have considered such a remark unbearably pompous. He said, "Because I'm too lazy", to which I replied, "Well, I'm too lazy not to."

To many in the statistical world "Student" was regarded as a statistical adviser to Guinness's brewery; to others he appeared to be a brewer devoting his spare time to statistics. I have tried to show that though there is some truth in both of these ideas they miss the central point, which was the intimate connexion between his statistical research and the practical problems on which he was engaged. I can imagine that many think it wasteful that a man of his undoubted genius should have been engaged in industry, yet I am sure that it is just that association with immediate practical problems which gives "Student's" work its unique character and importance relative to its small volume. On at least one occasion he was offered an academic appointment, but it is almost certain that he would not have been a successful lecturer, though perhaps a good individual teacher; nor is it likely that his research work would have flourished in more academic circumstances; his mind worked in a different way.

The work in connexion with barley breeding carried out by the Department of Agriculture in Ireland, in which Messrs Guinness took a prominent part, enabled "Student" to get that first-hand experience of yield trials and agricultural experiments generally which contributed so largely to his great knowledge of the subject. He did not merely sit in his office and calculate the results, but discussed all the details and difficulties with the Department officials, and went round all the experiments before harvest, when a "grand tour" is annually carried out by the Department, the brewery, and sometimes statisticians or others interested from England or abroad. As well as the work carried out at the actual cereal station near Cork, three or four varieties of barley are grown in

$\frac{3}{4}$ or 1 acre plots at ten farms representing all the principal barley-growing districts of Ireland, so a visit to all of them entails a fairly comprehensive inspection of the crops.

"Student" took a great deal of interest in this work from the beginning and correspondence shows that he discussed the results of these tests with Karl Pearson at great length when he went to study with him at University College in 1906.

In the last ten years or so of his time in Ireland he played a leading part in these investigations, and thus had a perhaps unique opportunity of following experimental varieties from sowing through growing and harvest to malting and brewing results, and also of carrying out or supervising all the relevant mathematical work. At one time he also made some barley crosses in his own garden, and accelerated their multiplication by having one generation grown in New Zealand during our winter. These crosses were known as Student I and II, and have now been discarded as failures, the inevitable fate of the large majority. With characteristic self-effacement he was the first to point out that they were not worth going on with.

He also made frequent visits to Dr E. S. Beaven, whose work on barley breeding is well known, and discussed every aspect of yield trials with him. These visits were undoubtedly very useful, and although Dr Beaven is never tired of protesting that he is no mathematician and does not understand "magic squares" or "birds of freedom", which he prefers to the more orthodox expressions, he has a vast experience of agricultural trials and is very quick to see the weak point of any experiment.

In spite of the quantity of work "Student" did he was never in a hurry or fussed; this was largely due to the absence of lag when he turned his mind to a new subject; unfortunately others were not always equal to this. He would ring one up on the phone and plunge straight into some subject which might have been discussed some days previously. The slower-witted listener would probably lose the thread of his discourse before realizing what it was about and would ignominiously have to ask him to begin again. I have many times seen him hard at it on a Monday morning, but at first meeting it was always "How did the sailing go?" "Well, did you catch any fish?", and he would recount any notable event of his own week-end before plunging into the very middle of some subject. I never heard him say "I'm busy".

"Student" had many correspondents, mostly agricultural and other experimenters, in different parts of the world. He took immense pains with these and often explained points to them at great length when he could easily have given a reference. His letters contain some of his clearest writing, and the more difficult points are often better elucidated than in his published papers.

Karl Pearson emphasized the fact that a statistician must advise others on their own subject, and so may incur the accusation of butting in without adequate knowledge. "Student" was particularly expert at avoiding any such disagreement; usually he was such an enthusiastic learner of the other's subject that the fact that he was giving advice escaped notice.

The reader will by now have realized that "Student" did a very large quantity of ordinary routine as well as his statistical work in the brewery, and all that in addition to consultative statistical work and to preparing his various published papers. It might thus be thought that he could have done nothing else but eat and sleep when at home; this, however, was far from being the case, and he had a great many domestic and sporting interests. He was a keen fruit-grower and specialized in pears. He was also a good carpenter, and built a number of boats; the last, which was completed in 1932, and on whose maiden voyage I had the honour to be nearly frozen to death, was equipped with a rudder at each end by means of which the direction and speed of drift could be adjusted—an advantage which will be readily appreciated by fly-fishermen. This boat with its arrangement of rudders was described in the *Field* of 28 March 1936. In his carpentry he showed preferences analogous to his mathematical ones previously mentioned; he disliked complicated or specific tools, and liked to do anything possible with a pen-knife. On one occasion, seeing him countersinking screw-holes with a pocket-knife, I offered him a proper countersink bit which I had with me, but he declined it with some embarrassment, as he would not have liked to explain or perhaps could not have explained why he preferred using the pen-knife. Out of doors he was an energetic walker and also cycled extensively in the pre-war period. He did a lot of sailing and fishing. For his last boat he had a most unconventional sail, which cannot be exactly described under any of the usual categories; it was illustrated in the *Field* article referred to above.

In fishing he was an efficient performer; he used to hold that only the size and general lightness or darkness of a fly were important; the blue wings, red tails and so on being only to attract the fisherman to the shop. This view was more revolutionary when I first heard it than it is now. He was a sound though not spectacular shot, and was well above the average on skates. Until the accident to his leg in 1934 he was quite a regular golfer, and once went round a fairly difficult course in 85 strokes and $1\frac{1}{4}$ hours by himself. He used a remarkable collection of old clubs dating at least from the beginning of the century. In the last few years since his accident he took up bowls with great keenness, and induced many other people to play as well. One of his last visits to Ireland was with a team which he had organized at the new brewery at Park Royal.

On top of all this he knew as much as most people of the affairs of the world in general and of what was going on about him. It became very difficult to imagine how he found 24 hours in any way a sufficient length for the day. His wife certainly organized things so that the minimum amount of time was wasted, but even so few people could approach such activity in quantity or diversity.

In personal relationships he was very kindly and tolerant and absolutely devoid of malice. He rarely spoke about personal matters but when he did his opinion was well worth listening to and not in the least superficial.

In the summer of 1934 he had a motor accident and broke the neck of his femur. He had to lie up for three months, of course working at statistics, and was a semi-cripple for a year. This was particularly irksome for such an active man, as was the sheer unnecessariness of the accident, for he ran into a lamp-post on a straight road, through looking down to adjust some stuff he was carrying; but with great hard work and persistence he eventually reduced the disability to a slight limp.

At the end of 1935 he left Ireland to take charge of the new Guinness brewery in London, and I saw comparatively little of him after that. The departure from Ireland of "Student" and his family was a great loss to many who had experienced their hospitality.

His work in London was necessarily very hard and accompanied by all the vexations inevitably associated with a big undertaking in its first stages, before any settled routine has been established; nevertheless, he still found time to continue his statistical work and wrote several papers.

His death at the comparatively early age of 61 was not only a heavy blow to his family and friends, but a great loss to statistics, as his mind retained its full vigour, and he would undoubtedly have continued to work for many more years.

I am painfully conscious of the inadequacy of this sketch, which cannot hope to convey more than a faint impression of his unique personal quality to those who did not know him, but it will have served its purpose if it helps any readers to grasp the essential unity and directness of the personality which lay behind such widely varied manifestations.

(2) "STUDENT" AS STATISTICIAN

By E. S. PEARSON

For many years after the publication of his first paper in *Biometrika*, in 1907, the name of "Student" was associated in statistical circles with an atmosphere of romance. Those who knew him only through his written contributions must often have wondered who was this unassuming man, content to remain anonymous, who wrote so clearly and simply on so wide a range of fundamental topics. To those of us who came into touch with him personally, the knowledge that "Student" was W. S. Gosset did not altogether dispel that romantic impression. Here, in London, he would pay us visits from time to time at the old Biometric Laboratory on his way to Euston station to catch the Irish mail;

he would be wearing the grey flannel trousers that were a tradition of his Wykehamist schooldays and carrying a rucksack on his back. And then after a short hour's talk, perhaps on statistical subjects, perhaps on his garden experiments in cross-breeding, he would be off again to that wild Ireland where, in the "bad times", we had heard that gunmen were to be found hiding behind his hedges or even searching his house for arms. We had heard too of great exploits by members of his family of an entirely non-statistical character, of their boat-building and of their construction of a pair of water-skis which they used for walking over Kingstown Harbour.

My one short winter visit to Gosset's house at Blackrock, a few miles outside Dublin, would hardly by itself have cleared away this element of myth or made me appreciate fully the sterling values that lay beneath that friendly and unassuming exterior. We talked very little about statistics during my stay, and the strongest impressions remaining are of a morning spent among the immense vats and varied smells of the brewery; of drives out of town on misty evenings through the badly lit Dublin suburbs in that old, high two-seater Model-T Ford of his, christened "The Flying Bedstead"; of the warm hospitality of his fellow-brewers; and of a Saturday in the snow-covered Wicklow Mountains when, letting his folk go off to test the more exciting slopes, he patiently tried to teach me to ski on a short stretch of mountain road.

My real understanding of Gosset as a statistician began, as no doubt for many others, when I joined that wide circle of his scientific correspondents. Perhaps to the majority of these he has stood as the friend who, with a greater mathematical knowledge, helped them to understand the statistical approach to experimental problems. In my own case the position was a little different, as his endeavour was always to temper my mathematical reasoning with sane common sense. I can think of no other statistician who would have shown that interest and forbearance over many years to a young man who was continually posting to him the results of half-finished investigations for comment and criticism. In looking back through this correspondence I realize more clearly now than I could ever have done at the time what its value to me has been, and I can see how many of his ideas scattered through these letters have since almost unconsciously become part of my own outlook. I think this must be true also in the case of other persons with whom he corresponded, so that one can say that the last thirty years' progress in the theory and practice of mathematical statistics owes far more to "Student" than could be realized by a mere study of his published papers.

One of the striking characteristics of these papers, also of course evident in correspondence, was the simplicity of the statistical technique he used. The mean, the standard deviation and the correlation coefficient were his chief tools; hardly adequate for treating specialized problems it might be thought; yet how extremely effective in fact in his skilled hands! There is one very simple and illuminating theme which will be found to run as a keynote through much of his work, and may be expressed in the two formulae:*

$$\sigma^2_{x+y} = \sigma^2_x + \sigma^2_y + 2\rho\sigma_x\sigma_y, \qquad \text{......(1)}$$

$$\sigma^2_{x-y} = \sigma^2_x + \sigma^2_y - 2\rho\sigma_x\sigma_y. \qquad \text{......(2)}$$

Perhaps we may count as one of his big achievements the demonstration in many fields of the meaning of that short equation (2); as he wrote in 1923 (**11**, p. 273, but with modified notation):

> The art of designing all experiments lies even more in arranging matters so that ρ is as large as possible than in reducing σ^2_x and σ^2_y.

It is a simple idea, certainly, but I cannot doubt that its emphasis and amplification helped to open the way to all the modern developments of analysis of variance, and there may be some who have felt that where this technique runs a risk of defeating its ends by over-elaboration is just where that simple maxim has been set on one side. Recently I came across a short passage in a letter to a friend in Australia which refers to this theme and illustrates Gosset's own humorously modest outlook on his own contributions. He had just received a good deal of criticism of a paper he read in March 1936 before the Industrial and Agricultural Research Section of the Royal Statistical Society (**21**), particularly because of his advocacy of the half-drill strip method of agricultural experiment. This is essentially a method of comparison whose efficiency depends on maximizing correlation by taking the difference between the yields of neighbouring strips of the two varieties or treatments compared. He wrote:

> Meanwhile I...enclose the rough proof of what I said at the Statistical. You will gather from that that I am not in the fashion....Some years ago an American referred to difference treatment as "Student's" method and, though at the time I referred it to Noah, I am beginning to think that there is something in the name.†

Another point which must be borne in mind in gaining a real understanding of Gosset's character and outlook is that all his most important statistical work was undertaken in order to throw light on problems which arose in the analysis of data connected in some way with the brewery. The subject of statistics was in no sense a whole-time job for him, nor, on the other hand, was it his hobby as it might perhaps be described in the case of W. F. Sheppard; he undertook theoretical investigations only when he or his colleagues were faced with difficulties which needed solution along statistical lines. Rapid if less accurate methods appealed to him because in much heavy routine work it was a question of finding such methods or of making no attempt at statistical treatment. He was in no hurry to see his results in print, and several of his papers in *Biometrika* were written in response to an editorial request rather than on his own initiative. In two cases at least, which I shall refer to below, he was using methods in the brewery ten years before publication was undertaken. He was indeed the ideal

* σ_x, σ_y, σ_{x+y} and σ_{x-y} are the standard deviations of x, of y, of $x+y$ and of $x-y$ respectively, and ρ is the coefficient of correlation between x and y.

† See (**14**, p. 709).

servant of his firm, and part of the value of his life's work would need to be recorded in a history of progress gained by scientific research in industry rather than in the pages of *Biometrika*.

Yet in spite of the fact that only a small part of his time was taken up with statistics, Gosset had a wonderful power of "getting there first" before the more professional statisticians. Perhaps it was because his greater detachment meant a continual freshness of mind. It is this characteristic, as well as those others I have mentioned, that I shall try to bring out in my description of his work in the following pages.

EARLY STATISTICAL INVESTIGATIONS

Gosset became one of the brewers of Messrs Arthur Guinness Son and Co., Ltd., in 1899. The firm had shortly before initiated the policy of appointing to their staff scientists trained either at Oxford or Cambridge, and these young men found before them an almost unexplored field lying open to investigation. A great mass of data was available or could easily be collected which would throw light on the relations, hitherto undetermined or only guessed at in an empirical way, between the quality of the raw materials of beer, such as barley and hops, the conditions of production and the quality of the finished article. With keen minds playing round the interpretation of these data it was almost inevitable that before long the need was realized of some understanding of the theory of errors. No doubt during the first few years of his appointment Gosset was mainly occupied with learning the routine work of his job, but once this knowledge had been gained it was natural that he, as the most mathematical of the younger brewers, should give his attention to the question of error theory. He seems to have made use of the following books: G. B. Airy, *Theory of Errors of Observations*; Lupton, *Notes on Observations*; M. Merriman, *The Method of Least Squares*.

By 1904 he had made himself sufficiently familiar with the subject to draw up a *Report* on "The Application of the 'Law of Error' to the work of the Brewery". This document, dated 3 November 1904,* opens with some paragraphs which set out in simple terms a case for the introduction of statistical method in large-scale industry. They are worth quoting since they might be put before many a board of directors to-day with just as much cogency as they were put 34 years ago in Dublin:

> The following report has been made in response to an increasing necessity to set an exact value on the results of our experiments, many of which lead to results which are probable but not certain. It is hoped that what follows may do something to help us in estimating the Degree of Probability of many of our results, and enable us to form a judgment of the number and nature of the fresh experiments necessary to establish or disprove various hypotheses which we are now entertaining.

* I am extremely grateful to the firm for giving me permission to see and quote from this and other records available in their Dublin brewery.

When a quantity is measured with all possible precision many times in succession, the figures expressing the results do not absolutely agree, and even when the average of results, which differ but little, is taken, we have no means of knowing that we have obtained an actually true result, and the limits of our powers are that we can place greater odds in our favour that the results obtained do not differ more than a certain amount from the truth.

Results are only valuable when the amount by which they probably differ from the truth is so small as to be insignificant for the purposes of the experiment. What the odds should be depends:

(1) On the degree of accuracy which the nature of the experiment allows, and

(2) On the importance of the issues at stake.

It may seem strange that reasoning of this nature has not been more widely made use of, but this is due:

(1) To the popular dread of mathematical reasoning.

(2) To the fact that most methods employed in a Laboratory are capable of such refinement that the results are well within the accuracy required.

Unfortunately, when working on the large scale, the interests are so great that more accuracy is required, and, in our particular case, the methods are not always capable of refinement. Hence the necessity of taking a number of inexact determinations and of calculating probabilities.

The *Report* then introduces the error curve and discusses some of its properties. The curve is written in Airy's form

$$y = \frac{1}{c\sqrt{\pi}} e^{-x^2/c^2}, \qquad \text{......(3)}$$

where c is the modulus. The method is given for estimating c from a sample of n observations, by calculating (*a*) the mean deviation, (*b*) the mean square deviation (dividing by $n-1$), and using the appropriate correcting factors. It is stated that (*b*) gives a better value "in proportion 114/100".* A numerical example is given and it is suggested that both methods (*a*) and (*b*) should be used to check one another. There is next some discussion given to what was then clearly a most important practical problem in the brewery: the size of sample needed to make the odds that the mean lay within desired limits sufficiently large. Chauvenet's criterion for the rejection of extreme observations is quoted, as well as the modulus of the estimate of c (obtained by the mean square process), namely $c/\sqrt{(2n)}$.

All this is simply Airy or Merriman put by Gosset into the form most useful for his fellow brewers. What, however, shows a flash of his own insight is the use which he makes of Airy's theorems on the "Error of the result of the addition (or subtraction) of fallible measures". Thus if

$$W = X \pm Y \pm Z \pm \text{etc.}, \qquad \text{......(4)}$$

* This is the ratio of the sampling variances of (*a*) the mean deviation, and (*b*) root mean square deviation estimates of c, in large samples. I do not know from what source Gosset obtained these figures. The full value of the standard error of the mean deviation for samples of any size from a normal population was first derived, I believe, by Helmert (1876), but Gosset could not have known of this paper.

and E, e, f, g, etc., are the probable errors (or alternatively the moduli or the mean errors) of W, X, Y, Z, ... respectively, then Airy gives the law

$$E^2 = e^2 + f^2 + g^2 + \dots \qquad \dots\dots(5)$$

Gosset had noticed in certain cases he had met with that the result $E^2 = e^2 + f^2$ did not hold, as it should according to this law, for both $W = X + Y$ and $W = X - Y$. In other words he found that if W, X and Y are measured from their means there was very considerable difference between *Sum* $(X + Y)^2$ and *Sum* $(X - Y)^2$. He concluded that when this was the case it was a sign of the existence of a correlation between the variables. Thus he was feeling his way towards the fundamental relations (1) and (2) of p. 212 above, but he had not yet been introduced to the correlation coefficient.

The concluding remarks of the *Report* are interesting:

We may point out that, although the proof of the law (of Error) rests on higher mathematics, the application of it only demands quite simple algebra. We have been met with the difficulty that none of our books mention the odds, which are conveniently accepted as being sufficient to establish any conclusion, and it might be of assistance to us to consult some mathematical physicist on the matter.

This last difficulty was repeated in the summary which contains the sentence:

Explains that we have no information of the degree of probability to be accepted as proving various propositions, and suggests referring this question to a mathematician.

It is curious perhaps that Gosset should have felt at first that a mathematician was needed to solve this particular problem, which is just the point which the mathematician would now consider that the practical man must answer.* As we shall see in a moment he changed his view, but it seems to have been uncertainty on this question which led almost at once to that important contact between Gosset and Karl Pearson. A minute of March 1905 added to the printed *Report* indicates that arrangements for this meeting are to be made.

The interview was arranged through Vernon Harcourt, a chemistry don at Oxford whose pupil Gosset may have been and who perhaps got into touch with Pearson through Weldon, who was then Professor of Comparative Anatomy at Oxford. The opportunity for a meeting came about 12 July 1905 when Pearson was spending his long vacation at East Ilsley in Berkshire and Gosset bicycled over from his father's house at Watlington, preceded by a list of questions from which the following paragraphs are taken:

(1) *My original question and its modified form.* When I first reported on the subject, I thought that perhaps there might be some degree of probability which is conventionally treated as sufficient in such work as ours and I advised that some outside authority should be consulted as to what certainty is required to aim at in large scale work. However it would appear that in such work as ours the degree of certainty to be aimed at must depend

* I have, however, heard of another very recent case where an industrialist considered that it was the mathematical statistician's job to suggest the appropriate odds to use.

on the pecuniary advantage to be gained by following the result of the experiment, compared with the increased cost of the new method, if any, and the cost of each experiment. This is one of the points on which I should like advice.

(2) *Another problem.* I find out the P.E. of a certain laboratory analysis from n analyses of the same sample. This gives me a value of the P.E. which itself has a P.E. of P.E.$/\sqrt{2n}$. I now have another sample analysed and wish to assign limits within which it is a given probability that the truth must lie. E.g. if n were infinite, I could say "it is 10 : 1 that the truth lies within 2·6 of the result of the analysis". As however n is finite and in some cases not very large, it is clear that I must enlarge my limits, but I do not know by how much.

(3) *What is the right way to establish a relationship between sets of observations?* I use the following method when endeavouring to establish a relationship between sets of observations, but I have reason to suppose that it is not a good way and would like criticism on my method and advice as to the proper way. Suppose observations A and B taken daily of two phenomena which are supposed to be connected. Let A_1, A_2, A_3, etc. be the daily A observations and let B_1, B_2, B_3, etc. be the daily B observations. (I reduce the B observations if necessary or increase them by multiplying by a constant so that the P.E. of the A and B is about the same.) Then I form two series A_1+B_1, A_2+B_2, etc. and A_1-B_1, A_2-B_2, etc. and find the P.E. of each of the new series. If they are markedly different, it is clear (sufficient observations being taken) that the original series A and B are connected and proceed to attempt to find it quantitively. I cannot however at present find the P.E. of my results, nor can I be quite sure how great a difference between the P.E.'s of the sum and difference series is necessary to shew the connection.

(4) *What books would be useful?* When you talk with me you will doubtless find out many other points on which I require enlightenment and could perhaps recommend me some books on the subject.

The solution of "*another problem*" was to be given $2\frac{1}{2}$ years later in Gosset's paper on "The probable error of a mean" (2). The method described in paragraph (3) is interesting. I do not know exactly how Gosset attempted to measure the relationship quantitively, but if, as would seem natural, he compared the difference between $\Sigma(A+B)^2$ and $\Sigma(A-B)^2$ with their average, then by adjusting the scale so as to make the P.E.'s of A and B approximately the same, he had secured a maximum value for this ratio, and therefore presumably minimized the risk of overlooking a relationship. For

$$\frac{\Sigma(A+B)^2-\Sigma(A-B)^2}{\frac{1}{2}\{\Sigma(A+B)^2+\Sigma(A-B)^2\}}=\frac{4r_{AB}\,\sigma_A\,\sigma_B}{\sigma_A^2+\sigma_B^2},$$

which, for a given value of r_{AB}, has a maximum value of $2r_{AB}$ when $\sigma_A=\sigma_B$. One feels that, given a little more time, with his unerring instinct for reaching the best solution, Gosset would have found for himself Galton's correlation coefficient, just as he was later to rediscover Poisson's limit to the binomial and Helmert's distribution of a squared standard deviation.

Among Pearson's rough jottings written down for Gosset at the interview is the basic formula that he needed,

$$\sigma_{A\pm B}^2=\sigma_A^2+\sigma_B^2\pm 2r\sigma_A\sigma_B$$

(with the letter r doubly underlined), the probable error formula for r and also references to a number of papers on the theory of statistics.

Gosset was a quick learner; the immediate results of this visit include a *Supplement* to the brewery *Report* of 1904, from which I have quoted, and a second *Report* on correlation dated 30 August 1905. In both of these the influence of new ideas received from Pearson is evident. The *Supplement* contains a warning that distributions may not always be normal, although in small sample problems "it is practically convenient to use a curve...which has been thoroughly investigated, of which the values have been tabled, and which in the majority of cases describes them 'within the error of random sampling'". His colleagues are also advised to use the standard deviation and not the mean error. The *Report* is headed "The Pearson Co-efficient of Correlation", and describes, with a numerical example, the method of calculating this coefficient, r, as well as the use of the regression straight line for prediction.

This idea of correlation, which in origin is of course Galton's rather than Pearson's, has more than once during the past fifty years brought with it a stimulus leading to fresh discovery. The conception, presented with all its novelty to minds which had hitherto only considered the perfect relationship of the physicist as a relationship which could be scientifically handled, has seemed to provide a key to the solution of a host of problems. The inspiration which Galton's discussion of correlation in his *Natural Inheritance* gave to Weldon and Pearson in the early nineties has often been referred to and, now, the introduction of the new ideas opened out fresh avenues of research to both Gosset and his colleagues. The crude method which Gosset had invented of examining the difference between $\Sigma(A+B)^2$ and $\Sigma(A-B)^2$ could be abandoned. It became possible to assess with precision the relative importance of the many factors influencing quality at different stages in the complicated process of brewing, and before long the methods of partial and multiple correlation were mastered and applied.* The *Reports* circulated within the brewery constantly quote correlation coefficients and their probable errors, while Gosset's rough notebooks of this date contain numerous correlation tables. Apart from the actual calculation of r, the idea of arranging data in a two-way table was possibly novel and certainly illuminating to the brewers.

It seems, however, to have been at once obvious to Gosset that the methods developed by Pearson and his co-workers for handling the large samples met with in biometric inquiries would probably need modification when applied to the problems of the brewery. In his *Report* on correlation of August 1905 he notes that "correlation coefficients are usually calculated from large numbers of cases, in fact I have only found one paper in *Biometrika* of which the cases are as few in number as those at which I have been working lately". He was dealing at this time with all the possible correlations between a number of characters for which 31 observations were available; in another problem only 10 observations

* A *Report* of Gosset's of June 1907 applies multiple correlation to prediction. The mathematical Appendix dated 27 September 1906 is stated to have been read through by Karl Pearson.

could be used. He gives a reason which, though faulty, is extremely interesting, for doubting the validity of the probable error formula for r in small samples. Thus, if r is an observed correlation from a sample of n individuals, he takes the ratio

$$\frac{\text{Deviation of } r \text{ from zero}}{\text{Probable error of } r} = \frac{r}{0{\cdot}6745\,(1-r^2)/\sqrt{n}} \qquad \ldots\ldots(6)$$

as a measure of the significance of the correlation, remarking that if the ratio is greater than $2\frac{1}{2}$ the odds are about 20 : 1 on the existence of a real relationship. He then says that if n be very small "I expect a larger ratio is required", and illustrates this by supposing that $r = 0{\cdot}9$, $n = 4$, when the probable error calculated as in (6) becomes 0·064 and the ratio is 14. "Yet", he remarks, "no one would claim any certainty from four experiments."

If we are asking whether an observed r is consistent with sampling from a population in which the correlation, say ρ, is zero, then the appropriate probable error is approximately $0{\cdot}6745/\sqrt{n}$ and not the value used in (6). Thus in Gosset's example the ratio is really 2·7 and not 14; as he was afterwards to show, it was not the standard error that was seriously at fault in testing significance when dealing with small samples, but the assumption of normality. For $n = 4, \rho = 0$, the distribution of r is rectangular. The faulty reasoning involved in the interpretation of equation (6) has been used again and again in statistical literature; the reason that in 1905 the difficulty had not caught the attention of the workers at the Biometric Laboratory was that they were dealing with large samples and, for these, the error involved is of relatively small consequence. It was Gosset, "naughtily" playing about with absurdly small numbers,* who stumbled on the inconsistency, although not at first understanding its reason. Here perhaps we may see the first illustration of the tremendous gain in clear thinking that has followed in statistics from an approach to the subject from the small-sample end. Also this is one of the many occasions on which Gosset was first on the spot.

There were other difficulties in application that he was already turning over in his mind. For instance, he wished to obtain a combined measure of the correlation between two characters measured on several varieties of the barley used for malting and he considered the possibility of taking deviations from variety means. "I hope to find out the limitations of this device at some later date", he reported. "I am using it and similar devices pretty freely.... "

A point which may be of interest to industrial statisticians to-day is that the practical brewer of thirty years ago, as the practical engineer to-day, was objecting to the introduction into his reports of the statistician's term *population*, yet was unable to suggest an appropriate substitute. A footnote to the word population ran as follows: "This appears to be a general statistical term to

* Writing to Gosset on 17 September 1912 on the subject of the standard deviation, not correlation, Pearson remarked that it made little difference whether the sum of squares was divided by n or $(n-1)$, "because only naughty brewers take n so small that the difference is not of the order of the probable error!"

express a number of things or people of the same kind. We have tried to find a word in common use to express this, but have failed."

The *Report* closes with a characteristic piece of sound advice:

> It must be borne in mind, however, that the better the instrument the greater the danger of using it unintelligently: it is more important than ever to think carefully in what way any connection may have arisen accidentally, and, more especially, any semi-constant variation must be treated with particular care.
>
> Statistical examination in each case may help much, but no statistical methods will ever replace thought as a way of avoiding pitfalls, though they may help us to bridge them.

THE YEAR IN LONDON, 1906–7, AND THE WORK ON SMALL SAMPLES

Following a general practice of the brewery, Gosset was sent away from Dublin for a year's specialized study. He spent the greater part of this time either working at or in close contact with the Biometric Laboratory, where he arrived at the end of September 1906. During the year which had elapsed since he first met Karl Pearson he must have given a great deal of time and thought to the application of current statistical methods to the type of experimental and routine data analysed in the brewery. He was now anxious to obtain Pearson's opinion on the work he had already done and to ask his advice on a number of unsolved questions. Probably he had already realized that the most important problem on which he required further information was the behaviour of frequency constants in small samples. In a letter written to a friend at the brewery on 30 September 1906, just after his arrival, he outlines, however, only a modest programme:

> Then he [K. P.] proposes to give me a room to work in, that I should attend his lectures, and become as far as possible accustomed to the calculations, etc., of his department. I had a long talk with him, and told him the lines I had been going on in the Hops..., and he seemed to consider that I had been over most of the ground, but points soon cropped up which showed him the necessity for going deeper. I think that from what he said I am more or less on the right lines so far; perhaps when the reports have been considered you might let me have a copy of each of them, to ask about anything which may have occurred to me by then about them. I think he would be very willing to give us advice on any points which crop up.

The first problem which he took up was of considerable practical importance in one department of the brewery activities: the question of the sampling error involved in counting yeast cells with a haemacytometer. In his paper (1) published early in 1907 he derived afresh Poisson's limit to the binomial distribution, namely,

$$e^{-m}\left\{1+m+\frac{m^2}{2!}+\dots+\frac{m^r}{r!}+\dots\right\}, \qquad \dots\dots(7)$$

and showed by a comparison of the series with four sets of experimental results that it did represent well the observed distribution of cell counts in an investigation carried out under carefully controlled conditions. The paper should be read in conjunction with another that he wrote on the same subject twelve years later (9).

2	2	4	4	4	5	2	4	7	7	4	7	5	2	8	6	7	4	3	4	91
3	3	2	4	2	5	4	2	8	6	3	6	6	10	8	3	5	6	4	4	94
7	9	5	2	7	4	4	2	4	4	4	3	5	6	5	4	1	4	2	6	88
4	1	4	6	3	9	3	5	8	2	9	5	3	9	5	5	2	4	3	4	88
4	1	5	9	3	4	4	6	6	5	4	6	5	5	4	3	5	9	6	4	98
4	4	5	10	4	4	3	8	3	2	1	4	1	5	6	4	2	3	3	3	79
3	7	4	5	1	8	5	7	9	5	8	9	5	6	6	4	3	7	4	4	110
7	5	6	3	6	7	4	5	8	6	3	3	4	3	7	4	4	4	5	3	95
8	10	6	3	3	6	5	2	5	3	11	3	7	4	7	3	5	5	3	4	103
1	3	7	2	5	5	5	3	3	4	6	5	6	1	6	4	4	4	6	4	84
4	2	5	4	8	6	3	4	6	5	2	6	6	1	2	2	2	5	2	2	77
5	9	3	5	6	4	6	5	7	1	3	6	5	4	2	8	9	5	4	3	100
2	2	11	4	6	6	4	6	2	5	3	5	7	2	6	5	5	1	2	7	91
5	12	5	8	2	4	2	1	6	4	5	1	2	9	1	3	4	7	3	6	90
5	6	5	4	4	5	2	7	6	2	7	3	5	4	4	5	4	7	5	4	94
8	4	6	6	5	3	3	5	7	4	5	5	5	6	10	2	3	8	3	5	103
6	6	4	2	6	6	7	5	4	5	8	6	7	6	4	2	6	1	1	4	96
7	2	5	7	4	6	4	5	1	5	10	8	7	5	4	6	4	4	7	5	106
4	3	1	6	2	5	3	3	3	7	4	3	7	8	4	7	3	1	4	4	82
7	6	7	2	4	5	1	3	12	4	2	2	8	7	6	7	6	3	5	4	101
96	97	100	97	85	100	74	88	115	86	102	96	106	103	105	87	84	92	75	84	1872

Page from Gosset's notebook containing the analysis of haemacytometer counts. (Left-hand page.)

			Calc			χ^2
0	0		3.704	3.712	−4	4.00
1	20	20	17	17.37	+3	.53
2	43	86	41	40.65	+2	.10
3	53	159	63	63	−10	1.59
4	86	344	74	74	+12	1.95
5	70	350	70	69	0	
6	54	324	54	54	0	
7	37	259	36	36	+1	.03
8	18	144	21	21	−3	.43
9	10	90	11		+1	.09
10	5	50	5		0	
11	2	22	2		0	
12	2	24	1		+1	1.00
	400	1872				9.72

$(1 - vp)$ $(P + Q) = 1$

$1 + \frac{1}{1} + \frac{1}{1.2} + \frac{1}{1.2.3} + \dots$

$5^2 = 25$

about .64

from $(.9525 + .0475)^{98.53}$ $\chi^2 = 9.28$

Average number 4.68

Calculating the number of mean cases on formula $\frac{N}{\sqrt{2\pi nq}}$ & putting $q = 1$ for this purpose we get the number to be 73.8 (interpolation) a fair agreement between 86 @ 4 } 4.68 about 75 or 76; 70 @ 5 }

$400\left(\frac{399}{400} + \frac{1}{400}\right)^{1872}$ $\left(\frac{1872}{400}\right) \times 400$ $1 + \frac{1872}{400}$ $1 + 4.68$

$Me^{-m}\left\{1 + m + \frac{m^2}{1.2}\right.$

Page from Gosset's notebook containing the analysis of haemacytometer counts. (Right-hand page.)

Plate II

The derivation of the limiting form of the binomial was not in itself an achievement of any special difficulty; the series has been obtained independently from time to time by a number of investigators. But it was characteristic of "Student's" flair or, as he himself would have said, luck that when he had a practical problem to solve he should go straight to the correct solution; and that because it was a fundamental type of biological problem his research should have been of much greater value in the field of applied statistics than von Bortkiewicz's work, illustrated by fitting the Poisson series to suicides of German women and deaths of Prussian soldiers from the kicks of a horse.

I have reproduced in facsimile in Plate II two pages from Gosset's notebook containing the rough working for this paper. The experimental data are those of the series IV (see his p. 357). They are quoted also as an example of a Poisson distribution by R. A. Fisher in *Statistical Methods for Research Workers* (1938, p. 58). The left-hand page contains the 400 individual yeast cell counts and the resulting frequency distribution and histogram; the right-hand page shows the calculation of the mean, m, as well as the theoretical series and the derivation of χ^2. The expression $N/\sqrt{(2\pi mq)}$ (or $N/\sqrt{(2\pi nq)}$), where q is put equal to unity, is an approximation to the frequency in the group containing the mean. In the notes, Gosset seems to have reached this result by a rather lengthy method, but it can be obtained by putting $r=m$ in the general term of series (7) and using the first order term in Stirling's approximation to $m!$ No reference to this comparison was made in the published paper. A few figures, which are in pencil in the note-book, appear to be in Pearson's hand; e.g. the theoretical frequencies 3·712, 17·37 and 40·65 as well as the three terms of the Poisson series at the bottom of the page. They were jottings made no doubt by K. P. on one of his daily "rounds" of the laboratory.

A good part of the work on Gosset's second paper on "The probable error of a mean" (**2**) was also carried out during his year in England; with it is closely associated his third paper on the "Probable error of a correlation coefficient" (**3**), as both were supported by the same piece of experimental sampling. I have already referred to Gosset's doubts regarding the distribution of r in small samples; since in the brewery work a mean value had often to be estimated from eight or ten determinations he also felt uneasy about the applicability to such work of accepted theory regarding the distribution of the mean and the standard deviation. A letter written on 12 May 1907 to a colleague in Dublin shows him to be in the middle of his investigation. After dealing with some points about the significance of differences* he adds:

> Herewith my answer to your questions. I hope it is quite clear, but I am afraid I rather increase the difficulties when I try to explain anything as a rule.

* There is a reference to that long-standing difference of opinion regarding n and $n-1$, in the following sentence: "When you only have quite small numbers, I think the formula we used to use for the P.E. ($\sqrt{\{\Sigma(x^2)/(n-1)\}}\times 0\cdot 6745$) is better, but if n be greater than 10 the difference is too small to be worth taking the extra trouble." Here K. P. and Airy were in disagreement.

What I have written on the back is true for large samples, and approximately so for small, and is the accepted theory. My work on small numbers may or may not modify it. We shall know later....

I go up to K.P.'s lectures from here [The Ousels, Tunbridge Wells] and on other days work at small numbers: a greater toil than I had expected, but I think absolutely necessary if the Brewery is to get all the possible benefit from statistical processes.

There could be no better illustration than these last sentences of the way in which Gosset's best work was called forth in the service of his firm.

The contents of the paper on the probable error of the mean are too well known to require more than a brief summary. Starting with a sample of n observations, $x_1, x_2, \ldots, x_n$, from a normal population with standard deviation σ and mean at the origin for x, Gosset obtained the sampling moments of $s^2 = \Sigma(x-\bar{x})^2/n$, where $\bar{x}$ is the sample mean. He showed that these moments corresponded exactly with those of a Pearson type III curve and hence inferred that the curve representing the sampling distribution of s^2 must almost certainly be

$$y = \text{constant} \times \sigma^{-n+1}(s^2)^{\frac{1}{2}(n-3)}\, e^{-ns^2/2\sigma^2}. \qquad \ldots\ldots(8)^*$$

He then showed that the correlation coefficient between $\bar{x}^2$ and s^2 was zero and, making the assumption (which does not necessarily follow though in fact it is true in this case) that this meant that $\bar{x}$ and s were absolutely independent, he deduced the probability distribution of $z = \bar{x}/s$ as

$$p(z) = \text{constant} \times (1+z^2)^{-\frac{1}{2}n}. \qquad \ldots\ldots(9)$$

He considered the properties of this curve,† gave a table of its probability integral for $n=4$ to 10 and examined its approach to a normal curve with standard deviation $1/\sqrt{(n-3)}$. He next compared the distributions (8) and (9) with the results of a sampling experiment for the case $n=4$ and finally illustrated the use of his results on four examples.

When two years ago the question of the photographic reissue of the paper had been suggested to meet a continued demand for offprints, Gosset wrote to me describing it as now "rather a museum piece". That is true, though perhaps in a different sense than he meant. It is a paper to which I think all research students in statistics might well be directed, particularly before they attempt to put together their own first paper. The actual derivation of the distributions of s^2 and z, or of $t=\sqrt{(n-1)}\,z$ in to-day's terminology, has long since been made simpler and more precise; this analytical treatment need not be examined carefully, but there is something in the arrangement and execution of the paper which will always repay study.

In the first place, in the Introduction and Conclusions we find an excellent illustration of Gosset's wise advice given to a beginner in the art of composition: "First say what you are going to say, then say it and finally end by saying that

* That this result had previously been derived by Helmert (1876), English-speaking statisticians were quite unaware till many years later.

† There are some minor errors in §§ IV and V of the paper.

you have said it."* The main part of the paper, the "saying it", is divided clearly into headed sections. The adequacy of the assumptions on which the mathematical theory rests is tested by a piece of experimental sampling; this test being satisfactorily passed, computed tables required for application are given and finally a number of well-chosen examples illustrate the purpose of the inquiry.

Before considering some other notable features of the paper and attempting to assess its influence on later work, it is important to see just what was the main purpose of the inquiry that its author had in mind. As usual with him, this was simple and practical. Having n observations, he wished to know within what limits the mean of the sampled population—the "true result" of the 1904 *Report*—probably lay. His solution involved a tacit introduction of the method of inverse probability, but I do not think he ever tried to put this into precise terms.† Thus the last sentence on the first page of the paper runs as follows:

> The usual method of determining the probability that the mean of the population lies within a given distance of the mean of the sample, is to assume a normal distribution about the mean of the sample with a standard deviation equal to $s/\sqrt{n}$, where s is the standard deviation of the sample, and to use the tables of the probability integral.

The results of the present investigation meant to Gosset that he could now assume in small samples a z-distribution for the population mean about the sample mean, the scale now being the sample standard deviation, s. In his examples he uses the z tables, not to test the hypothesis that the population mean is zero or has some other specified value, but to find the odds that this mean lies within specified limits, e.g. between 0 and ∞, that is to say is positive. Take for instance his *Illustration* I (pp. 20–1); the average number of hours of sleep gained by ten patients treated with *D. hyoscyamine hydrobromide* is $\bar{x} = 0{\cdot}75$ while the standard deviation is $s = 1{\cdot}70$. If we regard the population mean, say ξ, to be distributed about the sample mean 0·75 in the z-form, with a standard deviation of s, it follows that the chance that $\xi > 0$ is the proportionate area under the z-curve between the ordinate at

$$z = \frac{0 - 0{\cdot}75}{1{\cdot}70} = -0{\cdot}44$$

and ∞. This is the same as the chance that $z < +0{\cdot}44$, which interpolation in his tables in the column $n = 10$ shows to be 0·887. He therefore argued that the odds are 0·887 to 0·113 that the population mean ξ is positive, i.e. that the soporific will

* The advice was not originally Gosset's. Writing in 1934 he says: "This is a rule which we owe to A. J. (I think at second hand)." He then quotes the rule and adds, "It does make things so much easier for everybody concerned, besides which 'what I tell you three times is true'"; the last words are those of the Bellman in *The Hunting of the Snark*.

† In his paper on the correlation coefficient written in the same year (**3**, p. 302) Gosset states definitely that a knowledge of the *a priori* probability distribution of the population correlation coefficient, R, is needed in order to determine "the probability that R...shall lie between any given limits".

on the average give an increase of sleep. While a somewhat loosely defined conception of inverse probability seems to underlie the argument, it will be seen that as far as the practical consequences go, Gosset had reached a result which we can hardly improve on 30 years later. It is true that, using the idea of the fiducial or confidence interval, some of us would word our statement of limits and probabilities a little differently so as to avoid any appeal to inverse probability, but as practical statisticians we must, I think, admit that our conclusions would be identical.

There are some other features of the paper which are interesting historically. Gosset remarks on p. 13 that before he succeeded in solving the problem analytically, he had endeavoured to do so empirically. The sampling experiment which he carried out for this purpose involved the drawing of 750 samples of 4 by means of shuffled slips of cardboard, from W. R. Macdonell's (1901) correlation table containing the distribution of height and middle-finger length of 3000 criminals. As far as I know this was the first instance in statistical research of the random sampling experiment which since has become a common and useful feature in a large number of investigations where precise analysis has failed. The results of this same experiment were used by Gosset in a number of later papers. On p. 16 he draws attention to a difficulty in the application of Pearson's χ^2-test of goodness of fit which was later to lead to R. A. Fisher's modification in terms of degrees of freedom. On p. 19 he gives reasons for believing that even when the population sampled is not normal the sampling distribution of z will be very little modified; this was a prediction which experimental and theoretical investigations carried out in recent years have confirmed.

Finally we may note the introduction of a difference in notation to distinguish between sample and population characters, viz. s for the sample and σ for the population standard deviation. The need for this distinction seems obvious to us to-day, but it is interesting to notice that it was only when attention was directed to the problem of small samples that statisticians grasped the clarification resulting from this innovation.

As the theory of mathematical statistics has developed, the significance of "Student's" test has been elaborated from many angles and deeper meanings associated with it than its author had ever dreamed of. This is a common feature of scientific progress, but as Neyman very appropriately remarked on a recent occasion (1937, p. 142): "The role of a rigorous scientific theory is frequently very modest and is reduced to explaining to the practical man—and this sometimes with a certain difficulty—how good is what he himself knew to be good long ago." To understand the reason for the historical importance that has rightly been associated with this paper, it is not however necessary to discuss the abstract conceptions of the mathematical statistician and their relation to forms of critical regions in hyperspace; it can be explained much more simply than that. As Gosset wrote on the second page of the paper, referring to the inadequacy for certain purposes of the statistical technique available in 1908:

> There are other experiments, however, which cannot easily be repeated very often; in such cases it is sometimes necessary to judge of the certainty of the results from a very small sample, which itself affords the only indication of the variability. Some chemical, many biological, and most agricultural and large scale experiments belong to this class, which has hitherto been almost outside the range of statistical inquiry.

It is probably true to say that this investigation published in 1908 has done more than any other single paper to bring these subjects within the range of statistical inquiry; as it stands it has provided an essential tool for the practical worker, while on the theoretical side it has proved to contain the seed of new ideas which have since grown and multiplied an hundredfold.

The sampling experiment used to test the accuracy of the theoretical distributions of s^2 and z was also planned to throw light on the distribution of the correlation coefficient r, in very small samples. In this second problem (**3**) Gosset was forced to rely much more on his empirical approach than before, since the mathematical solution lay beyond his powers. In suggesting the probable form of the distribution of r when sampling from a population in which the two variables were uncorrelated (i.e. $R=0$)* he could get no clue from known values of moments as in the case of s^2. He started from the following basis: (*a*) the distributions must be symmetrical about $r=0$ and be limited within the range -1 to $+1$; (*b*) he had available the distributions of r found from his experiment for 745 samples of 4 and 750 samples of 8; (*c*) of these, he noticed that the former was approximately rectangular.

As in the case of s^2, his training at the Biometric Laboratory naturally suggested that he should try to use a Pearson curve for the unknown distribution; a type II curve was the only one suitable, and therefore in his own simply expressed phrase, "working from $y=y_0(1-x^2)^0$ for samples of 4 I guessed the formula

$$y=y_0(1-x^2)^{\frac{1}{2}(n-4)}\text{''}. \qquad \text{......(10)}$$

He then showed that for $n=8$ this formula represented his empirical sampling distribution very well, and pointed out that the result agreed with large sample theory, since the standard deviation $\sigma_r=1/\sqrt{(n-1)}$ would equal Pearson and Filon's value of $(1-R^2)/\sqrt{n}$ when $R=0$ and $n\to\infty$. He also gave the correct limiting result, which he had been able to establish for any R, when $n=2$, suggesting that this might furnish a clue for the distribution when $n>2$. It was a brilliant piece of guessing and all the more striking because of the forceful way in which the supporting evidence was marshalled.

In the case where the population correlation, R, was not zero Gosset provided three empirical sampling distributions for the cases $R=0{\cdot}66$ and $n=4$, 8 and 30.

* He used R for the population correlation; the notation, ρ, seems to have been first used by H. E. Soper (1913).

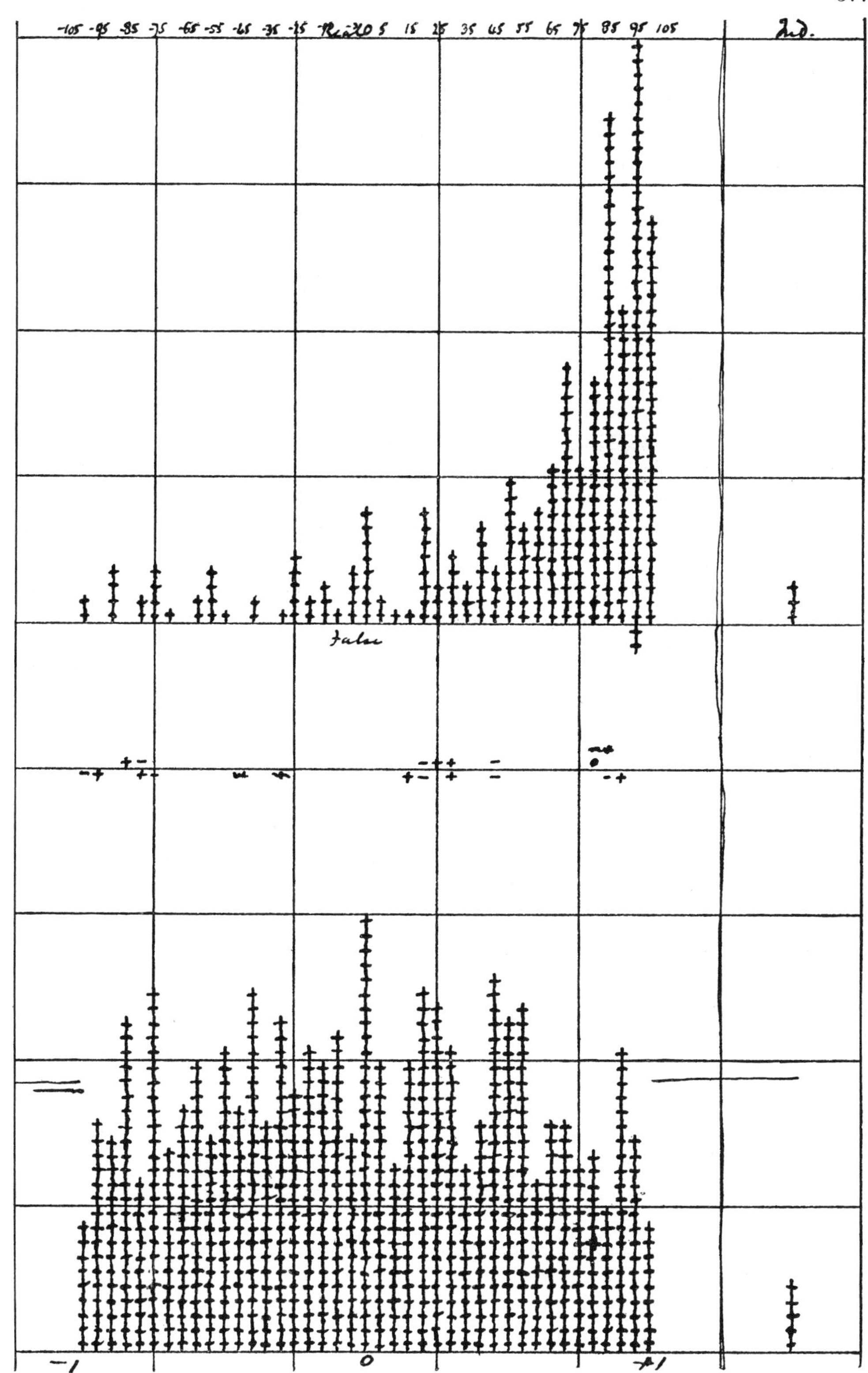

Distribution of the correlation coefficient in samples of 4, tabled in Gosset's notebook.

Above, $R = 0{\cdot}66$; *below*, $R = 0$.

Plate III

He also set out very clearly the conditions which his work showed must be satisfied by the true distribution. "I hope", he concluded, "they may serve as illustrations for the successful solver of the problem". Six years later R. A. Fisher was able to demonstrate the substantial accuracy of all Gosset's predictions both in the r and the z paper.

In the notebook containing the original samples of 4 from Macdonell's correlation distribution, there are given what I think must be the original distributions built up by Gosset as he tabled his calculated values of r. Two of these are shown in facsimile in Plate III ($n=4$, $R=0{\cdot}66$ and $n=4$, $R=0$). It is hard to believe that Gosset did not experience a very pleasurable excitement as these distributions gradually took shape on the paper, for he was exploring a region entirely unmapped and the discovery of the rectangular distribution in the case when $R=0$ must have been a complete surprise.*

One of the curious things that must strike us now about these two papers of Gosset's (**2**, **3**) is the small influence that their publication had for a number of years on current statistical literature and practice. The z-test was used in the brewery at once, but I think very little elsewhere for probably a dozen years. Perhaps because he realized that it showed how little reliability could be placed on a correlation coefficient based on small numbers, Gosset does not seem to have recommended the use of the r-test even to his colleagues and he made no tables of the probability integral for the distribution (10). I have come across, however, one reference to the work in a letter of 3 April 1912 to E. S. Beaven, in which the following remarks occur:

By the way, don't be *too* cock-a-hoop about your 0·95 correlation with 7 cases. Such a thing might occur more than once in a hundred trials of 7 cases, even if there were no correlation. (I haven't got tables to evaluate

$$\int_{\sin^{-1}0\cdot95}^{\frac{1}{2}\pi} \cos^4\theta\, d\theta \Big/ \int_{0}^{\frac{1}{2}\pi} \cos^4\theta\, d\theta,$$

but you get that fraction of N at each end 0·95 or over in N trials); and I guess its about 2% at each end.† All the same it seems very reasonable to suppose that it is right.

From Gosset's point of view, he had developed the tools which he needed for practical application in Dublin and he was not primarily interested in their wider use. If Pearson failed to realize the importance of the work and did not assimilate the results into current practice and teaching, it was because he too was mainly interested in what appeared to be of value in the research investigations of his laboratories. To him all small sample work was dangerous and should be avoided. But it would be wrong to suppose that there was a lack of sympathy between the two; except at a far later stage when opposite views over z found their way into print, Pearson's attitude towards Gosset's small sample

* Yet Mrs Gosset, who was helping him at the time, writes: "Whatever thrill he may have got out of that experiment he showed nothing whatever of it, and his amanuensis never realized that there was anything original about it!"

† Gosset was wrong here. The fraction is actually 0·001.

work was one of humorous protest, well conveyed in the quotation I have given about "naughty brewers" who take n too small (p. 368 above). The readiness with which he would talk to Gosset over his problems and at times refer to him on matters of difficulty shows how highly he rated his ability and insight. Although Gosset launched off along independent lines of investigation directly he had mastered the elements of statistical theory, it is clear that he owed a great deal to the early guidance that he received in London. In the first place he had that very great advantage of being freed for a year from his official duties and of spending that time in close contact with persons who were enthusiasts in the study of statistics. Although, as he wrote at a later date, "I am bound to say that I did not learn very much from his [K. P.'s] lectures; I never did from anyone's and my mathematics were inadequate for the task", he obtained from the Biometric Laboratory a number of things which were not to be found in Airy or Merriman: the theory of correlation, the χ^2-test, and above all Pearson's system of frequency curves. It is doubtful for instance if he could have reached the distribution of s^2, and hence that of z, if he had not had available for use Pearson's type III curve.

After his year in London was over Gosset kept in close touch with Pearson for 29 years, and to his intimate friends would speak with admiration of his teacher. Some sentences which he spoke at the opening meeting of the Industrial and Agricultural Research Section of the Royal Statistical Society in November 1933 were composed, I know, with this aspect of the relationship between professor and student in mind:

> Another point arises from the peculiar nature of statistics. It is impossible to apply statistical methods to industry or anything else unless one has a certain amount of intelligent experience as a background. That works both ways. The practical man has to go and talk to his Professor partly in order that the Professor himself should share his experience. ...The whole art of statistical inference lies in the reconciliation of random mathematics with biassed samples. Every new problem has some fresh kind of bias and might contain some new pitfall. The only way not to fall into these pitfalls is to talk over the problem with some intelligent critic; and so the practical man, if he is not entirely foolish, talks over his problems with the Professor, and the Professor does not consider himself to be a competent critic unless he has had some experience of applying the statistics to industry and has learned the difficulties of that application.

MISCELLANEOUS PAPERS, 1909–21

Before considering the very important part that Gosset played in the development of agricultural experimentation, it is desirable to give a brief account of six papers on a variety of subjects which were published in *Biometrika* between 1909 and 1921.

(i) The first of these papers on "The distribution of the means of samples which are not drawn at random" (**4**, 1909) dealt with one aspect of that theme which, as I have already mentioned, runs through so much of his work. He had realized at an early date how frequently there existed a correlation between successive observations either in time or space. Thus if x and y are two contiguous observations it would follow that

$$\sigma^2_{x+y} > \sigma^2_x + \sigma^2_y > \sigma^2_{x-y}.$$

Hence if x and y were successive duplicate chemical analyses of the same quantity their mean would be less reliable than we should expect on the usual theory of random sampling. On the other hand were x and y the yields from plots of two different cereals which were to be compared, by placing the plots side by side in space, the difference $x-y$ would be more reliable than on the classical error theory. In this paper he considers the distribution of the mean not of two but of n observations, so selected that they are correlated, i.e. more like one another than individuals randomly selected from the population. It is the problem of fraternities which Pearson had termed homotyposis in his biometric work. Gosset gave the second, third and fourth moments of the sample mean, the second having the value

$$M_2 = \frac{\sigma^2}{n}\{1+(n-1)\rho\}, \qquad \text{......(11)}$$

where σ is the population standard deviation of x and ρ the correlation between the x's in a sample, which Fisher has termed the intraclass correlation. From the values of the third and fourth moments he deduced that in general it was likely that the distribution of the mean would tend to normality less rapidly than when $\rho=0$.

From the practical point of view he was concerned to warn the chemist that "repetition of analyses in a technical laboratory should never follow one another, but an interval of at least a day should occur between them. Otherwise a spurious accuracy will be obtained which greatly reduces the value of the analyses".

(ii) The next paper (**6**) published in 1913 dealt with "The correction to be made to the correlation ratio for grouping", an investigation no doubt connected with Pearson's work (1913) on the same subject published in the same number of *Biometrika*.

(iii) Volume x of *Biometrika* (1914) contains a short note on "The elimination of spurious correlation due to position in time or space" (**7**). In this, Gosset showed that the difference correlation method used by F. E. Cave (1904) and R. H. Hooker (1905) could be extended to differences of higher order than the first. This paper was the basis of later investigations on the variate difference correlation method.

(iv) In 1917 (**8**) Gosset published an extension of his tables of the probability integral of z; the range covered now ran from $n=2$ to $n=30$. In the introductory remarks he again gave advice "as to the best way of judging the accuracy of physical or chemical determinations". He wrote:

After considerable experience, I have not encountered any determination which is not influenced by the date on which it is made; from this it follows that a number of determinations of the same thing made on the same day are likely to lie more closely together than if the repetitions had been made on different days. It also follows that if the probable error is calculated from a number of observations made close together in point of time much of the secular error will be left out and for general use the probable error will be too small. Where then the materials are sufficiently stable, it is well to run a number of determinations on the same material through any series of routine determinations which have to be made, spreading them over the whole period.

(v) Gosset's paper of 1919 (**9**) on "An explanation of deviations from Poisson's law in practice" answered some questions regarding the relation of this series to the positive and the negative binomial raised by Lucy Whittaker (1914) in a paper published five years earlier from the Biometric Laboratory. Since the rather severe criticisms of the latter paper directed against the applications of the Poisson law made by Bortkiewicz and Mortara might have discouraged its use in other directions, Gosset pointed out that the object of his own earlier paper (**1**) was to give the user of the haemacytometer a guide to the error of his count. From this first practical point of view it made little difference whether, theoretically, the better fitting distribution was a positive or negative binomial, although as a further point it was of interest to consider what such departures implied if the data were sufficient to establish them.

(vi) The final paper (**10**) of this group on "An experimental determination of the probable error of Dr Spearman's correlation coefficients", was written in the first instance for reading at one of the early meetings (13 December 1920) of the newly formed Society of Biometricians and Mathematical Statisticians. Gosset had many years before realized the value of the method of rank correlation in assessing quickly the order of relationship between two short series of numbers. Probably while working at the Biometric Laboratory he had developed the proof quoted by Pearson (1907, p. 13), that the standard error of the coefficient

$$\rho = 1 - \frac{6\Sigma D^2}{n(n^2-1)} \qquad \ldots\ldots(12)$$

is $1/\sqrt{(n-1)}$, in the case of independence in the population. In a Report written in 1911 for his colleagues in the brewery he illustrated the use of the method and gives what is substantially the correction for "ties" described in the present paper of 1921. Apart from the publication of this correction, the paper is of interest because Gosset again made use of his sampling experiment of 1907. For the 375 samples of 8 from a population having correlation 0·66 he calculated both of Spearman's rank correlation coefficients, in their raw and corrected form and, in the case of his 100 samples of 30 added Sheppard's estimate of correlation obtained from a median fourfold division. He uses these results to make a number of comparisons between the methods, in particular paying regard to the amount of additional sampling needed if one of these more rapid methods of "assay" is to give as reliable an estimate of the population correlation coefficient as that obtained from the usual product-moment formula. He concludes by suggesting to mathematicians a problem which has still remained unsolved, that of determining the sampling distribution of the rank coefficient of equation (12) above, in random samples from a bivariate normal population, in which the correlation is not zero.

The application of statistical method to agricultural plot experiments

It is a feature commonly noticeable in the advance along any new line of scientific inquiry that the first steps in that progress are made hesitatingly and with difficulty, accompanied by much trial and error; and then after many years of what seems, looking back, to have been a painfully slow advance to an obvious goal, a stage is reached where the way forward has been almost cleared so that the introduction, perhaps, of some new tool or some fresh personality leads to a rapid advance into fresh country. In later years the casual student may well attribute the beginning of an epoch to that moment of rapid advance, partly because few records of the earlier struggle have found their way into print and partly because the later workers themselves have hardly realized the amount of thought that has gone into the creation of ideas which have formed the groundwork of their own further progress.

The history of the introduction of statistical methods in the planning and interpretation of agricultural experiments provides an illustration of these points. The large extension of technique with the accompanying stimulus to scientific planning which followed R. A. Fisher's introduction of the methods of analysis of variance in the years following 1923, may have caused the present-day statistician to overlook the essential pioneer work of the preceding years, without which it is certain that the later advance would have been impossible.* It therefore seems appropriate to take this opportunity of giving rather special attention to this aspect of Gosset's contribution to statistics and to do so by following out the gradual stages by which he advanced from simple beginnings to the analysis of a balanced block experiment.

A number of persons contributed to this early work and, as is often the case when methods of attack are in an imperfect or trial stage, ideas were worked out in correspondence or by word of mouth rather than in print. The brewery, as a very large consumer of barley, was naturally interested in agricultural problems and in particular in certain large-scale experiments undertaken in Ireland under the supervision of the Irish Department of Agriculture. Gosset was not, however, concerned with giving advice in these experiments till a

* Fisher himself has on many occasions paid a warm tribute to the help he received both from "Student's" published work and from correspondence and discussion.

number of years after he had specialized in statistics, and I think his first real interest in agricultural work arose from his contact with E. S. Beaven, who as a maltster was from time to time in Dublin on official business. Beaven had started experimental work in the nineties and about 1905 approached Gosset for an interpretation of apparently anomalous results, afterwards seen to be due to interference, that he found in comparing the yields of two varieties of barley in his "cage" at Warminster. From that date until Gosset's death there was a continuous flow of correspondence between them in which ideas were exchanged and thrashed out, and the more mathematical approach of the younger man was influenced by the practical experience of his older friend.

It will be noticed that three out of Gosset's four illustrations in the paper on the probable error of the mean (2) deal with agricultural topics; the data were taken from published accounts of Woburn farming experiments and Gosset shows how, by taking appropriate differences and using his z-test, a more precise interpretation of such results could be obtained than had hitherto seemed possible. Beaven was in touch with the agricultural work both at Rothamsted and Cambridge and it was no doubt owing to his report of Gosset's keen interest in these problems that both of those classical papers by Wood & Stratton (1910) and by Mercer & Hall (1911), dealing with the analysis of what we now term uniformity trial data, passed through Gosset's hands before publication. The first was only "an affair of a day or two's glancing at" after which he "made one or two suggestions, most of which were quite rightly turned down as being too refined for the purpose".* But in the second case he "brooded over the paper for months", and made suggestions which were incorporated, as well as adding an Appendix (5). If we compare the two statistical contributions, that of Stratton to the first paper and that of Gosset to the second, it is possible, I think, to see without difficulty the latter's special contribution to the subject. Stratton is following the approach of the classical theory of errors, which he had learnt and applied as an astronomer; he shows that variation in plot yields can be represented by the error curve and hence that the results of that theory regarding the probable error of a mean are applicable. These results are used to show the relation of size and number of plots (or animals) to the reliability of the results. No reference is made to "Student's" paper of 1908.

Gosset, writing his Appendix a year later, brings to the problem the added insight that he has gained from an understanding of correlation theory and from much discussion of the Warminster results with Beaven. He shows how it is possible to bring the changing fertility level or "patchiness" of the experimental field into service (*a*) by scattering the varieties to be compared in small plots over the field, and (*b*) then taking as the statistical variable for analysis the difference between the characteristics of two varieties on neighbouring plots. Thus the standard error, by way of formula (2), p. 362 above, can be very much

* These quotations come from a letter of 4 June 1922 from Gosset to Beaven.

reduced. The illustration which he gives deals only with the case of two varieties A and B, and at this date he had probably not thought out a technique for dealing with more comparisons.

There is another point of difference that may perhaps be noted; Wood and Stratton by raising the question, "What is the probable error of a single field experiment?" seemed to suggest that it might be possible to determine a single value, σ, which it would be appropriate to apply to future experiments of a given type. Gosset however emphasized a rather different idea. He writes (**5**, p. 130):

> But, it will be asked, why take all this trouble? The error of comparing plots of any given size has been found by the authors of the paper, and all that has to be done is to apply this knowledge to the particular set of experiments.
>
> The answer to this is that there is no such thing as the absolute error of a given size of plot. We may find out the order of it, be sure perhaps that it is not likely to be less than (say) 5 per cent. nor more than 15 per cent.... but the error of a given size of plot must vary with all the external conditions as well as with the particular crops upon which the experiment is being conducted, *and it is far better to determine the error from the figures of the experiment itself; only so can proper confidence be placed in the result of the experiment.**

His own z-distribution was available, if the number of observations was scanty.

If the field were divided into m pairs of plots and x_i and y_i were the yield, say, of varieties A and B on contiguous ith plots, then Gosset's test for a difference in yield may be summarized as follows:

Write $d_i = x_i - y_i$ and $\bar{d} = \sum_i d_i/m$.

Calculate the ratio
$$z = \frac{\bar{d}}{\sqrt{\left\{\sum_i (d_i - \bar{d})^2/m\right\}}}, \qquad \text{......(13)}$$

and if $m \leqslant 10$ refer this to the z-tables (**2**, p. 19). Otherwise, if $m > 10$, since z has a standard deviation of $1/\sqrt{(m-3)}$, refer $z\sqrt{(m-3)}$ to Sheppard's tables of the normal probability integral.

In the years 1912 and 1913 at Beaven's suggestion plot experiments of similar design, each comparing eight varieties of barley, were carried out at three centres, viz. Warminster, Cambridge and Ballinacurra in Co. Cork. The experiments were carried out in cages, and there were twenty replications of each of the eight varieties in square-yard plots. The arrangement of the varieties in a "chess-board" pattern was effectively what we should now term balanced; a plan of one of the schemes has been shown in Gosset's paper of 1923 "On testing varieties of cereals" (**11**, p. 277) and I have reproduced a portion of this below, only adding some thicker rules to separate the different sets of eight plots.

Beaven suggested that the results might be analysed by using as a statistical

* The italics are mine.

variable the difference between (1) the yield on a plot of A, say, and (2) the mean yield for the eight varieties (including A) on the 9-plot area in which this A-plot lay at the centre.* This was a rough and ready procedure but, as Gosset pointed out, owing to correlation there would be difficulty in the statistical interpretation. The method which he preferred was a very natural extension of his difference method advocated in the case where there were only two varieties. He could still clearly use that method to compare any two of the eight varieties,

<table>
<tr><td>E
230·1</td><td>B
249·3</td><td>G
312·2</td><td>D</td><td>A</td><td>F</td><td></td></tr>
<tr><td>D
255·9</td><td>A
222·6</td><td>F
218·7</td><td>C</td><td>H</td><td>E</td><td></td></tr>
<tr><td>C
265·6</td><td>H
205·0</td><td>E
246·7</td><td>B</td><td>G</td><td>D</td><td></td></tr>
<tr><td>B
265·9</td><td>G
236·7</td><td>D
295·8</td><td>A</td><td>F</td><td>C</td><td>H</td></tr>
<tr><td>A
236·5</td><td>F
210·4</td><td>C
291·1</td><td>H
223·9</td><td>E</td><td>B</td><td>G</td></tr>
</table>

Fig. 1.

say A and D, taking the corresponding pair of plots from each set of eight, and differencing the character measured, although the plots would not now be generally contiguous. This would mean that changes in soil fertility, etc. would make the comparison less accurate than before,† but that could not be helped if eight varieties were to be compared in a single experiment in place of two. He saw, however, that it was possible to compensate to some extent in another direction for this loss in accuracy, by getting a single combined estimate of error from all the $\frac{1}{2}n(n-1) = 28$ possible sets of differences between $n = 8$ varieties, a method which he described as "hotchpotching" the comparisons. The reasoning which he used in reaching his result may be set out as follows:

Let there be n varieties each repeated m times and denote by $d_{uv \cdot i}$ the difference obtained from the ith comparison of the uth and vth varieties ($i = 1, 2, \ldots, m$) and by $\bar{d}_{uv}$ the mean of these m differences. Thus in Fig. 1, if u and v stand for varieties A and D, respectively, then

$$d_{uv \cdot 1} = 236{\cdot}5 - 255{\cdot}9 = -19{\cdot}4, \quad d_{uv \cdot 2} = 222{\cdot}6 - 295{\cdot}8 = -73{\cdot}2, \text{ etc.}$$

To obtain a common estimate of the standard deviation of differences, say σ, proceed now, he argued, as follows: (1) calculate the $\frac{1}{2}n(n-1)$ possible values of $s_d^2 = \sum_i (d_{uv \cdot i} - \bar{d}_{uv})^2/m$; (2) multiply each by a factor $m/(m-1)$ so that its

* One variety would appear twice in this mean and its yield must be suitably weighted.

† Gosset at a later date made comments on this point and on the assumption involved in getting a pooled estimate of standard errors that might differ; see (**11**, pp. 285 and 282).

expectation becomes σ^2; (3) sum these quantities and divide by their number. Thus the final estimate of σ^2 becomes

$$s^2 = \frac{2 \sum_{u,v} \sum_{i} (d_{uvi} - \bar{d}_{uv})^2}{n(n-1)(m-1)}. \qquad \text{......(14)}$$

As I shall explain later, this is exactly the estimate which would now be used, only it would be calculated in a more direct manner. The division of Beaven's plots into sets of eight which I have shown in Fig. 1, would to-day be termed a division into blocks (though the blocks are not similar in shape), and the arrangement of the different varieties within a block would be called balanced rather than random. Thus already in 1912 Beaven and Gosset together had gone a long way towards reaching one form of the present-day experimental technique.

Having obtained the estimate s^2 of (14), Gosset was then able to consider the significance of the difference between any pair of varieties by calculating the ratio

$$x = \frac{\bar{d}_{uv}\sqrt{m}}{s}, \qquad \text{......(15)}$$

and referring to Sheppard's tables.* His method was to place the eight varieties in order of magnitude of the character under consideration and, by applying the test as a foot-rule to selected differences, draw reasoned conclusions as to the existence or absence of real variety differences. A test (R. A. Fisher's z-test) which would determine whether as a whole the eight variety means differed significantly would clearly have been useful, but sound common sense could make the difference test yield reliable results.

This method was applied to the English and Irish chess-board results; the computation was lengthy and many pages of a large notebook of Gosset's are filled with the calculations. G. U. Yule carried out the Cambridge computations in consultation with Gosset. But, however laborious the work, the conclusions obtained from the analysis combined with results of large scale tests played an important part in securing the steady improvement that was being effected in the quality of Irish grown barley.

It is perhaps of historical interest to note a more general formula that Gosset was using at this time to obtain a common estimate of standard deviation from data classified into a number of groups with possibly different means.† The formula would not now be regarded as satisfactory, but it illustrates well the slow progress of the human mind to its final goal.

Suppose that N observations of a variable x are divided into n groups of unequal size, that x_{ti} is the ith observation in the tth group; further that m_t is

* The common estimate, s^2, of (14) is based on so many observations that Gosset probably had not considered whether $\bar{d}_{uv}/s$ could be referred to the z-distribution.

† I have taken the expression from a letter of 1912 to Beaven.

the number and $\bar{x}_t$ the mean in that group. Then Gosset took as an estimate of a supposed common within-group variance, σ^2, the expression

$$s^2 = \frac{1}{N} \sum_{t=1}^{n} \frac{m_t}{m_t - 1} \sum_{i=1}^{m_t} (x_{ti} - \bar{x}_t)^2. \qquad \text{......(16)}$$

Since the expectation of $\sum_i (x_{ti} - \bar{x}_t)^2$ is $(m_t - 1)\,\sigma^2$ and $N = \sum_t m_t$ it will be seen that the expectation of s^2 is σ^2. Except in the case where m_t is the same for every group, which was the case he was concerned with in the chess-board analysis, the factors weighting the sums of squares are not, however, those which we now know give an estimate of σ^2 having minimum sampling error. When however $m_t = m$ his estimate assumed the correct form

$$s^2 = \frac{1}{n(m-1)} \sum_t \sum_i (x_{ti} - \bar{x}_t)^2. \qquad \text{......(17)}$$

Had he applied formula (16) to the chess-board problem in a case where the number of plots was not the same for all varieties, his final estimate would have been less satisfactory.

During the war period of 1914–19 the analysis of the chess-board results was discontinued. In 1920 Gosset took over responsibility for the statistical aspects of the barley experiments conducted at a number of centres by the Irish Department of Agriculture, and this made him particularly interested in the possibilities of Beaven's new half drill strip method of arrangement. Correspondence with Beaven is full of discussion of the possibilities of this method and of the best way of analysing the results. At the same time he was in touch with R. A. Fisher who was beginning to turn his great mathematical powers to similar problems at Rothamsted.

The next reference I can find to the chess-board analysis is early in 1923, when Beaven had asked Gosset to explain again the procedure he had used ten years before. The final lap of the long passage to an "analysis of variance" is of sufficient historical and personal interest to place on record. On 29 March 1923 Gosset writes:

> I enclose a note on the chess-board error. I was using the formula before the war and see no reason to repent of it. I am writing Fisher asking him to look it over and if necessary criticize.

The method given is that which I have described above, involving the calculation of the $\frac{1}{2}n(n-1)$ squares of differences. It was naturally a lengthy procedure, and I find a brief note of Beaven on the papers, after working through an example: "Conclusion (if any possible) from above is that P.E. with chess-boards might be guessed at almost as well as calculated." It needed a "Student" with his facility for doing calculations in spare moments on the back of an envelope to cope with such computations. But the author of the method himself was not content and on 9 April in the second half of a letter started on the 6th, he writes again to Beaven:

Since writing the above I have had a vision on the subject of chess-board error and enclose a rough proof of my new method. I have written to Yule asking him whether he is in fact working at chess-board error and enclosing a similar proof. If he is *not* I shall be inclined to write it up and shall ask your leave to use the No. 1 chess-board of 1913 as an illustration. If he *is*, he has doubtless got something as good or better, and he can put mine in the W.P.B.

To use my new method with 15 plots, each of 8 varieties (1) find the square of the S.D. of the whole 120 plots, Σ^2; (2) after calculating the averages of the eight varieties, find the square of the S.D. of these eight figures, σ_8^2; (3) after calculating the averages of the fifteen groups of eight, find the square of the S.D. of these fifteen figures, σ_{15}^2. Then the P.E. of the error of a comparison should be

$$0{\cdot}6745\sqrt{\frac{2\times 8(\Sigma^2-\sigma_8^2-\sigma_{15}^2)}{120-8-15}}.*$$

In calculating the S.D.'s do not use the $(n-1)$ divisor.

The "rough proof" of the method which he enclosed was as follows: it will be seen to be on similar lines to that given in the paper "On testing varieties of cereals" (**11**, pp. 282–3) except for the omission of the term $-\sigma_e^2/mn$ referred to in the published paper, which resulted in a divisor of $mn-m-n$ instead of $mn-m-n+1$.

Memorandum

Let m plots of each of n varieties be chessboarded. There will be m groups each containing one of each of the n varieties. If Σ^2 be the variance of the nm plots, it may be considered to be composed of three parts which as a first approximation may be taken as uncorrelated:

(1) The real differences between the varieties, σ_v^2,

(2) The errors common to each group of n, σ_c^2,

(3) The remaining casual errors, σ_e^2.

Of these the last is the only part that affects the comparison of varieties since the differences which we intend to measure compose (1), and (2) is eliminated by the process of chess-boarding.

It remains to find the best estimate of (1), (2) and (3) given Σ^2, the averages of the n varieties, and those of the m groups.

Now if σ_n be the S.D. of the averages of the n varieties

$$\sigma_n^2=\sigma_v^2+\sigma_e^2/m,\dagger$$

and if σ_m be the S.D. of the averages of the m groups

$$\sigma_m^2=\sigma_c^2+\sigma_e^2/n.$$

Also

$$\Sigma^2=\sigma_v^2+\sigma_c^2+\sigma_e^2.$$

* This is the P.E. of the difference between two means of fifteen plots. It must be squared and multiplied by $m=15$ to get into the form of (18) below. [E. S. P.]

† The expression on the right-hand side should have been $\sigma_v^2+\sigma_e^2(n-1)/mn$; this is equal to the expectation of σ_n^2. Similar corrections to the σ_e^2 term are required in the next two equations. [E. S. P.]

Hence $$\Sigma^2 - \sigma_n^2 - \sigma_m^2 = \sigma_e^2\left(1 - \frac{1}{n} - \frac{1}{m}\right),$$

therefore $$\sigma_e^2 = \frac{mn(\Sigma^2 - \sigma_n^2 - \sigma_m^2)}{mn - m - n}.$$

Whence the others follow, and the error of a comparison between a pair of varieties is

$$\sqrt{\frac{2}{m}}\,\sigma_e = \sqrt{\frac{2n(\Sigma^2 - \sigma_n^2 - \sigma_m^2)}{mn - m - n}}.$$

In the next letter to Beaven of 20 April Gosset writes:

Now as to chess-board error. About a week after I sent the proposed simplified method to you and Yule, I got a note from Fisher via Somerfield giving the same method in rather more technical language. Next I got a reply from Yule saying that the method was new and giving it his blessing more or less, and finally I got a p.c. from Fisher this morning saying that the divisor should be $mn - m - n + 1$ not $mn - m - n$. Anyhow the thing seems to have some weight behind it now.

It should give the same result as my original method....

That the agreement between the two results depends on the identity*

$$\frac{2\sum\limits_{u,v}\sum\limits_{i}(d_{uv\cdot i} - \bar{d}_{uv})^2}{n(n-1)(m-1)} = 2 \times \frac{\sum\limits_{u}\sum\limits_{i}(x_{ui} - \bar{x})^2 - m\sum\limits_{u}(\bar{x}_u - \bar{x})^2 - n\sum\limits_{i}(\bar{x}_i - \bar{x})^2}{(n-1)(m-1)} \quad \text{......(18)}$$

was shown by Fisher in the letter Gosset quotes in the footnote to p. 283 of his paper (**11**). The expression on the left-hand side is taken from formula (14) above, while that on the right represents the estimate of the sampling variance of the difference between two single plot yields obtained by the usual analysis of variance method.

Fisher's application of the method was given in a joint paper with W. A. Mackenzie on "Studies in crop variation", received by the *Journal of Agricultural Science* on 20 March 1923 and published in July. The theory was illustrated on an experiment with potatoes "planted in triplicate on the 'chess-board' system"; the arrangement of the plots was not so well balanced as in Beaven's chess-board and as yet no question of randomization was considered. The paper contained what was I think the first published arrangement of numerical data in an analysis of variance table (then described as analysis of variation), and a method was given of testing for the significance of the treatment (or variety) sum of squares, taken as a whole.

"Student's" paper (**11**) was read before the Society of Biometricians and Mathematical Statisticians on 28 May 1923 and published in *Biometrika* in the following December. In obtaining the formula of the memorandum even with the slip which no doubt he would later have found out himself, and in the description of the method of procedure given to Beaven, he had so evidently after long searching reached the essential conception of breaking up a total sum

* In this notation $d_{uv\cdot i} = x_{ui} - x_{vi}$ or is the difference between uth and vth varieties in the ith block. $\bar{x}_u$, $\bar{x}_i$ and $\bar{x}$ are the variety, the block and the grand mean respectively. There are n varieties and m blocks.

of squares into parts* that I feel his achievement should be put on record. As we have seen, in his modest way he was ready to have his results thrown into the waste paper basket, if another statistician could improve on his work! Whether his mathematics could ever have shown unaided that if no variety differences existed:

(1) the expressions $\Sigma^2 - \sigma_n^2 - \sigma_m^2$ and σ_n^2 of his memorandum were independent,

(2) were each distributed in a modified form of the distribution he had discovered in 1908,

(3) gave a ratio whose distribution law was a Pearson type VI curve;

all this is doubtful. But, as he would have said himself, why speculate, these further results were derived by Fisher; the problem was therefore solved and a new chapter opened.

The 1923 paper (**11**) contains much else of interest besides this handling of the chess-board type of experiment. It starts with an historical survey of the development of experiments aiming at the comparison of cereals and concludes with a critical discussion of the half drill strip method. The simple theme which I have referred to on many occasions runs through the whole and takes form in a final concluding sentence:

It is shown that methods (2) [chess-board] and (3) [half drill strip] depend for their accuracy on the fact that the nearer two plots of ground are situated, the more highly are the yields correlated, so that we are able to increase the effect of the last term of the equation

$$\sigma_{A-B}^2 = \sigma_A^2 + \sigma_B^2 - 2r_{AB}\sigma_A\sigma_B$$

(where A and B are the varieties to be compared) by placing the plots to be compared with one another as near together as possible.

Later papers

In his later papers Gosset tended to avoid, as far as possible, the introduction of mathematics and he would ask his friends to regard him as a non-mathematician. Thus he forwarded his paper on the Lanarkshire milk experiment (**17**) to Karl Pearson with the words:

I hope you will find it interesting, though its chief merit to the likes of me (that there are no mathematics in it), will hardly commend it to you.

Or again, writing to me in 1926 regarding the original χ^2 paper (Karl Pearson, 1900) he remarked:

I have now read the χ^2 paper in *Phil. Mag.* **50**. It may be divided into three parts, one that I can follow as a man who could cut a block of wood into the rough shape of a boat with his penknife might appreciate a model yacht cut and rigged to scale, the second I can

* His original approach to statistics through Airy's book made this a natural way of regarding things; see the formula (5) I have quoted above. There are points in Gosset's proof in (**11**, p. 282) also reminiscent of Airy, *Theory of Errors of Observations* (1875, p. 46).

only compare to a conjuring trick of which I haven't got the key (such for example as the transformation to polar co-ordinates on p. 158) and lastly quite a small part which I think I can understand.

When at last, after the war, an increasing number of men trained as mathematicians began to turn their attention to statistics, it was not perhaps surprising that one whose mathematical training had ceased with Oxford Mods. in the nineties should refuse to regard himself as a mathematician. Besides, the increasing responsibilities of his work as a brewer left him little time or inclination to follow out in detail the continuous elaboration of the theory of mathematical statistics. As a result, in his relatively rare publications he tended to concentrate on simple exposition of the function of statistical method. The best examples of such work are:

(1) The paper on "Errors of routine analysis" of 1927 (**15**) which develops more fully a theme he had touched on before (**4** and **8**), and shows how some recent theoretical work on the distribution of "range" in small samples might be made to give a useful working tool for the analyst.

(2) Two admirable papers on the use of statistical methods in agriculture, both unfortunately rather inaccessible to the ordinary student: "Mathematics and Agronomy", 1926 (**14**), and the article on "Yield Trials" in *Baillière's Encyclopedia of Scientific Agriculture* 1931 (**16**).

This recession from the mathematical approach of his earlier papers had other consequences. In the first place it meant that during a period of rapid advance in statistical technique there was available, for almost anyone in need of advice, a statistician of great practical experience and unusual insight, whom the inquirer could be sure would not be carried away by the fascination of any mathematical model into allowing abstract theory to step beyond its proper sphere. On the other hand there were certain disadvantages; Gosset's avoidance of a mathematical statement of his case sometimes, as in his last two papers (**21**), (**22**), made it difficult for others to grasp an idea or method which probably was clear enough in his own mind. The theory of probability is based on mathematics, and beyond a certain point there are dangers in introducing it into practice without a precise mathematical statement of the assumptions underlying the method of procedure.

If we return to 1923, it is clear that Gosset welcomed with enthusiasm the new methods that R. A. Fisher was developing. The neatness of the arrangement of calculations in an analysis of variance table for example, appealed to him. It brought to the rather laborious calculation methods of his own a simplification whose value he was quick to realize. The introduction of t as the ratio of a deviation to an estimate of its standard error, in place of his own criterion z, and the use of degrees of freedom, appealed to him at once because of the greater generality; as a result he calculated extended values of the probability integral of t to replace his old z tables and published these in 1925 (**13**) in conjunction with a theoretical contribution of Fisher's. In print and in correspondence he emphasized the importance of randomness. "The experiments", he wrote in 1926 (**14**, p. 711), "must be capable of being considered to be a *random* sample of the population to which the conclusions are to be applied. Neglect of this rule has led to the estimate of the value of statistics which is expressed in the crescendo 'lies, damned lies, statistics'."

This paper of 1926 contains perhaps the extreme limit to which he ventured in allowing the toss of a coin or a die to decide the arrangement of plots in an agricultural experiment. On the last page (p. 719) he suggests the arrangement of four varieties in an 8×8 square, in which two plots of each variety are to fall in each row and each column. Subject to this restriction the arrangement was to be obtained in a random manner.

He must soon, however, have realized the disadvantages of such a procedure. If A, B, C and D represent the varieties, a possible if unlikely run of luck might lead to the following pattern of plots in one corner of the square:

A	*A*	*C*	*D*	*C*
A	*A*	*C*	*D*	*B*
B	*C*	*A*	*A*	*D*
C	*C*	*C*	*B*	*A*

Fig. 2.

Should this chance juxtaposition of many A-plots happen to coincide with a "fertility summit" or "depression" in the field, the resulting statistical analysis of plot yields might easily attribute a characteristic to the variety A which it did not possess. His practical mind could not accept such a state of affairs. To know in advance that if an experiment was carried out with a particular pattern of plots there was quite a chance that it would be misleading, and to continue with this pattern—this was a course he was not prepared to follow. It was no compensation to be told that in the long run, if the verdict of the random toss was accepted and the 5% significance level of mathematical tables used in the statistical analysis, then misleading results would be obtained only 5 times in 100. In his own words (**22**, p. 366):

It is of course perfectly true that *in the long run*, taking all possible arrangements, exactly as many misleading conclusions will be drawn as are allowed for in the tables, and anyone prepared to spend a blameless life in repeating an experiment would doubtless confirm this; nevertheless it would be pedantic to continue with an arrangement of plots known beforehand to be likely to lead to a misleading conclusion.

His withdrawal from the out and out randomization position is illustrated in his article of 1931 on "Yield Trials" (16). Here he speaks of the Latin square arrangement as ideal in the types of experiment for which it is suited, because it combines the elements of balance and randomness, but he is critical of the randomized block arrangement because of the risk involved of getting misleading results. He gives the following illustration of a balanced or equalized block design which he had recommended to a horticultural correspondent, comparing ten treatments with five replications:

G	*H*	*E*	*C*	*A*	Block I
F	*D*	*J*	*B*	*I*	
H	*J*	*D*	*F*	*E*	Block II
B	*G*	*I*	*A*	*C*	
E	*I*	*A*	*G*	*D*	Block III
J	*B*	*C*	*H*	*F*	
C	*F*	*B*	*I*	*J*	Block IV
A	*E*	*G*	*D*	*H*	
D	*A*	*F*	*J*	*B*	Block V
I	*C*	*H*	*E*	*G*	

Fig. 3.

In this example the assignment of treatments to plots in Block I is random, but each successive block has its arrangement more and more controlled, so that (i) each of the five columns contains one plot only of the ten varieties, (ii) *A*, *D*, *E*, *F* and *J* occur in the top row of their block three times and in the lower row twice, while for *B*, *C*, *G*, *H* and *I* the position is reversed, an arrangement as nearly balanced as possible for an odd number of blocks.

In advocating the introduction of this element of balance, he did not consider that the random element could be dispensed with; but he believed that if a regular pattern was used to equalize the more probable variations in fertility there were still sufficient complications to leave the residual variations random enough to justify from the practical point of view the application of probability theory. It was here that he disagreed and was eventually forced into open controversy with R. A. Fisher and the Rothamsted school.

This is not the place to enter into detail regarding the nature of this controversy, which resulted in Gosset's last paper published a few months after his death (22). It is however well to emphasize that his attitude was closely related to the type of agricultural problem with which he had had most experience, the development of improved strains of barley. In such a case as this he saw that success was only likely to result from a comparison of two or more strains in a number of years and in a number of different localities. Small scale investigations must be followed by others in which the technique conformed as far as possible to ordinary agricultural practice. In each case some experimental plan was needed which would give the yields, let us say, of variety A and variety B on the experimental area with as little error as possible, that is to say freed from bias such as might be introduced by changes in fertility, patches of weed, etc. Provided that the error of the difference (yield of A − yield of B) could be kept low, he was satisfied with a knowledge of its probable upper limit and did not mind if he was told that the ratio of this difference to the estimate of its standard error in a particular experiment could not be referred with mathematical precision to a table of probabilities. He was interested primarily in the behaviour of the difference from farm to farm and year to year, and experience had shown him, beyond any possibility of doubt, that small scale balanced plot experiments followed by larger scale tests with the half drill strip method of Beaven's, the purpose of which any intelligent farmer could understand, had achieved remarkable success in the improvement of barley. If it were argued that fully randomized experimental designs would have achieved the same or better results he would not have denied this dogmatically, but he felt doubtful on the point because his perusal of reports on such experiments showed to his mind an unduly high proportion of inconclusive results. He would also have added that with the staff, the ground and other facilities available in the investigations for whose planning he was responsible, fully randomized designs could not have been carried out. This was his attitude in writing the Statistical Society paper of 1936 (21).

In his final paper (22) he attacked his critics on their own ground by pointing out that in the experiment at a single station a balanced arrangement of plots in blocks was on the whole more likely to detect variety differences than a random arrangement when those differences were really large and therefore important, although for small differences the reverse would be true.

The ultimate decision on these points can hardly be expected as yet; it will come in time, perhaps after 10 or after 20 years, when there has been ample opportunity for the practical experimenter, freed from the weight of authority, from fear of mathematics on the one side and from the fascination of a new technique on the other, to judge from accumulated experience what methods have been most worth while having regard to the results they have achieved.

In addition to these papers on agricultural subjects a brief reference may be given to some other published work of the last few years:

(1) A paper on "The Lanarkshire milk experiment", 1931 (17); his suggestion

that the experiment should be repeated on a more precise but far less expensive scale by using pairs of twins involves a characteristic introduction of his paired difference plan.

(2) Two papers on certain implications of F. L. Winter's selection experiments with maize 1933 (**19**) and 1934 (**20**). The plant breeder's problem of improving varieties of cereal by continued selection had long been of interest to him in connexion with barley and in these papers he discusses the bearing of these experimental results upon evolutionary theory.

(3) A number of short but suggestive contributions to the discussion of papers read before the Industrial and Agricultural Research Section of the Royal Statistical Society (see references on p. 249 below).

Extracts from letters

I have spoken more than once of Gosset's correspondence; the professional statistician, whether he be attached to a university or research station, receives and expects to receive appeals for advice which will continue to increase through life as his circle of contacts grows. But with Gosset the position was somewhat different; to provide advice to correspondents all over the world was in no way part of his job. Yet he gave that help unstintingly and unless it could be described as brewery business, he gave it out of his own time. Advice as to how to plan a particular experiment, or explanations of misunderstood points in statistical theory, while of extreme value at the time to the individual who receives them are rarely of interest to the general reader. Nevertheless, I believe that a few quotations from letters will add to the record of Gosset's personality by showing something of his patience, his practical mind, his suggestiveness and his characteristic freedom of expression.

The first quotations are taken from a long letter written to me in 1926. At that time I had been trying to discover some principle beyond that of practical expediency which would justify the use of "Student's" ratio $z=(\bar{x}-m)/s$ in testing the hypothesis that the mean of the sampled population was at m. Gosset's reply had a tremendous influence on the direction of my subsequent work, for the first paragraph contains the germ of that idea which has formed the basis of all the later joint researches of Neyman and myself. It is the simple suggestion that the only valid reason for rejecting a statistical hypothesis is that some alternative hypothesis explains the observed events with a greater degree of probability. The second part of the letter probably put into my mind the very extensive plan of sampling from non-normal populations which we carried out in the Department of Statistics at University College during the next few years.

Letter I

From a letter of W. S. G. to E. S. P., dated 11 May 1926.

In your large samples with a known normal distribution you are able to find the chance that the mean of a random sample will lie at any given distance from the mean of the population. (Personally I am inclined to think your cases are best considered as mine taken to the limit n large.) That doesn't in itself necessarily prove that the sample is not drawn randomly from the population even if the chance is very small, say ·00001: what it does is to show that if there is any alternative hypothesis which will explain the occurrence of the sample with a more reasonable probability, say ·05 (such as that it belongs to a different population or that the sample wasn't random or whatever will do the trick) you will be very much more inclined to consider that the original hypothesis is not true.

I can conceive of circumstances, such for example as dealing a hand of 13 trumps after careful shuffling *by myself*, in which almost any degree of improbability would fail to shake my belief in the hypothesis that the sample was in fact a reasonably random one from a given population.

* * * * * * *

I'm more troubled really by the assumption of normality and have tried from time to time to see what happens with other population distributions, but I understand that you get correlation between s and m with *any* other population distribution.

Still I wish you'd tell me what happens with the even chance population ▭ or such a one as △: it's beyond my analysis.

* * * * * * *

If Student is wrong it is up to you to give us something better. You see one must experiment and frequently it is quite out of the question, from considerations of cost or of impossibility of duplicating conditions in the time scale, to do enough repetitions to define one's variability as accurately as one could wish. It's no good saying "Oh these small samples can't prove anything". Demonstrably small samples *have* proved all sorts of things and it is really a question of defining the amount of dependence that can be placed on their results as accurately as we can. Obviously we lose by having a poor definition of the variability but *how much do we lose?*

Letter II, with its enclosure which, for reasons I have forgotten, was never published, was written shortly after K. P. had made an editorial comment on "Sophister's"* (1928) interpretation of the distribution of "Student's" ratio in samples from a non-normal population. It had been found that in such cases the distribution of t was asymmetrical, but that the distribution of $|t|$ (or of t^2) followed very closely the standard normal-theory form, i.e. if the distribution of t was curtailed on one side of the origin this was balanced by a corresponding extension on the other side. The letter also refers to a suggestion of bringing up at a meeting of the International Statistical Institute the question of differentiating between the symbols used for probable error and standard error.

* "Sophister" like "Mathetes" was the *nom de plume* of a disciple of "Student". The particular sampling investigation in question had been sketched out by Gosset and myself before "Sophister" came to spend a year in the Biometric Laboratory.

Letter II

Holly House,
Blackrock,
Co. Dublin.
May 18th, '29.

Dear Pearson,

I was rather amused to see your letter open with an apology for delay in writing as I have for some time been acutely conscious that I have been in arrears. However, last things first.

(i) I agree that Z's second suggestion though sound is not workable. Your idea of raising the question at Warsaw seems to me to suggest the right way of getting to work. I think they should raise the question on the grounds (*a*) that $\pm$ is being used in two senses, (*b*) that the prob. error is no longer the slightest use to anyone and (*c*) that as the tables are in terms of the S.D. a simple notation such as : or ; or anything of the sort is required.

(ii) I fancy you give me credit for being a more systematic sort of cove than I really am in the matter of limits of significance. What would actually happen would be that I should make out P_t (normal) and say to myself "that would be about 50 : 1; pretty good but as it may not be normal we'd best not be too certain", or "100 : 1; even allowing that it may not be normal it seems good enough" and whether one would be content with that or would require further work would depend on the importance of the conclusion and the difficulty of obtaining suitable experience.

One so often finds that the importance (and even occasionally the direction of the result) of varying one factor, change from experiment (or experience) to experiment according to the accompanying variations in other factors, that it often doesn't pay to make too certain of any one result.

E.g. You may have two varieties of barley one of which will give the best yield in one season or place while the other will win in another season or place; hence we have to sample places and seasons widely rather than aim at being meticulously accurate at all places sampled: there must be economy of effort.

* * * * * * *

Lastly I am enclosing a short note in reply to the Editorial footnote. Probably you are going to say all that is at all useful in it in your next paper, and in any case I haven't the least intention of indulging in a controversy, so suppress it unless you think it will clear up our position. All the same I think it is a pity to let the thing go by default without any comment.

Yours v. sincerely,
W. S. Gosset.

Suggested Note for "Biometrika"

17th May, 1929.

In his footnote on page 422 of Sophister's paper the Editor asks, "Supposing 50 per cent of prisoners tried for murder were acquitted and the remainder found guilty should we be right *in the long run* to drop the trial and toss up for judgment?" This, if I may say so, is hardly what Sophister proposes to do. If I may deal first with the Editorial analogy the position is rather, "The evidence before the court is such that the chances are even that the prisoner committed the murder". Doubtless if more evidence were forthcoming we should know more about it; as it is, an English Court will acquit, though the inexorable Justice of Shan Tien would condemn the prisoner to piecemeal slicing, unless of course sufficiently weighty evidence for the defence could be imperceptibly introduced within the Mandarin's sleeve. But, seriously, a better illustration can be drawn from the practice of

Insurance where in the first place the premium is calculated on the Healthy Male table and, I suppose, originally this was the only basis after a medical examination. But the material which supplied the experience for the H.M. table can be subdivided into various classes, by professions and occupations, by stature or eye colour, total abstainers or moderate drinkers and so forth, which further investigation may find to have expectations of life which do not accord with the table. The life expectation of some of these classes is probably taken into consideration by the Companies—I doubt whether a Lion Tamer, however healthy, could insure at the ordinary rates—but no company, as it well might, charges a lower rate of premium for the descendants of centenarians or a higher for orphans; they are most unfairly lumped together just as Sophister proposes to do with his samples from unknown populations. In effect he says, "This small sample is from an unknown population, which *may* be normal; it probably is not far from normal; if it *is* normal we use the table justly, if it is anormal but symmetrical we can still use the table with sufficient accuracy; even if it is skew, about which we cannot be sure—much less about the direction of the skewness—we shall in the long run draw much the same proportion of correct inferences as if it were normal." Admittedly our ignorance of the nature of the population introduces an element of uncertainty which no sensible person will ignore when using the tables, but recent work, and not least Sophister's, shows that this uncertainty, while not altogether negligible, is much less than we had any right to expect.

Student.

The suggestion in Letter III of 1932 ultimately led to the production of tables of percentage limits of the ratio of (*a*) range in a sample of *n* observations to (*b*) an independent estimate of standard deviation, which are to be published shortly in *Biometrika*. From the beginning of his analysis of the results of the chess-board experiments, Gosset had wondered how best to judge what differences among variety means were significant. While the ratio of (*a*) the difference between any two means selected at random to (*b*) the estimate of standard error could be referred to "Student's" distribution or, if desired, the significance of the set as a whole could be judged by Fisher's *z*-test, it was not possible to treat selected differences in either of these ways. In the article in *Baillière's Encyclopedia* (**16**, p. 1358) he refers to a method suggested by Fisher of taking the differences between individual variety yields and the mean yield. He felt however that a knowledge of the probability levels of "studentized" range would in addition be very useful; on this could be based a rough test of the kind he had suggested in his paper on "Errors of routine analysis" (**15**, p. 161).

Letter III

St. James's Gate,
Dublin.
Jan. 29th, '32.

Dear Pearson,

Many thanks for your letter and enclosure: as I am at the moment

"The Cook and the Captain bold
And the mate of the Nancy brig",

I have handed all the lot to Mathetes till such time as I can get a chance of dealing with it which should be sometime next week.

I have been meaning to write to you for some time re the proposals for the use of range and sub-range which I made in my last letter to you. Of course there is a serious crab which

I had at one time recognised and then forgotten in that the thing would have to be "Studentised": the only measure of the S.D. is provided by a limited number of degrees of freedom. Whether one could get an approximate correction for this with moderately small numbers by reducing still further the degrees of freedom or whether it would be necessary as Fisher suggested when I mentioned the matter to him (he was here lecturing) to dive into the depths of hyperspace to produce the jewel I am not clear, but obviously something would have to be done about it.

* * * * * * *

Yrs. v. sincerely
W. S. Gosset.

Letters IV and V of 1936, which Dr Beaven has kindly allowed me to reproduce, deal with the interpretation of the results of half drill strip barley experiments carried out at six stations in England; the two varieties compared were Plumage Archer and Beaven's 35/7. The second letter followed a reply from Beaven discussing the position in terms of betting on two horses, whose form varies on different courses. The argument illustrates Gosset's outlook on the function of large scale experiments to which I have already referred.

Letter IV

From a letter of W. S. G. to E. S. B., dated 8 January 1936.

If you derive the S.E. from a set of 10 strips at one station, you are sampling "comparisons between plots grown at a certain station in the weather of 1935" and can draw the appropriate conclusion, e.g. that at Sprouston it is quite certain that Beaven's 35/7 would have beaten Plumage Archer in any sound arrangement of plots in 1935.

When however you regard the six stations as a small sample of the barley land of England you can very nearly draw the conclusion that Beaven's 35/7 would on the average have beaten Plumage Archer if compared all over the barley land of England in 1935.

The chance that so favourable a result would have happened if there were really no difference between them is only 1/38, i.e. the odds are 37 to 1 against it's happening. This is very nearly significant but as you know, what odds are to be considered significant is a matter of convention—or taste.

Naturally, in calculating the S.E. (not really an *error* at all) of the second conception where the variation from Station to Station depends as much, (or much more...than), on the differential response to weather and soil as on the soil errors taken account of in each station, one takes no particular account of the S.E.'s at the individual stations: one merely rejoices because the Half Drill Strip method has largely eliminated the errors due to soil position and left us mainly the differential response aforesaid, which would have affected the result to a greater or less extent in every field of barley-growing England and which we have assumed that we have sampled by the six results which we have examined.

I hope I have made the distinction clear between the S.E. of the result at one station, which is rightly derived from the plots grown at that station but which only enables us to judge whether the result is significant for that station, and the S.E. of the whole series, derivable only from the six mean results of the six stations but which enables us to make an estimate of the result of comparing the barleys "everywhere", where "everywhere" represents the whole extent of country that may properly be considered to be sampled by the six stations.

Letter V

Davan Hollow,
Denham,
Bucks.
14. 1. 36.

Dear Beaven,

I don't think your analogy is quite exact: this is mine.

The two horses 35/7 and P.A. are known to vary somewhat from day to day and also to be very much affected by the particular course on which they are running.

They have raced ten times at Sprouston and 35/7 has won every time by amounts varying from one furlong to two furlongs. At Sprouston then you may lay longish odds on 35/7. At Cambridge they raced ten times and on the average 35/7 won by 50 yds, the amounts varying from 270 yds in favour of 35/7 to 170 in favour of P.A. You would not therefore bet very heavily on 35/7 at Cambridge. At four other places 35/7 beat P.A. on average by various amounts. What odds is to be given on another hitherto untried course?

You are surely as much influenced by the narrowness of the margin at Cambridge as by the width of it at Sprouston: the new course may resemble the one with just as much likelihood as the other and may even as far as you can see favour P.A. rather than 35/7, since your knowledge of the difference between courses rests on only six cases.

Furthermore a new method of training may reduce the variation so that the Sprouston results may lie between $1\frac{1}{4}$ and $1\frac{3}{4}$ furlongs and the Cambridge between 160 yds in favour of 35/7 and 60 yds in favour of P.A., without altering very much* the odds on a series of races on a new course, since the chief source of variation remains the reaction of the horses to the courses and not the day to day variation which alone is measured by the variation on a single course.

* * * * * * *

Yours v. sincerely
W. S. Gosset.

* But since the smaller day to day variation prevents an accidentally high or low value of mean obscuring the real value of the course there is a better chance of getting the right odds—not of getting higher odds.

Letter VI was written at the time when Gosset was putting together his last paper (**22**).

Letter VI

Dart Cottage,
Postbridge,
Devon.
19. iv. 37.

Dear Pearson,

Many thanks for yours of 10th; I feel I'm rather wasting your time but as long as you ask questions you must expect to get answers. You have given my reason for not changing the level of significance *viz.* that while balancing certainly *tends* to produce a lower real error and consequently higher calculated error one cannot say how much one has succeeded in any particular case. I therefore content myself with pointing out that the tendency is beneficial, not only are the cases missed of comparatively little value but one actually gets more conclusions of real value.

* * * * * * *

Now I was talking about Cooperative experiments and obviously the important thing in such is to have a low real error, not to have a "significant" result at a particular station. The latter seems to me to be nearly valueless in itself. Even when experiments are carried out only at a single station, if they are not mere five finger exercises, they will have to be part of a series in time so as to sample weather and the significance of a single experiment is of little value compared with the significance of the series—which depends on the real error not that calculated for each experiment.

But in fact experiments at a single station *are* almost valueless; you can say "In heavy soils like Rabbitsbury potatoes cannot utilise potash manures", but when you are asked "What *are* heavy soils like Rabbitsbury?" you have to admit—until you have tried elsewhere—that what you mean is "At Rabbitsbury etc." And that, according to X may mean only "In the old cow field at Rabbitsbury". What you really want to find out is "In what soil and under what conditions of weather do potatoes utilise the addition of potash manures?"

To do that you must try it out at a representative sample of the farms of the country and correlate with the characters of the soil and weather. It may be that you have an easy problem, like our barleys which come out in much the same order wherever—in reason—you grow them or like Crowther's cotton which benefitted very appreciably from nitro-chalk in seven stations out of eight, but even then what you really want is a low real error. You want to be able to say not only "We have significant evidence that if farmers in general do this they will make money by it", but also "we have found it so in nineteen cases out of twenty and we are finding out why it doesn't work in the twentieth". To do that you have to be as sure as possible which *is* the 20th—your real error must be small.

* * * * * * *

Tedin:* Somerfield sent me the number and I have just had time to glance at it. T. put down three kinds of patterns of Latin Squares (5 × 5) on various uniformity trials. There were

Two Knight's moves:

A	*B*	*C*	*D*	*E*
D	*E*	*A*	*B*	*C*
B	*C*	*D*	*E*	*A*
E	*A*	*B*	*C*	*D*
C	*D*	*E*	*A*	*B*

Two Diagonals:

A	*B*	*C*	*D*	*E*
E	*A*	*B*	*C*	*D*
D	*E*	*A*	*B*	*C*
C	*D*	*E*	*A*	*B*
B	*C*	*D*	*E*	*A*

and a number of randoms.

Of course all Latin squares are "balanced" but one wouldn't care too much for the "Diagonal" arrangement and the Knight's move would, I think, be preferred to all others. In conformity with this Tedin found a slight tendency for the Knight's move to give a low actual and a high calculated error while the diagonal tends to give a high actual and a low calculated error. The whole thing is not worth worrying about but is interesting as an illustration of what actually happens when we depart from artificial randomisation: I would Knight's move every time!

Yours
W. S. G.

P.S. Beaven after all got some slight ailment which prevented his being in the chair for Bartlett's paper: I proposed the vote of thanks....I was heard without enthusiasm but there were no cat calls!

Such are my impressions of Gosset and of his work. Others will have different views on the relative importance of his many contributions to statistics; on his rightness or wrongness. The experimentalist will have seen him in a different light from the mathematician; his personal friends will have realized aspects of his character which his correspondents could not see. But all who have known him will agree that he possessed almost more of the characteristics of the perfect

* A reference to the paper by O. Tedin (1931).

statistician than any man of his time. They will agree, too, on the essential balance and tolerance of his outlook, and on that something which a friend of his schooldays has described as an "immovable foundation of niceness" which made him through life the same friendly dependable person, quiet and unassuming, who worked not for the making of personal reputation, but because he felt a job wanted doing and was therefore worth doing well.

BIBLIOGRAPHY OF "STUDENT'S" PAPERS

(1) 1907. "On the error of counting with a haemacytometer." *Biometrika*, **5**, 351.

(2) 1908. "The probable error of a mean." *Biometrika*, **6**, 1.

(3) 1908. "Probable error of a correlation coefficient." *Biometrika*, **6**, 302.

(4) 1909. "The distribution of the means of samples which are not drawn at random." *Biometrika*, **7**, 210.

(5) 1911. Appendix to paper by W. B. Mercer and A. D. Hall on "The experimental error of field trials." *J. Agric. Sci.* **4**, 128.

(6) 1913. "The correction to be made to the correlation ratio for grouping." *Biometrika*, **9**, 316.

(7) 1914. "The elimination of spurious correlation due to position in time or space." *Biometrika*, **10**, 179.

(8) 1917. "Tables for estimating the probability that the mean of a unique sample of observations lies between $-\infty$ and any given distance of the mean of the population from which the sample is drawn." *Biometrika*, **11**, 179.

(9) 1919. "An explanation of deviations from Poisson's law in practice." *Biometrika*, **12**, 211.

(10) 1921. "An experimental determination of the probable error of Dr Spearman's correlation coefficients." *Biometrika*, **13**, 263.

(11) 1923. "On testing varieties of cereals." *Biometrika*, **15**, 271.

(12) 1924. Note by "Student" with regard to his paper "On testing varieties of cereals." *Biometrika*, **16**, 411.

(13) 1925. "New tables for testing the significance of observations." *Metron*, **5**, 105.

(14) 1926. "Mathematics and Agronomy." *J. Amer. Soc. Agron.* **18**, 703.

(15) 1927. "Errors of routine analysis." *Biometrika*, **19**, 151.

(16) 1931. Article on "Yield Trials" in *Baillière's Encyclopedia of Scientific Agriculture*. **2**, 1342.

(17) 1931. "The Lanarkshire milk experiment." *Biometrika*, **23**, 398.

(18) 1931. "On the 'z' test." *Biometrika*, **23**, 407.

(19) 1933. "Evolution by selection. The implications of Winter's selection experiment." *Eugen. Rev.* **24**, 293.

(20) 1934. "A calculation of the minimum number of genes in Winter's selection experiment." *Ann. Eugen., Lond.*, **6**, 77.

(21) 1936. "Co-operation in large-scale experiments." A discussion opened by W. S. Gosset. *J.R. Statist. Soc.* Suppl. **3**, 115.

(22) 1937. "Random and balanced arrangements." *Biometrika*, **29**, 363.

A FEW SHORTER CONTRIBUTIONS

(*a*) *Letters to "Nature"*.

29 November 1930, **126**, 843: "Agricultural Field Experiments".
14 March 1931, **127**, 404: "Agricultural Field Experiments".
5 December 1936, **138**, 971: "The half drill strip system of Agricultural Experiments".

(*b*) *Contributions to discussions at meetings of the Industrial and Agricultural Research Section of the Royal Statistical Society.*

J.R. Statist. Soc. Suppl. (1934), **1**, 18; (1936), **3**, 173; (1937), **4**, 89, 170.

REFERENCES TO PAPERS BY OTHER AUTHORS

Cave, F. E. (1904). *Proc. Roy. Soc.* **74**, 403.
Helmert, F. R. (1876). *Astr. Nachr.* **88**, S. 122.
Hooker, R. H. (1905). *J.R. Statist. Soc.* **68**, 696.
Macdonell, W. R. (1901). *Biometrika*, **1**, 219.
Mercer, W. B. & Hall, A. D. (1911). *J. Agric. Sci.* **4**, 107.
Neyman, J. (1937). *Lectures and conferences on mathematical statistics.* Graduate School of U.S. Dept. of Agriculture, Washington.
Pearson, K. (1900). *Phil. Mag.* **50**, 157.
Pearson, K. (1907). *Drapers' Company Research Memoirs.* Biometric Series, **4**.
Pearson, K. (1913). *Biometrika*, **9**, 116.
Soper, H. E. (1913). *Biometrika*, **9**, 91.
"Sophister" (1928). *Biometrika*, **20a**, 389.
Tedin, O. (1931). *J. Agric. Sci.* **21**, 191.
Whittaker, L. (1914). *Biometrika*, **10**, 36.
Wood, T. B. & Stratton, F. J. M. (1910). *J. Agric. Sci.* **3**, 417.

SOME EARLY CORRESPONDENCE BETWEEN W. S. GOSSET, R. A. FISHER AND KARL PEARSON, WITH NOTES AND COMMENTS

By E. S. PEARSON

SUMMARY

Letters or extracts from letters which passed between W. S. Gosset, R. A. Fisher and Karl Pearson during the years 1912–20 are reproduced. They throw light on the start of Fisher's statistical career. In the notes accompanying the correspondence, attention is drawn to the problems in estimation and significance testing which were brought to the front at that date by the unusual character of the sampling distribution of the correlation coefficient. An early disagreement between Pearson and Fisher on estimation through minimizing χ^2 and maximizing likelihood is covered.

1. INTRODUCTION

In going through some of my father's correspondence of over 50 years ago, I came across two letters to him from W. S. Gosset ('Student') written in 1912 and three letters from R. A. Fisher written in 1916 which seemed of special interest because of some fresh light which they threw on Fisher's entry on his career as a statistician. To these could be added a letter of Gosset's to Fisher of September 1915. This is the first of a series of letters (1915–36) from Gosset to Fisher reproduced a few years ago at Sir Ronald's suggestion and given a limited circulation. The letters showed how at this very early stage, when he was scarcely more than an undergraduate, Fisher was already feeling his way among problems of distribution and estimation along lines which he was to develop and enlarge within the following years.

It then occurred to me that some letters of the same period from my father to Fisher might also have been preserved. Through the courtesy of Dr E. A. Cornish I was indeed able to see copies of some eighteen letters from Pearson to Fisher, written during the years 1914–20. These had been preserved in the collection of Fisher's papers now in Cornish's charge in Adelaide. It was clear that certain of these letters, when studied in conjunction with the published articles of that date, were of such historical interest that rather full reproduction was justified. I am grateful to Arthur Guinness Sons and Co. Ltd, acting on behalf of Gosset's family, and to the University of Adelaide, in respect of the Fisher material, for permission to reproduce.

For purposes of comment, the correspondence falls into three groups: (i) the letters from Gosset; (ii) the exchange of correspondence between Fisher and Pearson, mainly concerning the papers on the distribution of the correlation coefficient; (iii) further correspondence between Fisher and Pearson arising from Dr Kirkstine Smith's article on fitting theory to data by minimizing χ^2.

2. Letters from W. S. Gosset

12th September 1912 Woodlands, Monkstown,
Co. Dublin

Dear Pearson,

I am enclosing a letter which gives a proof of my formulae for the frequency distribution of $z\,(=x/s)$, where x is the distance of the mean of n observations from the general mean and s is the S.D. of the n observations. Would you mind looking at it for me; I don't feel at home in more than three dimensions even if I could understand it otherwise.

The question arose because this man's tutor is a Caius man whom I have met when I visit my agricultural friends at Cambridge and as he is an astronomer he has applied what you may call Airy to their statistics and I have fallen upon him for being out of date. Well, this chap Fisher produced a paper giving 'A new criterion of probability' or something of the sort. A neat but as far as I could understand it, quite unpractical and unserviceable way of looking at things. (I understood it when I read it but its gone out of my head and as you shall hear, I have lost it). By means of this he thought he proved that the proper formula for the S.D. is

$$\sqrt{\frac{\Sigma(x-m)^2}{n}} \quad \text{vice} \quad \sqrt{\frac{\Sigma(x-m)^2}{n-1}}.$$

This, Stratton, the tutor, made him send me and with some exertion I mastered it, spotted the fallacy (as I believe) and wrote him a letter showing, I hope, an intelligent interest in the matter and incidentally making a blunder. To this he replied with two foolscap pages covered with mathematics of the deepest dye in which he proved, by using n dimensions that the formula was, after all

$$\sqrt{\frac{\Sigma(x-m)^2}{n-1}}$$

and of course exposed my mistake. I couldn't understand his stuff and wrote and said I was going to study it when I had time. I actually took it up to the Lakes with me—and lost it!

Now he sends this to me. It seemed to me that if it's all right perhaps you might like to put the proof in a note. It's so nice and mathematical that it might appeal to some people. In any case I should be glad of your opinion of it.

. . . .

[The rest of the letter is concerned with tuberculosis death-rates, a matter which W. S. G. was already in correspondence about with K. P.]

Yours very sincerely,
W. S. Gosset

September 1912 Woodlands, Monkstown,
Co. Dublin

Dear Pearson,

Since I wrote to you Fisher's first letter has turned up. It is nearly as incomprehensible to me as the other, but shows signs of supplying the missing links in the argument. I am sending it to you in case it should interest you.

If I'm the only person that you've come across that works with too small samples you are very singular. It was on this subject that I came to have dealings with Stratton, for in a paper setting up to teach agriculturalists how to experiment he had taken as an illustration a sample of 4! I heard about it, wrote to the man whom I supposed to be writing the paper with him and he forwarded my letter to the guilty pair. They sent me their papers to correct the day before the proofs were sent in and I mitigated some of it! A high handed proceeding, but all for the good of the cause. The next paper on the subject was by A. D. Hall and came to me much earlier. I think I sent you a copy.

. . . .

[The remainder of the letter is concerned with the possible use of variate differences in eliminating the effect of secular trends when studying the relations between two variables.]

Yours v. sincerely,
W. S. Gosset

15. 9. 1915

Holly House,
Blackrock, Co. Dublin

Dear Mr Fisher,

Very many thanks for the copy of your paper in *Biometrika*. When I first saw it I nearly wrote to thank you for the kind way in which you referred to my unscientific efforts but my natural indolence, joined to the fact that our last correspondence petered out owing to my lack of courtesy and mathematics, allowed me to let the opportunity slip.

I am very glad that my problem is a step nearer solution. I never really liked Soper's approximation though of course it was colossal but there still remains the determination of the probability curve giving the probability of the real value (for infinite population) when a sample of x has given r. Of course this would have to be worked out for two or three *a priori* probabilities and if otherwise convenient I would try* $y = y_0(1-x)^{\frac{1}{2}(m-4)}$ (giving m the values of 3, 4 and 6 in succession) as the *a priori* distribution of the probability of x being the real value of r.

But of course anything almost would do if it gave an integrable expression.

I don't know if it would interest you to hear how these things came to be of importance to me but it happened that I was mixed up with a lot of large scale experiments partly agricultural but chiefly in an Experimental Brewery. The agricultural (and indeed almost any) Experiments naturally required a solution of the mean/S.D. problem and the Experimental Brewery which concerns such things as the connection between analysis of malt or hops, and the behaviour of the beer, and which takes a day to each unit of the experiment, thus limiting the numbers, demanded an answer to such questions as 'If with a small number of cases I get a value r, what is the probability that there is really a positive correlation of greater than (say) ·25?'

After my own investigations I got of course a rough idea of the thing which has been very useful to me but I should still like a full solution.

Yours v. sincerely,
W. S. Gosset

3. Notes on Gosset's letters

Gosset's two letters to Pearson are of interest for several reasons. They show that Fisher had derived 'Student's' distribution by an appeal to n dimensional space, at least by September 1912, within a few months of taking his Cambridge tripos examination and that he was put in touch with Gosset by the astronomer F. J. M. Stratton, Fellow of Gonville and Caius College. That this early contact was made is consistent with Fisher's remarks in his obituary article on Gosset (1939), though the account given by Mahalanobis (1938, 1964) had seemed to suggest that it was not until after he had drafted his paper on the correlation coefficient, finally published in 1915, that Fisher became aware of Student's work. It will be noted that Gosset suggested to Pearson that Fisher's proof might be published as a nice mathematical solution of his own results. Perhaps the proof was in rough form; at any rate the result only appeared as part of the correlation paper of 1915.

Fisher's paper to which Gosset refers in the first letter is clearly that with the title 'On an absolute criterion for fitting frequency curves' (Fisher, 1912). As the volume is dated May 1911–April 1912, Fisher had written the paper while still an undergraduate at Cambridge. In it he suggests that the parameters of frequency curves should be estimated by maximizing what he later termed the likelihood function; he illustrated the method on the normal probability function, obtaining the estimates

$$m = \bar{x}, \quad \sigma^2 = \Sigma(x-m)^2/n.$$

There is no record of why he changed the divisor to $n-1$ in a later letter. In his paper of 1920, when comparing estimates of σ^2 from the mean absolute deviation, σ_1^2, and the mean

* Reference to the original MS shows that Gosset definitely wrote $(1-x)$, but it is possible that $(1-x^2)$ was intended. E. S. P.

squared deviation, σ_2^2, he still used the divisor n, but added: 'Both σ_1 and σ_2 may be adjusted by means of appropriate functions of n so as to make the mean value of each of them obtained from a number of samples agree with the true value, but this for the moment is immaterial'.

Gosset's remarks about himself in his second letter to Pearson confirm what was already known (see E. S. Pearson, 1938,† p. 390) about his connexion with those two pioneer papers on sampling of plots in the field, by Wood & Stratton (1910) and Mercer & Hall (1911). It is very likely that Fisher's appointment to Rothamsted in 1919 owed something to Gosset's links with the agriculturalists.

Gosset's letter to Fisher of 1915 appears to have been the earliest to be preserved out of a long correspondence. No doubt the previous correspondence which he says 'petered out' was that referred to in the letters to Pearson. As in his paper on 'The probable error of a mean' (1908*a*), Gosset was still looking for a solution of problems of inference in terms of prior probabilities.

4. Correspondence between Pearson and Fisher on the distribution of the correlation coefficient

The next series of letters is almost entirely concerned with Fisher's derivation of the sampling distribution of the correlation coefficient and of further work flowing from this. The correspondence which has survived is rather one-sided as there are very few of Fisher's answers, but it seems clear that as a result of discussion certain additions were made to the original draft. At the same time Pearson emphasized the importance of illustrating the theory with rather extensive numerical work. The first letter I quote in full.

September 26, 1914 — Galton Laboratory, University College

My dear Sir,

I could not arrange a meeting for today or tomorrow, because the whole of my week's work is now devoted to non-Laboratory business connected with the war, and Saturday and Sunday are a continuous crush of neglected duties. Indeed, your paper was the fourth I received this week end, asking for 'immediate' consideration.

I have only had time to run very hurriedly through your paper, but it appears to me of very great interest, and I congratulate you very heartily on getting out the actual distribution form of r. If you will let me have the paper till the end of next week, I will write more at length to you on the subject. But if the analysis is correct which seems highly probable, I should be delighted to publish the paper in *Biometrika*. The occurrence of frequency curves with finite range and finite terminal ordinates is of much interest, and I should like to see your paper extended with graphs of some of the curves, and tracing as n increases the change of the frequency form towards a normal distribution. Also the distortion in the normal distribution for n, say 50, when r is large, say ·85. I take it your formulae would be more troublesome to handle than Mr Soper's* approximations, in which case for values of n say 10 to 25 it would be of great value to show that Mr Soper's were sufficiently approximate for practical purposes.

Please let me have, if you are going into camp, your address, that I may write further to you next week.

I am, Yours very faithfully,
Karl Pearson

P.S. Of course the values of the third and fourth moments of the curve with a view to finding β_1 and β_2, and testing the rapidity of the approach with n increasing to normality, would be of great value. I don't think, myself, that values of r for n less than 10 ought *ever* to be considered, but tables of the distribution of r for $n = 10$ to 25, say, and for $\rho = 0{\cdot}1$ to $1{\cdot}0$ by tenths would be of special value. I fear, however, that the war may keep you, like others of us from our own work!

* This is the paper of H. E. Soper (1913). E. S. P.

†Paper No. 24 in this volume

The next letter followed a week later.

October 3, 1914 7 Well Road, Hampstead, N.W.

My dear Sir,

I have now read your paper fully and think it marks a distinct advance and is suggestive in character. I shall be very glad to publish it, as I feel sure it will lead to developments. I wish you had had the leisure to extend the last pages a little. I will have the numerical results verified.

. . . .

As I have said I should like to see some attempt to determine at what value of n and for what values of ρ we may suppose the distribution of r practically normal.

I am just completing the proofs of the current number of *Biometrika*, but your paper shall appear in the next issue. Will you let me have some permanent address, which will reach you?

I am, Yours very faithfully,
Karl Pearson

P.S. I wonder if you have come across my nephew Reginald Sharpe who is also in the Inns of Court, and I think now at Berkhamsted.

The next three letters from Pearson (written on 18 October 1914, 30 January 1915 and 1 February 1915) are concerned with matters arising from the paper. The second letter is particularly interesting because it illustrates the stage which the problem of small-sample estimation of an unknown parameter value had reached at that moment. The inherent difficulty of this problem in the case of the correlation coefficient certainly resulted in Fisher's introduction of the z transformation (1921) and possibly ideas stimulated by the discussions of 1915 ultimately bore fruit in the concepts of fiducial limits and fiducial probability (1930).

In the 1915 paper Fisher had introduced the transformations

$$t = r/(1-r^2)^{\frac{1}{2}}, \quad \tau = \rho/(1-\rho^2)^{\frac{1}{2}},$$

where r and ρ are respectively the sample and population correlation coefficients. He was able to derive quite simple expressions for the first four moments of t and in his first draft provided numerical tables of $\sigma^2(t)$ and $\beta_1(t)$ for selected values of τ and the sample size, n. Later, in proof, he added a similar table of $\beta_2(t)$. The transformed variable was still some way from being normally distributed and its variance was not independent of τ. However, it seemed likely that t would have a more stable distribution than r, although this could not be fully established until the moments of r were known. Pearson's doubts about the practical value of the transformation are shown in the following extracts from a much longer letter.

January 30, 1915 7 Well Road, Hampstead, N.W.

Dear Mr Fisher,

I have delayed answering your letter about the paper because I wanted to think over the problems involved, and just now I don't get much time to think....

I have rather difficulties over this τ and t business—not that I have anything to say about it from the theoretical standpoint—but there appear to me difficulties from the everyday applications with which we as statisticians are most familiar. Let me indicate what I mean.

A man finds a correlation coefficient r from a small sample n of a population; often the material is urgent and an answer on the significance has to be given at once. What he wants to know, say, is whether the true value of $r(\rho)$ is likely to exceed or fall short of his observed value by, say ·10. It may be for instance the correlation between height of firing a gun and the rate of consumption of a time fuse, or between a particular form of treatment of a wound and time of recovery. The data are very sparse and an answer has to be prompt. If n be considerable he knows the p.e. of r and can use tables of the probability integral to determine the degree of accuracy of his result. Failure only comes in when (i) n is small or (ii) r is near unity, when the Gaussian cannot be applied. What we want to know is (i) how soon the Gaussian can be applied and (ii) failing that what is the real or approximate distribution of frequency.

. . . .

For example, suppose that $\rho = \cdot 30$, and I want to find what is the chance that in 40 observations the resulting r will lie between $\cdot 20$ and $\cdot 40$. Now what we need practically are the β_1 and β_2 for $\rho = \cdot 30$ and $n = 40$, and if they are not sufficiently Gaussian for us to use the probability integral, we need the frequency curve of r for $\rho = \cdot 30$ and $n = 40$ to help us out. That is, we need a table of frequencies for ρ going say by $\cdot 1$ and $n = 6$ or 8 up to 40. Now I do not clearly see that your table of τ values will help me to this ....

. . . .

Had I the graph of t I could deduce the graph of r, and mechanically integrate to determine the answer to my problem, but you have not got the ordinates of the t-curve and the practical problem remains it seems to me unsolved. It still seems to me essential (i) to determine β_1 and β_2 accurately for r... and (ii) determine a table of frequencies or areas (integral curve) of the r distribution curve for values of ρ and n which do not provide approximately Gaussian results. Of course you may be able to dispose of my practical difficulties, which do not touch your beautiful theory.

I am still inclined to tabulate the ordinates of the frequency curves for r, starting say, with $\rho = 0$, $\cdot 1, \cdot 2, \cdot 3, \ldots, \cdot 9$, and $n = 4, 8, 12, 16, \ldots, 40$. That is only 100 curves. I am quite aware that this will be very laborious, and can only be done by a trained calculator, who devotes his life to this sort of work, but I think it is the best thing to do ....These tables shall be taken in hand as soon as opportunity offers (my best man for this is at present at work on a table of the incomplete Γ-functions), unless you really want to do them yourself. I do not feel—subject to your reply—that the t-values you give will satisfy practical purposes, even had they been arranged for interpolation. I will publish, if you wish, your tables of σ^2, β_1 and β_2, but I think they lay us open to the criticism that they do not serve to answer the practical problems from which the whole discussion has arisen.

Believe me, Yours very faithfully,

Karl Pearson

Fisher's reply has not been found, but it may have been a letter, which unfortunately seems to have been lost through the disturbance of papers during the 1939–45 war, in which he welcomed the suggestion that computations should be put in hand at the Galton Laboratory. A brief letter of Pearson's of 4 February 1915 states that the paper has gone to press, that certain additions can be made in proof and that the Laboratory will proceed with calculating the frequency distributions for r. Proofs were sent to Fisher on 4 April, were returned before the end of the month, and the paper appeared in the May issue of *Biometrika*. Two letters written by Pearson in April report that some progress has been made on the computaion of the frequency distribution.

The next letter from Pearson is dated 4 November 1915 and reports.

I have at last got good formulae for *all* the moment coefficients of the r-frequency in very rapidly converging series, and hope to publish tables in the next number of *Biometrika* as well as ordinate and area tables. Soper's values are hardly exact enough before $n = 70$, and long after that the frequency curves fail to be normal, if r be over $\cdot 70$ or so. The problem is a very interesting one, and I only wish I could get the mode as simply as I have now got the mean.

There are no further letters available until May 1916, when we have two letters of Fisher's to Pearson as well as Pearson's replies. It is clear that Fisher had been continuing to work on the subject while the long drawn-out Galton Laboratory computations were in progress, but whatever he had done, he may have been discouraged from going further because of the extensive nature of the work completed in London.

May 10th, 1916

9 Horton Crescent,
Rugby

Dear Professor Pearson,

I am afraid I have been very slow about my paper on the probability integral. I have got it written out now, quite shortly as I have cut down on the unnecessary Mathematics. I think you will like the method of calculation in the last section, but of course I don't know what has been done. I have run out the first 27 values of R for $\theta = 90^\circ$ and 120° in no time, but its no good multiplying repeatedly by a long number without a machine. A table by degrees from 90°–180° should meet all requirements, and would save all double entry.

Yours very truly,
R. A. Fisher

May 13, 1916

7 Well Road, Hampstead, N.W.

Dear Mr Fisher,

I ought to have written to you before and told you that the *whole* of the correlation business has come out quite excellently. We have calculated all the frequency distributions from 3 to 25, 50, 75, 100 and 400. We have got good formulae for the moments, the odd ones are all given in terms of complete elliptic integrals and the even as you know in terms of $\cot\alpha$. We have calculated in most cases the β_1 and β_2 and the modal values and these constants ought to be completely tabled in another month. By 25 my curves* give the frequency very satisfactorily, but even when $n = 400$, for high values of ρ the normal curve is really not good enough. Soper's paper really failed because he took *the range* as a fixed quantum. The fit to the frequency is far better, if you simply fit from β_1 and β_2 without regard to the range, the *appreciable* frequency then always lies inside the theoretical range. Many rather new points have arisen from the whole investigation, especially some, I think, new mathematical identities.

I do not know whether and when the results will be published. There is great uncertainty about the future of *Biometrika*; the war has cut off the bulk of our continental subscribers who were the mainstay of the journal—far more important in a way than the English subscribers; and, as I alone am now responsible for the journal, and the deficit was very heavy last year, it will probably have to be suspended or transferred to other proprietors, possibly American. It had been paying its way quite well before the war and readers had no doubt thought it was run by the Cambridge Press, but they have only issued it for the proprietors on commission and decline to give aid when hard times come. I only mention this because it is quite uncertain whether I can publish your paper, and you may care to issue it *at once* elsewhere. Many thanks for your friendly notice of *Biometrika* in the Eugenics Education Society's journal. I don't think it will convert that body!

Yours very sincerely,
Karl Pearson

May 15, 1916

9 Horton Crescent,
Rugby

Dear Professor Pearson,

Your news about *Biometrika* is extremely serious. It seems incredible that such a publication should be in jeopardy for lack of foreign support. It would be a most terrible loss as well as an appalling indignity, if this country cannot support such an important and valuable school of research. Could not the Universities be induced to tide over the War? They have a particular obligation to support such a work, and although I suppose they are hard hit, their honour is at stake in such a matter. I hope you will not allow it to pass into other hands without allowing the widest publicity to the injury which would be done to English learning.

As regards my own paper, parts of it will evidently have to be rewritten with fuller knowledge of what you have done. I should be very glad if you could send me copies of proofs of such of this work, as is complete, if possible with numerical tables. Also my own paper when you have done with it. I could probably have worked more profitably, if I had been in closer touch with the Laboratory, although such collaboration is never easy.

I remain,
Yours very truly,
R. A. Fisher.

* That is, curves of the Pearson system. E. S. P.

May 18, 1916

Department of Applied Statistics,
University College, London, W.C.

Dear Mr Fisher,

Many thanks for your kindly letter about *Biometrika*....

. . . .

As regards our paper it is difficult to determine where, when or how it will be printed as matters are difficult at present with regard to all publication of scientific work. When it comes to type I will certainly send you proofs. With regard to your own paper it seems to me that it would be very nice to have a table of the probability integral, but it would undoubtedly mean on your lines a large amount of stiff work. I think it could be found for $n = 3$ to 25 and $\rho = 0$ to 1 by ·05 with two or three months work (r from -1 to $+1$ by ·05), but it would be a fairly long job, and I am not sure it could not be deduced as accurately from our completed tables of *ordinates* of the like cases. These ordinates enable the areas to be found either by quadrature, or by plotting and the integraph. My original idea was to use the latter method, but if you will work a table of the areas it would undoubtedly be preferable. Our work here has been *very* slow as I have only one man left on the staff and four women, and the man may be called up at any moment. Further we have been doing war work since last January, so that other things can only come in odd moments.

Yours very sincerely,
Karl Pearson

5. Comments on the Pearson–Fisher correspondence

This concludes the 1914–16 correspondence about the distribution of the correlation coefficient. I have reproduced it at some length because of its interest from a variety of angles. On the personal side it shows that at this period the relationship between the two men was a friendly one and the implication, which has been made in more than one account, and in particular by Mahalanobis (1964, p. 241) and Yates (1963, pp. 98–9), that the long 'Co-operative study' (Soper *et al.* 1917) containing tables of the ordinates and moments of r was published without Fisher's knowledge is far from correct. What seems likely, however, is that Fisher did not know until late in the day of §8 of the 'Co-operative study', headed 'On the determination of the "most likely" value of the correlation in the sampled population.'

This contained a criticism of what the authors supposed was Fisher's use in his §12 of a prior distribution for ρ in deriving an optimum estimate of this parameter. The misunderstanding arose because Fisher had not succeeded in making clear in his 1912 paper the meaning of his concept of what he later termed 'likelihood'. This may have been partly due to the fact that the maximizing criterion used, e.g. in his Note printed in §6 below, is said to be derived from the Principle of Inverse Probability. But the 1912 paper should have made clear that Fisher was not associating the term 'inverse probability' with any use of Bayes's Theorem, for he there questioned the propriety of integrating the elements $d\theta_1, \ldots, d\theta_r$ through the parameter space of the θ's. It took many years to get the new idea across, but in the present instance the lack of understanding was particularly unfortunate since both Fisher and Pearson, who presumably drafted the Co-operative Study paper, were equally opposed to the use of Bayes's Theorem in the r, ρ context. Had there been the 'closer touch' which Fisher spoke of in his letter of 15 May 1916 as 'profitable' but 'never easy', it is possible that the argument could have been thrashed out before publication. As it was, since papers were not readily typed in those days, there will only have been a single manuscript of the text and lengthy tables, so that Fisher probably knew nothing of the offending section until the paper was available in proof in 1917; owing to financial difficulties hanging over the journal, there was no issue of *Biometrika* between May 1916 and May 1917.

His answer to this criticism, dated Rothamsted, October, 1920, appeared in the first volume of *Metron* (1921). In this paper he took the opportunity of restating more clearly his concept of likelihood; he also derived the distribution of the intraclass coefficient of correlation, and introduced his z transformation for this and for the interclass coefficient. From the following letter of Pearson's to Fisher, it appears that the paper had been offered shortly before to *Biometrika*.

August 21, 1920

Galton Laboratory
University College

Dear Mr Fisher,

Only in passing through Town today did I find your communication of August 3rd. I am very sorry for the delay in answering it, but it is not my fault. During my holiday no journals or pamphlets are forwarded to me and your paper being enclosed in an envelope with a large 'Lawes Agricultural Trust' heading had been taken by the laboratory steward for an offprint and not forwarded.

As there has been a delay of three weeks already, and as I fear if I could give full attention to your paper, which I cannot at the present time, I should be unlikely to publish it in its present form, or without a reply to your criticisms which would involve also a criticism of your work of 1912—I would prefer you published elsewhere. Under present printing and financial conditions, I am regretfully compelled to exclude all that I think erroneous on my own judgment, because I cannot afford controversy.

I think it most important to obtain values for the p.e. of r especially for n subgroups of m in questions of homotyposis and inheritance. The Danes are working at this question, but I am in great hopes it may be done on broader lines which do not assume Gaussian distributions, although what you have done on that point is undoubtedly suggestive and important.

Yours sincerely,
Karl Pearson

On a different point, the sidelight thrown on the difficulties encountered by *Biometrika* under wartime conditions is interesting; financial support from several generous friends enabled publication to continue, but it was not until the publication was transferred in 1922 from the Cambridge University Press to the Biometric Laboratory that the position was finally stabilized.

But perhaps of most interest are the leads into the field of statistical inference which the papers and correspondence reveal. The problem of estimation in the case of the correlation coefficient, with its sampling distribution changing so rapidly in shape and dispersion with n and ρ, provided a challenge to statistical thinking and it is in seeking answers to such challenges that progress is made.

Pearson had stated the practical man's problem clearly at the beginning of his letter of 30 January 1915: the need to find a means of expressing the probability that the unknown population, ρ, fell within certain distances on either side of the observed sample, r. However, in going on to speak of the problem of finding limits for r, given ρ, he was really turning away from estimation to significance testing. But with his usual instinct for getting down to the fundamentals of a problem, however laborious the spade-work involved, he launched on the calculation of the basic numerical tables, without which the accuracy of any approximate estimating or testing procedure could hardly be established.

Gosset, in his letter to Fisher of 15 September 1915 quoted above, had also stated the estimation problem, and as in his paper of 1908 (*b*) had suggested, without stopping to question the philosophy of approach, a solution which involved the use of Bayes's Theorem. This of course, would not have appealed to Fisher.

Although he had already advocated in 1912 the use of what was to be later termed the method of maximum likelihood in estimating a single value for a parameter, it is unlikely that Fisher had yet given much thought to the question of what has been termed 'interval estimation.' He realized, however, the importance in this problem of some form of transformation of variable, and the transformation of r and ρ to t and τ, through the relations

$$t = r/(1-r^2)^{\frac{1}{2}}, \quad \tau = \rho/(1-\rho^2)^{\frac{1}{2}},$$

was a first attempt at a solution by this means. The concluding paragraph of the 1915 paper gives a lead to the next step, the transformation $z = \tanh^{-1} r$. When by 1920, he had completed his examination of this transformation he had gone a long way towards the solution of the immediate practical problem. For in so far as z is distributed normally about $\zeta = \tanh^{-1}\rho$ with a standard deviation of $1/\surd(n-3)$, it matters very little what inferential approach is used in determining limits for ρ, given r.

However, the z approximation was not completely adequate, as comparison with the tabulated results of the Co-operative Study showed, and there still remained the need for a principle of interval estimation not involving the use of Bayes's Theorem. This Fisher provided in his paper of 1930, in which he introduced his ideas of fiducial probability, taking as his one numerical example the estimation of a lower 5 % fiducial limit for ρ, given r, in a sample of four observations. Since in this single parameter problem there is no difference, from the practical man's point of view, between the fiducial limits of Fisher and the confidence limits of Neyman, we can see the final answer of Pearson's and Gosset's 1915 question in the four charts incorporated in David's (1938) *Tables*. These give for four probability levels either confidence limits for ρ, given r, or significance levels for r, given ρ, and were derived from tables of the probability integral of r, themselves based on the ordinates published 21 years before in the Cooperative Study.

6. MINIMUM χ^2 AND MAXIMUM LIKELIHOOD

A final pair of letters of 1916 shows the beginning of a long controversy in which the final words seem not yet to have been spoken. The May 1916 issue of *Biometrika* contained an article by Kirstine Smith 'On the "best" values of the constants in frequency distributions.' Fisher's reaction must have been prompt because, although the letter given below is undated, it was replied to at length by Pearson on 26 June 1916.

9 Horton Crescent,
Rugby

Dear Professor Pearson,

There is an article by Miss Kirstine Smith in the current number of *Biometrika* which, I think, ought not to pass without comment. I enclose a short note upon it.

I have recently completed an article on Mendelism and Biometry which will probably be of interest to you. I find on analysis that the human data is as far as it goes, not inconsistent with Mendelism. But the argument is rather complex.

Yours v. truly,
R. A. Fisher

Draft of Note

In your issue of May 1916 Miss Kirstine Smith proposes to use the minimum value of χ^2 as a criterion to determine the best form of a frequency curve; and proceeds to compare in a number of cases the distributions obtained by ordinary methods with those 'improved' by the use of χ^2. It should be observed that χ^2 can only be determined when the material is grouped into arrays, and that its value depends upon the manner in which it is grouped.

As an example I take the 53 first differences of C. A. Claremont's corrected death rates for diabetes. After determining m and σ by moments, the data was grouped in intervals of $\frac{1}{2}\sigma$ cut off on either side of the mean. The groups were then shifted through a distance of $\sigma/10$, four times in succession, so as to give 5 different groupings. The values of χ^2 obtained were:

groupings	1	2	3	4	5
χ^2	8·33	11·95	11·78	6·57	9·61

With this amount of variation due to grouping, it is evidently hopeless to use the precise minimum of χ^2 for any one group, as a criterion of the frequency constants. This conclusion is reinforced when we calculate the 5 values of $(\sigma/N)\,\partial\chi^2/\partial m$ which are not even all of the same sign.

grouping	1	2	3	4	5
$(\sigma/N)\,\partial\chi^2/\partial m$	·1014	·0718	·0975	−·0013	·0626

The Gaussian would have to be 'improved' by shifting the mean not only by different amounts, but in opposite directions, in the several cases.

There is nothing at all 'arbitrary' in the use of the method of moments for the normal curve; as I have shown elsewhere it flows directly from the absolute criterion ($\Sigma \log f$ a maximum) derived from the Principle of Inverse Probability. There is, on the other hand, something exceedingly arbitrary in a criterion which depends entirely upon the manner in which the data happens to be grouped.

June 26, 1916

Department of Applied Statistics,
University College

Dear Mr Fisher,

I am afraid that I don't agree with your criticism of Frøken K. Smith (she is a pupil of Thiele's and one of the most brilliant of the younger Danish statisticians). In the first place you have to demonstrate the logic of the Gaussian rule. I have the more right to ask a proof as I followed it in 1897, but very much doubt its logic now. In the next place your argument that χ^2 varies with the grouping is of course well known and is one of the modes of finding the best grouping. What we have to determine, however, is with *given* grouping which method gives the lowest χ^2. Frøken shows there is extremely little difference, but she can get better fits by making χ^2 a minimum, always on the hypothesis that χ^2 a minimum is a more reasonable thing to start from than P a minimum [? maximum]. I think the keynote to this is the footnote on p. 263. Now if you look at Frøken's illustrations you will see that no choice of grouping is possible in Illustrations III, IV and V. In II she takes the actual facts as given by Bessel and asks whether values can be chosen which give a better fit than Bessel's moment values. I can see nothing whatever valid in the argument that if another grouping were taken χ^2 would change. Data must be grouped in all series of astronomical and anthropometric observations, even if only owing to the limitation in reading accuracy.

It is clear to me that your true position for criticism must arise, not from saying that χ^2 a min. does not give a 'better value' for m and σ—it obviously must if you accept the χ^2 test—than the method of moments, but that you must demonstrate that the Gaussian method of making the ordinate of a certain contour of the multiple frequency-surface a maximum is more legitimate than making a minimum the chance of a series of observations as bad or worse than the observed series. I think the latter is the true test, not the Gaussian method. I frankly confess I approved the Gaussian method in 1897 (see *Phil. Trans.* Vol. 191, A, p. 232), but I think it logically at fault now.

If you will write me a defence of the Gaussian method, I will certainly consider its publication, but if I were to publish your note, it would have to be followed by another note saying that it missed the point, and that would be a quarrel among contributors.

Yours very sincerely,
Karl Pearson

P.S. Of course the reason I published Frøken Smith's paper was to show that by *another* test than the Gaussian, the method of moments gave excellent results, i.e. her second conclusion.

Whereas the controversy of the supposed use by Fisher of Bayes's Theorem in the estimation of ρ was an unnecessary one, there was in this second case a more fundamental point at issue. Neither writer was apparently aware at this stage that Edgeworth in 1908 had suggested one practical point in favour of maximum likelihood estimators, namely that their standard errors would be expected to be smaller asymptotically than those of alternative estimators.

It appears from the following letter that Fisher did submit another and no doubt fuller justification for his criticism of Kirstine Smith.

October 21, 1918

Department of Applied Statistics,
University College

Dear Mr Fisher,

Many thanks for your memoir* which I hope to find time for. I am afraid I am not a believer in cumulative Mendelian factors as being the solution of the heredity puzzle.

Also I fear that I do not agree with your criticism of Dr Kirstine Smith's paper and under present pressure of circumstances must keep the little space I have in *Biometrika* free from controversy which can only waste what power I have for publishing original work. I have been doing little but war work for the past few years and the issue of the Journal has been delayed or suspended. I should have returned the paper to you only in pressure of other matters it had wholly passed from my mind.

Believe me, sincerely yours,
Karl Pearson

It is easy to understand that Fisher must have felt a sense of frustration in having two of his papers refused on the grounds that they were controversial and how, as a result, he decided to turn to other means of publication. At the same time we can respect Pearson's reasons for refusal. The war had put a very heavy strain on him and he had seen the completion of his laboratory's research projects of 1914 postponed indefinitely. At over 60, with much of his own work unfinished he shunned controversy and he remembered well how 10–15 years before the need to reply to William Bateson's criticisms of biometric work had put too great a strain on his friend and colleague W. F. R. Weldon. There was deeply ingrained in him, as there was too in Fisher, an urge to reply to any expression of opinion which he believed to be wrong and perhaps harmful to the development of his subject. But at least the need to reply would be less compelling if the 'faulty' article was not in print in his own journal! The titanic battles which have from time to time been waged across the statistical field were perhaps enlivening to the onlookers, but they were very real and I think harmfully moving to the participants. History we may hope will forget them, but I have felt that these groups of letters are worth putting on record both for what they explain and also because they provide a contribution of some historical value.

References

David, F. N. (1938). *Tables of the Ordinates and Probability Integral of the Distribution of the Correlation Coefficient in Small Samples.* Cambridge University Press.

Edgeworth, F. Y. (1908). On the probable error of frequency constants. *J. R. Statist. Soc.* **71**, 381–97, 499–512, 652–78.

Fisher, R. A. (1912). On an absolute criterion for fitting frequency curves. *Messeng. Math.* **41**, 155–60.

Fisher, R. A. (1915). Frequency distribution of the values of the correlation coefficient in samples from an indefinitely large population. *Biometrika* **10**, 507–21.

Fisher, R. A. (1918). The correlation between relatives on the supposition of Mendelian inheritance. *Trans. R. Soc. Edinb.* **52**, 399–433.

Fisher, R. A. (1920). A mathematical examination of the methods of determining the accuracy of an observation by the mean error, and by the mean square error. *Mon. Not. R. Astr. Soc.* **80**. 758–70.

Fisher, R. A. (1921). On the probable error of a coefficient of correlation deduced from a small sample. *Metron* **1** (4), 1–32.

Fisher R. A. (1930). Inverse probability. *Proc. Camb. Phil. Soc.* **26**, 528–35.

Fisher, R. A. (1939). 'Student.' *Ann. Eugen.* **9**, 1–9.

* The reference is presumably to Fisher's 1918 paper on Mendelian inheritance. E.S.P.

MAHALANOBIS, P. C. Professor Ronald Aylmer Fisher (1938). *Sankhyā* **4**, 265–72. (1964). *Biometrics* **20**, 238–50.

MERCER, W. B. & HALL, A. D. (1911). The experimental error of field trials (with an Appendix by 'Student'). *J. Agric. Sci.* **4**, 107–32.

PEARSON, E. S. (1938). 'Student' as statistician. *Biometrika* **30**, 210–50.*

SMITH, K. (1916). On the 'best' values of the constants in frequency distributions. *Biometrika* **11**, 262–76.

SOPER, H. E. (1913). On the probable error of the correlation coefficient to a second approximation. *Biometrika* **11**, 328–413.

SOPER, H. E., YOUNG, A. W., CAVE, B. M., LEE, A. & PEARSON, K. (1917). A cooperative study. On the distribution of the correlation coefficient in small samples. Appendix II to the Papers of 'Student' and R. A. Fisher. *Biometrika* **11**, 328–413.

STUDENT, (W. S. GOSSET) (1908*a*). The probable error of a mean. *Biometrika* **6**, 1–25.

STUDENT (W. S. GOSSET) (1908*b*). Probable error of a correlation coefficient. *Biometrika* **6**, 302–10.

WOOD, T. B. & STRATTON, F. J. M. (1910). The interpretation of experimental results. *J. Agric. Sci.* **3**, 417–10.

YATES, F. (1963). Ronald Aylmer Fisher, 1890–1962. *Biographical Memoirs of Fellows of the Royal Society*, pp. 91–129.

**Paper No. 24 in this volume*

GEORGE UDNY YULE

GEORGE UDNY YULE, 1871–1951

By M.G. KENDALL

THE YULE family has roots reaching down deep into Scottish history. William Yule, the grandfather of George Udny Yule, was born in 1764, nearly two centuries ago. Although he published little, he was widely known for his extensive oriental learning, particularly his scholarship in Persian and Arabic. Of his three sons, Robert (1817–1857) was killed in action at Delhi, commanding the 9th Lancers during the Indian Mutiny. George (1813–1885) became distinguished for his administrative work in India, for which he was knighted. Henry, the youngest (1820–1890), who also received the honour of knighthood, was a man of many parts, whose edition of Marco Polo's travels is a definitive work exhibiting all the patient scholarship characteristic of his family.

George Udny Yule, the subject of this memoir, and one of several members of his family to bear the name, was the son of Sir George Udny Yule and the nephew of Sir Henry. The literary and administrative traditions of his family were strong behind him and were to influence him throughout his life; but through him they found expression in a new medium. He was born on February 18th, 1871, at Beech Hill, a house at Morham, near Haddington in Scotland, which was destroyed in 1944 by an aircraft crashing on it. (His second name derives from an ancestor of the family of Udny of that ilk.) After schooling at Winchester he proceeded at the age of 16 to study engineering at University College, London. There he spent three years as an undergraduate (1887–1890), and a further two years (1890–1892) in engineering workshops. But he seems to have felt that engineering was not his *métier* and in 1892 went for a year to Bonn, where he embarked on research into electric waves under Hertz. His first published papers (1893*a*, *b*, *c*, 1895*a*) were based on this work, but experimental physics failed to hold him any more than had engineering and he never again wrote on either subject. It does not appear, in fact, that this early training left a permanent imprint on his habits of thought. One would not suspect an engineering background behind his mature work; the only point at which it exerted some influence was in his careful and expert draughtsmanship and his preference for diagrammatic representation.

In the summer of 1893, at the age of 22, he returned to London, and was promptly offered a demonstratorship at University College by Karl Pearson, who was then Professor of Applied Mathematics and had known Yule as a student. Yule accepted the post, discovered in Pearson an inspiring teacher, and before long was himself making fundamental contributions to the theory of statistics (1897*a*,*c*, 1899). His long association with the Royal Statistical Society began with his election to fellowship in 1895; at his death he had been a fellow for 56 years. About this time he decided to make statistics his life's work and his career was firmly founded.

Although Yule was given the title of Assistant Professor at University College, the salary of such a post in those days was scarcely a living wage. Early in 1899 he left the College for secretarial work with an examining body (the Department of Technology of the City and Guilds of London Institute). This helped to provide bread and butter, but his interest in statistical work was undiminshed, and he continued to publish numerous papers on association and correlation (1900*a*, 1901, 1903). Moreover, his relationship with University College was not severed. Between 1902 and 1909 he held, in addition to his post at the City and Guilds Institute, the Newmarch Lectureship in Statistics. During this period he gave an annual course of lectures on statistical method which formed the basis of his

Introauction to the Theory of Statistics, the first edition of which was published in 1911. During his lifetime it ran to fourteen editions, and was to make his name known and respected all over the scientific world.

In the meantime he became (in 1907) an honorary secretary of the Royal Statistical Society, an office which he held for twelve years. The Society awarded him its highest honour, the Guy Medal in gold, in 1911. It was during the period from 1900 to 1912 that his basic work on correlation and association reached its peak (1907*a*, 1912); but his work was always practical, and these theoretical studies were accompanied by contributions to various economic and sociological subjects (1906*a*,*c*, 1907*b*, 1909, 1910*a*) as well as to Mendelian inheritance (1902*a*, 1907*c*, 1914).

The year 1912 was a turning-point in Yule's career. The University of Cambridge offered him a new post of lecturer in statistics. He accepted, and was duly appointed from October, 1912. There then began his long association with St. John's College, of which he became a member in 1913 and a Fellow in 1922, and where he resided for the rest of his life except during the last years of illness. He was to hold his University post, which was raised to the status of readership, until reaching the age of 60.

The first world war interrupted his academic career and Yule spent four years (1915–1919), first as statistician to the Army Contracts Department of the War Office and then as Director of Requirements with the Ministry of Food. He never spoke to me of these years with any affection, but he must have performed his exacting duties efficiently for his work at the Ministry of Food was recognized by a C.B.E. in 1918.

In 1919 he returned to Cambridge. The war years had not been unproductive of theoretical research, and two papers (1915*a*, 1920*a*) written in collaboration with his lifelong friend Major Greenwood are notable contributions to methodology. The next ten years were to see the full expression of his genius: the papers on time-correlation (1921, 1926, 1927), in which he introduced the correlogram and laid the foundations of the theory of autoregressive series. Further honours came to him naturally, as his reputation was consolidated. In 1922 he was elected a Fellow of the Royal Society, and from 1924 to 1926 was President of the Royal Statistical Society. Various foreign societies elected him to membership and his book was officially translated into the Czech language.

During the 1920's Yule developed a keen interest in motoring,and scandalized some of his fellow dons by the speed at which he travelled. This led him on to an interest in flying, but yielding to some pressure he postponed learning to fly until his retirement in 1930. Then he found that, being over sixty, he was an unacceptable risk as a pilot, and no private company could teach him to fly one of its planes because of the lack of insurance cover. "So of course", he said to me afterwards "I had to buy my own plane". He did so and acquired a pilot's flying certificate A in 1931, a feat of which he was secretly rather proud, as well he might be. Ten years later he and I were sitting together in the Fellows garden at St. John's when a bombing raid was in preparation. The sky was throbbing with aircraft circling round Cambridge. Yule threw his head back and laughed in his hearty way. "I was just thinking", he said, "that I am licensed to fly every one of those things and I haven't the first notion about the controls of any of them".

However, his flying career was soon brought to an end. In 1931 his heart, which had never been very robust, gave serious trouble (there was, I think, a partial heart-block), and for the rest of his life he was a semi-invalid, alert enough in mind, but inactive in in body, climbing the flight of stairs to his rooms only with some difficulty. This irked him considerably, and reacted to some extent on his work. He continued to give some lectures, mainly on vital statistics, as college lecturer until 1940, when he finally gave up teaching; but apart from the paper on vital statistics (1934) he produced very little between 1931 and 1938. He began to feel that the new developments in his subject had overrun him. The death of Karl Pearson in 1936 affected him deeply. "I feel", he wrote to me, "as though the Karlovingian era has come to an end, and the Piscatorial era which succeeds it is one in which I can play no part". In particular he refused to revise any more editions of the "Introduction", and it was a purely chance meeting between us in 1935 which led him to ask me to take over future revisions.

It was during this period that he enlivened an argument about the modern theory of small samples by some Latin verses, from which the following is an extract:

Nonne hoc mirificum?
Sicut sanctum templa
Attrahunt statisticum
Parvula exempla

Multas horas disputat
Mente laborante
Multas horas computat
Machina crepante

Agitat memoria
Multum verbum bonum
Machina scriptoria
Spargit mirum sonum

Plagula novissima
Tandem terminata
"Omnia clarissima"
Clamat "enodata"

Ai! inundor symbolis
Verbis longis tundor
Occaecor parabolis
Juppiter, confundor!

O Exemplum Parvulum
Sero te amavi
Tecum lusi paululum
Mentem fatigavi

O aenigma lepidum
Nova pulchritudo
Vae! Ardorem tepidum
Vicit senectudo.

There is a good deal of the real Yule in this *jeu d'esprit*. The facility with medieval Latin; the sly references to the Dies Irae, and to St. Augustine's famous invocation to the Deity ("Sero te amavi, pulchritudo tam antiqua et tam nova, sero te amavi"); the legitimate scepticism of a practical statistician for the monstrous regiment of mathematicians; the genuine regret of a man who lived to see his subject opening up new pathways along which he could not hope to tread: and most of all, perhaps, the revealing fact that he felt regret rather than resentment.

I like to think that the publication of the revised "Introduction" in 1937 gave him a new lease of life. The first four editions of the book (1911–1917) had totalled only 2,750 copies, and the first ten editions (1911–1935) amounted to 12,250 copies, equivalent to an average sale of about 500 copies a year. The eleventh edition, of 3,000 copies, was exhausted in less than two years, and by 1950 the revised version (11th to 13th editions) had sold about 15,000 copies – an annual rate of more than double that of the earlier years. Since the war, in fact, the rate of disposal has been four times the pre-1937 figure. The increasing popularity of the book did a good deal to counteract Yule's feeling of being left behind by modern developments. He professed to be astonished that the work fulfilled his earlier hope that it would be useful to new generations of students, but he was undoubtedly greatly pleased and comforted.

However that may be, he resumed work and soon began to make further characteristically original contributions. Between 1934 and 1939 his publications were slight; they read like the work of a man who was tidying up his desk in preparation for retirement from the subject. But at long last the philological interests of his family began to show themselves. He became interested in doubts cast on the attribution of the *De Imitatione Christi* to Thomas à Kempis, and was led on to study the statistical characteristics of an author's style. His earliest attempts in this field were concerned with sentence length (1939*b*), and these alone were almost sufficient to dismiss a number of claimants to the *Imitatio* such as Gerson. His main work, however, related to the occurrence of words (principally nouns), and his researches found expression in his last book (1944) on *The Statistics of Literary Vocabulary*. As by-products he wrote a note on a textual emendation of Milton's *Areopagitica* (1943*a*) and two papers on errors in copying manuscripts (1946, 1947). These, I believe, were his last published works.

And so the man who has contributed as much as anyone to the true science of statistics, who began his career with a paper in the *Proceedings of the Royal Society* on interference phenomena in electric waves passing through an electrolyte, closed it with a study in a theological journal on the dating of families of manuscripts. He completed a concordance to the prayer-book version of the psalms, and was greatly disappointed at failing to find a publisher for it; but all further labour came to an end as his health grew steadily worse. His heart became less and less equal to its task and he spent the last two and a half years of his life in nursing homes, walking a little, reading a little, corresponding a little, but conscious that his powers were failing, and waiting, not always patiently, for the end. It came in his eighty-first year on June 26th, 1951, in the Evelyn Nursing Home at Cambridge.

A great deal of Yule's contributions to the advancement of statistics cannot come to light; they reside in the stimulus he gave to his students, the discussions he held with his colleagues on a host of subjects, notably agriculture and demography, and the advice he freely tendered to all who consulted him, for he was always a most approachable man. Of his published work, also, the value of some of his contributions has been lost to view in sheer virtue of their success; for example, his work on correlation and regression is now such standard practice that only the student of history would consult the original papers. When all this is said, however, there remains enough in his bibliography to illustrate amply the breadth of his vision and the originality of his treatment. Fundamental ideas abound in his work, and are usually put forward in such a cautious way that the reader does not always appreciate their importance. It was Yule who gave, in the "Introduction," formulae on correlated sums which are still being rediscovered by students of systematic sampling. It was Yule who invented the correlogram, though he did not invent the name; and likewise it was he who developed the autoregressive series, though again, another invented the name. It was Yule who cut through several pages of Pearsonian algebra to point out that the sampling formulae for partials must be of the same form as those of total correlation coefficients in normal variation, and hence paved the way for Fisher's derivation of the distribution of partial correlations. Only in one respect has his name been attached to a statistical concept, the so-called Yule process; and if I had not chanced to mention it in casual conversation a few weeks before his death he would have died in ignorance of the of the fact.

The three requirements in a good detective, according to Sherlock Holmes, are knowledge, powers of observation, and powers of deduction. All these Yule had in abundance. He was not an expert mathematician, but his mathematics were always equal to their task. He possessed, in addition, extraordinary insight and a balanced judgement which earned him the respect and admiration of all who knew his work. Apart from a clash with Karl Pearson – which was scarcely to be avoided by any of his generation – his writings are exceptionally free from controversy and the personal embitterments which mar so much of the statistical literature of the past thirty years. It was not that Yule did not possess a temper or a set of decided opinions; he kept them, as he kept all his faculties, under control in the interests of the cause of science.

In character he was kindly, gentle and genial. His wide knowledge of many subjects and his love of an apposite story made him the best of companions. His correspondence was a delightful mixture of shop, anecdote and commentary on things in general, as the following random extracts will show:

"I began to keep a commonplace book many years ago, filled with quotations of rude things people have said about statistics. I gave it up as they became less and less imaginative . . . ".

"Isn't it extraordinary how difficult it is to get a sample really random? Every possible precaution, as it may seem, sometimes fails to protect one. I remember Greenwood telling me that, in some experiments done by drawing differently coloured counters from a bag, there seemed to be a bias against one particular colour. On testing, they concluded that this colour had given the counters a slightly greasy surface, so that it tended to escape the sampler's fingers . . .".

"Egon Pearson sent me a few days ago a folder advertisement of *Who's Who*. The inside was headed 'People in the News' and there was given seven specimen biographies, Anthony Eden, Sir Patrick Hastings, General Smuts, Sir John Reith, Ellen Wilkinson, ME!!! and Sir Basil Zaharoff. Isn't this fame? . . . If any reporters do come to a paper of mine at the Stat. Soc. they soon rise up and walk out gently with bowed heads, moaning like the wind in the keyhole . . .".

"I enclose a paper by Raymond Pearl. Here contraceptive methods appear to have brought about a reduction of some 20 per cent. for married white women – but an *increase* of some 14 per cent. among married negresses! . . . It interests me that you are sceptical as to the truth of the usual view that contraception depresses the birthrate. Willcox of the U.S.A. is about the only man I can recall who agreed with me. Almost all the others jeered . . .".

"The college is suffering from a frightful insult from the A.R.P. authorities. They have put up a notice outside our back gate pointing to the Cam and saying 'Static Water' . . .".

"Measurement does not necessarily mean progress. Failing the possibility of measuring that which you desire the lust for measurement may, for example, merely result in your measuring something else – and perhaps forgetting the difference – or in your ignoring some things because they cannot be measured . . . To my mind Freudian psychology made more progress in a few years than measurement-psychology had made for decades. Mendelism again meant more progress than Biometry . . .".

"Gosset came in to see me the other day. He is a very pleasant chap. Not at all the autocrat of the *t*-table . . .".

A man of Yule's age has the misfortune of seeing many of his friends and contemporaries procede him to the grave. Some are left to mourn him; far larger is the number of younger men who knew him first as instructor and then as friend and will always remember him as one of the ablest, kindliest and most lovable of men.

Bibliography

1893*a*. "On interference phenomena in electric waves passing through different thicknesses of electrolyte", *Proc. Roy. Soc.*, **54**, 96.

1893*b*. "Ueber den Durchgang elektrischer Wellen durch Elektrolytenschichten", *Wied. Ann.*, **50**, 742.

1893*c*. "On the passage of electric wave-trains through layers of electrolyte", *Phil. Mag.*, **36**, 531.

1895*a*. "On the passage of an oscillator wave-train through a plate of conducting dielectric", *Phil. Mag.*, **39**, 309.

1895*b*. "On a simple form of harmonic analyser", *Phil. Mag.*, **39**, 367. (A short report of the discussion at the Physical Society is in the *Electrician* for March 22nd.)

1895*c*. "On the correlation of total pauperism with proportion of out-relief. I. All ages", *Econ. Jl.*, **5**, 603.

1896*a*. "Notes on the history of pauperism in England and Wales from 1850, treated by the method of frequency curves, etc", *J.R.S.S.*, **59**, 318.

1896*b*. "On the correlation of total pauperism with proportion of out-relief. II. Males over 65", *Econ. Jl.*, **6**, 613.

1897*a*. "On the significance of Bravais' formulae for regression, etc., in the case of skew correlation", *Proc. Roy. Soc.*, **60**, 477.

1897*b*. "Note on the teaching of the theory of statistics at University College (London)", *J.R.S.S.*, **60**, 456.

1897*c*. "On the thory of correlation", *J.R.S.S.*, **60**, 812.

1897*d*. "Anti-vaccination statistics," Jenner Soc. (Practically a reprint of remarks made at the Stat. Soc. discussion on a paper by A. Milnes.) *J.R.S.S.*, **60**, 608.

1897*e*. "Statistics of small-pox and vaccination", *Public Health*, **9**, 324 (cf. above).

1899. "An investigation into the causes of changes in pauperism in England, etc.", *J.R.S.S.*, **62**, 249

1900*a*. "On the association of attributes in statistics", *Phil. Trans.*, **194A**, 257.

1900*b*. With Miss M. Beeton and Karl Pearson. "On the correlation between duration of life and the number of offspring", *Proc. Roy. Soc.*, **67**, 159.

1901. "On the theory of consistence of logical class-frequencies and its geometrical representation", *Phil. Trans.*, **197A**, 91.

1902*a*. "Mendel's Laws and their probable relations to intra-racial heredity", *New Phytologist*, **1**, 193.

1902*b*. "Variation in the number of sepals in *Anemone nemorosa*," *Biometrika*, **1**, 307.

1903. "Notes on the theory of association of attributes in statistics", *Biometrika*, **2**, 121.

1904. "On a convenient means of drawing curves to various scales", *Biometrika*, **3**, 469.

1905. "The introduction of the words 'statistics', 'statistical' into the English language", *J.R.S.S.*, **68**, 391.

1906*a*. "On the changes in the marriage- and birth-rates in England and Wales during the past half century", *J.R.S.S.*, **69**, 88.

1906*b*. (With R.H. Hooker.) "Note on estimating the relative influence of two variables upon a third", *J.R.S.S.*, **69**, 197.

1906*c*. (With H.D. Vigor.) On the sex-ratios of births in the registration districts of England and and Wales, 1881–90, *J.R.S.S.*, **69**, 576.

1906*d*. "On a property which holds good for all groupings of a normal distribution of frequency for two variables, with applications to the study of contingency-tables for the inheritance of unmeasured qualities", *Proc. Roy. Soc.*, **77A**, 324.

1906*e*. "On the influence of bias and of personal equation in statistics of ill-defined qualities", *J. Anthrop. Inst.*, **36**, 325. Abstract in *Proc. Roy. Soc.*, **77A**, 337.

1907*a*. "On the theory of correlation for any number of variables, treated by a new system of notation", *Proc. Roy. Soc.*, **79A**, 182.

1907*b*. "Statistics of production and the Census of Production Act (1906)", *J.R.S.S.*, **70**, 52.

1907*c*. "On the theory of inheritance of quantitative compound characters on the basis of Mendel's Laws", *Report 3rd Int. Conf. on Genetics* (1906), 140.

1909. "The applications of the method of correlation to social and economic statistics", *Bull. de l'Inst. Int. de Statistique*, **18**, liv. 1. 537, 265, and *J.R.S.S.*, **72**, 721.

1910*a*. "On the distribution of deaths with age when the causes of death act cumulatively", *J.R.S.S.*, **73**, 26.

1910*b*. "On the interpretation of correlations between indices or ratios", *J.R.S.S.*, **73**, 644.

1911. Obituary: Sir Francis Galton, *J.R.S.S.*, **74**, 314.

1912. "On the methods of measuring association between two attributes", *J.R.S.S.*, **75**, 579.

1914*a*. (With M. Greenwood.) "On the determination of size of family and of the distribution of characters in order of birth, etc.", *J.R.S.S.*, **77**, 179.

1914*b*. "Fluctuations of sampling in Mendelian ratios", *Proc. Camb. Phil. Soc.*, **17**, 425.

1914*c*. (With F.L. Engledow.) "The determination of the best value for the coupling-ratio from a given set of data", *Proc. Camb. Phil. Soc.*, **17**, 436.

1914*d*. (With T.B. Wood). "Statistics of British feeding trials and the starch equivalent theory", *J. Agric. Sci.*, **6**, 233.

1915*a*. (With M. Greenwood.) "The statistics of anti-typhoid and anti-cholera inoculations, and the interpretation of such statistics in general", *Proc. Roy. Soc. Med.* (Epidemiology), **8**, 113.

1915*b*. "Crop production and price: a note on Gregory King's law", *J.R.S.S.*, **78**, 296.

1917. (With M. Greenwood.) "On the statistical interpretation of some bacteriological methods employed in water analysis", *J. of Hygiene*, **16**, 36.

1920*a*. (With M. Greenwood.) "An enquiry into the nature of frequency distributions representative of multiple happenings, etc.", *J.R.S.S.*, **83**, 255.

1920*b*. "The fall of the birth-rate". (Pamphlet) Camb. Univ. Press.

1920*c*. "A note on Mr. King's method of graduation and its relation to graphic method", *J. Inst. Act.*, **52**, 135.

1921 "On the time-correlation problem, with especial reference to the variate-difference correlation method", *J.R.S.S.*, **84**, 497.

1922*a*. "On the application of the x^2 method to association and contingency tables, with experimental illustrations", *J.R.S.S.*, **85**, 95.

1922*b*. (With Dr. J.C. Willis.) "Some statistics of evolution and geographical distribution in plants and animals, and their significance". (Linnean Soc., February 2nd), *Nature*, February 9th, **109**, 177.

1923*a*. "The progeny, in generations $F_{.12}$ to $F_{.17}$ of a cross between a yellow-wrinkled and a green-round seeded pea: a report on data afforded by experiments initiated by the late A.D. Darbishire, M.A., in 1905, and conducted by him until his death in 1915", *J. of Genetics*, **13**, No. 3, 255.

1923*b*. "The laws of probability and their meaning", *Annals of Botany*, **37**, 541.
1924*a*. "A mathematical theory of evolution, based on the conclusions of Dr. J.C. Willis", *Phil. Trans.*, **213**B, 21.
1924*b*. "The function of statistical method in scientific investigation", *Medical Research Council Industrial Fatigue Research Board*, Rep. 28, H.M.S.O.
1924*c*. "Some life-table approximations". *Proceedings of International Mathematical Congress*, Toronto.
1925. "The growth of population and the factors which control it" (Presidential Address), *J.R.S.S.*, **88**, 1.
1926*a*. "Why do we sometimes get nonsense correlations between time-series? A study in sampling and the nature of time-series" (Presidential Address), *J.R.S.S.*, **89**, 1.
1926*b*. (With F.L. Engledow.) "The principles and practice of yield trials" (Reprinted from the *Empire Cotton Review*, Empire Cotton Growing Corporation, Vol. III, Nos. 2 and 3. Revised edition 1930).
1927*a*. "On a method of investigating periodicities in disturbed series, with special reference to Wolfer's sunspot numbers", *Phil. Trans.*, **226**A, 267.
1927*b*. "On reading a scale", *J.R.S.S.*, **90**, 570.
1931. In memory of the Rev. William Cecil, M.A., sometime Fellow of Magdalene College and Fellow of the Cambridge Philosophical Society, *Proc. Camb. Phil. Soc.*, **27**, Pt. I.
1933*a*. "Note on the number of Jews in Germany", *J.R.S.S.*, **96**, 478.
1933*b*. (With E.L. Collis.) "The mortality experience of an occupational group exposed to silica dust, compared with that of the general population and an occupational group exposed to dust not containing silica", *J. of Industrial Hygiene* (U.S.A.), **15**, Pt. 6, 395.
1934. "On some points relating to vital statistics, more especially statistics of occupational mortality", *J.R.S.S.*, **97**, 1.
1935. Commemoration Sermon, St John's College. *The Eagle*, **49**, 7.
1936*a*. "On a parallelism between differential coefficients and regression coefficients", *J.R.S.S.*, **99**, 770.
1936*b*. "Karl Pearson", *Royal Society*, Obituary Notices, **2**, 73.
1938*a*. "A test of Tippett's random sampling numbers", *J.R.S.S.*, **101**, 167.
1938*b*. "Notes of Karl Pearson's lectures on the theory of statistics, 1884–96", *Biometrika*, **30**, 198.
1938*c*. "On some properties of normal distributions, univariate and bivariate, based on sums of squares of frequencies", *Biometrika*, **30**, 1.
1939*a*. "John Wallis, D.D., F.R.S.", Notes and Records of the Royal Society, April, **2**, 74.
1939*b*. "On sentence-length as a statistical characteristic of style in prose, etc.", *Biometrika*, **30**, 363
1941. "Note on the statistical theory of accidents with special reference to the time-factor", *J.R.S.S. Supplement*, **7**, 91.
1943*a*. "The word 'muing' in Milton's *Areopagitica*", *Rev. English Studies*, 19.
1943*b*. Obituary of Sir Henry Howard, *Cambridge Review*, November, and *The Eagle*.
1944. Obituary of R.H. Hooker, *J.R.S.S.*, **107**, 74.
1945. "On a method of studying time-series based on their internal correlations", *J.R.S.S.*, **108**, 208.
1946. "Cumulative sampling: a speculation as to what happens in copying manuscripts", *J.R.S.S.*, **109**, 44.
1947. "Puyol's Classes A and B of texts of the 'De Imitatione Christi' ", *Recherches de Théologie ancienne et médiévale*, **14**, 65.

Books

(1) 1911. *An Introduction to the Theory of Statistics*. Charles Griffin & Co., Ltd., London. 2nd edn. 1912. 3rd edn. 1916, 4th edn. 1917, 5th edn. 1919, 6th edn. 1922, 7th edn. 1924, 8th edn. 1927, 9th edn. 1929, 10th edn. 1932.

The book was then re-written by M.G. Kendall and re-set in a larger format, the new edition appearing under their joint names: 11th edn. 1937, 12th edn. 1940, 13th edn. 1945, 14th edn. 1950.

A very handsomely produced official translation into Czech was published in 1926, and an unauthorized Polish translation in 1921. Translations have also appeared in Spanish and Portuguese.

(2) 1944. *The Statistical Study of Literary Vocabulary*. Cambridge University Press.

KARL PEARSON

KARL PEARSON, 1857–1957

Being a Centenary Lecture by J. B. S. HALDANE, delivered at University College London on 13 *May* 1957*

We are met here to-day to celebrate the centenary of the birth of Karl Pearson. To me, at least, this means that I am glad that Karl Pearson was born, that I think the world is better because he was born.

A greater man than any of us said

> The evil that men do lives after them
> The good is oft interrèd with their bones.

Let us begin, therefore, with some criticisms. And then let us study not only those of Pearson's contributions to science and culture which are widely known, but perhaps some also which should be disinterred and brought once more into the light of day.

As Pitt first stated, and Acton restated more precisely, all power corrupts. It is impossible to be a professor in charge of an important department, and the editor of an important journal, without being somewhat corrupted. We can now see that in both capacities Pearson made mistakes. He rejected lines of research which later turned out to be fruitful. He used his own energy and that of his subordinates in research which turned out to be much less important than he believed. It is, however, very easy to say what any one ought to have done fifty years ago!

But this criticism can be, and has been, pushed much further. It is said that Pearson espoused a fundamentally false theory of heredity, and therefore of evolution, and that as a consequence his work was not merely useless, but actually retarded progress. Had Pearson become dictator of British research on heredity and evolution, this might have been true. Fortunately he did not. I believe that his theory of heredity was incorrect in some fundamental respects. So was Columbus' theory of geography. He set out for China, and discovered America. But he is not regarded as a failure for this reason. When I turn to Pearson's great series of papers on the mathematical theory of evolution, published in the last years of the nineteenth century, I find that the theories of evolution now most generally accepted are very far from his own. *But* I find that in the search for a self-consistent theory of evolution he devised methods which are not only indispensable in any discussion of evolution. They are essential in every serious application of statistics to any problem whatever. If, for example, I wish to describe the distribution of British incomes, the response of different individuals to a drug, or the results of testing materials used in engineering, I must start off from the foundation laid in his memoir on 'Skew variation in homogeneous material'. After sixty-three years I shall certainly take some short cuts through the jungle of his formulae, some of which he himself made in later years. Very few ships to-day follow Columbus' course across the Atlantic.

Let me put the matter in another way. Anyone reading the controversy between Pearson and Weldon on one side, and Bateson and his colleagues on the other, which reached its

* The author is quite aware that he has repeated himself in a way which would be unjustified had the material been put together for an article, but which is justifiable in an oration. On the other hand, he considers that it is seldom desirable to hack what was designed to be spoken into a form suitable for reading.

culmination about fifty to fifty-five years ago, might have said 'I do not know who is right, but it is certain that at least one side is wrong'. In fact both were right in essentials. The general theory of Mendelism is, I believe, correct in a broad way. But we can now see that if Mendelism were completely correct, natural selection, as Pearson understood it, could not occur. For the frequency of one gene could never increase at the expense of another, except by chance, or as we now put it, sampling errors. It is just the divergence between observed results and theoretical expectations, to which Pearson rightly drew attention, which gives Mendelian genetics their evolutionary importance.

After this preamble I pass to my main task. Pearson's connexion with this College began when he was nine years old, and was sent to University College School, where he remained for seven years. He left at sixteen and obtained a scholarship at King's College, Cambridge, at eighteen. As an undergraduate he studied mathematics and was third wrangler in 1879. He had already shown something of his future mettle by a successful refusal to attend divinity lectures. In spite, or perhaps because, of this independence of spirit he became a fellow of King's in 1880. He spent about a year in the universities of Heidelberg and Berlin, attending lectures on philosophy and Roman law as well as physics and biology. However, the most striking effect of his German year was to interest him in mediaeval and Renaissance German literature, especially the development of ideas on religion and the position of women. At about this time he began to spell his Christian name with a K instead of a C. This may have been a homage to German culture. It may have been a special homage to Karl Marx, for we know that he later lectured on Marx, and his daughter tells me that when in Germany the police once searched his rooms, and he considered that one of Marx's books was the most subversive of the documents which they found there.

In 1880 he began the study of law in London, and was called to the bar in 1881. This may have been a tribute to his father, who was a Q.C., or a means of ensuring a livelihood in future, more probably both. He also published his first books, *The New Werther** and *The Trinity, a Nineteenth Century Passion Play*. Both were anonymous, and had they been signed, would certainly have prejudiced their author's chance of appointment in many institutions, perhaps even in the Infidel College, which suffers from occasional outbreaks of respectability. For both attack Christian orthodoxy.

It was at this period of his life that he lectured on Marx to small audiences in London, on the 'Ethic of Freethought' at South Place, and to the Sunday Lecture Society on 'Matter and Soul'.

In 1884, at the age of 27, he was appointed to the Chair of Applied Mathematics and Mechanics in this College. He had only published two small papers on rather academic problems of applied mathematics. His first publication after his appointment was *The Common Sense of the Exact Sciences*, by his illustrious predecessor in this College, W. K. Clifford. His next was even more surprising. It was written in German, entitled *Die Fronica: Ein Beitrag zur Geschichte des Christusbildes in Mittelalter*, and published at Strassburg. So far as I know it was the first contribution made by a professor of this College to the history of art. It is interesting to see that he regarded this as a worthy topic of academic study. May I hope that, now that we have a Chair of this subject, our Professor may comment on Pearson's contribution to it.

He was clearly a very successful and thorough teacher of applied mathematics, mainly

* For the bibliography of Pearson's works, and for much else, I rely on E. S. Pearson's invaluable memoir (Cambridge, 1938). In this lecture I have not even mentioned some of his books.

to students of engineering. He edited de Saint-Venant's work on the theory of elasticity, and wrote the second part of Todhunter's *History of the Theory of Elasticity*. His radical activities continued. In 1885 he joined 'The men's and women's club', a small body devoted to 'the free and unreserved discussion of all matters in any way connected with the mutual position and relation of men and women'. As, in *The Ethic of Freethought*, Pearson defended the view that unmarried women should be allowed sexual freedom, it is not surprising that legends arose, and still exist, as to this club.* In fact Karl Pearson married one of its members, Miss Sharpe. To-day it is quite normal for a couple to discuss human sexual physiology before marriage. Seventy years ago it was regarded as grossly improper, and all kinds of accusations were made against those who did so. I have not the faintest doubt that in fact the male members of the club were far less promiscuous than most of their contemporaries. If to-day association with prostitutes is generally regarded as degrading, while seventy years ago it was generally condoned and not rarely approved, we owe it largely to men like Karl Pearson.

The Ethic of Freethought was published in 1888, and is a collection of lectures and essays, some of which had been reprinted as pamphlets. It is, in essence, a religious book. Pearson defined religion as 'the relation of the finite to the infinite'. 'Hence', he continued, 'all systems of religion are of necessity half truths.' The most scholarly part of the book deals with the history of religious systems, particularly in Germany. He believed that such a study was part of the duty of an educated man or woman. I read a few sentences. 'By studying the past I do not mean reading a popular historical work, but taking a hundred, or better fifty, years in the life of a nation, and studying thoroughly that period. Each one of us is capable of such a study, though it may require the leisure moments, not of weeks, but of years. It means understanding, not only the politics of that nation during those years; not only what its thinkers wrote; not only how the educated classes thought and lived; but in addition how the mass of the folk struggled, and what aroused their feeling and stirred them to action. In this latter respect more may be learnt from folk-songs and broadsheets than from a whole round of foreign campaigns.'

The book is largely a record of its author's search for truth among religious systems. One chapter is devoted to the mystic Eckehart, and was the first introduction of that remarkable thinker to the British public. Of all the systems examined there can be no doubt that that of Spinoza appealed most deeply to Pearson; and he devoted another chapter to demonstrating Spinoza's debt to Maimonides. If I may be allowed to express a regret which is in no sense a criticism, it is that Pearson's acquaintance with Indian philosophy was confined to translations of Hinayana Buddhist scriptures. I think that he would have recognized more kindred spirits in such ancient Hindu thinkers as Yajnavalkya and the great anonymous humanist whose words are preserved in the first section of the Bṛhadaranyaka Upanishad.

It is a little surprising that the title page does not mention the author's professorship at University College. Perhaps his senior colleagues thought that such a mention would have got him into trouble.

If, in 1890, one had had to pass judgement on Pearson, it might have run as follows,

* [As might be expected, the critics tended to fix on only one side of the picture of the ideal relationship between the sexes in a socialist state which Pearson elaborated in his lectures on "The Woman's Question" (1885) and "Socialism and Sex" (1886), afterwards published in *The Ethic of Freethought*. Ed.]

'He is a first-rate teacher of applied mathematics, and a scholarly compiler of the work of more original men. He has a knowledge of literature and art most unusual in a professor of mathematics. He is somewhat of a radical, but he is only thirty-three years old. He will settle down as a respectable and useful member of society, and may expect a knighthood if he survives to sixty. He will never produce work of great originality, but the College need not be sorry to have appointed him.' Had this judgement been correct, we should not be here to-day.

In 1890 two events occurred which, in my opinion, shaped the course of Pearson's future life. He applied for, and received, the lectureship in Geometry at Gresham College; and W. F. R. Weldon succeeded Lankester in the chair of zoology at University College. At Gresham College he could lecture on what he pleased. His first set of lectures developed into *The Grammar of Science*, his main contribution to philosophy. Later series dealt with 'The Geometry of Statistics', and 'The Laws of Chance'. But since the discussion of probability and statistical method in the first edition of *The Grammar of Science* is superficial, we may take it that in 1891 he had not considered the subject seriously. He certainly did so in later years. I have little doubt that the stimulus to do so came largely from Weldon.

The Grammar of Science is a very remarkable book. Pearson claimed that material objects were merely a conceptual shorthand used to describe regularities in our sense-impressions. This idea is hard to develop, if only because our language is in terms of material objects such as eyes and brains. He did not in fact develop it without some self-contradiction, at least on the verbal level. But he did so, in my opinion, with much less self-contradiction than contemporaries such as Mach and Avenarius. He must be regarded as one of the founders of the important school of logical positivism.

I can well remember the impression which his book made on me when I first read it about 1909. If it is less impressive to modern readers, this is probably because physical theories have changed profoundly, a fact which would in no way have surprised or distressed its author. I do not personally think that Pearson's philosophical views are correct. Nevertheless, a man who first states an important doctrine clearly, even if it is subsequently rejected, is a moment in the thought process of humanity. We can best see whether Pearson did this by listening to the judgement of one of his adversaries. In 1908 Vladimir Ilyitch Lenin wrote *Materialism and Empirio-criticism*. This was an attack on people who, in his words, or rather those of his translator, 'under the guise of Marxism were offering something incredibly muddled, confused and reactionary'.

Now Lenin disagreed strongly with Pearson, and claimed, in my opinion correctly, to have found self-contradictions in his arguments. Nevertheless, he found him vastly clearer than other Machians. Let me read a few of Lenin's sentences. 'The philosophy of Pearson, as we shall repeatedly find, excels that of Mach in integrity and consistency' (p. 119).* 'The Englishman, Karl Pearson, expresses himself with characteristic precision, "Man is the creator of natural law".' (p. 221). And finally (p. 243) Lenin described him as 'This conscientious and scrupulous foe of materialism'. Unfortunately, I do not know how precise is this translation from the original Russian. But praise of this kind from an opponent is in my opinion worth a great deal more than either the assent of uncritical disciples or the patronizing acknowledgements of successors who claim to have improved on Pearson's

* The references are to the pagination in Vol. 11 of Lenin's *Selected Works*. London, Lawrence and Wishart, 1939.

treatment of the subject. The only other contemporary British opponent of materialism to whom Lenin was equally polite was James Ward. I cannot help thinking of Dante's treatment of Saladin, who was, of course, in hell, but so far from suffering from heat, cold, or other torments, was housed in a noble castle. Whatever may be the fate of Pearson's philosophy in his own country, *The Grammar of Science* is assured of attentive reading in those states where Leninism is orthodox.

To go back to Pearson's own views, I quote three sentences from the *Grammar*, which I think illustrate the strength and the weakness of Pearson's approach to science. 'The unity of all science consists alone in its method not in its material.' 'No physicist ever saw or felt an individual atom. Atom and molecule are intellectual conceptions by aid of which physicists classify phenomena and formulate the relationship between their sequences.' The strength is shown by the fact that the distributions, which Pearson worked out to describe Weldon's measurements of populations of crabs, will equally well serve to describe populations of stars, manufactured goods, durations of life, incomes, barometer readings, and so on. The weakness is shown by the fact that physicists have, during this century, seen individual atoms, or rather atomic nuclei, by the tracks which they make when moving rapidly. Pearson's philosophy discouraged him from looking too far behind phenomena. It was, I think, for this reason, that he never accepted Mendelian genetics, although the *Treasury of Human Inheritance* and his own monograph on albinism contain plenty of evidence in its favour.

His later series of Gresham Lectures dealt with statistics and probability, particularly with graphical methods of representing distributions. I have no doubt that they were written partly as a result of the questions which Weldon began putting to him soon after his arrival in University College. But his full answer to these questions is to be found in the great series of memoirs on the Mathematical Theory of Evolution which were published in the *Philosophical Transactions* of the Royal Society between 1893 and 1900. It is not too much to say that the subsequent developments of mathematical statistics are largely based on Pearson's work between 1893 and 1903. Perhaps we shall be helped to estimate its importance by an exercise in hypothetics. What would have been the effect on Pearson had Bateson obtained the Jodrell Chair of Zoology in place of Weldon? And what would have been the effect had our College contained an economist or engineer interested in what is now called Quality Control? Although Bateson was as interested as Weldon in animal variation, he was more concerned with exceptions, and with discontinuous, or as Pearson and Lee called it in 1899, exclusive inheritance. I doubt if Bateson would have put his questions in a form which would have aroused Pearson's interest. If he had done so, they would probably have discovered what is now called Mendelism. For Bateson, before reading Mendel's paper, did not realize the necessity of dealing with large samples, which Pearson certainly did.

If an economist or technologist had interested him in the variation of manufactured goods, he would have had to deal, as he did, with skew variation. He would presumably have used correlation to measure the likeness between the products of the same craftsman or machine as he in fact used it to measure the likeness between the children of the same parents. Perhaps in 1901 he might have founded a journal *Technometrika* not wholly unlike *Biometrika*. He would almost certainly have invented some, at least, of the statistical methods now used in industrial quality control. He might perhaps have added 1% or so to the industrial productivity of Britain in the early years of this century.

The papers to which I refer are hard to read because Pearson reached his conclusions by algebraical and arithmetical methods which are now seen to be needlessly laborious. Many of them have since been simplified. As a humble tribute to Pearson I have, as I believe, simplified the first of them, which deals with the dissection of a skew frequency distribution into two normal distributions. By an elementary transformation I have thrown his rather formidable nonic equation into a form which allows numerical tabulation, and this tabulation is now under weigh in the electronic laboratory of the Indian Statistical Institute. I hope that as a result, the method will be available to statisticians less pertinacious than Karl Pearson.

Commenting on a particular passage in (I think) one of Beethoven's works, a German musical critic remarked 'Hier ist Titanenthum Pflicht'. (Here titanicity is a duty). Karl Pearson attacked Olympus by piling Ossa on Pelion rather than by seeking an easy path. If we, his successors, have made statistical theory relatively easy, and much of Pearson's mathematics are no longer used, we should remember that we are treading in the footsteps of an intellectual titan.

The germs of many later developments in mathematical statistics are to be found in these papers. Thus, in Contribution III to the theory of evolution, Pearson discussed 'the best value of the correlation coefficient' based on a given sample. He decided on the value which maximized the chance of obtaining the observed sample. This method was developed by Edgeworth as 'the method of maximum credibility', and by Fisher as 'the method of maximum livelihood'. In the succeeding paper, with Filon, Pearson developed it further. Critics have asked why he did not generalize it. I think one possible answer is as follows. The expression 'the best' is unfortunately seldom applicable to statistical estimates. The best for one purpose is not usually the best for another. I think Pearson realized this. Some of his successors have not.

In 1900 Pearson attacked the problem of curve fitting. Having fitted the best available curve to a series of data, for example, the numbers of human beings whose heights were in intervals such as 70–71 in., he asked what was the probability that a sample from a population truly represented by his curve should fit it as badly as, or worse than, the sample in question. If the chance was 1 in 3, there is no good reason to doubt the validity of the theory on which the curve is based. If the chance was 1 in 300, the theory is almost certainly wrong, though it may be a useful approximation for some purposes. But the question arises 'what is a bad fit?' Is 38 a worse fit to an expected number of 30 than 4 to an expected number of 10? (It is not!) And how are we to combine these in an overall estimate of badness of fit? Pearson solved this problem by the invention of the function of observations called χ^2, which increases as the fit becomes worse.

This has turned out to be an immensely powerful tool, and is used on a huge scale. To take one example, in the last number of the *Journal of Genetics*, at least fifty-three values of χ^2 were calculated by three different authors. But now comes the curious and characteristic fact. None of these authors used χ^2 as a test of curve-fitting, and it is very rarely so used. It is used as a test of agreement with hypothesis wherever the hypothesis is tested by counting individuals. And it is used, as Pearson pointed out that it might be used, to discover whether a number of sets of data agree with the same unknown hypothesis. For example, if the total of a number of families contains about 17 % of a particular type we may have had no reason beforehand to expect 17 % rather than any other frequency. But we can use χ^2 to determine whether some of the families have a proportion which diverges more from 17 % than could reasonably be accounted for by chance. If not, but only if not, we can justifiably pool the data. In this case χ^2 is said to be used as a test of homogeneity.

This is, of course, a commonplace with great human achievements. The wheel was invented for use in chariots, carts, and so on. But to-day most wheels are used, not for the support of vehicles, but for power transmission. Perhaps the majority of wheels in England are inside watches. It was absolutely characteristic of Karl Pearson that his intellectual inventions were often extremely general. He obtained a solution of a problem which was of such generality that it had entirely unexpected applications.

But for this very reason it often had a limited applicability to the problem for which it was originally designed. In the last few years many experiments have been done on artificial selection of quantitative characters, particularly in *Drosophila* and mice. Their results during the first few generations are often much as Pearson would have expected. But after this they diverge very greatly. In spite of this they are best described by the use of the mathematical tools which Pearson first applied to such problems, that is to say by describing changes in the moments of character distributions, and simple functions of them such as standard deviations and correlations. One can only defeat Pearson intellectually with the weapons which he himself forged. If I may be allowed to quote William Blake,* Pearson's main service to humanity was

> In all his ancient strength to form the golden armour of science
> For intellectual war.

About the same time he began not only to use data collected by Galton and others, on man and other animals, but to collect his own. Among the important biological results of this period were the demonstration that fertility is inherited both in our own species and in race-horses. As an example of his thoroughness I mention his measurements of the same human bones after various periods of wetting and drying, which never changed their length by as much as 1 %, though they did change it.

It was probably through Weldon that he came to know Galton. This very remarkable man had, among other things, invented the recognition of criminals by finger prints, and psychoanalysis (as may be seen from pages 185–207 of *Inquiries into Human Faculty*). Much of Pearson's work in the '90's was a development of the notions used by Galton in his *Natural Inheritance* in 1889. However, there is no reason to think that Pearson's one serious excursion into practical psychology owed anything to Galton. This is described in a pair of papers published in 1899 and 1902 alleged to be on the mathematical theory of errors of judgement, but in fact incorporating a series of measurements made on the same material by Dr Alice Lee, Dr Udny Yule, and Pearson, and by Lee, Dr Macdonell, and Pearson. Each observer had, of course, a characteristic bias and a characteristic spread round the mean. But what was utterly unexpected was the discovery that the errors made by two observers varied in Pearson's words, sympathetically. In fact in one series, Lee and Macdonell showed a high correlation, Pearson being independent. He attributed this to 'the influence of the immediate atmosphere'. Others might have attributed it to telepathy.

Some of his finest work at this time was with Lee, Bramley-Moore and Beeton on the inheritance of human fertility and longevity. I cannot say more for the value of this work than that I could find no better data on which to base a theory which I published in 1949,

* *Vala*, or *The Four Zoas* (End of last Night).

and which I venture to think explains some results which Pearson found surprising at the time, though, of course, he published them, and did his best to explain them.

In 1901 the first volume of *Biometrika* was published, partly no doubt because the Royal Society, although it had awarded him its fellowship in 1896 and its Darwin medal in 1898, objected to publishing advances in mathematical and biological knowledge in the same paper! *Biometrika* has not fulfilled what Galton, in its first number (p. 9) stated to be the primary object of biometry, namely 'the discovery of incipient changes in evolution which are too small to be otherwise apparent'. The reason for this failure is simple. The mean rate of increase in tooth length during the evolution of the horse since the Eocene is now known to have been about 4 % per million years. Such evolution could not be detected in a human lifetime. But the aims stated in the editorial introduction, presumably the joint work of Pearson, Weldon and Davenport, were fulfilled. In particular the first number contained a paper by Weldon on variation in snails whose importance he did not live to realize. He found that natural selection in a snail species weeded out extremes, reducing the standard deviation of a metrical character without affecting the mean. We now know that this centripetal selection is very common. Had Weldon lived longer he would presumably have discovered this, and the whole history of biometry would have been very different.

In 1903 Pearson's Department received a grant of £500 from the Drapers' Company, and these grants, at the rate of £500 per year, continued till 1932. In 1903 this sum was worth about £3000 or more to-day, and went partly in the payment of Dr Lee and other computers, partly for instruments, and partly for printing.

I have no idea how Pearson obtained this money. We may be sure that he did not either flatter rich men or promise to improve the national health and intelligence in their lifetimes. Perhaps Galton had the ear of some rich acquaintances. Perhaps too, at that time our ruling classes were less permeated than now with the ferocious contempt for the pursuit of knowledge for its own sake, which is voiced in the Archbishop of Canterbury's sermon of March 24, 1957. To-day it is not hard to get money for research which may have economic, military or hygienic advantages. It is extremely hard to do so for the mere search for truth.

About this time Pearson began the series of papers on human biology for which he is best known in some quarters, and the majority of which, I think, were joint work. Even where his name did not appear on papers, I think our chairman will agree that nothing was published from his laboratory without his *imprimatur*, and some of such work must at least briefly be considered here.

Many of these papers are as fresh to-day as when they were written. To take an example, in my opinion nothing since written on human craniometry has in anyway superseded Pearson's and Davin's great memoir of 1924. Some of this work was, at the time, of inestimable value. I think particularly of the *Treasury of Human Inheritance*. This is still indispensable. Nevertheless, we now know that it is possible to distinguish between conditions (for example, haemophilia, Christmas' disease, Owren's disease, and so on) which were inevitably classed under a single category by the writers of the *Treasury*. The more polemical writings of this period are of less value to-day, as Pearson doubtless realized when he called a series 'Questions of the Day and of the Fray'. The Fray in question was a many-sided contest. On the one hand, Pearson and his colleagues attacked those who underestimated the importance of heredity, including those who exaggerated the harm done by parental alcoholism. But they also attacked those who oversimplified it, including many Mendelians and many eugenists. Other attacks were on statistical data alleged to prove the value of immunization to diseases. These attacks were fully justified. Mental defect is certainly not a Mendelian character. As Pearson and Jaederholm showed, the distribution of intelligence quotients in defectives is the tail of a nearly normal frequency distribution. Forty-three years later we can say a great deal more about it. We can say, for example, that phenylketonuria is a chemically definable character inherited as a Mendelian recessive, and accounting for perhaps 1 % of certifiable mental defect. But the mental defect of phenylketonurics is graded, and a few of them are stupid, but not sufficiently so to be classed as feeble-minded. In fact the diagnosis of phenylketonuria enabled Penrose to dissect the distribution of human intelligence quotients into two very different but still overlapping distributions. The notion that such a dissection is possible was Pearson's first contribution to biometry.

Again, one series of memoirs was entitled *Studies in national deterioration*. This is a polemical title. And it is a fact that as regards most measurable characters the nation has not deteriorated. It may have done so as regards its 'nature' or inborn capacities. I think that if Weldon had lived Pearson would have realized the ubiquity of centripetal selection, and that in fact both the most successful and the least successful members of society were breeding more slowly than those a little below the median. It is, however, easy to be wise after the event. Moreover, Pearson and his colleagues were completely right in one respect. Even if, in spite of his predictions, the nation has improved in some measurable directions, it would have improved more if, say, a million children who were born to unskilled labourers had been born to skilled workers, teachers, and the like.

No such criticism is possible of the mathematical tables which he edited, and in whose computation he played a large part. Their utility was well shown by the fact that 'pirated' editions of them were soon published in America. The subsequent development of statistics is largely based on them. Even the advent of the electronic computer has not yet superseded them. They were published from 1914 to 1934, and the *Tables of the Incomplete Beta-function*, published in Pearson's seventy-eighth year, were his last, but not his least, contribution to science. It appears that no one has yet discovered how to use an electronic computer as efficiently as Pearson used his teams of devoted, painstaking, and remarkably accurate, lady assistants.

In 1911 Galton died, and left funds for the endowment of a Chair of Eugenics, of which Pearson became the first occupant. At last he was able to give up the teaching of applied mathematics to engineers and physicists, and in the next year the present laboratory was begun. Fortunately it was completed by 1914 though it was commandeered as an annexe to the hospital, and he did not get into it till after the war. From 1914 till 1919 he did very little but war work, first for the Board of Trade and later on calculation of trajectories of artillery. When in 1920 the Department of Applied Statistics was formally opened, he was sixty-three years old, and he had, among other things, to develop a new course of lectures and practical work. In 1923 he began a series of papers which combined biometrical and historical research. He was able to measure the skulls of a number of distinguished men and compare them with contemporary portraits. In the course of this work he played the detective, and reconstructed the murder of Lord Darnley, second husband of Queen Mary of Scotland; and his comments on the history of the Reformation in Scotland are well worth reading. To the same period belongs his great life of Galton, which involved much historical research.

I have devoted this lecture entirely to Pearson's published works. If he could hear it I believe his main criticism would be that I have said too little about his fellow workers; for much of his work was in collaboration. He had a wonderful gift for inspiring loyalty in his colleagues, of which more will be said to-day by others. I myself only met him frequently in the last years of his life, and can merely say that he was most gracious to me, though my outlook on many biological questions was very different from his own. He resigned his Chair in 1933, but published one book and at least three scientific papers before his death in 1936.

It remains to say a few words about what Samuel Butler would have called his life after death, the results which are still accruing from his original thought. To begin with, all subsequent statistical work is based on the foundations which he laid. If we sometimes find it more convenient to speak of a variance ratio where he would have used a coefficient of correlation this does not mean that his work on homotyposis is obsolete. If his system of frequency distributions is less used for biological data than he might have hoped, yet Gosset, Fisher and others have found that they describe exactly the distributions of many statistical estimates based on finite samples.

At University College his work is being carried on by three professors.* Under his son the Department of Statistics has become the leading teaching department in that subject in Britain, and the new Biometrika Tables, to take only one example of its work, continue his father's great tradition. Prof. Fisher, who succeeded him as Professor of Eugenics, did very great services to statistics in simplifying and rendering more accurate a number of statistical procedures; and, by the application of methods which owed much to Pearson's great memoir on homotyposis, made agricultural experiment an exact art. To mention only one of Fisher's contributions to eugenics, he established a laboratory for human serology, and his interpretation of the human *Rh* antigens has had a considerable influence on the prevention and cure of a very serious human congenital disease. His controversies with his predecessor were perhaps inevitable between two men each so determined to defend the truth as he saw it. Under Prof. Penrose the Department has swung back towards Pearsonian methods. If I may mention two researches which I believe would have delighted Pearson particularly they are the work of Karn and Penrose on infantile mortality as a function of birth weight, which measured natural selection in man with an accuracy which he would have envied, and that of Penrose on abnormality in the offspring as a function of parental age and maternal parity, a subject which Pearson had broached in his 1914 memoir on the handicapping of the first-born.

My own department of biometry has not been so fortunate. I was Professor of Genetics till 1937. I should not have accepted the Weldon Chair had I not been promised accommodation for Biometrical work. Owing to the war, and for other reasons, this promise was not kept. I have been unable to carry out the duties of this chair adequately. In my opinion the best biometric work of the last twenty years has been carried out by Teissier and Schreider in France, and by Mahalanobis and his colleagues in India. In a few months in India I have been able to start new lines of biometrical research. I think particularly of the work of S. K. Roy, now I hope in press, in which he took up the problem of homotyposis where Pearson left it in 1903. A study of some 60,000 flowers from three different plants (to be compared with Pearson's 4443 capsules from 176 poppy plants) has shown that individual plants not only have their characteristic means, but their characteristic standard devia-

* Perhaps I should also include the Professors of Astronomy and Art History, and those of Engineering.

tions, and that both of these alter in a characteristic manner during a season. I believe that the opportunities for Biometric research are now better in India than in Britain, and for this reason among others I have thought it my duty to migrate there. To quote Karl Pearson's most loved poet

> In Vishnu-land what avatar?

Whatever the fate of Pearsonian biometry in Britain, I believe that it will live and flower in India.

To me at least there seems to be an element of hypocrisy about the present celebrations. I believe that we should be honouring Pearson more effectively if, to take one example out of many possible, we ensured that the College Library possessed copies of all his works, placed where students could consult them, than by making speeches and eating food. I mention this particular example as I have been trying, without the faintest success, to secure such accessibility for at least ten years.

Pearson's work for free thought and the emancipation of women has been successful in this country, if not always as quickly so as he hoped seventy years ago. The history of art is now taught in this College. His work for socialism has not been as successful here as he hoped. Nor would he have approved of many features of the socialistic systems of the Soviet Union and China. Here again I believe his real heirs are to be found in India, where the editor of *Sankhyā*, the Indian Journal of Statistics, is also the principal planner of the approach to socialism under the second Five-Year Plan.

I fully realize that I have not done justice to my subject. The task set me was an impossible one. No one man now alive could do justice to the breadth of Karl Pearson's interests and achievements. But I thank you for joining with me in celebrating the memory of this great man.

RONALD AYLMER FISHER
1890–1962

RONALD AYLMER FISHER, 1890–1962

By M. G. KENDALL

Ronald Aylmer Fisher was born in 1890 at East Finchley, the youngest of seven children (There was an eighth, Fisher's twin, who did not survive.) His family were for the most part in business, his father being an auctioneer, his maternal grandfather a solicitor, but several of them struck out original lines for themselves and their intellectual interests were strong. Fisher himself, even as a child, began to show the precocity of the typical mathematician, and his gifts were encouraged by the excellence of the teaching he received in his school at Stanmore, and later at Harrow. From an early age his eyesight was weak. Some of his disciples have attributed his ability to analyse complicated situations in his head to the fact that he was prohibited, in his youth, from reading by artificial light. But this I only half believe. Powers such as Fisher's ability to visualize complicated geometrical relations are born, not made, although they may be strengthened by circumstance.

When Fisher left Harrow the family finances were not in a very strong state. Fortunately he won an entrance scholarship to Gonville and Caius College, Cambridge, where he was an undergraduate from 1909 to 1912. He graduated as a Wrangler in 1912, with distinction in the optics papers of what was then schedule B. A fourth year was spent in reading mathematical physics under Jeans and Stratton. His scientific interests at Cambridge, however, also included biometry and genetics and little as he might, in later life, have been inclined to acknowledge the fact, he must have been influenced by Karl Pearson's papers on the mathematics of evolution.

I think it was by way of the combination of observations in astronomy that his interests in probability distributions were first aroused. His first paper (1912) on an absolute criterion for fitting frequency curves, contains the germ of his later work on likelihood. He used to read *Biometrika* over lunch and before long a paper by H. E. Soper on the correlation coefficient prompted him to attempt some work of his own on distribution theory. Shortly before the First World War broke out he sent Pearson a rough draft of his famous paper (appearing in 1915) in which he obtained the distribution of the correlation coefficient in normal samples.

He began his working life in the office of the Mercantile and General Investment Company from 1913–15 but this was clearly only a matter of earning his bread, not a vocational call. In the First World War his eyesight debarred him from military service and from 1915 to 1919 he spent four years teaching in public schools. He cannot have been very happy in this work but the years were not wasted. His ideas were taking shape. He himself was maturing. His 1915 paper made him many acquaintances in the statistical world, notably Gosset, with whom he corresponded regularly, until the latter's death in 1937. And in 1917 he married Ruth Guinness, a doctor's daughter, who was to bear him nine children.

Despite his slender list of publications in 1919, he already had a reputation in quarters where it mattered, and soon had to make a major decision. Pearson offered him a post at University College. At almost the same time Sir John Russell offered him a post at Rothamsted. Fisher chose Rothamsted on the grounds that he would have greater opportunities for original research. There was never a happier appointment. Over the next fourteen years Fisher established Rothamsted as one of the Holy Places of the statistical world. He himself flourished in the agricultural environment to a degree which could hardly have been predicted, deriving inspiration from every branch of the Experimental Station's work and contributing to studies of crop yields, rainfall, bacterial counts, genetics and, above all, of field trials, while developing the theoretical side of statistics at an amazing rate. He founded a unit, which, under his friend and successor Frank Yates, remains one of the great statistical training centres of the world.

When Fisher started work on his 1915 paper he had not, apparently, seen Gosset's *Biometrika* paper of 1908 but when he did, new lines of development in distribution theory opened up immediately. In 1920 he published the distribution of the mean deviation in normal samples (obtained by one of his geometrical arguments), following it by a brilliant series of papers in which he obtained the distribution of regression coefficients, correlation ratios and multiple regression coefficients in the null case (1922), 'Student's' t (1925), partial correlation coefficients (1924) and the variance ratio (equivalent to Fisher's z) (1924). Concurrently he set out his ideas on estimation and inference in two basic papers (1922, 1925).

The twenty years from 1920 to 1940 were a period of astonishing productivity. Fisher's bibliography for that period contains about 120 titles, apart from three important books and a set of tables. For a long time it seemed that there was no branch of theory which he did not touch, and none which he touched that he did not materially advance. Some of his work, on analysis of variance and on experimental design, flowed in a smooth stream of development. But every now and then he would branch off on to new topics, some of them never again to be taken up. For example the paper with Tippett (1928) on extreme values, the paper (1929) on tests of significance in harmonic analysis, the paper (1929) on sampling cumulants of k-statistics, are typical Fisherian pioneer work. His pre-eminence among his contemporaries became a commonplace. Not to refer to some of his work in any theoretical paper written about this time was almost a mark of retarded development; nor did such a reference often have to be forced.

His distinction was soon officially recognized. His Sc.D. dates from 1926. He was a Fellow of Gonville and Caius from 1921 to 1927 and again from 1943 until his death (he was President of the College from 1956 to 1959). He was elected F.R.S. in 1934, received the Darwin medal in 1948 and the Copley medal in 1955 from the Royal Society, a gold medal from the Royal Statistical Society in 1945, and was President of the Society from 1952 to 1954. Honours came to him in abundance and in 1952 he was created a Knight, which was a source of pleasure to all theoretical statisticians as a well-merited tribute to their leader.

By 1925 Fisher had sketched out the essentials of his ideas on statistical inference concerning likelihood and estimation (fiducial inference was still to come). In that year Oliver and Boyd published the first edition of a book which made statistical history, *Statistical Methods for Research Workers*. It is not an easy book. Somebody once said that no student should attempt to read it unless he had read it before. Fisher had no gifts of exposition, even of his own ideas, and rarely set out explicitly the assumptions on which he was working. But no one who has wrestled through this book has doubted that the effort was worth it. It remains, after nearly 40 years, one of the standard texts of our subject.

In the meantime Fisher's relations with Karl Pearson and his group had degenerated almost into a personal feud. At precisely what point the trouble began it is difficult to say. I have heard several stories but it hardly seems worth while trying to verify them. The personalities being what they were, a breach sooner or later was inevitable. As things turned out, it happened sooner, and the two were on bad terms not long after 1920. Fisher never again published in *Biometrika*. He himself told me that when he was writing *Statistical Methods for Research Workers* he applied to Pearson for permission to reproduce Elderton's table of chi-squared and that it was refused. This was perhaps not simply a personal matter because the hard struggle which Pearson had for long experienced in obtaining funds for printing and publishing statistical tables had made him most unwilling to grant anyone permission to reproduce. He was afraid of the effect on sales of his *Tables for Statisticians and Biometricians* on which he relied to secure money for further table publication. It seems, however, to have been this refusal which first directed Fisher's thoughts towards the alternative form of tabulation with quantiles as argument, a form which he subsequently adopted for all his tables and which has become common practice.

What is so surprising is the intensity of feeling which was generated on Fisher's side. Karl, after all, was 65 years old in 1922, still vigorous and productive but no longer generating new statistical ideas. Fisher could well have been content to let his theoretical contributions be their own revenge. His tanh transformation for the correlation coefficient took most of the point out of the Cooperative Study, produced by a group at University College in 1917; his claims concerning the efficiency of curve fitting and his corrections to Pearson's notions about the degrees of freedom of χ^2 in contingency tables were accepted and upheld; he solved a whole field of distribution problems which were beyond Pearson's capacity, including the sampling distribution (or, at least, the moments) of Pearson's measure of skewness and kurtosis. This would have been enough for most people. But it was not enough for Fisher.

He may have thought he had been badly treated by a few more senior authorities. He had some right to complain—at least, he did complain—when Bowley, moving a vote of thanks at the paper which Fisher read to the Royal Statistical Society in 1935, seemed to be making fun of some of the exposition in a rather unkind way. But it is remarkable how, even at the height of his achievement and reputation, when Karl had been dead for twenty years, Fisher wrote as if the wounds were still smarting. In his last book (1956) he gives a sketch of K. P. and renews his attack 'on the whole corpus of Pearsonian writings'. And the strangest thing of all is that the faults of which he accused Pearson were ones which can be detected in Fisher himself. 'He reminds me', said a Dutch colleague, 'of one of those artists who, whenever they paint a portrait, paint a self-portrait.'

Egon Pearson, with admirable restraint, refused to tread on the coat which was so often trailed around the statistical world. His collaboration with Jerzy Neyman had started in 1926 or 1927, but it was not until after the latter had received a permanent appointment in Egon's new Department of Statistics at University College, in 1935, that fresh fuel was added to the flames—a result perhaps inevitable when Fisher, Neyman and Pearson felt compelled, under the same roof, to lecture along the lines they believed to be right. Neyman had been developing a theory of confidence intervals in Poland about the same time that Fisher (1930) put forward his ideas on fiducial inference, and Fisher thought, or professed to think, that Neyman had taken his ideas—and spoilt them. Neyman replied to Fisher's criticisms of K. P. in far from deferential terms. And so the feud moved on from likelihood versus Bayes to fiducial versus confidence intervals and took on a new lease of life. Neyman's transfer to Berkeley, California, removed some of the tension in London, but the two went on for the next twenty-five years firing everything they had at each other's positions without making the least impression.

If we strip off the controversy, the irrelevancies and the mathematics from these Titanic arguments we are nevertheless left with a fundamental difference of approach to scientific inference which, to understand Fisher's mode of thought, it is worth while setting out in some detail. In one of his earliest papers (1922) Fisher repudiated Bayes's postulate and propounded the likelihood principle instead. It appears to me that, at this point, his ideas were not very well thought out. Certainly his exposition of them was obscure. But, in retrospect, it becomes plain that he was thinking of a probability function $f(x, \theta)$ in two different ways: as the probability distribution of x for given θ, and as giving some sort of permissible range of θ for an observed x. To this latter he gave the name of 'fiducial probability distribution' (later to be known as a fiducial distribution) and in doing so began a long train of confusion; for it is not a probability distribution to anyone who rejects Bayes's approach, and indeed, may not be a distribution of anything. Fisher nevertheless manipulated it as if it were, and thereafter maintained an attitude of rather contemptuous surprise towards anyone who was perverse enough to fail in understanding his argument.

Now it is quite possible to give formal expression to one's intuitive feelings about how likely certain values of θ are when a given sample is observed. But to do so one requires the explicit statement of certain postulates to replace those which are rejected. Fisher never stated these basic assumptions and, to maintain his position, had to impose what looked like very dogmatic assertions, for example, that only sufficient estimators could give fiducial intervals. It was not surprising that even his supporters misunderstood him. And to those statisticians who adhered to inference based solely on a frequency probability theory, the fiducial approach was not merely unnecessary but misconceived. (It also failed to solve estimation problems except in the very limited class of case where sufficient estimators exist.) The position on both sides has been restated *ad nauseam*, without much attempt at reconciliation or, as I think, without an explicit recognition of the real point, which is that a man's attitude towards inference, like his attitude towards religion, is determined by his emotional make-up, not by reason or by mathematics.

More recently the ground of controversy has moved still farther, though the basic point remains the same. Under the influence of Neyman and Wald (who also came under the Fisherian anathema, though posthumously), there has been a strong movement in the U.S.A. to regard inference as a branch of decision theory. Fisher would have maintained (and in my opinion rightly) that inference in science is not a matter of decision, and that, in any case, criteria for choice in decision based on pay-offs of one kind or another are not available. This, broadly speaking, is the English as against the American point of view. We shall see a lot of water under the bridge before this conflict is resolved, if indeed it can ever be resolved. Not wishing to be controversial about it, I propound the thesis that some such difference of attitude is inevitable between countries where what a man does is more important than what he thinks, and those where what he thinks is more important than what he does.

Let us return to those of Fisher's achievements which have a universally accepted value. The year 1928 saw another triumph. For some time, having given in 1922 the distribution of the multiple correlation coefficient in the null case, he had been thinking about the problem of finding the distribution in the non-null (normal) case. His geometrical powers again came into play and, by an argument which few of his statistical contemporaries were able to understand, he derived the distribution in closed form.

In 1929 Fisher published the paper on sampling moments. This has always seemed to me the most remarkable paper he ever wrote. It forms the basis of most subsequent work on the subject. Thiele had previously introduced semivariants, but defined his sample values by formulae which were structurally the same as those obtaining for parent values; and hence missed the essential simplification of k-statistics. Craig also had put forward the suggestion that new sample functions were required to replace the sample-moments. It was left to Fisher to define those new functions as the sample symmetric functions whose mean values were the cumulants, and to show how their cumulants could be obtained by combinatorial methods.

How Fisher thought of these results is a mystery. It was a mystery to Wishart, who was closely working with him at the time; and Fisher himself was never able to explain it to me in later years. What actually happened (so Wishart told me) was that Fisher was on a train journey when he thought of the combinatorial rules which express the cumulants of k-statistics in terms of parent cumulants. He put them to Wishart (without extensive notes) on reaching his destination. A group at Rothamsted, including Wishart and Hotelling, then proceeded to work out a number of formulae by the combinatorial method. When Fisher came to write the paper, however, he had forgotten how he arrived at the basic results. In some peculiar intuitive way he must have perceived the essential structure of k-statistics and the reason why they lead to his bi-partitional functions. But afterwards, so far as I was able to discover, the illumination had faded and his 1929 paper concludes with an obscure section which is no proof at all. Authors occasionally have a mystical feeling, on looking back, that some of their work was written through them, not by them. If this were held to be true of mathematicians, the 1929 paper would be a good case for consideration in support of the belief.

By 1930 Fisher's methods were beginning to inspire other workers in multivariate analysis. Wishart, in 1931, generalized from the two-dimensional case to obtain the distribution of dispersions in multivariate normal variation. Hotelling in 1932 produced the T-distribution and in 1936 put canonical correlation analysis on a firm theoretical footing. Wilks in 1932 generalized variance analysis to the multivariate case. The Indian School, Mahalanobis, R. C. Bose and S. N. Roy, were obtaining significantly important results. The pace was quickening, but the maestro was still in the lead. In 1939 he produced his last great contribution to distribution theory—the joint distribution of the characteristic roots of a dispersion matrix. His paper is typically condensed and his proof has one or two important lacunae, but any doubts about the correctness of the result were put at rest by a matrix proof by Hsu and independent proofs by Roy and by Mood.

In some of his earlier papers, especially the one (1923) on wheat yields at Rothamsted, Fisher had clearly perceived the importance of orthogonal components in the theory of explanatory variables. The simplicity of the fundamental idea, and what Fisher built from it, are wonderful examples of the creative mind at work. He was led, following the work of Lexis, Yule and Gosset, to the analysis of variance, and, in turn, to his great work on experimental design. The first edition of *The Design of Experiments* was published in 1935. It is another landmark in the development of statistical science. The experiment on tea-tasting with which it opens has, I suppose, become the best-known experiment in the world, not the less remarkable because, as he once told me, Fisher never performed it. There can be few books, if any at all, which have done so much to raise standards in scientific technology.

In 1933 Karl Pearson retired and his department at University College was, in effect, split into two. Fisher left Rothamsted to become professor of Eugenics. Egon Pearson took charge of statistics and became Professor in 1935. (It is remarkable that the greatest statistician in the world never held a chair in statistics.) Fisher's flow of publications was uninterrupted. A further series of papers appeared on the logic of inductive inference. In

1936 he published the first of his papers on discriminant analysis. His powers in combinatorial algebra found expression in papers on solutions of problems in enumerating Latin squares and incomplete blocks. Even the Second World War failed to check him, and at one stage he had a row with the police for refusing to allow him entrance to his own laboratory, which was in a danger area.

Finally, in 1943, he became professor of Genetics at Cambridge. His work in theoretical statistics was practically complete, though he often returned to it to argue some point of inference with a heretic. His work in genetics continued. I am not competent to discuss it but I am told that, of itself, it would have been enough to create a major reputation.

There should, perhaps, be mentioned two of his later publications, the pamphlet on *Smoking and Cancer* (1959) and the book on *Statistical Methods and Scientific Inference* (1956). Personally I wish he had never written either. Whatever the rights and wrongs of the cancer-smoking controversy, Fisher's intervention did no good to statistics in general or to this subject in particular, and the less said about it the better. His book on Scientific Inference, apart from being marred by the usual provocative attacks on his opponents, added very little to what had already been said. He still possessed the magic touch, as witness the paper (1953) on dispersion on a sphere; but if we had to sacrifice any of his writings, these two books would have a strong claim to priority.

Somehow, one never thought of Fisher as in retirement, but eventually he laid down his professorial duties and, to the general surprise, decided to live in Australia. His death at Adelaide in 1962, due to embolism following some complicated intestinal trouble, was a shock, for he had always enjoyed good health and showed no signs of flagging even at the age of 70. But I think it was such a death as he himself would have chosen. One cannot imagine that dynamic personality slowly disintegrating.

The portrait by Mrs Tintner at the beginning of this article reproduces faithfully the jaunty appearance of his later years. In character he was, let us admit it, a difficult man. Among his wide circle of admirers he accepted homage with fair grace. With most other people he seemed to feel that his genius entitled him to more social indulgence than they were willing to concede; in short he could be, and not infrequently was, gratuitously rude. In his written work he rarely acknowledged any kind of indebtedness to other statisticians and mentioned them, if at all, only to correct or to castigate. In fact, to win written approbation from him in his later work one had to have been dead for some time. And yet, some of his unpublished correspondence corrects the impression of petulant intolerance which so often crept into his scientific publications. He argued randomization versus systematic design in experimentation with Gosset over a period of years with even temper. And I can personally testify to the fact that, to those who did not cross him, he was affable enough.

It may be many years before an objective critical appraisal can be made of Fisher's work in its entirety. But certain things are incontestable. He raised his subject to a new level of achievement. His mathematical work on distribution theory is unrivalled for its scope and power. He created a new science of experimental design. His work on scientific inference, though possibly of less permanent value, at least stimulated a good deal of the thinking which has been done on that subject throughout the world. As with all true statisticians, his theoretical work went hand in hand with practical enquiries of the most far-reaching importance. He was a leavening factor such as we are unlikely to see again. With all his human failings, he was a truly great statistician.

In compiling the bibliography I have had valuable help from Mr J. A. King, Librarian of the Royal Statistical Society, Mr Michael Healy of the Rothamsted Experimental Station, Mr Douglas Grant of Messrs Oliver and Boyd and Professor J. H. Bennett and Professor E. A. Cornish of the University of Adelaide. It may be of interest to record in this connexion that a complete edition of Fisher's work is under consideration.

The portrait by Leontine Tintner is in the possession of the Department of Statistics of the University of North Carolina, and I am indebted to her for permission to reproduce it and to Dr Gertrude M. Cox for a photograph of it.

M. G. KENDALL

Bibliography of R. A. Fisher

1. *Books*

Statistical Methods for Research Workers (1925). Thirteenth edition, 1958. Edinburgh: Oliver and Boyd. (Also published in French, German, Italian, Japanese and Spanish.)

The Genetical Theory of Natural Selection (1930). Oxford: Clarendon Press. Second edition, 1958. New York: Dover Publications.

The Design of Experiments (1935). Seventh edition, 1960. Edinburgh: Oliver and Boyd.

Statistical Tables for Biological, Agricultural and Medical Research (1938) (with F. Yates). Sixth edition, 1963. Edinburgh: Oliver and Boyd.

The Theory of Inbreeding (1949). Edinburgh: Oliver and Boyd.

Statistical Methods and Scientific Inference (1956). Second edition, 1959. Edinburgh: Oliver and Boyd.

2. *Scientific papers, reviews, letters to journals, etc.*

1912 On an absolute criterion for fitting frequency curves. *Messeng. Math.* **41**, 155–60.

1913 Applications of vector analysis to geometry. *Messeng. Math.* **42**, 161–78.

1914 Some hopes of a eugenist. *Eugen. Rev.* **5**, 309–15.

1915 Cuénot on preadaptation: A criticism. (With C. S. Stock.) *Eugen. Rev.* **7**, 46–61.
Frequency distribution of the values of the correlation coefficient in samples from an indefinitely large population. *Biometrika*, **10**, 507–21.
The evolution of sexual preference. *Eugen. Rev.* **7**, 184–92.

1916 Biometrika. *Eugen. Rev.* **8**, 62–4.

1917 Positive eugenics. *Eugen. Rev.* **9**, 206–12.

1918 The correlation between relatives on the supposition of Mendelian inheritance. *Trans. Roy. Soc. Edinb.* **52**, 399–433.

1919 The causes of human variability. *Eugen. Rev.* **10**, 213–20.
The genesis of twins. *Genetics*, **4**, 489–99.

1920 A mathematical examination of the methods of determining the accuracy of an observation by the mean error, and by the mean square error. *Mon. Not. R. Astr. Soc.* **80**, 758–70.

1921 Some remarks on the methods formulated in a recent article on 'the quantitative analysis of plant growth'. *Ann. Appl. Biol.* **7**, 367–72.
Studies in crop variation. I. An examination of the yield of dressed grain from Broadbalk. *J. Agric. Sci.* **11**, 107–35.
On the probable error of a coefficient of correlation deduced from a small sample. *Metron*, **1** (4), 1–32.

1922 Darwinian evolution by mutations. *Eugen. Rev.* **14**, 31–4.
On the interpretation of χ^2 from contingency tables, and the calculation of *P*. *J. R. Statist. Soc.* **85**, 87–94.
On the mathematical foundations of theoretical statistics. *Phil. Trans.* A, **222**, 309–68.
New data on the genesis of twins. *Eugen. Rev.* **14**, 115–17.

1922 The evolution of the conscience in civilised communities: In special relation to sexual vices. *Eugen. Rev.* **14**, 190–3.

On the dominance ratio. *Proc. Roy. Soc. Edinb.* **42**, 321–41.

Biological studies of *Aphis rumicis* Linn: reproduction on varieties of *Vicia faba*. (J. Davidson; with a statistical appendix by R. A. F.) *Ann. Appl. Biol.* **9**, 135–45.

The goodness of fit of regression formulae and the distribution of regression coefficients. *J. R. Statist. Soc.* **85**, 597–612.

The correlation of weekly rainfall. (With Winifred A. MacKenzie.) *Quart. J. R. Met. Soc.* **48**, 234–45.

The systematic location of genes by means of crossover observations. *Amer. Nat.* **56**, 406–11.

The accuracy of the plating method of estimating the density of bacterial populations, with particular reference to the use of Thornton's agar medium with soil samples. (With H. G. Thornton and W. A. MacKenzie.) *Ann. Appl. Biol.* **9**, 325–59.

1923 Statistical tests of agreement between observation and hypothesis. *Economica*, **3**, 139–47.

Studies in crop variation. II. The manurial response of different potato varieties. (With W. A. MacKenzie.) *J. Agric. Sci.* **13**, 311–20.

Note on Dr Burnside's recent paper on errors of observation. *Proc. Camb. Phil. Soc.* **21**, 655–8.

1924 The theory of the mechanical analysis of sediments by means of the automatic balance. (With S. Odén.) *Proc. Roy. Soc. Edinb.* **44**, 98–115.

The conditions under which χ^2 measures the discrepancy between observation and hypothesis. *J. R. Statist. Soc.* **87**, 442–50.

The elimination of mental defect. *Eugen. Rev.* **16**, 114–16.

A method of scoring coincidences in tests with playing cards. *Proc. Soc. Psych. Res., Lond.*, **34**, 181–5.

The influence of rainfall on the yield of wheat at Rothamsted. *Phil. Trans.* B, **213**, 89–142.

The biometrical study of heredity. *Eugen. Rev.* **16**, 189–210.

The distribution of the partial correlation coefficient. *Metron*, **3** (3–4), 329–32.

On a distribution yielding the error functions of several well known statistics. *Proc. Int. Congr. Math., Toronto* (1924), 805–13. (Published 1928.)

1925 Note on the numerical evaluation of a Bessel function derivative. (With P. R. Ansell.) *Proc. Lond. Math. Soc.* **24**, 54–7.

Theory of statistical estimation. *Proc. Camb. Phil. Soc.* **22**, 700–25.

Sur la solution de l'équation intégrale de M. V. Romanovsky. *C.R. Acad. Sci., Paris*, **181**, 88–9.

The resemblance between twins, a statistical examination of Lauterbach's measurements. *Genetics*, **10**, 569–79.

Applications of 'Student's' distribution. *Metron*, **5** (3), 90–104.

Expansion of 'Student's' integral in powers of n^{-1}. *Metron*, **5** (3), 109–12.

1926 On the random sequence. *Quart. J. R. Met. Soc.* **52**, 250.

Bayes's theorem and the fourfold table. *Eugen. Rev.* **18**, 32–3.

Eugenics: can it solve the problem of decay of civilizations? *Eugen. Rev.* **18**, 128–36.

On the capillary forces in an ideal soil; correction of formulae given by W. B. Haines. *J. Agric. Sci.* **16**, 492–505.

Periodical health surveys. *J. State Med.* **34**, 446–9.

The arrangements of field experiments. *J. Min. Agric.* **33**, 503–13.

Variability of species. (With E. B. Ford.) *Nature, Lond.*, **118**, 515–16.

1927 On the existence of daily changes in the bacterial numbers in American soil. (With H. G. Thornton.) *Soil Sci.* **23**, 253–9.

The actuarial treatment of official birth records. *Eugen. Rev.* **19**, 103–8.

Biological measurements: Report of Committee appointed to draw up recommendations for the taking and presentation of biological measurements, and to bring such before persons or bodies concerned. *Rep. Brit. Ass.*, Sect. D, Leeds (1927), 1–13.

On the distribution of the error of an interpolated value, and on the construction of tables. (With J. Wishart.) *Proc. Camb. Phil. Soc.* **23**, 912–21.

Studies in crop variation. IV. The experimental determination of the value of top dressings with cereals. (With T. Eden.) *J. Agric. Sci.* **17**, 548–62.

1927 The effect of family allowances on population. (With W. H. Beveridge, V. H. Mottram, H. N. Brailsford, J. L. Cohen and J. Murray.) Conference on family allowances. *Proc. Lond. Sch. Econ.* (1927), 7–11.

On some objections to mimicry theory: statistical and genetic. *Trans. Ent. Soc. Lond.* **75**, 269–78.

1928 The possible modification of the response of the wild type to recurrent mutations. *Amer. Nat.* **62**, 115–26.

Limiting forms of the frequency distribution of the largest or smallest member of a sample. (With L. H. C. Tippett.) *Proc. Camb. Phil. Soc.* **24**, 180–90.

Further note on the capillary forces in an ideal soil. *J. Agric. Sci.* **18**, 406–10.

Income-tax rebates: The birth rate and our future policy. *Eugen. Rev.* **20**, 79–81.

The estimation of linkage from the offspring of selfed heterozygotes. (With Bhai Balmukand.) *J. Genet.* **20**, 79–92.

The differential birth rate: New light on causes from American figures. *Eugen. Rev.* **20**, 183–4.

Two further notes on the origin of dominance. *Amer. Nat.* **62**, 571–4.

Maximum- and minimum-correlation tables in comparative climatology. (With T. N. Hoblyn.) *Geogr. Ann. Stockh.* **3**, 267–81.

The general sampling distribution of the multiple correlation coefficient. *Proc. Roy. Soc.* A, **121**, 654–73.

On a property connecting the χ^2 measure of discrepancy with the method of maximum likelihood. *Int. Congr. Math.*, VI, Bologna (1928), 95–100.

Triplet children in Great Britain and Ireland. *Proc. Roy. Soc.* B, **102**, 286–311.

Correlation coefficients in meteorology. *Nature, Lond.*, **121**, 712.

The variability of species in the Lepidoptera, with reference to abundance and sex. (With E. B. Ford; app. by E. H. Poulton.) *Trans. Ent. Soc. Lond.* **76**, 367–84.

1929 A preliminary note on the effect of sodium silicate in increasing the yield of barley. *J. Agric. Sci.* **19**, 132–9.

Studies in crop variation. VI. Experiments on the response of the potato to potash and nitrogen. (With T. Eden.) *J. Agric. Sci.* **19**, 201–13.

'The balance of births and deaths' by R. Kuczynski; 'The shadow of the world's future' by Sir G. H. Knibbs. (Reviews.) *Nature, Lond.*, **123**, 357–8.

The over-production of food. *The Realist*, **1**, 45–60.

Statistics and biological research. *Nature, Lond.*, **124**, 266–7.

Tests of significance in harmonic analysis. *Proc. Roy. Soc.* A, **125**, 54–9.

The evolution of dominance; reply to Professor Sewall Wright. *Amer. Nat.* **63**, 553–6.

The sieve of Eratosthenes. *Math. Gaz.* **14**, 564–6.

The statistical method in psychical research. *Proc. Soc. Psych. Res. Lond.* **39**, 189–92.

Moments and product moments of sampling distributions. *Proc. Lond. Math. Soc.* (2), **30**, 199–238.

1930 The distribution of gene ratios for rare mutations. *Proc. Roy. Soc. Edinb.* **50**, 205–20.

Mortality amongst plants and its bearing on natural selection. *Nature, Lond.*, **125**, 972–3.

Biometry and evolution. *Nature, Lond.*, **126**, 246–7.

Note on a tri-colour (mosaic) mouse. *J. Genet.* **23**, 77–81.

The evolution of dominance in certain polymorphic species. *Amer. Nat.* **64**, 385–406.

Inverse probability. *Proc. Camb. Phil. Soc.* **26**, 528–35.

The arrangement of field experiments and the statistical reduction of the results. (With J. Wishart.) *Imp. Bur. Soil Sci. Tech. Comm.* no. 10.

The moments of the distribution for normal samples of measures of departure from normality. *Proc. Roy. Soc.* A, **130**, 16–28.

Genetics, mathematics, and natural selection. *Nature, Lond.*, **126**, 805–6.

1931 Pasteurised and raw milk. (With S. Bartlett.) *Nature, Lond.*, **127**, 591–2.

The evolution of dominance. *Biol. Rev.* **6**, 345–68.

The biological effects of family allowances. *Family Endowment Chronicle*, **1** (3), 21–5.

The sampling error of estimated deviates, together with other illustrations of the properties and applications of the integrals and derivatives of the normal error function. *Brit. Assoc. Adv. Sci., Math. Tables*, **1**, 26–35.

The derivation of the pattern formulae of two-way partitions from those of simpler patterns. (With J. Wishart.) *Proc. Lond. Math. Soc.* (2), **33**, 195–208.

1932 A new series of allelemorphs in mice. (L. C. Dunn with an appended note by R.A.F.) *Nature, Lond.*, **129**, 130.

The genetical interpretation of statistics of the third degree in the study of quantitative inheritance. (With F. R. Immer and O. Tedin.) *Genetics*, **17**, 107–24.

Family allowances in the contemporary economic situation. *Eugen. Rev.* **24**, 87–95.

The social selection of human fertility. *The Herbert Spencer Lecture*, 1932. Oxford: Clarendon Press.

Inverse probability and the use of likelihood. *Proc. Camb. Phil. Soc.* **28**, 257–61.

The bearing of genetics on theories of evolution. *Science Progr.* **27**, 273–87.

Inheritance of acquired characters. *Nature, Lond.*, **130**, 579.

The evolutionary modification of genetic phenomena. 6th *Int. Congr. Genet. Proc.* **1**, 165–72.

1933 Protective adaptations of animals—especially insects. *Proc. Ent. Soc. Lond.* **7**, 87–9.

Number of Mendelian factors in quantitative inheritance. *Nature, Lond.*, **131**, 400–1.

Statistical tables. *Nature, Lond.*, **131**, 893–4.

Selection in the production of the ever-sporting stocks. *Ann. Bot.* **47**, 727–33.

The concepts of inverse probability and fiducial probability referring to unknown parameters. *Proc. Roy. Soc.* A, **139**, 343–8.

On the evidence against the chemical induction of melanism in Lepidoptera. *Proc. Roy. Soc.* B, **112**, 407–16.

Mathematics of inheritance. *Nature, Lond.*, **132**, 1012.

Some modern statistical methods: their application to the solution of herring race problems. (H. J. Buchanan-Wollaston with a foreword by R.A.F.) *J. Cons. int. Explor. Mer*, **8**, 7–8.

The contributions of Rothamsted to the development of the science of statistics. *Annual Report, Rothamsted Exp. Stn* (1933), 43–50.

1934 Indeterminism and natural selection. *Philos. Sci.* **1**, 99–117.

Report of enquiry into the children of mental defectives. *Report of Departmental Comm. on Sterilisation*, 60–74.

Adaptation and mutations. *School Sci. Rev.* no. 59, 294–301.

Crossing-over in the land snail *Cepaea nemoralis* L. (With C. Diver.) *Nature, Lond.*, **133**, 834–5.

Professor Wright on the theory of dominance. *Amer. Nat.* **68**, 370–4.

The numbers of bacterial cells in field soils, as estimated by the ratio method. (H. G. Thornton and P. H. H. Gray; with appendix by R.A.F.) *Proc. Roy. Soc.* B, **115**, 540–2.

Probability, likelihood and quantity of information in the logic of uncertain inference. *Proc. Roy. Soc.* A, **146**, 1–8.

Crest and hernia in fowls due to a single gene without dominance. *Science*, **80**, 288–9.

The 6 × 6 Latin squares (With F. Yates.) *Proc. Camb. Phil. Soc.* **30**, 492–507.

The effect of methods of ascertainment upon the estimation of frequencies. *Ann. Eugen.* **6**, 13–25.

The amount of information supplied by records of families as a function of the linkage in the population sampled. *Ann. Eugen.* **6**, 66–70.

The use of simultaneous estimation in the evaluation of linkage. *Ann. Eugen.* **6**, 71–6.

Randomisation and an old enigma of card play. *Math. Gaz.* **18**, 294–7.

Two new properties of mathematical likelihood. *Proc. Roy. Soc.* A, **144**, 285–307.

The 'Viceroy' (*Basilarchia archippus* Cram.) mistaken for its model the 'Monarch' (*Danaus plexippus* Linn.). *Proc. Roy. Ent. Soc.* **9**, 97.

1935 The logic of inductive inference. (With discussion.) *J. R. Statist. Soc.* **98**, 39–82.

Some results of an experiment on dominance in poultry, with special reference to polydactyly. *Proc. Linn. Soc. Lond.* **147**, 71–88.

The calculation of the dosage–mortality curve. (C. I. Bliss; with appendix: The case of zero survivers, by R.A.F.) *Ann. Appl. Biol.* **22**, 164–6.

Eugenics, academic and practical. *Eugen. Rev.* **27**, 95–100.

On the selective consequences of East's (1927) theory of heterostylism in *Lythrum*. *J. Genet.* **30**, 369–82.

The detection of linkage with 'dominant' abnormalities. *Ann. Eugen.* **6**, 187–201.

The inheritance of fertility. Dr Wagner-Manslau's tables. (Ed. note.) *Ann. Eugen.* **6**, 225–51.

Dominance in poultry. *Phil. Trans.* B, **225**, 197–226.

Statistical tests. *Nature, Lond.*, **136**, 474.

The sheltering of lethals. *Amer. Nat.* **69**, 446–55.

1935 The mathematical distributions used in the common tests of significance. *Econometrica*, **3**, 353–65.
The detection of linkage with recessive abnormalities. *Ann. Eugen.* **6**, 339–51.
The fiducial argument in statistical inference. *Ann. Eugen.* **6**, 391–8.
Linkage studies and the prognosis of hereditary ailments. *Int. Congr. Life Assur. Med., Lond.* (1935).

1936 The 'coefficient of racial likeness' and the future of craniometry. *J. R. Anthrop. Inst.* **66**, 57–63.
Verification in mice of the possibility of more than fifty per cent recombination. (With K. Mather.) *Nature, Lond.*, **137**, 362.
Heterogeneity of linkage data for Friedreich's ataxia and the spontaneous antigens. *Ann. Eugen.* **7**, 17–21.
Has Mendel's work been rediscovered? *Ann. Sci.* **1**, 115–30.
Income-tax and birth-rate: family allowances. *The Times*, 30 April 1936.
Tests of significance applied to Haldane's data on partial sex linkage. *Ann. Eugen.* **7**, 87–104.
The use of multiple measurements in taxonomic problems. *Ann. Eugen.* **7**, 179–88.
A test of the supposed precision of systematic arrangements. (With S. Barbacki.) *Ann. Eugen.* **7**, 189–93.
The measurement of selective intensity. (A discussion on the present state of the theory of natural selection.) *Proc. Roy. Soc.* B, **121**, 58–62.
A linkage test with mice. (With K. Mather.) *Ann. Eugen.* **7**, 265–80.
Uncertain inference. *Proc. Amer. Acad. Arts Sci.* **71**, 245–58.
The half-drill strip system agricultural experiments. *Nature, Lond.*, **138**, 1101.
Curve fitting. *Nature, Lond.*, **138**, 934.

1937 The relation between variability and abundance shown by the measurements of the eggs of British nesting birds. *Proc. Roy. Soc.* B, **122**, 1–26.
Professor Karl Pearson and the method of moments. *Ann. Eugen.* **7**, 303–18.
The comparison of variability in populations having unequal means. An example of the analysis of covariance with multiple dependent and independent variates. (With B. B. Day.) *Ann. Eugen.* **7**, 333–48.
The wave of advance of advantageous genes. *Ann. Eugen.* **7**, 355–69.
On a point raised by M. S. Bartlett on fiducial probability. *Ann. Eugen.* **7**, 370–5.
W. F. Sheppard—The character of Sheppard's work. *Ann. Eugen.* **8**, 11–12.
Inheritance in man: Boas's data studied by the method of analysis of variance. (With H. Gray.) *Ann. Eugen.* **8**, 74–93.
Moments and cumulants in the specification of distributions. (With E. A. Cornish.) *Rev. Inst. Int. Statist.* **5**, 307–22.

1938 Presidential address. *Proc. Indian Statist. Conf.* (1938), *Sankhyā*, **4**, 14–17.
Statistical theory of estimation. *Calcutta University Readership Lectures.* Calcutta University Press.
Dominance in poultry—feathered feet, rose comb, internal pigment and pile. *Proc. Roy. Soc.* B, **125**, 25–48.
Local varieties and species—natural selection. *The Times*, 3 May 1938.
'The design and analysis of factorial experiments' by F. Yates. (Review.) *Nature, Lond.*, **142**, 90–2.
The mathematics of experimentation. *Nature, Lond.*, **142**, 442.
Quelques remarques sur l'estimation en statistique. *Biotypologie Bull.* **6**, 153–9.
On the statistical treatment of the relation between sea-level characteristics and high-altitude acclimatization. *Proc. Roy. Soc.* B, **126**, 25–9.
The statistical utilization of multiple measurements. *Ann. Eugen.* **8**, 376–86.

1939 'Student.' *Ann. Eugen.* **9**, 1–9.
The precision of the product formula for the estimation of linkage. *Ann. Eugen.* **9**, 50–4.
Citizens of the future: burden of a falling birth-rate: the case for family allowances. *The Times* 10 April 1939.
Selective forces in wild populations of *Paratettix texanus*. *Ann. Eugen.* **9**, 109–22.
The comparison of samples with possibly unequal variances. *Ann. Eugen.* **9**, 174–80.
The sampling distribution of some statistics obtained from non-linear equations. *Ann. Eugen.* **9**, 238–49.

1939 Blood groups in Great Britain. (With G. L. Taylor.) Copy of letter to editor. *Brit. Med. J.* (1939), **2**, 826.

Taste-testing the anthropoid apes. (With E. B. Ford and Julian Huxley.) *Nature, Lond.*, **144,** 750.

Surnames and blood-groups. (With Janet Vaughan.) *Nature, Lond.*, **144**, 1047.

A note on fiducial inference. *Ann. Math. Statist.* **10,** 383–8.

Stage of development as a factor influencing the variance in the number of offspring, frequency of mutants and related quantities. *Ann. Eugen.* **9**, 406–8.

The Galton Laboratory. *The Times*, 29 Sept. 1939; reprinted in *Science*, **90**, 436.

1940 The Galton Laboratory. *Science*, **91**, 44–5.

On the similarity of the distributions found for the test of significance in harmonic analysis, and in Stevens' problem in geometrical probability. *Ann. Eugen.* **10**, 14–17.

An examination of the different possible solutions of a problem in incomplete blocks. *Ann. Eugen.* **10**, 52–75.

Scandinavian influence in Scottish ethnology. (With G. L. Taylor.) *Nature, Lond.*, **145,** 590.

The quantitative study of populations in the Lepidoptera. I. *Polyommatus icarus* Rott. (With W. H. Dowdeswell and E. B. Ford.) *Ann. Eugen.* **10**, 123–36.

The estimation of the proportion of recessives from tests carried out on a sample not wholly unrelated. *Ann. Eugen.* **10**, 160–70.

The precision of discriminant functions. *Ann. Eugen.* **10**, 422–9.

Non-lethality of the mid factor in *Lythrum salicaria*. (With K. Mather.) *Nature, Lond.*, **146,** 521.

1941 The theoretical consequences of polyploid inheritance for the mid style form of *Lythrum salicaria*. *Ann. Eugen.* **11**, 31–8.

Average excess and average effect of a gene substitution. *Ann. Eugen.* **11**, 53–63.

The asymptotic approach to Behrens' integral, with further tables for the d test of significance. *Ann. Eugen.* **11**, 141–72.

The negative binomial distribution. *Ann. Eugen.* **11**, 182–7.

The interpretation of experimental fourfold tables. *Science*, **94**, 210–11.

1942 The polygene concept. *Nature, Lond.*, **150**, 154.

New cyclic solutions to problems in incomplete blocks. *Ann. Eugen.* **11**, 290–9.

The likelihood solution of a problem in compounded probabilities. *Ann. Eugen.* **11**, 306–7.

'The fundamental principles of mathematical statistics' by H. H. Wolfenden; 'Sampling methods in forestry and range management' by F. X. Schumacher & R. A. Chapman. (Reviews.) *Nature, Lond.*, **150**, 196.

'The advanced theory of statistics, Vol. 1' by M. G. Kendall. (Reviews.) *Nature, Lond.*, **52,** 431–2; *Camb. Rev.* **23**, Oct. 1943.

Polyploid inheritance in *Lythrum salicaria*. (With K. Mather.) *Nature, Lond.*, **150**, 430.

The theory of confounding in factorial experiments in relation to the theory of groups. *Ann. Eugen.* **11**, 341–53.

Some combinatorial theorems and enumerations connected with the numbers of diagonal types of a Latin square. *Ann. Eugen.* **11**, 395–401.

Completely orthogonal 9×9 squares. A correction. *Ann. Eugen.* **11**, 402–3.

1943 Note on Dr Berkson's criticism of tests of significance. *J. Amer. Statist. Ass.* **38**, 103–4.

The inheritance of style length in *Lythrum salicaria*. (With K. Mather.) *Ann. Eugen.* **12**, 1–23.

The birth rate and family allowances. *Agenda*, **2**, 124–33.

A theoretical distribution for the apparent abundance of different species. (Part 3 of a paper with A. S. Corbet and C. B. Williams.) *J. Anim. Ecol.* **12**, 54–8.

A sex difference in blood-group frequencies. (With J. A. Fraser-Roberts.) *Nature, Lond.*, **151,** 640.

1944 Mutation and the rhesus reaction. (With R. R. Race and G. L. Taylor.) *Nature, Lond.*, **153,** 106.

The experimental modification of dominance in Danforth's short-tailed mutant mice. (With S. B. Holt.) *Ann. Eugen.* **12**, 102–20.

Allowance for double reduction in the calculation of genotype frequencies with polysomic inheritance. *Ann. Eugen.* **12**, 169–71.

1945 G. L. Taylor, M.D., Ph.D., F.R.C.P. *Brit. Med. J.* (1945), **1**, 463.
A new rhesus antibody. *Nature, Lond.*, **155**, 542.
A new test for 2 × 2 tables. *Nature, Lond.*, **156**, 388.
The hereditary and familial aspects of exophthalmic goitre and nodular goitre. (L. Martin; with a genetical note by R.A.F.) *Quart. J. Med.* (new ser.), **14**, 207–19.
The logical inversion of the notion of the random variable. *Sankhyā*, **7**, 129–32.
The Indian Statistical Institute. *Nature, Lond.*, **156**, 722.
A system of confounding for factors with more than two alternatives giving completely orthogonal cubes and higher powers. *Ann. Eugen.* **12**, 283–90.
Recent progress in experimental design. In *L'application du calcul des probabilités de Coopération Intelletuelle*, 19–31 (1945).
Therapeutic use of Vitamin C. (With W. R. G. Atkins.)

1946 Rh gene frequencies in Britain. (With R. R. Race.) *Nature, Lond.*, **157**, 48–9.
The fitting of gene frequencies to data on *Rhesus* reactions. *Ann. Eugen.* **13**, 150–55.
A system of scoring linkage data, with special reference to the pied factors in mice. *Amer. Nat.* **80**, 568–78.
Testing the difference between two means of observations of unequal precision. *Nature, Lond.*, **158**, 713–14.

1947 Thomas Hunt Morgan, 1866–1945. Obit. Not. *Roy. Soc.* **5**, 451–4.
The analysis of covariance method for the relation between a part and the whole. *Biometrics*, **3**, 65–8.
The theory of linkage in polysomic inheritance. *Phil. Trans.* B, **233**, 55–87.
The renaissance of Darwinism. (Broadcast.) *Parents' Rev.* **58**, 183–7.
Spontaneous occurrence in *Lythrum salicaria* of plants duplex for the short-style gene. (With V. C. Martin.) *Nature, Lond.*, **160**, 541.
The spread of a gene in natural conditions in a colony of the moth *Panaxia dominula* L. (With E. B. Ford.) *Heredity*, **1**, 143–74.
Number of self-sterility alleles. *Nature, Lond.*, **160**, 797–8.
The sex chromosome in the house mouse. (With M. F. Lyon and A. R. G. Owen.) *Heredity*, **1**, 355–65.
Note on the calculation of the frequencies of *Rhesus* allelemorphs. *Ann. Eugen.* **13**, 223–4.
Development of the theory of experimental design. *Proc. Int. Statist. Confce*, **3**, 434–9.
The Rhesus factor: a study in scientific method. *Amer. Sci.* **35**, 95–102, 113.

1948 A quantitative theory of genetic recombination and chiasma formation. *Biometrics*, **4**, 1–13.
'Advances in Genetics, Vol. 1.' ed. M. Demerec. (Review.) *Sci. Progr.* **36**, 176–7.
Biometry. *Biometrics*, **4**, 217–19.
A twelfth linkage group of the house mouse. (With G. D. Snell.) *Heredity*, **2**, 271–3.
Genetics of style-length in *Oxalis*. (With V. C. Martin.) *Nature, Lond.*, **162**, 533.
Un résultat assez inattendu d'arithmétique des lois de probabilité. (With D. Dugué.) *C.R. Acad. Sci., Paris*, **227**, 1205–6.
Conclusions fiduciaires. *Ann. Inst. Poincaré*, **10**, 191–213.
Modern genetics. *Brit. Sci. News*, **1** (10), 2–4.

1949 What sort of man is Lysenko? *Soc. for Freedom in Sci., Occ. Pamphl.*, no. 9, 6–9.
A notation for the Lewis and Lutheran blood-group systems. *Nature, Lond.*, **163**, 580.
The quantitative study of populations in the Lepidoptera. II. *Maniola jurtina* L. (With W. H. Dowdeswell and E. B. Ford.) *Heredity*, **3**, 67–84.
Note on the test of significance for differential viability in frequency data from a complete three point test. *Heredity*, **3**, 215–28.
A preliminary linkage test with *agouti* and *undulated* mice. *Heredity*, **3**, 229–41.
A biological assay of tuberculins. *Biometrics*, **5**, 300–16.
The Report of the Royal Commission on Population. *Cambridge J.* **3**, 32–9.
A theoretical system of selection for homostyle *Primula*. *Sankhyā*, **9**, 325–42.
The linkage problem in a tetrasomic wild plant, *Lythrum salicaria*. *Proc. 8th Int. Congr. Genet.*, *Hereditas* (suppl. vol.), 225–33.

1950 A class of enumerations of importance in genetics. *Proc. Roy. Soc.* B, **136**, 509–20.
Polydactyly in mice. *Nature, Lond.*, **165**, 407, 796.
The significance of deviations from expectation in a Poisson series. *Biometrics*, **6**, 17–24.

1950 The 'Sewall Wright effect'. (With E. B. Ford.) *Heredity*, **4**, 117–19.
Creative aspects of natural law. 4th Eddington Memorial Lecture (1950). Cambridge University Press.
Gene frequencies in a cline determined by selection and diffusion. *Biometrics*, **6**, 353–61.
Sub-commission on statistical sampling of the United Nations. *Bull. Inst. Int. Statist.* **32**, 207–9.

1951 The hereditary and familial aspects of toxic nodular goitre (secondary thyrotoxicosis). (With Laurence Martin.) *Quart. J. Med.* (new ser.), **20**, 293–7.
A combinatorial formulation of multiple linkage tests. *Nature, Lond.*, **167**, 520.
Standard calculations for evaluating a blood-group system. *Heredity*, **5**, 95–102.
Statistics. *Scientific Thought in the Twentieth Century*. (Ed. by A. Heath.), 31–55.
Galton's 'Hereditary Genius'. (Review.) *Eugen. Rev.* **43**, 37.
'Inheritance in Dogs', by O. Winge. (Review.) *Heredity*, **5**, 149–50.

1952 Statistical methods in genetics. *Heredity*, **6**, 1–12.
Sequential experimentation. *Biometrics*, **8**, 183–7.

1953 The expansion of statistics. (Presidential address.) *J. R. Statist. Soc.* A, **116**, 1–6.
The variation in strength of the human blood group *P*. *Heredity*, **7**, 81–9.
The linkage of *polydactyly* with *leaden* in the house mouse. *Heredity*, **7**, 91–5.
Note on the efficient fitting of the negative binomial. *Biometrics*, **9**, 197–9.
Population genetics: Croonian lecture. *Proc. Roy. Soc.* B, **141**, 510–23.
Quantitative inheritance: Review of papers at colloquium held at Institute of Animal Genetics, Edinburgh University. *Heredity*, **7**, 293.
Sex differences of crossing-over in close linkage. (With W. Landauer.) *Amer. Nat.* **87**, 116.
Dispersion on a sphere. *Proc. Roy. Soc.* A, **217**, 295–305.

1954 The analysis of variance with various binomial transformations. *Biometrics*, **10**, 130–9.
A fuller theory of 'junctions' in inbreeding. *Heredity*, **8**, 187–97.
'Blood groups in anthropology', by A. E. Mourant. (Review.) *Brit. Med. J.* (1954), **2**, 1034.
Retrospect of the criticisms of the theory of natural selection. From *Evolution as a Process*, ed. by J. S. Huxley, A. C. Hardy and E. B. Ford, pp. 84–98.
The experimental study of multiple crossing-over. *Caryologia*, **6**, suppl., 227–31. (*Proc. 9th Int. Congr. Genet.* 1953.)

1955 Statistical methods and scientific induction. *J. R. Statist. Soc.* B, **17**, 69–78.
Science and Christianity. (Broadcast.) *The Friend*, **113**, 2 pp.
Double reduction at the rosy, or pink, locus in *Lythrum salicaria*. (With V. C. Fyfe.) *Nature, Lond.*, **176**, 1176.
On a test of significance in Pearson's *Biometrika* tables. *J. R. Statist. Soc.* B, **18**, 56–60.
New tables of Behrens' test of significance. (With M. J. R. Healy.) *J. R. Statist. Soc.* B, **18**, 212–16.
Blood-groups and population genetics. *Proc. 1st Int. Congr. Hum. Genet.* (1957); *Acta Genet. et Statist. Medica*, **6**, 507–9.
'Experimental design and its statistical basis', by D. J. Finney. (Review.) *Heredity*, **10**, 275.

1957 Space travel and ageing: a correspondence between Sir Ronald Fisher and W. H. McCrea, 1956. *Discovery*, **18**, 56–8, 174.
Comment on the notes by Neyman, Bartlett and Welch in *J. R. Statist. Soc.* B, **18**, 1956. *J. R. Statist. Soc.* B, **19**, 179.
Dangers of cigarette-smoking. *Brit. Med. J.* (1957), **2**, 43, 297–8.
The underworld of probability. *Sankhyā*, **18**, 201–10.
Method in human genetics. *Proc. 1st Congr. Hum. Genet.* (1957); *Acta Genet. et Statist. Medica*, **7**, 7–10.
Polymorphism and natural selection. *Bull. Inst. Int. Statist.* **36**, 284–9; *J. Ecol.* **46**, 289–93.
'Foundations of inductive logic', by R. F. Harrod. (Review.) *The Observer*, 27 Jan. 1957.
'Scientific Inference', by Harold Jeffreys. (Review.) *Camb. Rev.*, 18 May 1957.
The Autobiography of Charles Darwin, 1809–82. (Review.) *Nature, Lond.*, **182**, 71.

1958 Cigarettes, cancer, and statistics. *Centenn. Rev.* **2**, 151–66.
The nature of probability. *Centenn. Rev.* **2**, 261–74.
Lung cancer and cigarettes? *Nature, Lond.*, **182**, 108.
The discontinuous inheritance. (Broadcast.) *The Listener*, 85–7.
Cancer and smoking. *Nature, Lond.*, **182**, 596.

1958 Mathematical probability in the natural sciences. *Proc. 18th Int. Congr. Pharmaceut. Sci., Brussels*, 1958, 9 pp. Also in *Metrika*, **2**, 1–10; *Technometrics*, **1**, 21–29; *La scuola in Azione*, **20**, 5–19.

1959 *Smoking: the Cancer Controversy*. Edinburgh: Oliver and Boyd.
An algebraically exact examination of junction formation and transmission in parent-offspring inbreeding. *Heredity*, **13**, 179–86.
Natural selection from the genetical standpoint. *Aust. J. Sci.*, **22**, 16–17.

1960 The percentile points of distributions having known cumulants. (With E. A. Cornish.) *Technometrics*, **2**, 209–25.
On some extensions of Bayesian inference proposed by Mr Lindley. *J. R. Statist. Soc.* B, **22**, 299–301.
Scientific thought and the refinement of human reasoning. *J. Oper. Res. Soc. Japan*, **3**, 1–10.

1961 Possible differentiation in the wild population of *Oenothera organensis*. *Aust. J. Biol. Sci.* **14**, 76–8.
Sampling the reference set. *Sankhyā*, A, **23**, 3–8.
The weighted mean of two normal samples with unknown variance ratio. *Sankhyā*, A, **23**, 103–14.
A model for the generation of self-sterility alleles. *J. Theoret. Biol.* **1**, 411–14.

1962 Confidence limits for a cross-product ratio. *Aust. J. Statist.* **4**, 41.
The simultaneous distribution of correlation coefficients. *Sankhyā*, A, **24**, 1–8.
Enumeration and classification in polysomic inheritance. *J. Theoret. Biol.* **2**, 309–11.
Il ruolo del 'piano degli esperimenti' nella logica della inferenza scientifica. *La scuola in Azione*, **9**, 33–42.
Some examples of Bayes' method of the experimental determination of probabilities *a priori*. *J. R. Statist. Soc.* B, **24**, 118–24.
Self-sterility alleles. *J. Theoret. Biol.* **3**, 146–7.
The detection of sex difference in recombination values using double heterozygotes. *J. Theoret. Biol.* **3**, 509–13.

3. *Contributions to discussions of read papers*

1932 Biological principles in the control of destructive animals (M. A. C. Hinton). *Proc. Linn. Soc. Lond.* **144**, 124–5.

1933 Seedling mortality among plants (E. J. Salisbury). *Proc. Linn. Soc. Lond.* **145**, 100.

1934 On the two different aspects of the representative method (J. Neyman). *J. R. Statist. Soc.* **97**, 614–19.
Statistics in agricultural research (J. Wishart). *J. R. Statist. Soc. Suppl.* **1**, 51–3.
Discrimination by specification (B. H. Wilsdon). *J. R. Statist. Soc. Suppl.* **1**, 198.

1935 Statistical problems in agricultural experimentation (J. Neyman). *J. R. Statist. Soc. Suppl.* **2**, 154–7.
Complex experiments (F. Yates). *J. R. Statist. Soc. Suppl.* **2**, 229–31.

1936 Cooperation in large scale experiments (W. S. Gosset). *J. R. Statist. Soc. Suppl.* **3**, 122–4.

1937 On the value of Royal Commissions in sociological research (M. Greenwood). *J. R. Statist. Soc.* **100**, 403–4.

1939 Some aspects of the teaching of statistics (J. Wishart). *J. R. Statist. Soc.* **102**, 554–6.
Long-term agricultural experiments (W. G. Cochran). *J. R. Statist. Soc. Suppl.* **6**, 143–4.

1946 A review of recent statistical developments in sampling and sample surveys (F. Yates). *J. R. Statist. Soc.* **109**, 31.

1948 The validity of comparative experiments (F. J. Anscombe). *J. R. Statist. Soc.* A, **111**, 202–3.

1949 Some statistical problems arising in genetics (J. B. S. Haldane). *J. R. Statist. Soc.* B, **11**, 9–10.

1954 Interval estimation (E. C. Fieller, M. A. Creasy and S. T. David). *J. R. Statist. Soc.* B, **16**, 212–13.

1955 The Macmillan gap and the shortage of risk capital (Lord Piercy). *J. R. Statist. Soc.* A, **118**, 7–8.

The Neyman-Pearson Story: 1926-34

Historical sidelights on an episode in Anglo-Polish collaboration

E. S. PEARSON

1 Foreword

It has been suggested that I should contribute to this volume an account of my recollections of the period, starting just 40 years ago, when Jerzy Neyman first came to London and he and I began to wrestle with problems of statistical inference. I regarded this proposal at first with some hesitation, since our joint association was so complete that any account of it must be to a large extent autobiographical. But it is I suppose true to say that if the story of this episode in the history of statistics is to be put on record, it is likely to be more accurately done by the participants than by later searchers after we are gone. Of course, my account must be incomplete since I have only my own recollections and rough workings on numerous sheets of paper to go on, supplemented by a number of letters. The element of surprise in the presentation of this volume would, however, be defeated were I to ask Jerzy now to check my recollections. But if my record is at fault, I know he will correct it and forgive me!

2 The Biometric and Galton laboratories at University College

Jerzy came from Warsaw to London in the autumn of 1925, I think with a Polish Government Fellowship, to work for a year in my father's laboratories at University College. This year fell in the middle of what might be described as the third phase of Karl Pearson's long innings in the field of statistics and biometry. The first phase ended with W. F. R. Weldon's death in 1906, the second, with the intervention of the 1914–18 War. K.P. himself, though

68 on March 27, 1925, was still full of vigour and fresh ideas; he was giving both the 'first' and 'second-year' lecture courses in statistical theory; he suggested and supervised nearly all the research work of visiting graduate students; and at the end of 1925, with the assistance of Ethel Elderton, he launched his new journal, *Annals of Eugenics*. But the long run, spanning 40 years, of this foundation period of British mathematical statistics, was coming towards its end. In January 1926 two chapters in the history were closed with the deaths of William Bateson, the geneticist, and F. Y. Edgeworth, the statistician and mathematical economist. Of Bateson it is told how at the Cambridge meeting of the British Association for the Advancement of Science, in 1904, he had entered the lecture hall carrying a parcel containing all issues of *Biometrika* then published, and had thrown them down on the table with the remark: "I hold that the whole of that is worthless."* But had Weldon, the biologist, survived it is likely that it would have been realized much earlier that the approaches of biometricians and mendelians were complementary, not antagonistic. For 35 years Edgeworth had marched near the borders of the biometric land and now and again his lonely plough had cut right across the field. Yet he had always remained a friend.

Now, new ideas were stirring and although they had not as yet entered into the teaching programme at University College, some of the younger members of the staff, Oscar Irwin, John Wishart, and myself, were puzzling how to square the concept of likelihood and the new theory of estimation with the older approach we had learnt. In the summer of 1926 both Irwin and Wishart were to leave the College and, after an interval, to gain new experience with R. A. Fisher at Rothamsted.

The scope of the work of the Galton Laboratory is well summarized by the Foreword and the list of contents of the first number of the *Annals of Eugenics*. The more theoretical work which, for the purpose of grants was ascribed to the Biometric Laboratory, included several investigations to determine the moments of sample moments, both for infinite and finite populations; Neyman's first three papers published in *Biometrika* in 1925–26 were concerned with this field. The difficulties which were being encountered in the derivation of ever-lengthening algebraic expressions certainly

* Weldon, sitting beside Pearson, had remarked: "At least, we have got his subscription." But no, Bateson had brought the books from the library of the Cambridge Philosophical Society!

pointed to the need for the new technique which was to be introduced a few years later by Fisher. In another direction, L. H. C. Tippett's compilation of a long series of Random Sampling Numbers, which was ancillary to his work on the distribution of the range published in 1925,* opened up a whole new field for investigation: the empirical study of the sampling distributions of what we then termed frequency constants, but now would call statistics, in cases where the population sampled was not normal. From the earliest days when he developed his system of frequency curves, K.P. had, of course, laid stress on the point that statistical theory should not be limited to 'normal theory.' The advent of Tippett's Numbers made it relatively easy to explore in new directions where precise theory had failed to advance.

Into this Gower Street world came every year a small group of graduate students drawn from many countries. In two of his speeches (1926 and 1927) made at the Department's Galton dinner, held annually in January or February, we find K.P. referring to the group of this period:

> No workers in the laboratories are more welcome here than the post-graduates. It is they who keep us alive, bringing new ideas and new questions. This year, besides our own country, America, Japan, and Poland have been represented. If undergraduates teach a teacher to teach, it is equally certain that post-graduates teach a teacher to think, and this is especially the case when their training has been on different lines; they may reach *your* problem, but not by your route . . .

And again in 1927:

> They give us also wider international sympathies, although language differences present at times difficulties which can only be mastered with patience. In the past twelve months we have had students from Poland Spain, Canada, the United States, Yugoslavia, India, Japan, and China. We have learnt as usual much from them, not only as to what has been done in our science, but as to their social and national aspirations. On the whole I believe we gain more from them than we give . . .

If language presented difficulties to British teachers, communication must have been harder still between neighbouring desks in the post-graduate room where a Yasukawa from Tokyo sat next to a Neyman from Warsaw! I remember very little of Jerzy during his first two terms; he will have planned his research with K.P. and it was only when he attempted to carry out exercises on a Brunsviga

* Tippett completed two years of post-graduate work in the Department in June **1925.**

calculator, illustrating the theory given in K.P's lectures, that I should have had an official responsibility for him, as a demonstrator. It seems, however, from a note towards the end of his third *Biometrika* paper (1926b) that I helped him by correcting his English and suggesting that he include some diagrams and numerical tables to illustrate his study of the regression of the variance on the mean in samples from non-normal populations.

What I remember more clearly is a week-end which we spent together in the spring of 1926 at our family holiday cottage (the Old School House) at Coldharbour on Leith Hill in Surrey. It was then that I listened with fascination to an account of his early life in Russia and of the experiences which he had later undergone in the shadow of those disruptive forces, set in train throughout Central Europe by war and the Russian Revolution. In 1926 these events were near enough for the story to grip the imagination.

Perhaps through talks about his own paper (1926b) in which he drew "attention to a method of judging whether a sample is likely to have been taken from a population whose distribution is supposed to be known," I spoke to him towards the end of his stay in London about a very general statistical problem which I had for some time been puzzling round. I suggested that if he was interested we might collaborate in going further with the investigation. It was clear that this would have to be largely by post, for at the end of the university term, in June or July, he left England. His second year's Fellowship was to run in Paris. During the next eight years our communication was partly by letter—with occasions when a spurt of energy led to two or three letters a week, followed by long gaps when more demanding pressures intervened—partly in short holiday get-togethers when one or other of us crossed the Channel to meet in England, France, or Poland. It was only in 1934 that Jerzy came to London for a permanent appointment; by this time K.P. had retired and his Department of Applied Statistics had been split into two parts, under R. A. Fisher and myself. It was now my turn to be faced with administrative problems allowing much less time for doing freely what I wanted.

3 The first steps

To explain from my side the origin of our joint work, it is necessary to go back a little in time. No one can ever recall all the chance incidents, a paper read, a discussion half remembered, a letter received and destroyed, which may have put him on to a line of thought leading to a programme of action or research. It is probably true that there are moments in the history of a science when the time is ripe for certain questions to be raised and possible solutions worked out. In the middle 1920s the development of small-sample distribution theory, to a large extent as a result of Student's and R. A. Fisher's work, drew more than one statistician* into examining the philosophy of choice among statistical techniques. The need for critical discussion became apparent and the illustrative material on which to base discussion had become ample enough to make it worthwhile. If Fisher's theory of estimation did not seem to provide the complete answer, it was an incitement to further research.

In my own case there were two incidents which I know had an influence in determining the particular line of attack which the Neyman–Pearson theory was later to follow. One was the occasion of a paper read by E. C. Rhodes in 1924 before the informal Society of Statisticians and Biometricians and subsequently published in *Biometrika* with a following note by K.P. The other was a letter which W. S. Gosset ('Student') wrote to me in May 1926.

Rhodes's paper had the title "On the problem of whether two given samples can be supposed to have been drawn from the same population" and was concerned with alternative methods of assessing the significance of differences in data classified in a $2 \times k$ contingency table, the case of a 2×2 table being treated most fully. K.P's paper which followed (1924) set out the problem in rather more general terms. It was clear, he said, that there might be many procedures which could be applied with equal logical validity to test whether two samples had come from a common population; this being so, it was inevitable that the tests would not all place the observed data at the same level of significance. In such an event he suggested that the statistician should be guided in rejecting or accepting common origin by "the most stringent of these tests." By this he meant the test which, in a common terminology, gave the smallest tail-area P-value.

It was the suggestion of picking the most stringent test which raised many questions and doubts in my mind. Early in 1926 I had reached the stage of writing down a number of notes in which,

* For example in this case, among others, E. C. Rhodes and V. Romanovsky whose papers are referred to below. Apart from the 1924 paper, Rhodes published a second paper (1926) concerned with sampling from normal populations but this did not, I think, have any influence on our work.

to put it briefly, I set about exploring the multi-dimensional sample space and comparing what came to be termed the rejection regions associated with alternative tests. Could one find some general principle or principles appealing to intuition, which would guide one in choosing between tests?

Undoubtedly the rapid development in Fisher's hands of techniques capable of handling data in small samples called for a rethinking of the current philosophy of inference. The biometric school founded by Pearson and Weldon had obtained much of their data by sampling natural populations, not by controlled experiments. Emphasis had therefore been placed on the collection of large samples and this had determined the outlook on tests of significance. Theory had provided the standard errors in terms of population parameters of such differences as $\bar{x} - \mu$ (in the case of the mean of a single sample) or $s_1^2 - s_2^2$ (for the variances in two samples).* In so far as it was necessary to introduce sample estimates of variance into the standard errors, the result for large samples could still be assumed to provide a close approximation to the true values and the tests could be regarded as essentially equivalent to those which would have been carried out had the population variances been known.

Thus in large samples, in testing for significance of a shift in a mean value, the ratio $t = (\bar{x} - \mu)\sqrt{n}/s$ could be regarded as the best estimate available of the desired ratio $(\bar{x} - \mu)\sqrt{n}/\sigma$ and, as such, referred to the normal probability scale. If the sample was, however, quite small what could be regarded as anomalies began to be apparent. For example, a sample with a less divergent mean, $\bar{x}_1$, might well provide a larger value of t than a second sample with a more divergent mean, $\bar{x}_2$, simply because s_1 in the first sample happened through sampling fluctuations to be smaller than s_2 in the second. To someone brought up with the older point of view this seemed at first sight paradoxical.

To the statistical generation of today it may seem strange that there was any problem here to worry about, but looking back I realize that a re-orientation of outlook must for me at any rate have been necessary. It was a shift which I think K.P. was not able or never saw the need to make; as a result, controversy followed both

* It is interesting to speculate how far this attitude to the test ratio $(s_1^2 - s_2^2)/$(S.E. of the difference) for long obscured the fact that when estimates of variance were inserted in the S.E. the ratio became a function of s_1^2/s_2^2, so that for normal populations exact significance levels might be found by deriving directly the distribution of the variance ratio.

with Gosset and Fisher in connection with the use and interpretation of Student's t (or $z = t/\sqrt{n}$) and the goodness-of-fit χ^2.

Gosset often called both at University College and at Rothamsted when passing through London, either on his firm's business or to stay with his father who lived at Watlington under the Chilterns. It was natural that I should put my difficulties regarding both t and χ^2 before him. The earliest of my letters to him which has been preserved, dated April 7, 1926, began with the following paragraph.

> I was down last week in the middle of small samples at the Fruit Station at East Malling, Kent, which you probably know, of which my cousin G. R. Hatton is director. While wandering among apple plots I was suddenly smitten with a doubt as to what exact interpretation can be laid on your distribution of $z = \bar{x}/s$ or m/s. I have not really thought of the matter much before, but as it is a stepping stone from which so much small sample theory (particularly of Fisher's) starts, I feel the whole thing rather important and I should like to have your comments on my doubts...

I did not actually post the letter until May 5, after a holiday in Italy. Gosset's reply of May 11 was characteristically full of sound comments based on long practical experience. He found illuminating the presentation of the problem in the space of m ($= \bar{x} - \mu$) and s, with the density contours for $p(m, s \mid \mu, \sigma)$ sketched in, but emphasized that the essential character of Student's problem was that we had no possible means of connecting σ and s. At a later stage in his argument, he threw out some ideas which undoubtedly pointed a way to the resolution of my difficulties.

> In your large samples with a known normal distribution you are able to find the chance that the mean of a random sample will lie at any given distance from the mean of the population. (Personally I'm inclined to think your cases are best considered as mine to the limit n large.) That doesn't in itself necessarily prove that the sample is not drawn randomly from the population even if the chance is very small, say 0.00001: What it does is to show that if there is any alternative hypothesis which will explain the occurrence of the sample with a more reasonable probability, say 0.05 (such as that it belongs to a different population or that the sample wasn't random or whatever will do the trick) you will be very much more inclined to consider that the original hypothesis is not true.
>
> I can conceive of circumstances such for example as dealing a hand of 13 trumps after careful shuffling *by myself* in which almost any degree of improbability would fail to shake my belief in the hypothesis that the sample was in fact a (reasonably) random one from a given population.
>
> In the case of small samples if we have no other means of knowing anything about the population we cannot with any profit use that method to bet on the sample having been drawn from a given population but we can bet on the probability that the mean of the population shall lie within any

given distance of the known mean of the sample and we do so not by comparing areas but volumes.* Logically the nature of the inference is exactly the same but of course as we are more ignorant in the second case the odds we can lay are lower for the same number of cases; in fact your objections have been met by this lowering of the odds. ...

I'm more troubled really by the assumption of normality and have tried from time to time to see what happens with other population distributions but I understand that you get correlation between S and m with *any* other population distribution.

Still I wish you'd tell me what happens with the even chance population □ or such a one as △; it's beyond my analysis. ...

P.P.S. If Student is wrong it is up to you to give us something better. You see one must experiment and frequently it is quite out of the question —considerations of cost or of impossibility of duplicating conditions in the time scale—to do enough repetitions to define one's variability as accurately as one could wish. It's no good saying "Oh these small samples can't prove anything." Demonstrably small samples have proved all sorts of things and it is really a question of defining the amount of dependence that can be placed on their results as accurately as we can. Obviously we lose by having a poor definition of the variability but *how much do we lose?*

From then on a number of new ideas must have begun to take shape. The possibility of getting a mathematical entry into the problem by specifying the class of alternative hypotheses which should be accepted as 'admissible' for formal treatment; the difference between what we later distinguished as 'simple' and 'composite' hypotheses; the 'rejection region' in the sample space; the 'two sources of error.' These were points which we must have discussed during the autumn of 1926.

It was still necessary to find a principle for determining the choice among possible contours in the sample space in such a way that the hypothesis tested became 'less likely' and alternatives 'more likely' as a sample point moved outward across them. From rough notes, it seems that the idea of using the likelihood ratio criterion as a method of determining these contours took form in November 1926.

4 The likelihood ratio criterion and the sample space

4.1 *The first joint paper*

I am not sure what opportunity Neyman had had at this date to make a careful study of Fisher's 1922 *Phil. Trans.* paper, but his

* He was using one dimension for the ordinate of the probability density functions. The 'area' was therefore the tail area of the curve $y = p(m \mid \sigma)$ and the 'volume' the volume under the surface $y = p(m, s \mid \sigma)$ falling beyond the section along which $z = m/s$ was constant.—E.S.P.

first reaction to my suggestion of using the likelihood ratio criterion* was rather similar to that of other statisticians when faced in the 1920s with Fisher's use of likelihood in his theory of estimation. Jerzy thought that it involved a disguised appeal to Bayes' theorem and suggested that we might as well go the whole way by assigning a distribution of prior probabilities to the hypotheses of the 'admissible set.'

Personally, I had accepted Fisher's criticism of the use of methods of inverse probability, largely from not seeing how in practice meaningful prior distributions, linked with relative frequencies, could be associated with the unknown values of population parameters. In testing a statistical hypothesis, whether simple or composite, it seemed to me that the appropriate likelihood ratio picked out a system of contours in the sample space of the kind I had been looking for. The principle involved had an intuitional appeal and the first results achieved were straightforward and usable. Not only did the approach lead to many established tests, throwing some new light on them in the process, but also as time went on it pointed to several new procedures which were later incorporated into statistical methodology.

It was not long, I think, before Jerzy accepted the utility of the principle, but in our first joint paper (1928a) we agreed to keep the door open by tackling problems in a variety of alternative ways, one of which was based on an inverse probability approach. We were also careful to state that while this likelihood ratio procedure led to 'good' tests, we had no grounds for claiming them as 'best.' Indeed, we pointed out that acceptance of this or any other process of reasoning must be a matter of individual preference, depending ultimately on an appeal to 'the way we think.'

The paper which resulted from our discussions (1928a) went to press in March 1928 and was published in the July issue of *Biometrika.* I must have been working at the illustrative numerical material during 1927 while Jerzy was first in Paris, from where he returned in May to Poland. The paper was long and had some of the character of a research report, putting on record a number of lines of inquiry, some of which had not got very far. The way ahead was left open; for example, five different methods of approach were suggested for the test of the hypothesis that two different samples came from normal populations having a common mean! It was shown that the use of the λ-principle in the cases of sampling

* See Appendix on p. 475 for a definition.

from a rectangular and from an exponential population led to statistics analogous to Student's z (or t), whose sampling distributions took simple forms. A few sampling results were included from investigations in hand at University College, following Gosset's suggestion of examining the robustness of 'normal theory' tests. No doubt we could have produced a more concise paper had we waited longer, but the ideas had been under discussion for two years, we thought they were worth putting on record and K.P. accepted the paper for publication.

It has been asked why we did not at an earlier stage try to link our results with Fisher's theory of estimation and, in particular, explore the part played by sufficient statistics. Certainly we might have done this, but it should be remembered that we were searching for some general principles which could be used in deriving tests of statistical hypotheses whatever the probability density distribution in the sample space. From this point of view an independent attack had its advantages. A time for linking up came later.

Other questions have been asked as to why we followed certain lines of approach, and not others. Two of these may be mentioned:

(1) Why did we use tail-area probabilities, classing together all samples less likely to occur on the null hypothesis, H_0, than the sample observed? (Harold Jeffreys, at an early date.)
(2) Why did we not take the numerical value of the likelihood ratio itself as our criterion? (George Barnard, more recently.)

The answer to the first question is that this interpretation was not part of our approach; indeed in many cases, e.g. for Student's test, the contours used in the complete sample space cut right across those of equal probability density. At this stage we introduced the λ-principle as a working procedure in order to define a family of surfaces in the space, such that as a point moved outward across them and λ decreased from 1 towards 0, there would (speaking loosely) be alternatives to H_0 which became relatively more and more likely than H_0. One had then to decide at which contour H_0 should be regarded as no longer tenable, that is where should one choose to bound the rejection region? To help in reaching this decision it appeared that the probability of falling into the region chosen, if H_0 were true, was one necessary piece of information. In taking this view it can of course be argued that our outlook was conditioned by current statistical practice, but the criticism implied in (1) still does not strike me as relevant.

We certainly considered question (2) but soon realized that the value of λ itself (as defined in the Appendix on pp. 475) could provide no clear guide on which to base conclusions. In the k-sample problem mentioned below, for instance, the sampling distribution of λ under H_0 varies tremendously with the value of k in both shape and spread; it may be rectangular or J-shaped or bell-shaped. The refinement of working in terms of $-\log \lambda$ had not occurred to us and, if it had, we might not have found it meaningful.

4.2 *The χ^2 test for goodness-of-fit*

The interpretation of the χ^2 test for goodness-of-fit had been a subject of rather bitter controversy in the years 1922–24. Although this tended to be obscured in the course of argument, the point at issue really concerned the appropriate probability set to which to refer a unique value of

$$\chi_1^2 = \sum_{t=1}^{k} \frac{(n_t - \tilde{m}_t)^2}{\tilde{m}_t}$$

based on expected frequencies, $\tilde{m}_t$, calculated by some process of fitting theory to the data. K.P. maintained the view that the purpose of χ^2 was to test a simple hypothesis and that where the population values of m_t had to be estimated from the data, by $\tilde{m}_t$, we could still regard χ_1^2 as a good approximation to the measure of departure of sample from population that we were really seeking, viz. $\chi^2 = \sum (n_t - m_t)^2/m_t$.

In the problem discussed on p460 concerning the ratio $(\bar{x} - \mu)\sqrt{n}/s$, if the sample size was not too small it was in effect immaterial whether we regarded the ratio as appropriately referred to the normal or to Student's t-distribution. But if repeated samples, however large, were drawn from a fixed population and the values $\tilde{m}_t$ estimated each time by fitting to the data a frequency law containing c parameters, the resulting distribution of χ_1^2 would differ appreciably from that of χ^2. This was shown by Yule and Brownlee's sampling experiments and by Fisher's theory. In the latter case it was proved that in the large sample limit and if an 'efficient' method of fitting were used, the distribution of χ_1^2 would be that of the classical χ^2, with degrees-of-freedom $k - c - 1$.

It was clear that we had reached a division of ways in the conceptual approach to the theory of significance testing. But for the disagreement on interpretation and a natural resentment at the statement that the long established and practically useful method of fitting by moments was frequently 'inefficient,' I think that K.P. might have realized that he had himself already recognized the mathematical relevance of the reduction of degrees-of-freedom in two special cases: (i) in the comparison of two samples with frequencies classed in k-groups (1912); (ii) in dealing with what he had termed the problem of multiple and partial contingency (1916). But as it was he seemed never able to accept the newer viewpoint.

In the second of our joint papers published in the December issue of *Biometrika* (1928b), Neyman and I put the goodness-of-fit procedure into the setting of our likelihood ratio approach to testing a composite hypothesis. We gave more precise definitions of terminology and notation than in the earlier paper and were rather intrigued by the elegant way in which the situation could be presented geometrically in a k-dimensional sample space. The λ-contours for the composite hypothesis involved in testing whether an observed sample $(n_1, n_2, \ldots, n_k)$ was consistent with a given frequency law

$$p_s = f(s \mid \theta_1, \theta_2, \ldots, \theta_c), \quad s = 1, 2, \ldots, k \qquad \text{(A)}$$

with $\sum p_s = 1$, appeared as envelopes of the hyper-ellipsoidal contours associated with the simple hypotheses that the θ's had certain specified values. If the sample frequencies, n_s, were large enough, a single transformation would turn the hyper-ellipsoids into hyper-spheres, all of radius χ_1, and the population locus (A) into the intersection of a series of $k - c$ primes. Fisher's distribution of χ_1, as that of a χ with degrees-of-freedom $\nu = k - c - 1$, followed at once.

An advantage of this geometrical description was that it seemed to highlight the points where assumptions were being made in reaching the distribution of χ_1. The paper contained a number of experimental sampling results and an appendix of Jerzy's with a rigorous proof of the limiting form of the χ_1-distribution. About the same time W. F. Sheppard (1929) published a long mathematical paper filling in gaps in Fisher's very condensed proof.

The International Statistical Institute was to meet in Poland during August 1929 and through the preceding winter Jerzy had worked at a paper which he would have liked us to present jointly as No. III in the series "On the use and interpretation of certain test criteria for purposes of statistical inference." In this paper he showed that an inverse approach to testing goodness-of-fit for grouped data led, in the large sample limit, to the same χ_1^2 integral as the direct approach and it did so independently of the prior distribution of the estimated parameters. In this way the dual approach of our first paper in dealing with normal, rectangular, and exponential populations could be extended to the case of the χ^2-test. The work was entirely Jerzy's; it seemed to me more of mathematical interest than statistical and it had to be hurried to the printer to meet an I.S.I. date-line. He was disappointed when I refused to be rushed into appearing as a co-author. The paper was presented to the Conference with the title "Contribution to the theory of certain test criteria."

4.3 *The 'two-sample' and 'k-sample' papers*

A paper of different character was "On the problem of two samples" (1930) in which we applied the λ-principle in considering the question of whether two independent samples had been drawn from a common normal population. We had used as a starting point a paper of V. Romanovsky's (1928) in which four alternative criteria had been put forward for this purpose without, it seemed to us, any clear recognition that a test must be related to a background model adjusted to the questions put and the assumptions involved in answering them. Certainly this particular problem provided a text-book example of how to follow through with the likelihood ratio procedure. The t and F tests appeared as appropriate to answer certain questions while a third statistic emerged, involving the product of two likelihood ratios, when we asked whether both means and variances were the same in the populations sampled. The moments of this last statistic were found and it was seen to be distributed uniformly in the interval (0, 1) when the sample sizes increased indefinitely.

An extension of the investigation to the case of $k > 2$ samples was obviously called for. This appeared as a paper published in the same Polish journal in 1931. Here, if the population variances could be assumed the same and only the equality of means was in doubt, the λ-principle led to Fisher's analysis of variance test for the case of a one-way classification. The test for heterogeneity of variance which emerged was however novel. The likelihood ratio, raised to the power $2/N$ (where $N = \sum_{t=1}^{k} n_t$ was the total number of observations) was seen to be proportional to the ratio of the weighted geometric mean to the weighted arithmetic mean of the sample estimates of variance. The weights which came out from this approach were the sample sizes, n_t.

As in the case where $k = 2$, there was a similar ratio for testing the complete hypothesis that both means and variances in the

k sampled populations were the same. The sampling moments of the test criteria under the null hypothesis were obtained and it was shown that a good approximation to the sampling distribution was likely to be given by a beta distribution. In the limit we also pointed out that $-2 \log_e \lambda$ would tend to be distributed as χ^2, thus foreshadowing a later more general result of S. S. Wilks (1938). In 1936 B. L. Welch pointed out that by departing slightly from the λ-principle we could get similar results using other weights, w_t, and in 1937 Bartlett, approaching from another angle, reached weights which were the degrees-of-freedom ($w_t = n_t - 1$ in our problem) and derived a χ^2 approximation for the logarithm of the geometric/arithmetic mean ratio which worked very well, even with quite small samples.

4.4 *The laboratory in Warsaw. Difficulties of collaboration*

It is clear that our progress would have been a good deal quicker had we been able to meet more often. From the research point of view my life at University College in those days was what now seems ideal, with no administrative responsibilities and relatively little teaching. K.P., however, imposed rather strict limits on the vacation time of his staff and I had to choose between a working holiday and other commitments with family or friends.

In comparison, Jerzy had to struggle against many difficulties. In 1928 he succeeded in getting a small Biometric Laboratory established in Warsaw at the Nencki Institute for Experimental Biology. But the continuance of funds depended on the financial prospects of his country. "Certainly it [the creation of the Laboratory] is not yet sure, especially as our loan in America is not yet signed" he wrote on June 4, 1927. And four years later, on March 7, 1931:

> In town I am terribly busy in getting some job for the Lab. You may have heard that we have in Poland a terrific crisis in everything. Accordingly the money from the Government given usually to the Nencki Institute will be diminished considerably and I shall have difficulties in feeding my pups [research workers in the Lab.]

And again, on June 23, 1932:

> You seem to be a little annoyed with me: in fact you have some reasons as I do not answer properly your letters. This however is *really* not the result of carelessness or of anything which could be offensive. I simply cannot work; the crisis and the struggle for existence takes all my time and

energy. I am not sure that next year I shall not be obliged to take some job, I do not know where—in trade, perhaps, selling coal or handkerchiefs.*

Love from both,

J.N.

* I just examined about one dozen of asses, and am perhaps too pessimistic!

However, Jerzy and his 'pups' did produce quite a large output of work, much of it in connection with statistical methods in agriculture. All this was collected in the five numbers of *Statistica*, 1929–34, in which separates of papers published elsewhere were issued annually bound together in a printed wrapper. The most promising of these 'pups' was perhaps S. Kolodziejczyk whose 1935 paper in *Biometrika* on the testing of linear statistical hypotheses has formed the starting point of much later work. He was, alas, killed in action in September 1939, fighting desperately with his cavalry regiment against the advancing German tanks.

In the first few years after Jerzy had returned to Warsaw, we managed two meetings. In the summer of 1928 I spent about a fortnight with him and Lola in a small village on the south coast of Brittany, near Concarneau. I can recall a small hotel perched on the cliffs, a diet of shrimps and other sea foods, stretches of sand and warm rocks, and sessions at a café table on the beach when Jerzy scribbled algebra on paper blocks.

In 1929 I was in Poland for a much longer period. Of the I.S.I. Conference which had meetings at Warsaw, Poznan, and Krakow, the three capitals of a former partitioned Poland, I remember little, except that Corrado Gini attacked certain parts of Jerzy's paper with vigour. But memories of less statistical incidents remain. Of a poorly attended I.S.I. excursion, when J. W. Nixon (then at the International Labour Office), Jerzy and myself (both non-members) alone turned up to be escorted down a coal mine in Polish Upper Silesia and afterwards royally entertained at a banquet, with much vodka, by a group of Polish industrialists at Katowice. Of the old city of Krakow and of the broad stretch of the Vistula nearby, with cattle escorted by boys on horses standing knee deep in the shallows, leisurely drinking. Of a stay in the Tatra mountains and a climb of Świnica on the border with Czechoslovakia. Of a visit to a guest house, dedicated as a retreat for senior University research workers, among the pine woods at Mądralin, not many miles from Warsaw. Here, among ideal surroundings we got down to some concentrated work before I was finally taken to see the new port of Gdynia and then on to the 'free city' of Danzig, to sail home in a Polish ship. The shadow of Hitler had not yet fallen across Europe.

5 The best critical regions and the power function

The weeks spent together in Poland during the summer of 1929 must have thrown up a number of questions needing answers. Some months later, on March 8, 1930, Jerzy wrote: "I think we have to introduce a little more order into our work; we must fix a certain plan as we have a lot of problems already started and left 'in the wood.'" These problems included: (*a*) computation of the 'second type of error' for the *t*-test, or what we later termed the power function (for J.N.); (*b*) completion of the problem of *k* samples (I suspect this meant the numerical calculation of moments leading to some approximation to the percentage points of the test statistics), (work for E.S.P.); (*c*) further study of the robustness of test statistics already considered (sampling experiments to be carried out in both London and Warsaw); (*d*) test for the difference in means when two samples have been drawn from populations with different variances (for J.N.); and (*e*) a general problem in the calculus of variations (for J.N.).

Problem (*e*) turned out to be of fundamental importance for our work; where it might lead us we did not then know. Having established the usefulness of the λ-criterion, we realized that it was essential to explore more fully the sense in which it led to tests which were likely to be effective in detecting departures from the null hypothesis. So far we could only say that it seemed to appeal to intuitive requirements for a good test.

My own rather crude way of dealing with this was to start by making some explorations, using the random sampling results which were accumulating at University College. Table VIII of my joint paper with N. K. Adyanthaya, published in December 1929, gave what I suppose was the first comparison of power functions of two alternative tests. The null hypothesis was that the population mean had a specified value, say μ_0, while its standard deviation, σ, was unknown. The two test statistics were: (*a*) Student's $z = t/\sqrt{n} = (\bar{x} - \mu_0)/s$ and (*b*) the ratio of the deviation from the population mean of the mid-range to the semi-range, or $z' = (x_1 + x_n - 2\mu_0)/(x_n - x_1)$, where x_1 and x_n are the extreme observations in the sample.

It had already been shown (Neyman and Pearson, 1928a) that if the population was rectangular ($\beta_2 = 1.8$), application of the λ-principle picked out the z'-test, just as for a normal population ($\beta_2 = 3.0$) it picked out z. The sampling experiments included the cases of two symmetrical populations with $\beta_2 = 2.5$ and 4.1. Although the empirical power functions* were only based on 100 samples, with $n = 5$ and 10, they indicated the increasing effectiveness of Student's test over the mid-range/semi-range test as the population shape moved away from the rectangle towards greater and greater long-tailedness. While results were crude, they show that our thoughts were turning towards the justification of tests in terms of power.

Meanwhile Jerzy began to approach the problem in a far more fundamental way. It has been remarked that nearly always an important new scientific result is reached only after a process of slow groping and trial; what counts is persistence to carry on to the final rounded solution. Often, however, the scaffolding which has been needed to build this solution is not on record. The truth of this reflection struck me forcibly in re-reading Jerzy's letters to me written during 1930–31. He started his investigations with the hopeful reflection:

> If we can show that the frequency of accepting a false hypothesis is a minimum [that is the probability of detecting that H_0 is not true is a maximum—E.S.P.] when we use the λ-tests, I think it will be quite a thing! [February 1, 1930]

And then a little later (February 20, 1930) as he began to dig into the problem:

> At present I am working on a variation calculus problem connected with the likelihood method. The results already obtained are a vigorous argument in favour of the likelihood method. I considerably forget the variation calculus and until the present time I have only results for samples of two. But in all cases considered I have found the following:
>
> We test a simple hypothesis H_0 concerning the value of *some* character $a = a_0$, and wish to find a contour $\phi(x_1, x_2, \ldots, x_n) = c$ such that:
>
> (i) The probability $P(\phi, a_0)$ of a sample, Σ, lying inside the contour (which probability is determined by H_0) is $P(\phi, a_0) = \alpha$, where α is a certain fixed value, say 0.01 (this is for controlling the errors of rejecting a true hypothesis).
>
> (ii) The probability $P(\phi, a_1)$ determined by some *other* hypothesis H that $a = a_1 \neq a_0$ of the sample lying inside the same contour be a maximum.

* Actually, what was tabled was the percentage of samples for which a test at 5% level failed to establish significance, as the true mean shifted from μ_0 by steps of $\sigma/\sqrt{n}$.

Using such contours and rejecting H_0 when Σ is inside ϕ = const., we are sure that a true hypothesis is rejected with a frequency less than α, and that if H_0 is false and the true hypothesis is, say H', then *most often* the observed sample will be inside ϕ = const. and hence H_0 will be rejected.

I feel you start to be angry as you think I am attacking the likelihood method! Be quiet! In all cases I have considered the ϕ = const. contours are the λ-contours!

But the problem was not quite as simple as this. One would have liked the contour, ϕ = const., to be the same for all the hypotheses H belonging to what had been described formally as the 'admissible set' Ω. Were this the case it would determine a 'best critical region' (B.C.R.) leading to what later we termed a 'uniformly most powerful test.' Such a contour, it could be shown, was a λ-contour. But Jerzy soon realized that such a best contour frequently did not exist. Many questions had then to be followed up: under what conditions would a B.C.R. exist and how was it to be determined, particularly in the more difficult case of testing a composite hypothesis involving several nuisance parameters? What was the standing of the test derived from the λ-principle when there was no B.C.R.? What were suitable examples on which to illustrate theory?

These and other questions occupied him when time allowed in 1930–31. We were not able to meet during this period but when I returned after nearly six months in the United States in the autumn of 1931, he had made enough progress for it to be important for us to get together and put a further paper into shape. He was indeed already beginning to get on to the next stage of asking how to deal with certain defects in the λ-tests occurring when no B.C.R. existed.

In a letter of August 17, 1931 he illustrated one of these defects in the case where a single random sample, $x_1, x_2, \ldots, x_n$, is used to test the simple hypothesis, H_0, that the normal population sampled has a mean = μ and S.D. = σ.

It seems, he wrote, that if we use the λ-contours [which had been given in our first paper (1928a)—E.S.P.] to test the hypothesis, H_0, that $\mu = 0$, $\sigma = 1$, we shall accept more often when $\mu = 0$, $\sigma = 1.1$ (say) than when H_0 is true. In other words: *the true hypothesis will be rejected more often than some of the false ones.* I told Lola that we had invented such a test. She said: "good boys!" I should like to be wrong somewhere. If it is all correct, then the principle of maximum likelihood seems to lose its generality. Exceedingly interesting to know what are the conditions of its applicability. I think I shall send these notes as they are—you will see how useful it will be if you could come. The work would go ever so much quicker and better if we could talk.

I made a short visit to Warsaw just after Christmas 1931; we spent a little while in the quiet surroundings of Mądralin.* It was a period of concentrated discussion and writing which got us a long way towards rounding off our paper, "On the problem of the most efficient tests of statistical hypotheses" (1933a). My part was to help in shaping the material; sharpening the arguments; standardizing the notation and terminology; working on the illustrative examples of which there were eleven in all; and deciding on the best form for the diagrams. It must, I think, have been on this occasion that we decided to use the terms 'power function' and 'uniformly most powerful test.'

Final filling in of gaps and polishing the presentation was completed during the first few months of 1932 after we had separated. The paper was communicated to the Royal Society by K.P. on August 31 and, we had reason to believe, refereed by R.A.F. Publication occurred in February 1933.

Jerzy paid a short visit to London just before Christmas 1932. Official support was forthcoming to enable him: (*a*) to see foreign statisticians and get ideas for further work; (*b*) to collect information from England, France, and Germany on the methods of collection of data for the construction of sickness tables for the use of offices of social insurance. Unofficially he wrote:

> I am thinking of a present for Professor Pearson [K.P.]. Will he like to have a couple of hare ready for cooking? Going by sea it will be quite easy to bring them fresh. How is it with the etiquette? I suppose that such a present is fit to be given by a barbarian, but I do not know whether it will be a pleasure to digest the hare!

I have no record of my reply!

6 A return to prior probabilities

Our last joint paper before Jerzy came to London in 1934 was put together very speedily during a brief holiday which we spent in Paris in April 1933. Lola had been staying there often in connection with her painting work and I think he came to join her for a short break from teaching. My impression is that we met without any

* Not quite always so, though! Here is Jerzy writing on March 8, 1931: "Did you find the asymptotic expressions for the moments of λ_{H_1}? [for the k-sample paper—E.S.P.] I had little time and, besides, there is a poet here reading his poetries all the time, next door. I was too long ashamed to stop him. At present silence is reigning but I have to leave soon."

clear plan as to what we should work at as there were so many possibilities; on my arrival he suggested that it would be good to write a paper discussing "What statements of value to the statistician in reaching his final judgement can be made from the analysis of observed data, which would not be modified by any change in the probabilities *a priori*."

So we set down to this and had almost finished when we parted. This was the second time I was in Paris with the Neymans; the first had been in April 1927 and incidents from the two visits are no longer distinguishable. As always on such working holidays, statistics was closely mixed with other forms of entertainment and it is the highlights from the latter which I remember. The Tuileries Gardens in spring, restaurants on the left bank, the Impressionist pictures, a day's visit to Chartres....

Our paper was communicated by Udny Yule to the Cambridge Philosophical Society on May 31, refereed perhaps by Harold Jeffreys, and published in the *Proceedings* late that year (1933b).

7 Conclusion

In October 1933, after K.P's retirement, I had to take charge of a new Department of Statistics. That winter neither Jerzy nor I had much time for theoretical work and the Christmas break seemed too short to include some relaxation and a visit to Warsaw. We had, however, a number of problems in the air: there was the wide-open question of how to select a 'good critical region' when no 'best critical region' existed; also Jerzy was beginning to work on confidence intervals and regions and already doubting the wisdom of the particular approach which Fisher had taken to fiducial inference. In the spring of 1934 it became suddenly possible for me to find a temporary post for Jerzy in my Department. After a little reflection he came over gladly; the post became permanent in 1935.

I think that by 1934 we had found the answers that satisfied us to most of the more tractable problems. We had now been brought up against the hard fact that mathematical models which fit the observations of the real world will not respond beyond a point to a simple theory of statistical inference. Frequency distributions so often cannot be represented by mathematical functions which yield, for example, sufficient statistics or uniformly most powerful tests! Our joint work continued in London with many graduate students tackling a variety of problems, but the curtain had come down on the particular episode I set out to describe; our 8-year effort to carry out a joint programme of research across the face of Europe.

Sorting out the papers and letters which had been stored away, I can recapture some of the intellectual thrill of that time; the exhilaration which goes with the belief that one is chipping away along the fringe of the unknown. What value is or will be placed on our joint work suddenly seems relatively unimportant. It is the experience within oneself and the joint friendship which have really mattered in life. In this respect, at least, how extraordinarily lucky it was that I decided to launch some of my unsolved puzzles on that rather language-tied Research Fellow in the summer of 1926!

APPENDIX ON TERMINOLOGY USED IN CONNECTION WITH THE LIKELIHOOD RATIO CRITERION

The definitions given below follow the notation used in our *Phil. Trans.* paper (1933a).

The observations are represented by the values of n variates $x_1, x_2, \ldots, x_n$ defining a sample point Σ in the sample space W. H is a statistical hypothesis concerning the origin of the sample which is such as to determine the probability of occurrence of every possible sample represented in W.

$p_H = p_H(x_1, x_2, \ldots, x_n)$ is the probability density function associated with such a point Σ.

If H completely specifies this probability law it is termed a simple hypothesis. If the fundamental form of p_H is given but it involves the values of c unspecified parameters, say $\theta_1, \theta_2, \ldots, \theta_c$, it is termed a composite hypothesis having c degrees-of-freedom.

To place the problem in a mathematical setting it is supposed possible to define a set Ω of simple hypotheses which, for the problem at issue, are formally classed as 'admissible,' e.g. all univariate normal distributions with means, μ, and variances, σ^2, having any values whatever between the limits $-\infty, +\infty$ and $0, +\infty$, respectively.

The statistical hypothesis which we wish to use the sample to test, in other words the null hypothesis, will be termed H_0. If H_0 is simple it will clearly belong to Ω, e.g. it might specify that $\mu = 0$, $\sigma = 1.0$. If H_0 is composite it will be possible to specify a part of the set Ω, say ω, such that every simple hypothesis belonging to the subset ω will be a particular case of H_0. Thus H_0 might now be the hypothesis that $\mu = 0$, leaving σ unspecified.

The likelihood ratio is then obtained as follows. Consider the set A_Σ of probabilities p_H corresponding to a sample point Σ determined by different simple hypotheses belonging to Ω. We shall suppose that whatever the sample point the set A_Σ is bounded and denote by $p_\Omega(x_1, x_2, \ldots, x_n)$ the upper bound of the set. Then if H_0 is a simple hypothesis determining the probability density function p_0, the likelihood ratio associated with H_0 is defined as

$$\lambda = \frac{p_0(x_1, x_2, \ldots, x_n)}{p_\Omega(x_1, x_2, \ldots, x_n)}$$

If H_0 is composite, denote by $A_\Sigma(\omega)$ the sub-set of A_Σ corresponding to the set ω of simple hypotheses belonging to H_0 and by p_ω the upper bound of $A_\Sigma(\omega)$. Then the likelihood ratio associated with the composite hypothesis H_0 is

$$\lambda = \frac{p_\omega(x_1, x_2, \ldots, x_n)}{p_\Omega(x_1, x_2, \ldots, x_n)}$$

The use of the principle of likelihood in testing H_0 consists in taking for the critical region that determined by the inequality

$$\lambda \leqslant C = \text{const.}$$

To control the risk of rejecting H_0 when it is true, we choose a value C_α such that

$$P\{\lambda \leqslant C_\alpha \mid H_0\} = \alpha$$

In so far as it is impossible to give numerical expression to prior knowledge regarding the population parameters, the intuitional appeal of the use of the λ-principle might perhaps be summed up as follows. In testing a simple hypothesis we know that there is at least one and generally very many alternatives to H_0 among the admissible set for which the likelihood is increasingly greater than that of H_0 as $\lambda \to 0$. In testing a composite hypothesis there will be at least one and generally many alternatives in the sub-set $\Omega - \omega$ for which the likelihood is increasingly greater than that of any of the simple hypotheses included in ω.

The use of the λ-principle has been referred to so often in the text that it seemed desirable to include a definition in this appendix. The idea of a best critical region and a uniformly most powerful test is, I hope, set out sufficiently clearly for the present purpose in Neyman's letter of February 20, 1930 and the subsequent paragraphs (pp. 17–18) for a restatement not to be needed here.

REFERENCES

Bartlett, M. S. (1937). Properties of sufficiency and statistical tests. *Proc. Roy. Soc.* A, **160**, 268–82.

Fisher, R. A. (1922). On the mathematical foundations of theoretical statistics. *Phil. Trans.* A, **222**, 309–68.

Kolodziejczyk, S. (1935). On an important class of statistical hypotheses. *Biometrika*, **27**, 161–90.

Neyman, J. (1925). Contributions to the theory of small samples drawn from a finite population. *Biometrika*, **17**, 472–79.

Neyman, J. (1926a). Further notes on non-linear regression. *Biometrika*, **18**, 257–62.

Neyman, J. (1926b). On the correlation of the mean and the variance in samples drawn from an 'infinite' population. *Biometrika*, **18**, 401–413.

Neyman, J. (1930). Contribution to the theory of certain test criteria. *Bull. Int. Statist. Inst.*, **24**, No. 2, 44–86.

Neyman, J. and Pearson, E. S. (1928a, b). On the use and interpretation of certain test criteria for purposes of statistical inference. *Biometrika*, **20A**, Pt. 1, 175–240; Pt. 2, 263–94.

Neyman, J. and Pearson, E. S. (1930). On the problem of two samples. *Bull. Acad. Polonaise Sci. Lettres*, A, 73–96.

Neyman, J. and Pearson, E. S. (1931). On the problem of k samples. *Bull. Acad. Polonaise Sci. Lettres*, A, 460–81.

Neyman, J. and Pearson, E. S. (1933a). On the problem of the most efficient tests of statistical hypotheses. *Phil. Trans.* A, **231**, 289–337.

Neyman, J. and Pearson, E. S. (1933b). The testing of statistical hypotheses in relation to probabilities *a priori*. *Proc. Camb. Phil. Soc.*, **29**, 492–510.

Pearson, E. S. and Adyanthaya, N. K. (1929). The distribution of frequency constants in small samples from non-normal symmetrical and skew populations. *Biometrika*, **21**, 259–86.

Pearson, Karl. (1912). On the probability that two independent distributions of frequency are really samples from the same population. *Biometrika*, **8**, 250–54.

Pearson, Karl. (1916). On the general theory of multiple contingency with special reference to partial contingency. *Biometrika*, **11**, 145–58.

Pearson, Karl. (1924). On the difference and the doublet tests for ascertaining whether two samples have been drawn from the same population. *Biometrika*, **16**, 249–52.

Rhodes, E. C. (1924). On the problem whether two given samples can be supposed to have been drawn from the same population. *Biometrika*, **16**, 239–48.

Rhodes, E. C. (1926). The comparison of two sets of observations. *J. R. Statist. Soc.*, **89**, 544–52.

Romanovsky, V. (1928). On the criteria that two given samples belong to the same normal population. *Metron*, **7**, No. 3, 3–46.

Sheppard, W. F. (1929). The fit of a formula for discrepant observations. *Phil. Trans.* A, **228**, 115–50.

Tippett, L. H. C. (1925). On the extreme individuals and the range of samples taken from a normal population. *Biometrika*, **17**, 364–87.

Welch, B. L. (1936). Notes on an extension of the L_1 test. *Statist. Res. Mem.*, **1**, 52–56.

Wilks, S. S. (1938). The large-sample distribution of the likelihood ratio for testing hypotheses. *Ann. Math. Statist.*, **9**, 60–62.

APPENDIX

KARL PEARSON'S LECTURES ON THE HISTORY OF STATISTICS IN THE 17th AND 18th CENTURIES

These lectures were given in the Department of Applied Statistics at University College during a number of Sessions, starting in October 1921 and certainly continuing up till 1929. They were fully written up by Pearson in manuscript, but in certain years he repeated an earlier series of lectures. This was particularly so for the Graunt–Petty period and as a result there is sometimes more than one version in the MS. A few years after his death the late Mrs. M.V. Pearson spent a great deal of time in straightening out the MS., in checking most of the references and quotations and in having the lectures typed up to the beginning of the series concerned with the great French mathematicians Condorcet, D'Alembert, Lagrange and Laplace.

As far as is known the lectures are listed here in the order in which they were delivered. While to a large extent he provided the system of grouping himself, the "chapter headings" in the list are partly mine, and are introduced merely to give a provisional system of classification which may be helpful to anyone wanting to see at a glance the immense range which the researches covered.

E.S.P.

I INTRODUCTION. THE EARLY HISTORY OF STATISTICS.

II THE FOUNDING OF THE ENGLISH SCHOOL OF POLITICAL ARITHMETIC

- (1) John Graunt: 1620–1674; and Petty's claim to the authorship of *Observations on the London Bills of Mortality*
- (2) Edward Chamberlayne: 1616–1703
- (3) Graunt's *Natural and Political Observations on the London Bills of Mortality*
- (4) Sir William Petty: 1623–1687
- (5) Works of Petty

III CASPAR NEUMANN AND EDMUND HALLEY

- (1) The first real Life Table
- (2) Some account of the life of Edmund Halley: 1656–1742

IV EARLY SUCCESSORS OF SIR WILLIAM PETTY: THE POLITICAL ARITHMETICIANS

- (1) Sir Mathew Hale: 1609–1676
- (2) Gregory King: 1648–1712
- (3) Charles Davenant: 1656–1714

V FROM GRAUNT AND HALLEY TO THE ASSURANCE SOCIETIES

- (1) Introduction to the History of Statistics in the 18th century

(2) Condorcet's early work on the Theory of Testimony

(3) Condorcet's memoirs on the Theory of Probability

(4) Jean le Rond D'Alembert: 1717–1783
(5) Works of D'Alembert

(6) Joseph Louis Lagrange: 1736–1813
(7) Works of Lagrange

(8) Pierre-Simon Laplace: 1749–1827
(9) Works of Laplace